Flowering Plants & their Classification

Flowering Plants & their Classification

– Volume 2 –
Dicotyledons

by
Alfred Barton Rendle

BIOTECH

2019
BIOTECH BOOKS®

2019 Alfred Barton Rendle (b. 1865-1938)
Original Title: The Classification of Flowering Plants

ISBN 81-7622-151-1 (SET)
ISBN 81-7622-152-X (Volume 1)
ISBN 81-7622-153-8 (Volume 2)

Published by
BIOTECH BOOKS
4762-63/23, Ansari Road, Darya Ganj,
New Delhi-110002 (India)
Phone : 011 43003222
E-mail: biotechbooks@yahoo.co.in

PRINTED IN INDIA

PREFACE TO VOLUME II

No apology can be adequate for a delay of twenty years in the appearance of the second volume of this work. I much regret that increasing official and extra-official duties and responsibilities have allowed insufficient leisure for the continuous effort required. The volume was nearly finished in 1914, but has been considerably revised in the last few years during its completion.

As in the Monocotyledons the general arrangement is similar to that of Prof. Adolf Engler's *Syllabus der Pflanzenfamilien*, but with some, often considerable, difference in detail, especially in the arrangement and conception of the Orders of the first two Grades. On pp. 2 and 3 I have given reasons for maintaining a Grade Monochlamydeae. As in volume I some of the smaller families have been omitted and more space has been allotted to families represented in Britain.

In accordance with the *Rules of Nomenclature* formulated at the Vienna Congress in 1905 I have substituted the term Family for Natural Order, and have used the term Order for the higher grade (Series of volume I).

Again, as in the case of the earlier volume, I must express my obligation to the two great works on Systematic Botany, *Die natürlichen Pflanzenfamilien* and *Das Pflanzenreich*.

In the matter of illustration I trust that this volume will mark an improvement on the former. I am greatly indebted to the Trustees of the British Museum for permission to use a number of blocks from the *Flora of Jamaica* by Mr W. Fawcett and myself, now in course of publication. These (indicated as *From Flor. Jam.*) were drawn by Mr Percy Highley. For the drawings for the remaining blocks (except for a few repeated from volume I) I am indebted to Mr Percy Highley and my daughter, Phyllis Rendle.

Mr A. J. Wilmott of the Department of Botany, British Museum, has very kindly read the proofs and made some useful suggestions.

A. B. RENDLE

LONDON,
October 1925

NOTE

The method of production of the present reprint has allowed only some small textual alterations. A few notes have been added in an Appendix.

A. B. Rendle

CONTENTS

LIST OF ILLUSTRATIONS

DICOTYLEDONS

In germination the radicle grows out from the seed to form a primary root. In the great majority of cases the pair of seed-leaves (cotyledons) are carried above ground (epigeal) to form the first green leaves of the plant; they are drawn out of the seed partly by growth of the hypocotyl, partly by their own growth, the stem-bud being protected between them. In some cases they are hypogeal, remaining in the seed, where they act as storehouses of nourishment (see vol. I, 127). There is great variety in the form of the cotyledons, which generally bears no obvious relation to the form of the adult leaf; this form may be assumed by the leaves directly succeeding the cotyledons, or, especially in cases where the adult leaf is much divided or compound, the leaves immediately succeeding the cotyledon are simpler in form and shew transitional stages to the adult form. The Onagraceae afford a remarkable instance of the growth of the cotyledon to assume, by an intercalary development at the base, a form approaching that of the adult leaf; a still more remarkable growth of the same nature occurs in *Streptocarpus* (see family Gesneriaceae). The cotyledons of a pair are sometimes unequal, and in rare cases one is minute or aborted; in exceptional cases there are more than two cotyledons. When hypogeal the cotyledons are sometimes united so as to be indistinguishable (conferruminate)[1].

The habit of the adult plant is remarkably varied. A characteristic anatomical feature is the ring of open bundles in the young stem and the formation of a ring of cambium preliminary to secondary growth: in perennial woody plants exposed to regular alternation of seasons favourable and unfavourable to growth the secondary formation of wood in the stem takes the form of concentric "annual" rings. Secondary growth of other types, generally described as "abnormal," occurs in various families and is especially frequent in woody climbers. The details of the stem-anatomy shew considerable

[1] A detailed account of the forms of seedlings will be found in Lubbock, *A contribution to our Knowledge of Seedlings* (1892).

variety and various features may be characteristic of families or smaller groups of genera[1].

The main stem is sometimes unbranched, but generally forms a more or less widely branching system, the individual leaf being proportionately small. The characteristic tree-type with a spreading crown of branches contrasts with the unbranched palm-type of the Monocotyledons. The palm-type is rare in Dicotyledons but occurs in the Papaw and others, where a tall unbranched stem bears a terminal crown of large leaves.

The form of the leaf varies widely but a particular type is often characteristic of families or smaller divisions. With few exceptions the venation is obviously reticulate. The presence of a pair of bracteoles, placed laterally with regard to the bract, is characteristic of the group (see vol. I, 143).

There is also great variety in the form and arrangement of the parts of the flower, which may consist merely of one or more sporophylls protected by bracts, or of a definite perianth surrounding male or female sporophylls or both; further, the perianth may be differentiated into an outer, generally protective, green calyx, and an inner, generally white or coloured corolla. The most general type of arrangement is one of five alternating whorls, each whorl containing five or four members: calyx, corolla, two whorls of stamens, and carpels, in regular ascending order.

The seed generally contains a greater or less amount of endosperm which surrounds the embryo; in some cases perisperm is also present. In some groups the embryo completely fills the seed.

The following arrangement does not claim to be strictly phylogenetic. Various attempts have been made to construct a phylogenetic system of Angiosperms, but the results vary according to the point of view adopted[2]. In the arrangement adopted here the orders are grouped in three grades which

[1] For details of anatomy, see Solereder, *Systematic anatomy of the Dicotyledons* (English edition), 1908.

[2] See Hutchinson, *Families of Flowering Plants* (Macmillan, 1926).

correspond to grades of differentiation in the floral structure. The first grade includes orders with, on the whole, a comparatively simple type of flower; several orders represent very isolated groups, and while it is possible that some may be reduced forms, the affinities of which are to be sought in the higher grades, it is, on the other hand, possible to regard the members of this grade as representing lines of development from earlier extinct groups. It seems likely that development of the highly differentiated insect-pollinated dichlamydeous flower was preceded by numerous, so to speak, experimental stages arising from earlier, now long extinct, Angiosperms, and it is a tenable view that such stages are represented among the Monochlamydeae. In the last order, Centrospermae, the higher type of floral structure exists alongside of the simple monochlamydeous type.

In the second grade, Dialypetalae, the orders are arranged, so far as possible, in ascending sequence. In the first, Ranales, there is a marked indefiniteness in the number of members in the various series (calyx, corolla, androecium, gynoecium), the parts of which are generally free from each other and arranged in regular, frequently spiral, succession, on a convex floral receptacle. The grade includes several lines of development but the following general tendencies may be traced—diminution in the number of members in each series, and, in the higher groups, a tendency to their union, a whorled (cyclic) instead of a spiral arrangement of parts, a passage from hypogyny through perigyny to epigyny, and development of a zygomorphic from a regular flower. Comparison of the two extreme orders Ranales and Umbelliflorae well illustrates this.

The orders of Sympetalae represent still higher grades of floral development which have sprung from various dialypetalous groups. They illustrate further developments of the tendencies to floral complexity already noticed among the Dialypetalae and comprise at least three grades of development; first a pentacyclic group, with typically two whorls of stamens, which is the least removed from the Dialypetalae; secondly a higher group with a tetracyclic hypogynous flower; and thirdly a group, with a tetracyclic flower and an inferior ovary, in which the highest form of floral complexity is attained.

GRADE A. MONOCHLAMYDEAE

Flowers frequently unisexual, without a perianth, or with a simple perianth, which in the earlier orders is green and inconspicuous, in the later often coloured and attractive; in the highest developed families the perianth may be differentiated into an inner and an outer series. Stamens equal in number and opposite to the perianth-leaves or in greater number. Pollination in the earlier orders generally anemophilous, in the higher orders often entomophilous; course of the pollen-tube often endotropic, that is, otherwise than through the micropyle.

The plants of the earlier orders are exclusively woody plants; those of the later include also herbs.

Order 1. Salicales.
Family Salicaceae.

Order 2. Garryales.
Family Garryaceae.

Order 3. Juglandales.
Family I. Myricaceae.
,, II. Juglandaceae.

Order 4. Julianiales.
Family Julianiaceae.

Order 5. Fagales.
Family I. Betulaceae.
,, II. Fagaceae.

Order 6. Casuarinales.
Family Casuarinaceae.

Order 7. Urticiflorae.
Family I. Ulmaceae.
,, II. Urticaceae.
,, III. Moraceae.
,, IV. Cannabinaceae.

Order 8. Proteales.
Family Proteaceae.

Order 9. Santalales.
Family I. Santalaceae.
,, II. Loranthaceae.
,, III. Balanophoraceae.
,, IV. Cynomoriaceae.

Order 10. Aristolochiales.
Family I. Aristolochiaceae.
,, II. Rafflesiaceae.
,, III. Hydnoraceae.

Order 11. Polygonales.
Family Polygonaceae.

Order 12. Piperales.
Family I. Piperaceae.
,, II. Saururaceae.
,, III. Chloranthaceae.

Order 13. Centrospermae.
Family I. Chenopodiaceae.
,, II. Amarantaceae.
,, III. Nyctaginaceae.
,, IV. Phytolaccaceae.
,, V. Aizoaceae.
,, VI. Portulacaceae.
,, VII. Caryophyllaceae.

Order 1. *SALICALES*

Family. SALICACEAE

Flowers dioecious, in catkins, each flower subtended by a bract; bracteoles absent. Male flowers of two or more stamens; the female of two carpels united to form a one-

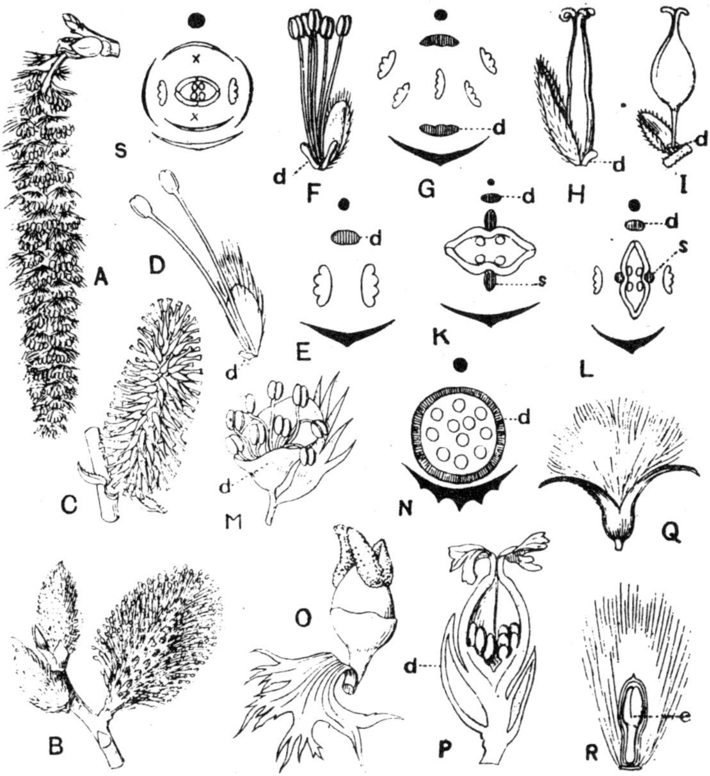

FIG. 1. A. Male catkin of *Populus tremula*. B. Male catkin of *Salix Caprea*. C. Female catkin of *S. Caprea*. D. Male flower of *S. Caprea*, viewed posteriorly (from the axis). E. Diagram of same. F. Male flower of *S. pentandra*. G. Diagram of same, a disc occurs both posteriorly and anteriorly. H. Female flower of *S. pentandra*. I. Female flower of *S. tetrasperma*. K, L. Diagrams of female and bisexual flowers of *S. Caprea*. M. Male flower of *Populus tremula*. N. Diagram of same. O. Female flower of *P. nigra*. P. Female flower of *P. tremula* in longitudinal section. Q. Open fruit of *S. Caprea*. R. Seed of *Salix* in longitudinal section. S. Theoretical floral diagram of *S. aurita* after Velenovsky. *d*, disc; *e*, embryo; *s*, stigma. D–R enlarged. (E, G, K, L, N after Eichler; F, H, I, M, R after Berg and Schmidt.)

celled ovary with parietal placentas bearing generally numerous anatropous ovules. There is no perianth but the axis is expanded beneath the sporophylls into a variously developed disc. Fruit a capsule dehiscing by two valves. Seeds very small with a thin testa; embryo straight, surrounded by a trace of endosperm.

Trees or shrubs with simple alternate stipulate leaves.

One family, Salicaceae, containing two genera, *Salix* (Willow), and *Populus* (Poplar), chiefly north temperate but extending into alpine and arctic regions where the plants become shrubby; extreme forms, such as *Salix herbacea* and *S. reticulata*, are small plants with slender creeping stems and branches on and beneath the ground sending up catkin-bearing twigs. A few species occur in the warmer parts of the world.

The adult plants shew considerable variety of habit dependent on the manner of branching. The branches are of two kinds: (1) vegetative, which go on growing at the tip (Poplars) or are closed by abortion of the terminal bud (most Willows), bearing axillary foliage-buds, and (2) dwarf-shoots, borne generally in the axils of the previous year's leaves, ending in a catkin and bearing either scale-leaves only or also foliage-leaves. The outermost bud-scale stands directly over the subtending leaf and is formed by union of two lateral scales, as is indicated by the presence of two main nerves, and especially also by the two axillary shoots which stand before the two portions of the bud-scale.

The simple scattered leaves vary in form from linear to cordate; individual species may shew great variety in leaf-form as in *Salix repens* or *Populus euphratica*. The stipules often fall early.

A constant anatomical character is the superficial origin of the periderm, which in *Salix* arises from the epidermis, in *Populus* from the next lower layer.

The catkins fall as a whole, after serving their purpose; the male after flowering, the female after bursting of the fruit. The flowers, which are sessile or have a shorter or longer stalk, stand each in the axil of a bract, which in the Willows is entire and hairy, in the Poplars lobed or cut.

The bracts on the axis of the inflorescence are arranged spirally or in alternating whorls. Except for the rare occurrence of bisexual flowers and the occasional appearance of male and female flowers in the same catkin the flowers are dioecious. Hegelmaier (1) states that they are formed in July of the year preceding that in which they open. The male flower contains in the Willows usually few stamens (1 to 8, rarely to 12), in the Poplars the stamens are usually numerous (4 to 40); the number is fairly constant in the same species; in *Salix incana* and *S. purpurea* they are united in a ring at the base. The female flower has two transversely placed carpels, rarely, in some Poplars, there are four. The two to four stigmas, which are generally sessile, more rarely borne on a style, are either lateral (above the midrib of each carpel) as in *Populus* and some species of *Salix*, or, as in other species of *Salix*, antero-posterior (above the ventral sutures of the carpels). The ovules are generally numerous and arranged on two parietal placentas. Van Tieghem finds a shorter inner integument in *Populus*, while in *Salix* only the outer integument is present. A similar difference occurs in the family Piperaceae where the ovule of *Piper* has two integuments, that of *Peperomia* only one.

Beneath the sporophylls (which are sometimes carried up beyond it on a stalk, as usually in the female flower of *Salix*) is a development of the axis which shews a variety of forms. In *Populus* it is an oblique cup, more developed in front than behind; in *Salix* it is often reduced to one or two small glandular teeth, which, in a few species, unite to form an inconspicuous ring; the reduction is generally greater in the female flower. In its size and early development as the first organ of the flower the cup-like structure in *Populus* contrasts with the less conspicuous structure in *Salix*; and this led Hegelmaier to suggest that the structures were not comparable in the two genera.

Eichler's hypothesis that the flowers of *Salix* are degenerate derivatives of a bisexual flower has later been developed by Velenovsky (2) who in abnormal late-flowering specimens of *S. aurita* found that in both male and female flowers the median gland opposite the bract (that is next the axis) was more or less split and displaced laterally, forming a pair of flat lanceolate scales. He interprets these as representing a pair of bracteoles (α and β). In the male flowers an anterior gland was also generally present (as occurs normally in *S. pentandra*) which always retained its scale-like form and never divided. This structure Velenovsky regards as representing the remains of a perianth, and compares the male flower with that of *Juglans* where the five similar scales represent a pair of bracteoles and three perianth-segments. Similarly the cup-like disc of *Populus* is interpreted as representing a perianth and a pair of bracteoles. The perianth-leaf is regarded as the remnant of a polymerous perianth and the reconstructed bisexual flower of *Salix* is represented as shewn in the diagram (fig. 1, S) which depicts a 2-merous flower with a theoretical posterior perianth-segment and the indication of a suppressed inner whorl of stamens. With this may be compared Eichler's theoretical flower of *Myrica*. The additional stamens which are often present in *Salix* are explained by doubling of the median stamens or the development of the inner series.

It is assumed that the original floral plan of *Populus*, *Myrica* and the Juglandaceae contained numerous stamens and several perianth-leaves from which the present-day flowers have been derived by reduction in various stages.

This comparison of the flowers suggests that the Salicaceae are most nearly related to the Myricaceae and Juglandaceae, and not to the Piperaceae alliance to which Van Tieghem (3) referred them on account of the presence of a second ovular integument (regarded as aborted in *Salix*) and the trace of endosperm which he finds in the seed in the form of a layer of oil-containing cells between the embryo and the seed-coat. A similar layer occurs in many other reputedly exalbuminous seeds.

The Poplars are wind-pollinated and have pendulous catkins, at least in the male; the Willows, on the other hand, with erect catkins, are pollinated by insects, many of which visit their flowers in the early spring, nearly ninety different species being recorded by Müller. The association of a large number of flowers in one inflorescence and the bright yellow colour of the anthers in the male flowers render the catkins conspicuous, especially among the bare twigs, when, as in many species, the flowers are developed before the leaves. The large store of nectar and pollen are an attraction to the insect at a time when nectar-secreting flowers are rare; many bees resort almost exclusively to Willows in search of food for their young (Müller). Even the tiny *Salix herbacea*, straggling over the bare rock of the high Alps, attracts, by its abundant supply of nectar, several insect-visitors in spite of its inconspicuous flowers.

The study of Willows is rendered very difficult by the occurrence in nature of hybrids, the production of which is facilitated by the prevailing dioecism. These hybrids may themselves hybridise, and Wichura (4) has produced by experiment unions of still higher complexity.

The two valves of the dehiscing capsule bend outwards or become spirally rolled. The seeds are very small and owing to the very thin testa become easily dried up. The time during which they remain capable of germination is accordingly short. On the other hand, the long silky hairs developed from the funicle, by which they are enveloped, effect their speedy distribution either by wind or by water and when stranded in sufficiently damp soil they will germinate in less than a day, and in a few days the seedling develops its first foliage-leaves. Rapid germination is rendered possible by the development of mucilaginous hairs on a raised girdle round the base of the hypocotyl and on the tip of the root while the embryo is still in the seed.

Plants are also very easily propagated by cuttings which root quickly in damp soil or water, forming trees or bushes in a short time.

Four centres of distribution of the Willows have been distinguished in the north temperate zone: (1) the district round Behring's sea, (2) Central Europe, (3) the Himalayas, and (4) Pacific North America. The tropical forms are widely distributed; in the Tertiary period they seem to have extended further north than at present. According to Nathorst's researches the arctic forms reached at the Glacial period into the more temperate regions. The genus *Populus* inhabits the temperate regions of both worlds, but does not reach the Arctic zone, though fossil remains indicate that it did so in Tertiary times; Eastern Asia and Atlantic North America are especially rich in species.

Of the two genera, *Populus* (Poplar) has broad leaves, pendulous catkins, at least in the male plant, a cup-like disc, numerous stamens and a lobed or cut bract. There are about 30 species, of which *P. alba* (White Poplar or Abele), *P. canescens* (Grey Poplar), *P. tremula* (Aspen) and *P. nigra* (Black Poplar) are British; *P. italica* (Lombardy Poplar) of doubtful origin has been regarded as a variety of the last. Fossil remains, chiefly leaf-specimens, closely resembling recent species, are known from Tertiary formations. Several species are planted as ornamental trees, or for screens, owing to their rapid growth. The young shoots and buds are sometimes rich in a balsam. The timber is soft and easily worked, but not durable: several species are used in paper-making. *Salix* (Willow) has generally narrow leaves, erect catkins, a disc reduced to one or two nectar-secreting glands, generally few stamens and an entire bract. There are about 160 species besides numerous hybrids. The great variability of the true species and the ease with which hybrids arise spontaneously render the genus one of the most difficult for study. Of our native species, *Salix alba* (White Willow) and *S. fragilis* (Crack Willow), when not pollarded, *S. triandra* and *S. pentandra* form large trees; *S. Caprea* (Palm or Goat Willow), *S. cinerea*, *S. aurita*, *S. repens*, and *S. lanata*, are generally shrubs; *S. viminalis*, the Common Osier, is a shrub with long straight flexible branches and very narrow leaves. *S. purpurea* (Purple Osier), also a shrub, is distinguished by its having the filaments of the stamens completely united. Fossil remains (including catkins as well as leaves) are frequent in Europe, America and the Arctic zone in Tertiary strata of different ages. Many species are valuable as timber-trees from their rapidity of growth and the production of light durable wood; the wood of *S. caerulea*, a hybrid between *S. fragilis* and *S. alba* occurring in the Eastern Counties, is used for cricket-bats. Some species, known as Osiers, are used for basket-making and wicker-work. The bark contains much tannin and a valuable medicinal glucoside, salicin. Several are ornamental trees such as *S. alba*, White Willow, and *S. babylonica*, Weeping Willow, a native of China. *Choisenia*, a genus widely distributed in Korea, has been described by Nakai (*Botan. Magazine*, of Tokyo, XXXIV, 67; 1920) as intermediate between *Populus* and *Salix*. It has pendulous male catkins and flowers without either a gland or a cup-like disc.

REFERENCES

(1) HEGELMAIER, F., in *Württemberg. Naturwiss. Jahresheft*, 1880, 204.

(2) VELENOVSKY, J. "Vergleichende Studien über die *Salix*-Blüte." *Beiheft. Botan. Centralbl.* XVII, 123 (1904).

(3) VAN TIEGHEM, PH. "Sur la structure de l'ovule et de la graine et sur les affinités des Salicacées." *Bull. Mus. Hist. Nat.* (Paris), VI, 197 (1900).

(4) WICHURA, M. *Die Bastardbefruchtung im Pflanzenreich, nachgewiesen an den Bastarden der Weiden*, Breslau, 1865.

Order 2. *GARRYALES*

Family. GARRYACEAE

The genus *Garrya*, formerly included in Cornaceae as an anomalous genus, has been removed from that family as it has a superior ovary, and apetalous or naked flowers arranged in catkins. It is regarded by Engler as the type of a distinct order Garryales, which he places among the amentiflorous orders next to Salicales. It comprises the single genus *Garrya*, with 15 species in California, the dry region of the south-west United States, and Mexico, and one in the West Indies.

The plants are dioecious evergreen shrubs, or sometimes trees, with opposite entire leathery leaves and small flowers in pendulous silky-hairy catkins. The flowers stand singly or two or three together in the axils of closely arranged opposite-decussate bracts; the male are stalked, the female sessile. The male have a simple four-leaved perianth, the leaves often remaining united at the tips and allowing the four stamens, which alternate with them, to protrude between them; the stamens are free, with long anthers which open lengthwise; in the centre of the flower is a small conical rudiment of an ovary. The female flowers are naked, consisting of a one-celled ovary bearing a pair of long spreading styles which

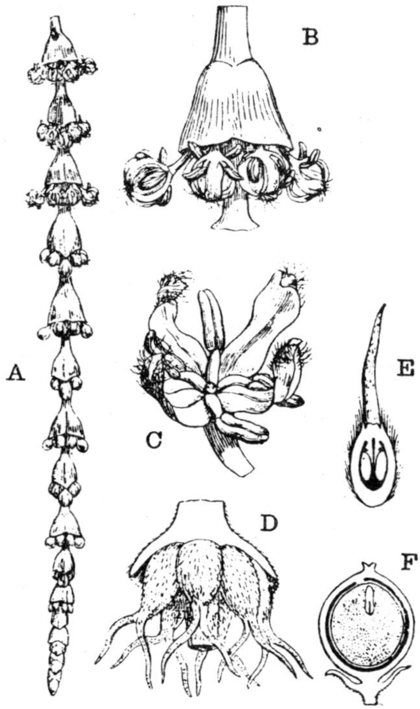

FIG. 2. A. Male catkin of *Garrya elliptica*. B. One pair of bracts of same from which protrude several flowers. C. Male flower. D. Portion of female catkin with bract cut away on one side exposing the flowers. E. Female flower in longitudinal section. F. Fruit in longitudinal section. B to F enlarged. (D, E after Engler; F after Baillon.)

bear stigmatic papillae on the inside. From the top of the ovary-cell hang two anatropous ovules, each with a single often incomplete integument; the micropyle is turned outwards and above it is an obturator or projecting thickening of the funicle. The fruit is an egg-shaped or roundish, 1- or 2-seeded berry crowned by remains of the styles. The seed has a membranous coat, and contains a very small straight embryo near the top of the copious fleshy endosperm.

Garrya elliptica is grown as an ornamental shrub, the long pendulous male catkins forming an attractive contrast with the green of the leaves.

Order 3. *JUGLANDALES*

Flowers unisexual, in catkins (at least the male), naked or with a simple scale-like perianth which is more or less united to the ovary in the female. Male flower with a variable number of stamens; female of two united carpels; ovary unilocular, containing a basal orthotropous ovule with usually one integument. Wind-pollination. Sometimes chalazogamic. Fruit a nut or drupe-like; often more or less enveloped by the persistent bracteoles. Embryo large, almost completely filling the seed, straight, with a superior radicle and two large cotyledons containing starch or oil.

Trees or shrubs with alternate simple or pinnately compound exstipulate leaves; rich in aromatic compounds.

Two families, Myricaceae and Juglandaceae.

Juglandales resemble Fagales in the arrangement of the unisexual flowers (at least the male) in catkins, which are here, however, simple inflorescences; the general plan of structure of the flowers is also somewhat similar, the presence in some genera of a pistil-rudiment in the male flower suggesting their derivation in each case from a group with bisexual flowers. In both orders the bracteoles play an important part in the protection and distribution of the fruit. The epigyny of the perianth in Fagales may be compared with the various stages of union of the perianth to the ovary in Juglandales. A constant and important distinguishing feature is found in the unilocular ovary of Juglandales with its single orthotropous ovule. The presence of aromatic compounds in the leaves is also distinctive.

Family I. MYRICACEAE

Flowers monoecious or dioecious, without perianth but protected by the bract and sometimes also by a pair of scale-like bracteoles. Male flowers with 2–20, often 4, stamens. Pistil of two carpels in the median plane bearing a pair of long thread-like stigmas. Fruit a small drupe with generally a wax-secreting exocarp. Embryo surrounded by a one-layered endosperm containing oil and proteid; cotyledons thick, plano-convex.

Shrubs or trees with flowers in short scaly spike-like catkins and simple rarely pinnately cut stiff leaves containing yellow aromatic resin-glands.

One genus containing 60 species. Widely distributed.

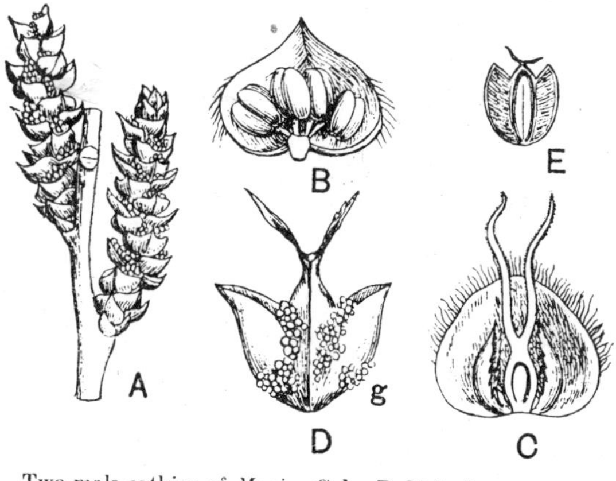

FIG. 3. A. Two male catkins of *Myrica Gale*. B. Male flower of same with subtending bract, enlarged. C. Female flower of *M. aspleniifolia* with subtending bract and pair of lateral bracteoles, enlarged. D. Fruit of *M. Gale*, enlarged; *g*, resin-glands. E. The same cut lengthwise (less enlarged than D) shewing the contained seed also in section. (A, B, D after Wettstein; C, E after Engler.)

Myrica Gale, Sweet-gale or Bog-myrtle, is a twiggy low-growing shrub, common on bogs and moors in the British Isles and occurring in Western and Northern Europe, North Asia and North America as far south as Virginia. The remaining species occur in the mountains of tropical Africa and in South Africa, the Canary Islands, on the Himalayas and the mountains of tropical Asia, in Eastern and North America, the West Indies and on the Andes of South America. In contrast with the restricted development of *Myrica* in Europe at the present day, fossil remains indicate a rich development during the Tertiary period, extending into the Arctic zone.

The leaves, which often shew a $\frac{2}{5}$ phyllotaxy, are stalked, and have an entire, toothed or more or less deeply cut, blade often shewing wide polymorphism on the same plant. On both the upper and lower epidermis, but generally more abundant on the lower, are shallow depressions containing yellow aromatic resin-glands. Root-tubercles similar to those of leguminous plants are frequent. The catkins are simple, as in *M. Gale*, or branched. In *M. Gale* and the American *M. aspleniifolia* they are arranged along short branches which cease growth after development of the stamens or of the fruits and perish after fall of the pollen or seeds.

This character is sometimes regarded as justifying the separation of these two from the remaining species, which then comprise the genus *Myrica*, the Sweet-gale becoming a distinct genus *Gale*, while the American species, in which alone stipules occur, comprise a third genus *Comptonia*. There are also differences in the fruits of the three groups (1).

The catkins arise from axillary buds; the bracts, which are small and scale-like, have, like the foliage-leaves, a $\frac{2}{5}$ spiral arrangement or one easily derived therefrom.

In some species male flowers appear below the female, forming androgynous catkins; bisexual flowers also occur occasionally. The stamens shew a wide variation in number, from two to twenty. Sweet-gale usually has four. They arise spirally on the floral axis which is quite short (*M. Gale*) or undergoes intercalary growth in length separating the filaments at different heights. When bracteoles are present (as in *M. cerifera*) these arise in precisely the same way as the stamens and follow the same phyllotaxy; their arrangement right and left in a plane at right angles to the bract is such as to form with the bract a perianth-like protection to the stamens. The anthers bear four pollen-sacs which open extrorsely by two longitudinal slits.

The female catkins develop more slowly than the male and are much smaller at time of flowering; in *M. Gale* the flowers are produced before the leaves. The ovary is formed at the apex of the floral axis, which bears below a pair of (sometimes three or four) lateral bracteoles, the subsequent development of which varies in the three sections (or genera).

Davey and Gibson (2) find that there is always a small proportion of monoecious plants in *Myrica Gale* which shew all gradations between the normal staminate and pistillate types, and that the sex of a bush or shoot may vary from year to year. They describe three grades of monoecism, namely, plants or shoots bearing staminate and pistillate catkins of the normal type, plants bearing androgynous catkins, and plants with shoots the bulk of whose catkins consist of bisexual flowers. The last-named are intermediate in size between the normal pistillate and staminate catkins. The bisexual flower contains a centrally placed ovary, similar to that of the normal female flower; round it are three or four stamens united to the base of the ovary and just below these on the very short floral axis are two minute lateral outgrowths corresponding to the bracteoles of the normal female flower. Both stamens and stigmas are functional and the flowers are proterogynous. They are capable of producing functional fruit.

Eichler had suggested that the flower was derived from a 2-merous hermaphrodite type: he regarded the additional bracteoles which sometimes occur in the female flower as representing a perianth.

E. M. Kershaw (3) finds that the single integument in the ovule of *Myrica Gale* is provided with a ring of vascular bundles; she compares this with the occurrence of vascular bundles in the ovule-integument of *Juglans* and *Juliania* and regards the phenomenon as indicating the primitiveness of the three families, since a ring of bundles also occurs in the integument in *Trigonocarpus* and other fossil seeds.

In *Gale* each bracteole until after fertilisation is represented only by a small lobule, but during development of the embryo they grow up round the fruit to form a pair of floats which become detached with the fruit and aid in its distribution; the pericarp is thin and tough.

In *Comptonia* the bracteoles develop with the ovary, and gradually envelop it forming a cupule around the nut-like fruit, which ultimately falls out.

In the restricted *Myrica* papillate emergences are developed on the ovary in spiral succession from below upwards. In the fruit, which is drupe-like, the hard mesocarp is covered with a parenchymatous exocarp, the development of which varies in different species. The papillae may become succulent, or covered with hairs, but in most species (e.g. *M. cerifera*) bear a thick white layer of wax. The fruits are spherical or ovoid, and vary from less than $\frac{1}{12}$ in. to $\frac{1}{3}$ in. in diameter. The bracteoles in this section remain rudimentary, or if developed dry up after the flowering period.

The seed has a single thin coat, to the inner face of which adheres an endosperm of a single cell-layer with oil and proteid contents. The rest of the seed is occupied by the embryo, which has a superior radicle and two large plano-convex cotyledons which contain starch and oil, and in germination are carried above the ground by growth of the hypocotyl.

REFERENCES

(1) CHEVALIER, A. "Monographie des Myricacées." *Mém. Soc. Sci. Nat. Cherbourg*, XXXII, 85 (1901).

(2) DAVEY, A. J. and GIBSON, C. M. "Note on the distribution of sexes in *Myrica Gale*." *New Phytologist*, XVI, 147 (1917).

(3) KERSHAW, E. M. "Structure and development of the ovule of *Myrica Gale*." *Annals of Botany*, XXIII, 353 (1909).

Family II. JUGLANDACEAE

Flowers monoecious. Male inflorescence a many-flowered catkin, each flower in the axil of a bract, with a pair of bracteoles and often a perianth of a few scale-leaves; stamens 3–40; a rudimentary pistil is sometimes present. Bract and bracteoles in the female flower free or more or less united with the ovary, as is also the inconspicuous usually 4-merous perianth; carpels two, median or transverse with a pair of stigmatic style-arms; rarely three carpels are present. Fruit a drupe or nut with a thin or fleshy exocarp; endocarp hard with two or four incomplete septa. Embryo with fleshy four-lobed corrugated or foliaceous cotyledons.

Trees with generally large compound imparipinnate aromatic leaves.

Six genera; about 40 species. Warmer parts of the north temperate zone extending into the tropics.

Germination is usually hypogeal. the large fleshy cotyledons remaining in the seed; but in *Pterocarya* the much-divided cotyledons are leaf-like and are carried up by growth of the hypocotyl to form the first green leaves of the plant.

The buds are protected by thick leathery or felted scales, the outer of which, especially in *Pterocarya*, sometimes bear a small pinnately cut rudimentary blade. They are borne somewhat above the leaf-axil; frequently there are two or three in a vertical series. The leaf-scars are large and bear three groups of vascular bundles.

The pendulous male catkins are borne laterally, generally on naked branches of the previous year. The flowers are subtended by a bract with which are united a pair of bracteoles and often several scale-like perianth-leaves; the floral axis is more or less elongated in the direction of the bract and bears an indefinite number of stamens; a central rudiment of a pistil is sometimes present. The short filament bears an erect anther which opens by two lateral longitudinal slits.

The male flower shews considerable variation in number of stamens and development of perianth. *Pterocarya* has a symmetrical four-leaved perianth the anterior member of which is the largest; the three hinder members may be more or less rudimentary or absent, the perianth consisting of three members or being reduced to the anterior one alone. The number of stamens varies between 8 and 16 and their arrangement seems to follow no definite rule; a rudiment of a pistil is sometimes present. Similarly in *Juglans* the number of perianth-leaves is variable, two to five. Eichler figures three as representative, but Nicoloff (5), following De Candolle (1), regards four as the typical form. In the lower flowers of the catkin there are as many as 20 stamens, but in the upper flowers the number is reduced to eight or six; there is no trace of a pistil.

In *Carya* the perianth is absent, and the stamens number

4 to 10, though sometimes reduced to three or two. In *Platycarya* neither bracteoles nor perianth are present; the stamens number 8 to 10; and a pistil is often represented by a rudiment. It would appear therefore that the male flower is derived from a bisexual flower with an indefinite number of stamens and a several-leaved perianth.

The female flowers are generally borne at the end of a leafy shoot of the current year and are few in number, as in

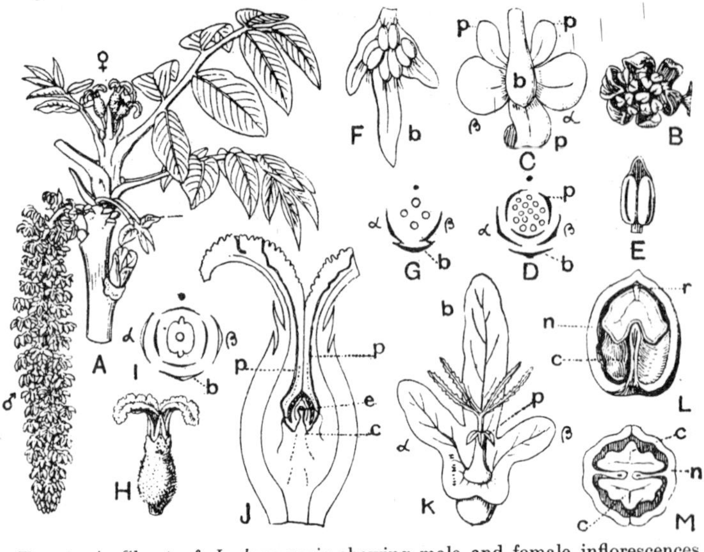

FIG. 4 A. Shoot of *Juglans regia* shewing male and female inflorescences. B. Male flower of same. C. The same viewed from behind shewing arrangements of floral leaves. D. Floral diagram. E. One stamen. F. Male flower of *Carya alba*. G. Floral diagram of same. H. Female flower of *Juglans regia*. I. Floral diagram of same. J. Longitudinal section of same to shew chalazogamy. K. Female flower of *Engelhardtia spicata*. L. "Nut" of *J. regia* cut lengthwise. M. The same cut across. *b*, bract; *α*, *β*, bracteoles; *p*, perianth; *p* (in J), pollen-tube; *e*, embryo-sac; *c*, chalaza; *n*, endocarp: *r*, radicle, *c*, cotyledons. A, L, M reduced, B, C, E, F, J, K enlarged. (A after Wossidlo; C, D, F, G, K after Engler; J after Nawaschin.)

Juglans and *Carya*; in *Pterocarya* and *Engelhardtia* they form many-flowered terminal (*Pterocarya*) or axillary (*Engelhardtia*) pendulous catkins. The female flower consists essentially of a bicarpellary ovary containing an erect ovule growing from the base and surrounded by an apparently single integument, and bearing a pair of styles with stigmas on the inner face.

According to Van Tieghem (7) the integument in *Juglans*, which is provided with a ring of vascular bundles, represents both inner and outer integuments. An additional partial outgrowth at the base of the ovule has also been described in *Juglans*. It consists of a pair of outgrowths from the placenta, apparently composed of a system of concrescent trichomes which grow until they abut on the carpellary walls and more or less fuse with them (6). It is through this tissue that the pollen-tube makes its way to the chalaza.

There has been considerable discussion on the structure and development of the female flower and the relation of the ovule to the carpels. Van Tieghem (7) states that the vascular supply for each of the floral leaves is given off from the stem-stele at the base of the flower, and the ovary is therefore superior; also that the ovule though apparently developed from the top of the floral axis is really an outgrowth from the carpels, one or both of which contribute to its vascular supply; Benson and Welsford (6) come to the same conclusions. Nicoloff (5), on the other hand, does not accept Van Tieghem's deduction from the course of the bundles and maintains that the ovule is a development from the top of the floral axis and that the carpels have no part in its formation.

The ovule is not developed at the time of pollination. The position of the carpels is sometimes median (*Juglans*, *Pterocarya*, *Engelhardtia*), sometimes transverse (*Carya*, *Platycarya*). The position of the stigmas also varies, and may correspond either with the edges or the midribs of the carpels. Karsten (2) states that at the time of pollination and later the union of the edges of the carpels is very often incomplete and even when they are united there is an open passage. The bract and pair of bracteoles are variously united with the ovary which is also crowned by a small, usually four-leaved, perianth. In *Carya* only the posterior perianth-leaf is present and in *Platycarya*, as in the case of the staminate flower, the perianth is altogether absent. In *Juglans* the bract is united with the ovary to a little above the middle and the bracteoles still higher, uniting to form a toothed rim (fig. 4, H): all trace of this is lost as the fruit ripens. In *Pterocarya* bract and bracteoles are almost free in the flower; in the fruiting stage the bract disappears, while the bracteoles, which have become carried up by the growth of the lower portion of the ovary, form a broad wing in the centre of which is the fruit. In *Engelhardtia* (fig. 4, K) bract and bracteoles are united with the lower half of the ovary and increase in the fruit to form a three-lobed involucre similar to that of the Hornbeam. In *Carya* they are united with the ovary to its apex. In *Platycarya* the bract is quite free, while the bracteoles are united with the ovary and develop in the fruit to form a pair of wings.

The flowers open in late spring after unfolding of the leaves and are wind-pollinated. *Juglans* (3), *Carya* (4) and *Pterocarya* (2) are chalazogamic. The pollen-tube grows down the conducting tissue of the style till near the ovary-cavity when it turns and passes down the ovary-wall (fig. 4, J). At a point a little below the funicle the tube curves upwards, passing through the chalaza to the embryo-sac; branching of the pollen-tube was observed. A micropylar canal was present but no pollen-tubes were found to enter it. The fruit is drupaceous, having a more or less succulent epicarp, in the formation of which the bract and bracteoles take part when coherent, and a woody or crustaceous endocarp (stone or "nut"). In *Carya* the epicarp opens regularly by four valves, in *Juglans* irregularly; otherwise the fruit is indehiscent.

The "nut" opens only on germination, splitting in *Juglans* and *Carya* into predetermined halves, the dividing lines corresponding with the middle lines (not the edges) of the carpels. In the interior of the nut are incomplete septa by which the seed, which conforms to the interior, becomes more or less lobed. In every case two so-called primary septa intrude from the line of union of the carpels and uniting in the lower part of the fruit to a varying height form a central column on the top of which the seed is supported (fig. 4, L); as the ovule is erect the radicle is in the upper part of the seed and the cotyledons become two-lobed to fill up the space on either side of the partition which crosses the lower portion of the fruit (*Juglans cinerea*, *Platycarya*, *Engelhardtia*). Frequently, however, two secondary septa are formed at right angles to the primary (and always less developed than these), dividing the chambers again so that the seed becomes four-lobed at the base, as in *Juglans regia* and *J. nigra*, and most species of *Carya* and *Pterocarya*. In *Juglans* and *Carya* there are also additional intermediate outgrowths which never reach the middle of the seed but correspond with shallow or deeper furrows on its contours. Symmetrically arranged hollows, characteristic of different species, are also often formed in the walls and septa by destruction of the tissue during development of the fruit.

Solereder remarks that the family is well characterised anatomically. A distinctive feature is the occurrence of peltate glands in all the species. Periderm development is superficial: and in the wood the medullary rays are not broad, the perforations of the vessels are mostly simple and wood-parenchyma is rather abundant. *Juglans* and *Pterocarya* have a chambered pith.

The family is found at the present day in the warmer parts of the north temperate zone and in tropical Eastern Asia to China and Japan. In North America it occurs mainly on the Atlantic side from Canada to Florida (*Juglans* and *Carya*), with two species of *Juglans* in California. The monotypic *Oreomunnea* is restricted to Central America and there is a species of *Juglans* in the West Indies (Jamaica). *Juglans* alone is found in both hemispheres. *J. regia* (Walnut) in prehistoric times grew wild in the western Mediterranean region, but is now native in Europe only in Greece whence it is found through Asia Minor to the Himalayas and Burmah. The allied *J. nigra* is native in the Atlantic United States. Fossil remains of leaves and fruits from Cretaceous and Tertiary strata indicate a former more northerly distribution of the family (represented by *Juglans*) extending as far as Greenland, Saghalien and Alaska, and also a greater community of representation between the eastern and western hemispheres; for instance, species of *Juglans* and *Carya* described from Tertiary beds in Europe closely resemble present-day American species. The former distribution of the family resembled that of Fagaceae at the present day in the northern hemisphere.

The seeds are rich in oil and several are edible; e.g. *Juglans regia* (Walnut), *J. nigra* (American Black Walnut) and species of *Carya* (Hickory, Pecan). *Juglans*, *Carya* and the Eastern Asiatic *Engelhardtia* are useful timber trees. *Pterocarya fraxinifolia* (Transcaucasia) is often cultivated as an ornamental tree. The family as a whole is rich in bitter principles and tannin.

REFERENCES

(1) C. De Candolle. "Mémoire sur la famille des Juglandées." *Ann. Sci. Nat.* (*Bot.*), sér. 4, XVIII, 5 (1862).

(2) Karsten, G. "Ueber die Entwickelung der weiblichen Blüthen bei einigen Juglandaceen." *Flora*, XC, 316 (1902).

(3) NAWASCHIN, S. "Ein neues Beispiel der Chalazogamie." *Botan. Centralbl.* LXIII, 353 (1895).

(4) BILLINGS, F. H. "Chalazogamy in *Carya olivaeformis*." *Botan. Gazette*, XXXV, 134 (1903).

(5) NICOLOFF, TH. "Sur le type floral et le développement du fruit des Juglandées." *Journ. de Botanique*, XVIII (1904), XIX (1905).

(6) BENSON, M. and WELSFORD, E. J. "Morphology of the ovule and female flower of *Juglans regia*." *Annals of Botany*, XXIII, 623 (1909).

(7) VAN TIEGHEM, PH. "Anatomie de la fleur femelle et du fruit du noyer." *Bullet. Soc. Botan. France*, XVI, 412 (1869).

Order 4. *JULIANIALES*

Family. JULIANIACEAE

Flowers dioecious, green, inconspicuous. Male flowers in catkins, or in simple or branched racemes, with a simple 3- to 9-partite perianth and an equal number of free stamens alternating with the perianth-segments; no pistil-rudiment. Female flowers three or four, collateral in an almost closed involucre, consisting only of a unilocular ovary and a deeply 3-partite exserted style. Ovule solitary, hemianatropous, with a single integument, and partially concealed by a much larger fleshy ventral development of the funicle. Fruit samaroid, compound and nut-like, enclosed in and adnate to the persistent involucre, and borne on the flattened wing-like stalk of the partial inflorescence. Endosperm absent; radicle long, applied to the edges of the large plano-convex cotyledons, which are epigeal in germination.

Small resin-containing trees or shrubs with deciduous alternate imparipinnate exstipulate leaves.

One family, Julianiaceae[1], with two genera—*Juliania*, with four species in Mexico, and *Orthopterygium*, a monotypic genus in Peru.

This small isolated group is placed between Juglandales and Fagales. It resembles the Juglandaceae in having alternate imparipinnate exstipulate leaves, in the dissimilarity of the male and female flowers, the broad stigmatic lobes of the style and the single integument of the ovule. The vegetative characters shew no resemblance to those of Fagales but the characters of the inflorescence and flower suggest a closer affinity with this order than with Juglandales. The male inflorescence and flowers and also the small globose pollen closely resemble those of species of *Quercus*, while the several female flowers arranged side by side

[1] See Hemsley, W. B. "On the Julianiaceae: a new natural order of plants." *Trans. Roy. Soc. Lond.* ser. B, CXCIX, 169–197 (1907).

in a closed persistent involucre recall strikingly what obtains in *Fagus* and *Castanea*. Julianiaceae differ in that the involucre remains closed in the fruit and the flattened nuts are adnate to its inner wall; they have a hard sclerenchymatous pericarp. Julianiaceae are also distinguished by the one-celled, one-ovuled ovary; but the seeds, as in Fagales, are exalbuminous and the cotyledons are epigeal in germination. In the remarkable structure of the ovule the family is unique. A resemblance also exists to the Anacardiaceae in the foliage, the presence of resin, the reduced unisexual flowers, the solitary exalbuminous seed and the form of the embryo; there is also a resemblance in anatomical characters.

Order 5. *FAGALES*

Flowers unisexual, in catkin-like generally compound inflorescences; the female flowers often a few together. Male flowers generally adnate to the bracts, sometimes with a simple inconspicuous perianth; female protected by the bract and bracteoles, sometimes with an inconspicuous superior perianth, of two or three (rarely six) carpels forming an ovary of as many cells, each with one or two pendulous ovules. Fruit a one-seeded nut associated with the persistent bract and bracteoles. Endosperm absent, embryo straight.

Trees or shrubs with simple alternate leaves, usually in two rows, with deciduous stipules.

Two families, Betulaceae and Fagaceae.

Family I. BETULACEAE

Flowers monoecious in unisexual catkin-like compound inflorescences. Perianth, when present, of small scale-leaves which vary in number and are free or united; superior in the pistillate flower. The male flowers adhere to their bract and contain 2–12 stamens which are generally divided; there is no rudiment of a pistil. The female flowers have a bilocular ovary with two styles; each chamber contains a pendulous anatropous ovule with one integument. Fruit with one seed; during its ripening the bracts undergo considerable development. The embryo with its large oil-containing cotyledons fills the seed.

Genera 6; species more than 100.

The family is to-day, as in past ages, characteristic of the north temperate regions of both hemispheres. In the Old World species of *Alnus* occur as far south as Bengal, while in the New World *Alnus acuminata* extends along the Andes from Mexico to Argentina; *Ostrya* and a species of *Carpinus* are found in Mexico.

Except in the Hazel (*Corylus*), where they remain in the seed and fruit, the cotyledons are epigeal in germination, forming a pair of small shortly-stalked green leaves. The foliage-leaves have generally a crenate or dentate margin and are spirally arranged on the primary axis, but on the branches become distichous, except in the largest section of Alders where the arrangement is one-third. Except in the Alders (*Alnus*) terminal buds are not found in the adult plant. The stipules serve only for bud-protection and fall when the leaves unfold.

In most species of *Alnus*, including the British representative, *Alnus glutinosa*, the buds are enclosed only by the pair of stipules of the first foliage-leaf. In *Betula* (Birch) the covering consists of two or three pairs of bladeless stipules; in *Carpinus* (Hornbeam) and *Corylus* (Hazel) these are preceded by two entire scales. The leaf-blade is generally plicate along the lateral nerves; in *Corylus* it is folded at the midrib with upwardly directed halves (conduplicate). The young branches and leaves in the bud bear glands in which a resinous secretion forms beneath the cuticle, and by the rupture of the outer layers of the latter is poured out on the surface. When the leaves unfold the glands sooner or later (*Alnus glutinosa*) disappear.

Periderm arises in the outermost layer of the cortex. In *Corylus* and *Betula* the cork-layers are alternately composed of wide and narrow cells; peeling of the bark is caused by the tearing of the former. Birch-bark owes its white appearance to the scarcity of cell-contents in the cork-layers.

The catkins terminate the growth of branches which bear below them foliage-leaves or at least bud-scales. In *Alnus*, species of *Betula*, and *Ostrya* the male catkins are developed at the ends of last year's leaf-shoots, and pass the winter unprotected by bud-scales. A tassel of catkins often results from their production also on short branches just below the end of the same shoot. In *Corylus*, where they are also naked, they are borne on lateral dwarf-shoots, which are usually arranged in a racemose manner, and similarly, but protected by bud-scales, in *Carpinus*, *Ostryopsis* and many species of *Betula*.

The female catkins of *Corylus* and *Carpinus* terminate leafy branches. In the former the branch does not elongate till the flowers have been pollinated, the red stigmas projecting in a tuft from the tip of the axillary bud. In *Betula* and some species of *Alnus* (§ *Alnaster*) they are borne on few-leaved dwarf-shoots, often with the addition of several lateral catkins forming a raceme. In the two other sections of *Alnus* (including the British species) the female catkins (solitary or in racemes) spring from a leaf-axil below the male, without bud-scales or foliage-leaves.

The bracts are arranged spirally on the axis of the catkin which is almost always a compound inflorescence, contrasting with the simple spike of the Salicaceae and Juglandales. In what may be considered the typical case, there is in the axil of each bract a group of three flowers forming a dichasium; the middle flower is, however, often absent; more rarely this alone is present, and the catkin becomes a simple spike.

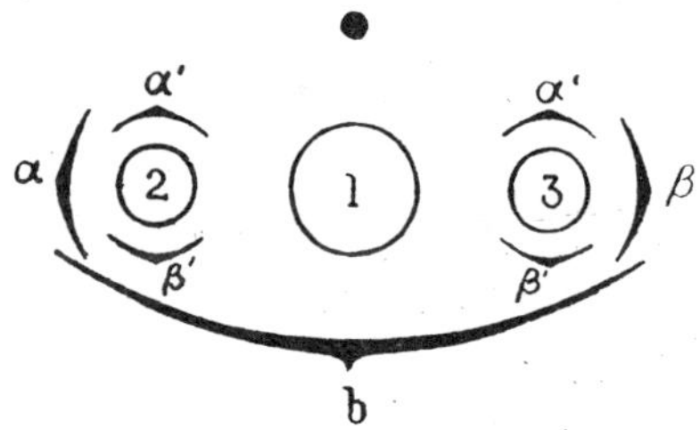

FIG. 5. Typical dichasium of Betulaceae *b*, bract subtending 1, central flower. *α*, *β*, lateral bracteoles subtending the pair of lateral flowers 2, 3; *α'*, *β'*, bracteoles of the lateral flowers.

The male flowers are more or less united with the bract and bracteoles. A perianth is absent or there is one of small scale-leaves with free or united parts. The number of stamens varies, the Birch has two, the Hazel four, the Hornbeam ten. Except in *Alnus* and *Ostryopsis* they are more or less divided, either the anther-halves only being separated or the filament being also more or less split lengthwise.

In the female flowers the perianth is absent or present and epigynous, consisting of a varying number of small scales. There are always two carpels, united to form a bicarpellary ovary; except in *Corylus*, they are placed transversely to the floral bract. At the period of pollination, which is effected by wind, only the two long styles have developed; the ovary cavity and ovules form later. Each of the two loculi contains a single pendulous anatropous ovule, with one integument.

Owing to the abortion of one of the two ovules the fruit is one-seeded. The pericarp is thin, or stout and hard as in the Hazel; it is crowned by the remains of the styles and the perianth, if present. The bracts become more or less changed during the ripening of the fruit. In the Birch and some Alders the fruit bears a pair of lateral wings which assist its dispersal. The embryo with its large oily cotyledons fills the seed.

As we have seen, the female flower consists of but little more than the pair of protruding styles at the time of pollination. Several months may elapse before fertilisation occurs, the pollen-tubes remaining at rest in the tissue of the style during the development of the ovary and ovules. On resuming growth, the pollen-tube (in *Betula*, *Alnus*, *Corylus* and *Carpinus*, the genera examined), having entered the ovary-cavity, does not penetrate the micropyle but passes directly into the tissue of the placenta, and entering the funicle passes up the raphe into the nucellus by way of the chalaza. In the Alder the route is an exceptionally roundabout one, the tube passing into the upper part of the nucellus and then bending down again to enter the embryo-sac (which is in the middle of the nucellus) at the upper end; a blind recurving branch is formed on the tube at its entrance into the nucellus.

In *Betula* and *Alnus* there is a single embryo-sac of the usual structure. *Corylus* and *Carpinus* on the other hand shew a resemblance to *Casuarina* in the formation of a sporogenous tissue from several axial strands in the nucellus. Several embryo-sacs are formed, one (*Corylus*) or more (*Carpinus*) of which send blind prolongations (caeca), resembling those described by Treub in *Casuarina*, down to the base of the nucellus. In *Carpinus* the pollen-tube enters the base of the caecum and passes up on its way to the definitive nucleus and the oosphere. The course of events in the ovule of *Carpinus* so closely resembles that in *Casuarina* as to have suggested a relationship between the two genera (1, 2, 3).

Two subfamilies may be distinguished; they are often regarded as distinct families but have much in common, including the chalazogamic route of the pollen-tube.

Subfamily I. CORYLOIDEAE. The male flowers are solitary, the inflorescence being a simple spike; they have no perianth. The female flowers have a perianth; the bract and pair of bracteoles unite in various ways in the fruiting stage. There are four genera. *Ostryopsis* has two species in Mongolia and China; *Ostrya* has seven species in Southern Europe, Asia Minor, Eastern Asia and North and Central America. Several species

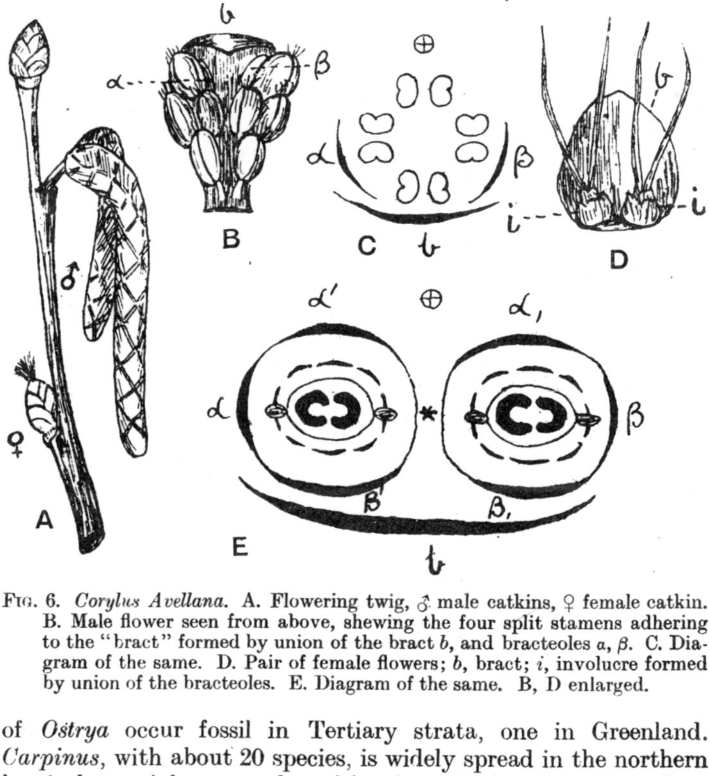

FIG. 6. *Corylus Avellana*. A. Flowering twig, ♂ male catkins, ♀ female catkin. B. Male flower seen from above, shewing the four split stamens adhering to the "bract" formed by union of the bract *b*, and bracteoles *a*, *β*. C. Diagram of the same. D. Pair of female flowers; *b*, bract; *i*, involucre formed by union of the bracteoles. E. Diagram of the same. B, D enlarged.

of *Ostrya* occur fossil in Tertiary strata, one in Greenland. *Carpinus*, with about 20 species, is widely spread in the northern hemisphere. A large number of fossil species have been described from the Tertiary, but most of these shew very slight differences from the Hornbeam (*C. Betulus*), a native of the southern half of England, where it forms a small tree with smooth light grey bark and white wood of close texture; it extends through Central and Southern Europe to Persia. In the bud the blade is plaited parallel to the lateral nerves. The female flowers, as in *Corylus*, are in pairs. The hard green fruit, about $\frac{1}{4}$ inch long, is borne on the base of a leafy trilobed bract which is formed by the union of the bract and bracteoles and aids in the distribution of the somewhat heavy fruit by the wind. *Corylus* has about eight species, with a range extending from Central and Southern Europe to Eastern Asia, and throughout temperate and North America. Fossils are known from the Tertiary, one species occurring in the Miocene within the Arctic zone. The Hazel (*C. Avellana*) is a widely spread European bush occurring in the British Isles as far north as Orkney. The asymmetrical leaves are plaited parallel to the midrib in the bud. The female flowers of *Corylus* differ from those of the other three genera in having the carpels placed medianly (fig. 6, E), not transversely, to the subtending bract, which in the fruit forms with the bracteoles a green irregularly cut cup surrounding the central nut. Bisexual flowers have been observed in the Hazel.

Subfamily II. BETULOIDEAE. The male flowers are in three-flowered dichasia in the axil of the bract, and have a perianth. The female flowers have no perianth; their bracts become united in the fruit with the bract of the inflorescence. Bisexual flowers occasionally occur. There are two genera, *Alnus* and *Betula*, both British.

Alnus (Alder). The structure of the inflorescence is indicated in the diagrams (fig. 7). In the male catkin each bract subtends a three-flowered dichasium, the lateral flowers of which have only one (the inner) bracteole apiece. Each flower has a perianth of four members, opposite which are the four entire stamens. The female inflorescence differs in the absence of the central flower and of a perianth. The main bract and the four bracteoles unite in each case to form the five-lobed "cone-scale," which in the fruit becomes hard and woody, the catkin resembling a small cone. The fruit is a nut with or without a wing. The genus contains 17 species of trees or shrubs and is generally distributed through the north temperate zone; two species occur in the Himalayas and further India, and one (*A. jorullensis*) extends from Mexico along the South American Andes as far as Argentina. *Alnus glutinosa*, the British species, grows in damp situations, ascending to 1600 ft. in the Highlands; it also occurs in Europe, reaching nearly to the Arctic circle, in North Africa and in West and North Asia.

Betula (Birch). The bracts of both male and female catkins subtend three-flowered dichasia; the lateral flowers have no bracteoles, while those of the median flower unite with the bract of the inflorescence forming a three-lobed scale. In the male flower the perianth is often reduced to the two median leaves, and there are only two, median, stamens, which are divided. The female flower is like that of the Alder. There are about 40 species of trees or shrubs, distributed through the north temperate zone, extending northwards beyond the Arctic circle, *B. glandulosa* occurring in Arctic America and *B. nana* in Kamschatka and Greenland. In the Old World the southern limit is reached in the Himalayan region and Southern China; in the New World in the Southern United States.

The genus is represented in Britain by two widely distributed species, *Betula alba* and *B. pubescens*, which often occur together and frequently hybridise. They are trees generally reaching 40 to 50 ft. in height with a silvery white peeling bark. *B. alba* has resinous glands but no hairs on the young branches, and acuminate leaves; *B. pubescens* has young branches softly hairy and leaves not acuminate. The form of the cone-scales is also different in the two. They occur in Europe and North Asia, *B. pubescens* attaining the most northerly latitude, often marking the limit of tree-growth.

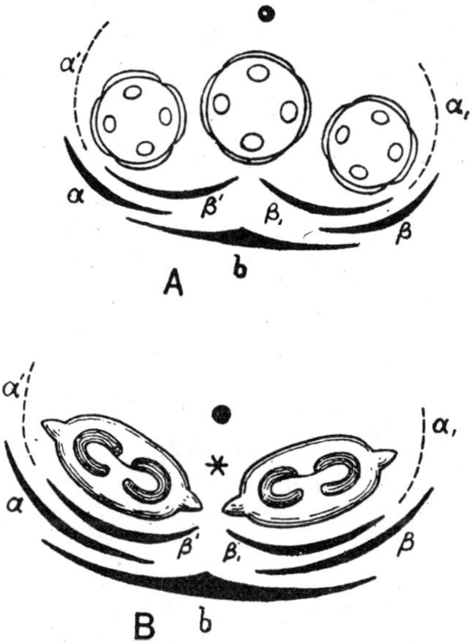

FIG. 7. A. Diagram of male dichasium of *Alnus*. B. Diagram of female dichasium. *b*, bract subtending the central flower (which is absent in B); α, β, lateral bracteoles subtending the pair of lateral flowers; α', β' and α_1, β_1, bracteoles of the lateral flowers of which α' and α_1 are suppressed.

Betula nana, a bush with shortly stalked roundish leaves, occurs locally on the mountains of Northumberland and in the Highlands, ascending to 2700 ft. Its range includes Arctic and Alpine North-west and Western Europe, North Asia and North America.

More than 30 fossil species of *Alnus* have been described, chiefly from Tertiary but also from Cretaceous beds. Many of the determinations are, however, extremely unsatisfactory, being based merely on leaves. *Alnus Kefersteinii* was a very widespread species in the Arctic regions in Miocene times, occurring in both Old and New Worlds. A still larger number of fossils are recorded as *Betula*. The genus was certainly present in the Tertiary, but doubtfully in the Cretaceous.

REFERENCES

(1) BENSON, M. "Contributions to the embryology of the Amentiferae, Part I." *Trans. Linn. Soc.* ser. 2 (Botany), III, 409 (1894).

(2) BENSON, M., SANDAY, E. and BERRIDGE, E. Part II. *Op. cit.* VII, 37 (1906).

(3) NAWASCHIN, S. "Zur Entwickelungsgeschichte der Chalazogamen. *Corylus Avellana.*" *Bull. Acad. Imp. Sci. St. Pétersbourg*, sér. 5, x, 375 (1899).

Family II. FAGACEAE

Flowers monoecious (except in *Nothofagus*), the male usually in catkin-like inflorescences, the female generally few. Male flowers solitary or cymose, often with a scale-like perianth; a pistil-rudiment is often present. Female flowers with an inferior generally tricarpellary ovary, bearing an inconspicuous six-leaved perianth; styles three; ovary generally three-celled with two pendulous anatropous ovules in each cell; ovule with two integuments. Fruit with one seed, invested at maturity with a characteristic cupule. The embryo fills the seed with its large cotyledons.

Genera 5; species about 400.

Except in the Beech, where they become aerial and green, the cotyledons are hypogeal in germination. As generally happens in such cases the leaves immediately succeeding the cotyledons are reduced to scales and the seedling devotes its energies to pushing its stem above the superincumbent decaying leaves or surrounding vegetation before producing its first foliage-leaves.

The plants are generally trees, often attaining a considerable size. A bushy growth is not infrequent, as in Holm Oak (*Quercus Ilex*), while some species of the south temperate genus *Nothofagus* have a dwarfed stem.

In *Quercus* (Oak) and *Pasania* (an important tropical Asiatic genus) the leaves are arranged spirally, generally with a divergence of $\frac{2}{5}$. In *Fagus* (Beech) and *Nothofagus* they are distichous, and the same arrangement holds on the less vigorous lateral shoots of *Castanea* (Chestnut). Many of the species are evergreen. A terminal bud is developed except on lateral shoots of the Chestnut. The stipules are concerned only with bud-protection. In *Nothofagus* the blade is generally flat, in *Fagus* folded along the lateral nerves, in *Quercus* and *Castanea* folded along the midrib with the halves convolute (*Castanea*) or superposed, convolute or revolute (*Quercus*).

The cork-cambium originates in the outermost layer of the cortex, and in *Fagus* persists throughout the life of the plant.

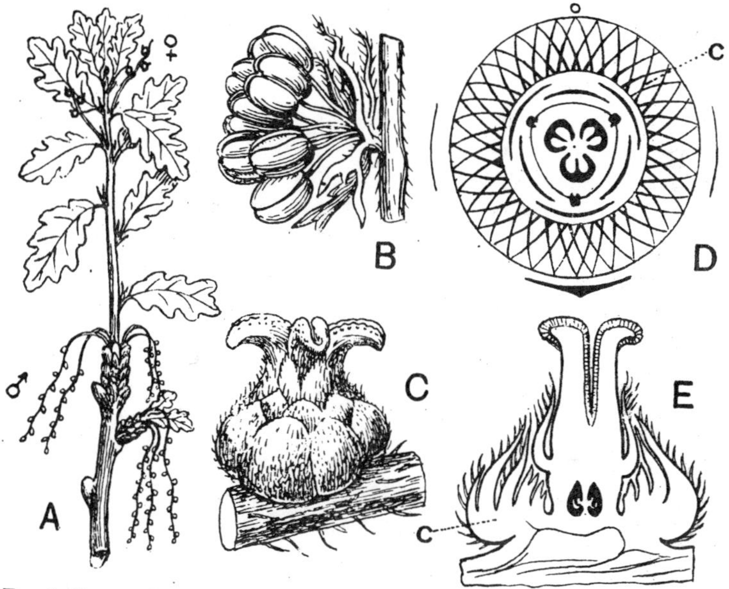

FIG. 8. *Quercus Robur*. A. Shoot bearing male, ♂, and female, ♀, inflorescences. B. Male flower. C. Female flower. D. Diagram of female flower. E. Longitudinal section of female flower; *c*, cupule. A reduced; B, C, E enlarged. (A, D after Eichler; B, C, E after Hempel and Wilhelm.)

Nothofagus has dioecious flowers; the remaining genera are monoecious; bisexual flowers occasionally occur. In *Nothofagus* and *Fagus* both male and female flowers stand solitary[1] or in dichasia in the axils of foliage-leaves; in other cases they form simple or compound catkins. In *Quercus* and in some species of *Pasania* and *Castanea* the male catkins are borne in the lower foliage- or scale-leaf-axils of the shoots

[1] Čelakovský regards the male "flower" as a condensed catkin, probably still with a terminal flower.

of the current year, the female in the upper leaf-axils of the stronger shoots. In most species of *Pasania* and *Castanea*, e.g. the Chestnut (*C. vulgaris*), the upper catkins are androgynous, bearing female flowers only at the base and male above. The male flowers of *Quercus* are solitary and have no bracteoles, in *Fagus* the bract also is absent; in *Castanea* and *Pasania* they are arranged in 3–7-flowered cymes with six bracteoles. The four to seven perianth-leaves are united at the base; the stamens are opposite and equal in number to them, or more numerous. In *Castanea* and *Pasania* the pistil is represented by a hairy protuberance.

The female flowers are in three-flowered dichasia in most species of *Nothofagus*, *Castanea* and many of *Pasania*. In *Fagus* the central flower is absent; in *Quercus* and in species of the other three genera the flowers are solitary. The solitary flowers and the flower-groups, except in *Pasania*, are surrounded by the *cupule*, a cup-like structure bearing numerous scales, and formed according to Eichler by the union of the bracteoles. Schacht, on the other hand, regarded it as an axial structure independent of the bracteoles, the scales representing true leaves. In favour of the latter view is urged the case of the three-flowered groups of *Pasania* (fig. 9, L), where each flower has a cupule resembling in structure and mode of development that of the Oak, and outside this the typical number (six) of bracteoles. The narrow scales at the base of the flower-group in *Fagus* (fig. 9, H, I) represent, on this view, the bracteoles; Eichler regarded them as merely the lowest scales of the involucral cupule. Čelakovský (1) also regards the cupule as a ring-like swelling of the axis beneath the flower bearing acropetally developed scales.

The female flower has an inferior tricarpellary ovary and a generally six-leaved perianth. In *Castanea* (§ *Eu-Castanea*), we find sometimes four or five, usually six carpels. The number of styles corresponds with that of the carpels. In *Castanea* and *Pasania* the stigma occupies the tip of the styles, in the other three genera the upper surface. The ovary is three-chambered in its lower part, bearing on each axile placenta a pair of pendulous ovules each with two integuments. In *Fagus* and *Castanea* this degree of development is reached before pollination but in *Quercus* and *Pasania* only subsequently. The pollen-tube enters the ovule through the micropyle.

Pollination is generally described as by wind-agency and this no doubt holds in *Quercus* and *Fagus*, where the inflorescence is pendulous. But in *Castanea* the large male catkins are directed obliquely upwards, are rendered conspicuous by the yellow perianth and anthers and have a strong scent; the pollen-grains also tend to cling together. The female flowers have a stiff style with a somewhat sticky stigma. These factors suggest entomophily, and the catkins are visited by honey-bees, flies and beetles. The male catkins of *Pasania* are also described as erect with styles and stigmas as in *Castanea*. Mr H. N. Ridley informs me that in Malaya the flowers of *Castanopsis* (which closely resembles *Castanea*) and *Pasania* have a strong scent and attract quantities of flies.

The development of the macrospore presents several interesting departures from the general type (2). In the three genera in which it has been studied, *Fagus*, *Castanea* and *Quercus*, certain resemblances to that of *Casuarina* and the Betuloideae have been discovered. In *Fagus* there is a considerable development of sporogenous tissue resembling Strasburger's description of *Rosa livida*, and occasionally more than one embryo-sac is produced. In *Castanea* and *Quercus* there is no suggestion of a sporogenous tissue, but in the former certain cells round the base of the macrospore, which have a similar origin to sporogenous cells of *Fagus*, become tracheides. These tracheides are of interest from the fact that they are developed also in *Casuarina* from cells of similar origin, which undoubtedly represent sterile sporogenous cells. In all the three genera the embryo-sac becomes elongated below into a caecum. In *Quercus* the pollen-tube also forms short blind branches before entering the micropyle.

In the development of the fruit all the ovules become aborted except one. The one-seeded nut has a leathery or hard pericarp. The familiar differences in appearance arise from the various developments of the cupule and the number of fruits which it surrounds. In *Castanea* it forms four spine-bearing valves enclosing the three nuts; in *Fagus* four bristly segments surround two triangular nuts; in *Quercus* the smooth or scaly cup more or less completely envelops a solitary fruit.

There are five genera, two of which, *Fagus* and *Quercus*, are British. *Fagus sylvatica*, the Beech, forms woods, especially on chalk and limestone. It grows as far north as 60° in Norway. and finds its north-eastern limit in a line drawn from there to the Crimea. In South Europe it grows only on the mountains, finding its upper limit in the Alps at about 5000 ft. A closely allied species occurs in Western Asia, two species occur in Japan, and one in the Eastern United States. Fossil remains, closely allied to the European and American species, occur in Cretaceous and Tertiary strata, and indicate a former wider distribution of the genus embracing California, Greenland, Spitsbergen and Iceland. The closely allied genus *Nothofagus* (with about 12 species) occurs in Antarctic South America, New Zealand and South Australia. *Quercus Robur* (Oak) ascends to 1350 ft. in the Highlands and spreads throughout Europe almost to the Arctic circle, and ascends to 3300 ft. on the Alps; it is native also in Asia Minor and on the Atlas Mts. of North Africa. The genus comprises more than half the family, containing more than 200 species, chiefly in the north temperate zone and the mountains of tropical Asia. It is absent from Central and South Africa, South America and Australasia. Many species are evergreen, such as Holm Oak (*Q. Ilex*) characteristic of the Mediterranean region and often planted in Britain, and the Cork Oak (*Q. suber*) a western Mediterranean tree, the bark of which yields the cork of commerce. The cupules of *Q. vallonea* and *Q. macrolepis* are used commercially for their copious tannin contents. A number of species have been described from Cretaceous and Tertiary formations. *Castanea vulgaris* (Chestnut), a native of the Mediterranean region, is widely cultivated. The genus contains, besides, two species in Eastern North America and one in China and Japan. Allied genera are *Castanopsis*, with about 25 species in tropical Asia, and *Pasania*, with about 100 species in Malaya and the warmer parts of the Pacific area.

REFERENCES

(1) Čelakovský, L. "Ueber die Blüthenstände der *Quercus ilicifolia* und die Eichel-cupula." *Oesterr. Botan. Zeitschrift*, XLIII, 272 (1893).

(2) Benson, M. See (1) p. 30.

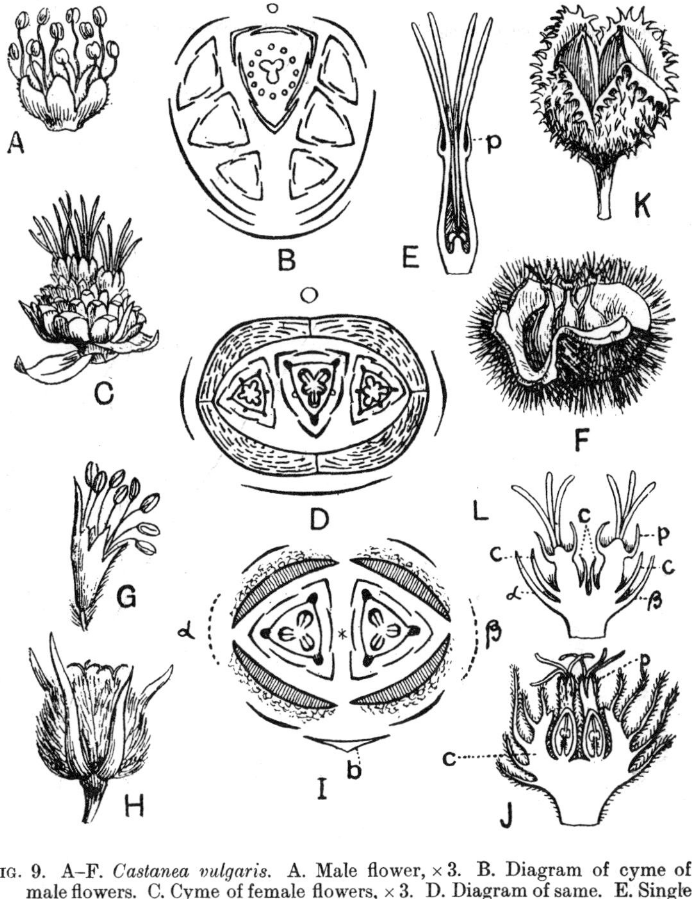

FIG. 9. A–F. *Castanea vulgaris*. A. Male flower, × 3. B. Diagram of cyme of male flowers. C. Cyme of female flowers, × 3. D. Diagram of same. E. Single female flower in longitudinal section, enlarged. F. Fruit, × ½. G–K. *Fagus sylvatica*. G. Male flower, × 3. H. Cyme of female flowers, × 1½. I. Diagram of same. J. Same in longitudinal section, enlarged. K. Fruit. L. Cyme of female flowers of *Pasania fenestrata* in longitudinal section. *b*, bract; *a*, *β*, bracteoles; *c*, cupule; *p*, perianth. (After Baillon, Eichler and Prantl.)

Order 6. *CASUARINALES*

Family. CASUARINACEAE

Flowers monoecious, extremely simple, in catkin-like spikes each in the axil of a bract and protected by a pair of lateral bracteoles. Male verticillate, consisting of a single stamen with a rudimentary perianth. Female consisting of a bicarpellary ovary bearing two long stigmas; ovary unilocular by suppression of the posterior cell, with generally two parietal ascending orthotropous ovules, only one of which develops to maturity. Wind-pollination; fertilisation chalazogamic. Fruit a one-seeded winged nut protected by the hardened bracteoles. Embryo large, filling the seed.

Shrubs or trees with xerophytic habit, mainly Australian. One family Casuarinaceae containing one genus *Casuarina*, with about 40 species.

The habit recalls that of *Ephedra* in the switch-like branches bearing alternating whorls of scale-like leaves which unite at the base to form a sheath. Germination is epigeal; the broad green sessile cotyledons are succeeded by a pair of much reduced leaves, followed by a number of decussating whorls of four leaves resembling those characteristic of the adult plant. The number of leaves in a whorl varies with the species from 4 to 12. The internodes are furrowed; the ridges contain green palisade parenchyma; the stomata are sunk in the furrows and protected by long protruding hairs. The branches are whorled.

The anatomical structure of the stem is peculiar (fig. 10, K). A transverse section of a young branch shews two rings of bundles which alternate. The wood contains vessels, tracheides, parenchyma and fibres.

The male flower-spikes are catkin-like and erect, generally at the ends of branchlets. The flowers are in whorls, each in the axil of a bract which with the other bracts at the node forms a sheath protecting the young flowers (fig. 10, B); there are a pair of bracteoles and two small median perianth-leaves of which rarely only the posterior is present. The single central stamen, which shews a tendency to split, has a filament at first short and a two-lobed anther which opens lengthwise.

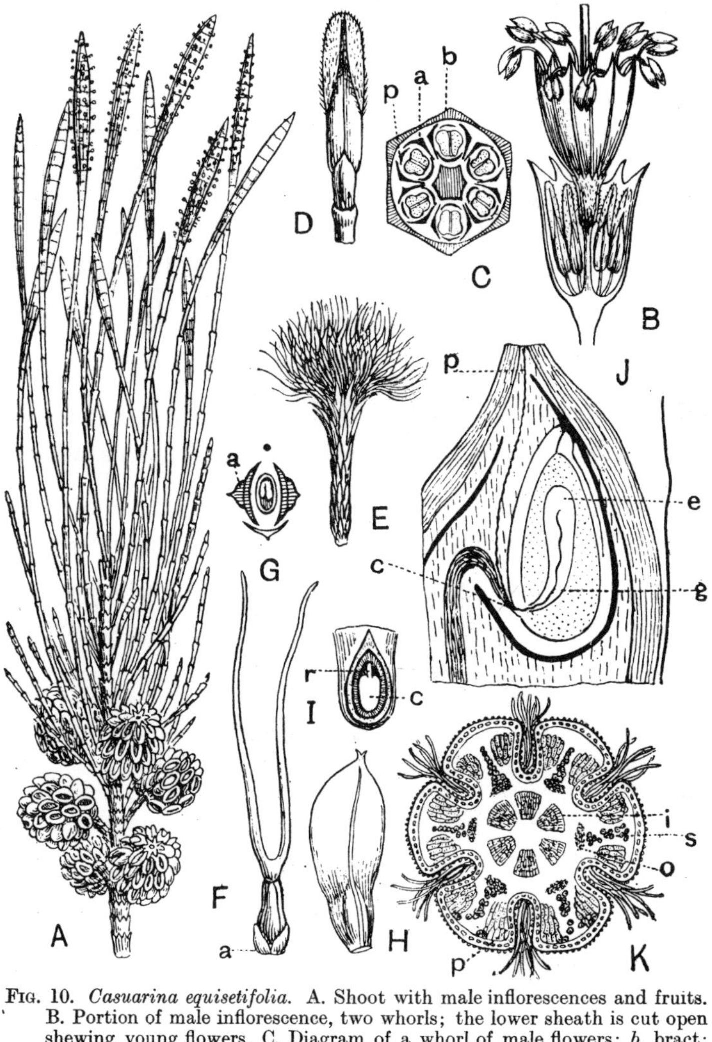

FIG. 10. *Casuarina equisetifolia.* A. Shoot with male inflorescences and fruits. B. Portion of male inflorescence, two whorls; the lower sheath is cut open shewing young flowers. C. Diagram of a whorl of male flowers; *b*, bract; *a*, bracteole; *p*, perianth. D. Male flower before elongation of the filament; the perianth-leaves are being carried up on the anther. E. Female inflorescence, with long, protruding stigmas. F. Female flower; *a*, bracteole. G. Diagram of same. H. Fruit. I. Lower portion of same with longitudinal section of seed; *r*, radicle; *c*, cotyledon. J. Portion of longitudinal section of an ovary shewing functional ovule and course of pollen-tube; *c*, chalaza; *e*, embryo-sac; *g*, caecum of embryo-sac; *p*, pollen-tube. K. Transverse section of a branchlet; *i*, *o*, inner and outer rings of bundles; *s*, sclerenchyma; *p*, palisade tissue. A, $\frac{2}{3}$ nat. size; the other figures variously enlarged. A after Poisson; C, D, G after Engler; J after Treub.

Before dehiscence the filament lengthens, the perianth which at first covers the anther is pushed off (fig. 10, D) and the open anther is then thrust above the sheath (fig. 10, B). The female flowers are crowded at the end of short lateral branches. Each stands singly in the axil of a bract with a pair of bracteoles and consists of a bicarpellary pistil in the median plane bearing two long stigmas. The stigmas are mature before the ovary with its ovules has begun to develop. The ovary is originally bilocular but becomes unilocular by suppression of the posterior chamber, and contains two parietal ovules each with two integuments; the functional ovule becomes connected with the base of the style by a bridge of tissue down which the pollen-tube (p) passes to the chalazal region (fig. 10, J).

The course of events preceding fertilisation shews remarkable deviations from the normal[1]. A many-celled sporogenous tissue is formed from which arise many embryo-sacs and also in some species equivalent spirally thickened tracheides. The embryo-sacs contain in the earlier stages the usual eight nuclei, but the antipodal cells are very transitory; the embryo-sacs form long tails (caeca) in the direction of and sometimes penetrating the chalaza. The pollen-tube passes from the tissue of the style down the raphe and enters the ovule by the chalaza, often passing up the caecum of an embryo-sac; the oblique position of the chalaza with regard to the raphe facilitates its entrance. The tube may branch at or before its entry into the ovule. Only one embryo-sac in an ovary is fertilised; "double fertilisation" occurs, and the oospore rests for some time during development of the endosperm. The development of the embryo is normal.

The bracteoles become woody and form five valves enclosing the compressed nut-like fruit which is expanded above into a membranous wing. The ripe catkins resemble small cones. The embryo fills the seed and is straight, with an upwardly directed radicle and a pair of large flat cotyledons.

Casuarina contains about 40 species and finds its chief development in Australia, but occurs also in south-east tropical Asia, the Mascarene Islands and the islands of the Pacific Ocean.

[1] Treub, M. "Sur les Casuarinées et leur place dans le système naturel." *Annal. Jard. Botan. Buitenzorg*, IX, 145 (1891). See also Rendle, A. B. "A new group of Flowering Plants." *Natural Science*, I, 132 (1892).

The wood is very hard, and that of *C. equisetifolia* (She-oak), which is widely cultivated in the warmer parts of the world, is known as iron-wood.

The foregoing orders have, if we except the catkin-formation (an adaptation to wind-pollination which occurs in other alliances) little in common, and, as has been pointed out, differ fundamentally in the important characters of ovary and ovule. *Casuarina* stands apart in its geographical distribution—it is a dominant Australian type—and in the morphology of its vegetative organs. It shews, however, a striking resemblance in the structure and development of ovary and pistil and the events associated with pollination to the Betulaceae (especially *Carpinus*) and it is possible to regard the male flower as an extreme case of reduction from the type of that family. The form of stem and leaf is obviously a highly adaptive one, but it is difficult to trace underlying this adaptation any suggestion of an affinity with the Fagales. *Casuarina* plays an important part in the system which derives the Angiosperms from the Gymnosperms. Wettstein[1] has used it as an illustration to support his theory of the development of the angiospermous flower from the inflorescence of an *Ephedra*-like ancestor. This, of course, implies that the genus is a primitive, not an extremely reduced, form.

The plants included in these orders are all woody plants with generally small simple leaves (large and compound in Juglandaceae), arranged alternately (except in *Casuarina*), and unisexual flowers arranged generally in catkin-like inflorescences. The flowers stand in the axil of a bract, with a pair of bracteoles (absent in Salicaceae) and frequently with an inconspicuous perianth which in the female flowers is superior. In some families a rudiment of a pistil is present in the male flower, and bisexual flowers occasionally occur. This supports the suggestion that they are derived from ancestors with bisexual flowers.

The male flower contains few to many stamens; in Salicaceae the number varies widely, and in Juglandales their number and arrangement are also indefinite and bear no relation to the perianth-scales which are, moreover, often absent. In Fagales the stamens are generally few and equal in number to and opposite the perianth-scales. In *Casuarina* one stamen only is present.

The female flower shews wide differences in the important characters of the structure of the ovary and ovule, but the presence of two united carpels, as indicated by a pair of stigmas, is general.

[1] *Handbuch der Systematischen Botanik*, II, 201.

except in Fagaceae, where there are three. The number, arrangement and attachment of the ovules are different in the several orders.

The number of ovule-integuments varies even in the same family (Salicaceae). Wind-pollination is very general. Departures from the normal structure and course of events in the interior of the ovule preceding and during pollination are frequent, and chalazogamy occurs in Juglandaceae, Betulaceae and *Casuarina*. The embryo is straight and, except in Salicaceae where it is small and incapable of passing through a resting stage, is large with fleshy cotyledons, which shew a remarkable development in the Juglandaceae.

The members of these orders are, except *Casuarina*, mainly mesophytic trees and shrubs which play a dominant part in the forest-vegetation of the northern hemisphere; and there is evidence that this was the case also in Cretaceous and Tertiary times, during which there was an extension into the Arctic zone; in the case of Juglandales the northern development is much more restricted at the present day. They pass, as mountain plants, within the tropics, where they are sometimes numerous, as *Quercus* and *Pasania* in tropical Asia, and occasionally reach the south temperate zone, as *Myrica* in South Africa, *Alnus*, which extends along the Andes of South America to Argentina, and notably *Nothofagus* which represents Cupuliferae in Antarctic South America, South Australia and New Zealand. They represent groups that have occupied large areas of the earth's surface for a very considerable time with not specially favourable means of distribution (compare *Quercus* with its heavy fruits and wide distribution right across the northern hemisphere).

The view most in accordance with our present knowledge is to regard them as the surviving representatives of more or less diverging lines of development from some early angiospermous type or types now extinct. The frequent presence of a pistil-rudiment in the male flower suggests a derivation from a type with bisexual flowers, probably with a simple inconspicuous bracteole-like perianth. Pollination took place presumably by wind-agency; the catkin-development represents a highly developed adaptation for anemophily and is associated with the separation of the sexes in two flowers. It is obviously a great advantage for pollen-dispersal by wind to associate the staminate flowers in lightly pendulous catkins. Salicaceae are of great interest as illustrating a development of entomophily in the catkin-type of inflorescence; a development which probably proceeded no further; at any rate we can trace no derivatives of this family among the existing families of Dicotyledons.

The complete absence of the herbaceous habit, the frequency of the chalazogamic method of fertilisation (which occurs also elsewhere, as in *Alchemilla* in Rosaceae), the marked interval between pollination and fertilisation, the wide distribution on the earth of the families from the earliest period at which we have any precise knowledge of angiospermic vegetation, and further the striking differences between the various orders, suggest that we are concerned with isolated remnants of relatively ancient groups which have no descendants among the more highly developed orders of our present-day flora.

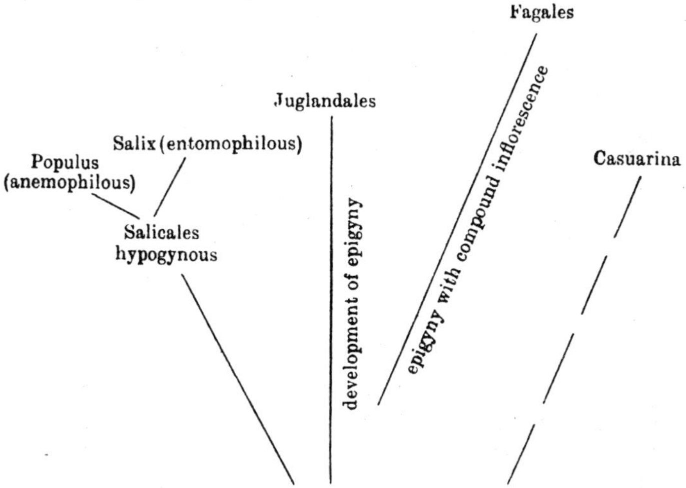

Suggested lines of development from a group (or groups) with a simple bisexual, hypogynous flower, anemophilous; the perianth perhaps not a general character, if present merely protective and bract-like; the protective function still shared with bracts.

Order 7. *URTICIFLORAE*

Flowers generally unisexual, small and regular, with a greenish inferior perianth of generally four to five more or less united leaves; rarely naked. Stamens equal in number and opposite to the perianth-leaves. There are one or two carpels, but the ovary is almost always unilocular and contains a solitary ovule. Flowers wind-pollinated; sometimes entomophilous. Fruit a nut or drupe, containing one seed; the embryo occupies the whole seed, or a fleshy or oily endosperm is present.

Herbs, shrubs or trees, with generally simple alternate stipulate leaves and cymose inflorescence. A large and widely distributed order containing four families which are sometimes, as in the *Genera Plantarum* of Bentham and Hooker, regarded as forming a single family divided into well-marked tribes. Other systematists prefer to combine these tribes under several families, an arrangement which seems more helpful and will be adopted here. Distinctive characters are found in the habit, presence or absence of latex, the inflorescence, the erect or bent position of the stamens in the bud, the position of the ovule, the nature of the fruit, and the seed. Cystoliths occur in the epidermis of many of the genera.

This order differs from the preceding in the remarkable diversity of habit, which includes herbs of various kinds (climbing in *Humulus*) as well as shrubs and trees. The inflorescence also shews considerable variety and in Moraceae includes remarkably specialised forms. The simple form of the flower recalls those of the preceding orders. The order may be regarded as derived from a simple angiospermous type.

Family I. ULMACEAE. Trees or shrubs without latex. Leaves simple, distichous, often oblique; stipules deciduous. Flowers often bisexual; perianth of four to seven leaves, stamens isomerous, erect in bud; styles two; ovary usually unilocular; ovule solitary, pendulous, anatropous. Fruit a nut or drupe. Embryo straight or curved. Endosperm generally absent.

Family II. URTICACEAE. Herbs, or sometimes woody, with simple stipulate leaves and often with stinging hairs; latex absent. Flowers unisexual, generally dimerous; stamens isomerous; filaments inflexed in the bud; style one; ovary unilocular; ovule orthotropous, erect. Fruit a nut or drupe. Embryo straight, surrounded by oily endosperm.

Family III. MORACEAE. Trees or shrubs with latex, rarely herbs. Leaves scattered, stipulate, often lobed. Flowers unisexual; perianth-leaves two to six; stamens isomerous; styles two; ovary generally unilocular, sometimes bilocular; ovule pendulous, more or less curved. Fruit a drupe, often enclosed in the succulent perianth. Embryo generally curved and surrounded by fleshy endosperm.

Family IV. CANNABINACEAE. Aromatic herbs with palminerved, more or less divided leaves and free persistent stipules; latex absent. Flowers dioecious; male with five perianth-leaves, and five stamens erect in bud; female with a small entire cup-like perianth; ovary unilocular with two stigmas and a pendulous curved ovule. Fruit a nut. Embryo curved or rolled. Endosperm scanty.

Family I. ULMACEAE

The plants are either trees, which may be very large, as in the Elm, reaching a height of 125 feet with a girth of 20 feet, or bushes, as in species of *Celtis* or *Trema*. The leaves are simple, often oblique, and arranged alternately in two rows on the branches. The stipules, which are either lateral or intra-petiolar, serve only for bud-protection and disappear with the unfolding of the leaves. Cystoliths occur in *Celtis* and allied genera but are absent in *Ulmus*.

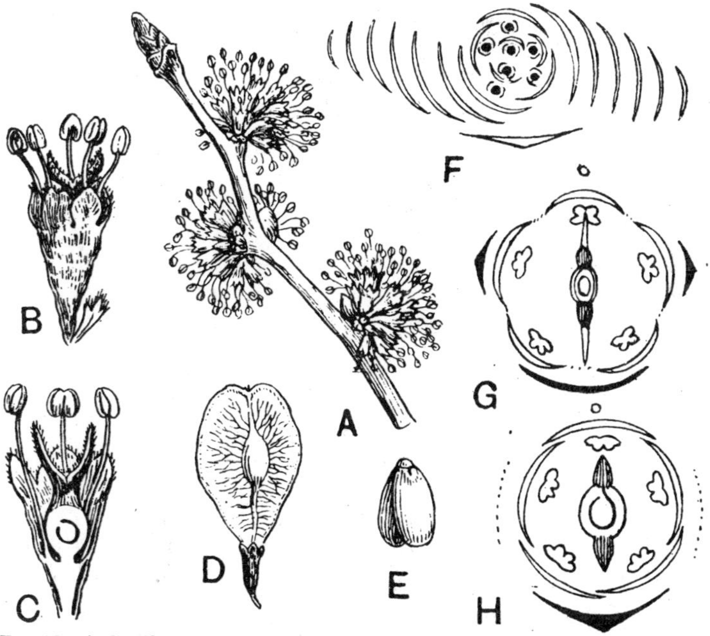

FIG. 11. A–G, *Ulmus campestris*. A. Flower-bearing twig. B. Flower. C. Same in longitudinal section. D. Fruit. E. Embryo. F. Diagram of inflorescence. G. Diagram of flower. H. Diagram of flower of *Celtis australis*. B–E enlarged. (A–D after Wossidlo; F, G, H after Eichler.)

The inflorescence of *Ulmus* forms scaly knobs borne in the axils of last year's leaves. On the outside are a number of empty bud-scales; flowers are borne in the axils of the inner scales. The arrangement of the scales is distichous throughout, as in foliage-buds, or passes gradually into a $\frac{2}{5}$ or $\frac{3}{8}$ arrangement above (fig. 11, F). In *Ulmus campestris* Eichler usually finds a single flower with a pair of bracteoles

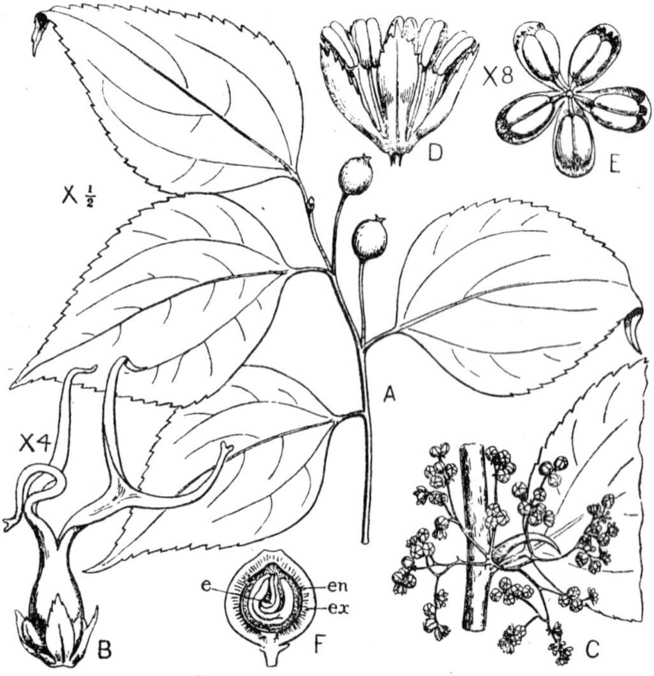

FIG. 12. *Celtis*. A. *Celtis trinervia*. Shoot with fruit. B. Female flower of *C. iguanaea*. C. Male inflorescence of same. D, E. Male flower of same. F. Fruit of *C. australis* cut longitudinally; *ex*, exocarp; *en*, endocarp; *e*, embryo. (From *Flor. Jam.*)

above each fertile bract-scale; in other species flowers or flower-bearing axes are developed in the axils of the bracteoles forming small dichasia. In *Celtis* the flowers are solitary or, especially in the male, in inflorescences (generally cymose), in the leaf-axils of shoots of the current year.

In the Elms (*Ulmus*) the flowers are bisexual; in the other genera mostly unisexual through more or less abortion of stamens or carpels. The perianth consists generally of four to seven sepaloid leaves which are free or more or less united. The stamens are equal in number and opposite to the perianth-leaves, or sometimes fewer; the filaments are erect and the anthers open by a lateral longitudinal slit. The pistil, which is often rudimentary in the male flowers, consists in the bisexual and female of two median carpels (sometimes transverse or oblique); occasionally in *Ulmus* the ovary is bilocular, usually, however, the posterior chamber is absent. A single anatropous or amphitropous ovule hangs from the apex of the chamber. The abortion of the hinder carpel does not extend to the style; there are two linear styles which bear stigmatic papillae on the upper surface.

The flowers contain no nectar and are wind-pollinated. Where both sexes are represented in an inflorescence the central flowers are generally female, and the later developing lateral flowers male or bisexual.

The fruit and seed afford characters for separating the family into two tribes, namely:

1. *Ulmeae*, including *Ulmus* and three small genera (two of which are monotypic[1]), with a broadly winged fruit, a flat seed, a straight embryo with generally flat cotyledons, and no endosperm.

2. *Celtideae* (nine genera), where the fruit is a more or less rounded drupe, and the embryo curved with folded or rolled cotyledons; endosperm is sometimes present.

These differences in the fruit involving different means of dispersal are of interest in connection with the geographical distribution of the two tribes. While *Ulmus* and its allies with the winged fruit occur on the mainland and neighbouring islands, the *Celtideae* which, with a frequently sweet pulp and hard endocarp, are adapted for dispersal by birds, have many representatives on oceanic islands. This illustrates the intermittent action of the wind as an agent in distribution: except in very violent storms it carries fruits and seeds for short distances only at one time, continually dropping and picking them up again, and is therefore not adapted for transport over wide stretches of water.

[1] A monotypic genus contains one species only.

The family is represented in Great Britain by one genus, *Ulmus*, the species of which have been very much confused. *U. campestris* (of British floras) is the English Elm, the tallest and most stately species and the commonest in the hedgerows and parklands of midland and southern England. It does not bear fertile seeds but is propagated by means of its copious suckers. It is not known as a native outside England. *U. sativa*, Small-leaved Elm, and *U. nitens*, Smooth-leaved Elm, occur mainly in the Eastern Counties. *U. glabra*, Wych Elm, a large tree with more or less arched branches, and large rough, almost sessile, acuminate leaves, is native throughout the British Isles. *U. hollandica*, Dutch Elm, with a short bole and wide-spreading, usually very long, lower branches, a hybrid, probably between *U. glabra* and *U. nitens*, is a hedgerow plant, producing numerous suckers. *U. minor* (*stricta*) is the Cornish Elm, pyramidal in outline with fastigiate branches and narrow leaves, abundant in Western Cornwall and rare elsewhere in Southern England; a variety *U. stricta* var. *sarniensis*, Jersey Elm, is limited to the south of England and probably not native; with the Cornish Elm it is commonly planted in avenues in the Channel Islands and some of our south-coast towns.

The family contains 13 genera with about 130 species which are distributed throughout the tropical and extra-tropical parts of the world wherever conditions favour the growth of woody plants. The northern limit in America is reached at 54° 30′ on the Saskatchewan, the most northerly station known for the American Elm (*Ulmus americana*); in Asia it reaches 58°, while in Western Europe *Ulmus glabra* occurs as far north as 66° 59′. In Miocene times, as indicated by fossil remains, *Ulmus* extended into Greenland and also westward to California where it is no longer found. *Celtis*, which has its northern limit at the present day in South Switzerland, extended at that period into Central Germany, fossil fruits having been found as far north as Frankfort-on-the-Main.

The largest genus, *Celtis*, contains about 60 species in the temperate and tropical zones, especially of the northern hemisphere. *C. australis* (Nettle-tree), a handsome tree of from 30 to 50 ft. in height, is a native of the Mediterranean region, and is also frequently planted in the south of France and Italy. *C. crassifolia* (Hackberry) is a fine forest tree abounding in the western United States.

Trema has 30 species in the tropics of both Old and New Worlds; *Ulmus*, 16 species in the temperate regions of the northern hemisphere and in the mountains of tropical Asia. The remaining genera are monotypic or contain only a few species.

Family II. URTICACEAE

The plants are generally herbaceous, as our nettles, or sub-shrubby, rarely forming a bush or tree, as in some tropical genera. The leaves are simple, with sometimes an alternate, sometimes an opposite arrangement; in the same plant they may be alternate on the stem and opposite in the inflorescence. With a few exceptions, for instance, our native Pellitory (*Parietaria*), stipules are present and their various positions afford marks for distinguishing genera. Thus they are attached on the base of the leaf-stalk, or stand separate at its side, or in the leaf-axil (intrapetiolar). In the last case the two stipules may be more or less completely united.

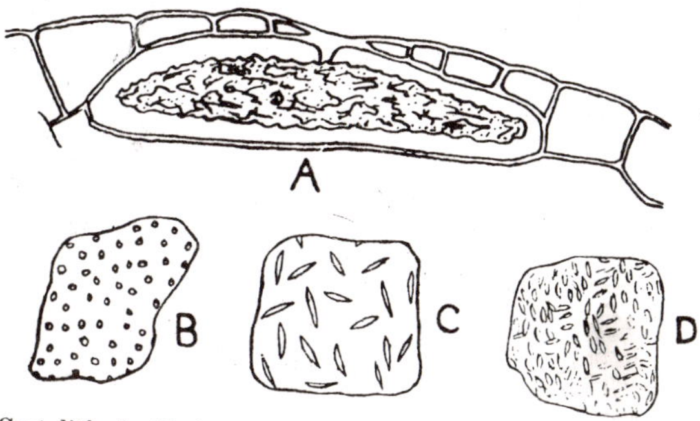

FIG. 13. Cystoliths in Urticaceae. A. Epidermis of *Boehmeria*, cut vertically, shewing one cell with a cystolith, highly magnified. B, C, D. Surface-views of pieces of leaf of species of *Laportea*, *Pellionia* and *Pilea* respectively, shewing various forms of cystoliths, magnified. (A after De Bary; B, C after Weddell.)

The milky latex so common in the next family is absent. The fibres of the bast are generally of considerable length, free laterally and firmly attached end to end and hence of great value for textile use. In Ramie grass (*Boehmeria nivea*) the individual fibre reaches nearly nine inches long, in the Stinging Nettle (*Urtica dioica*), three inches. Stinging hairs are often present on the stem and leaves. Cystoliths are commonly present in much enlarged epidermal cells, and vary in form and frequency in different species or groups of species (fig. 13). They are more or less evident in the dried leaf as punctiform or elongated markings.

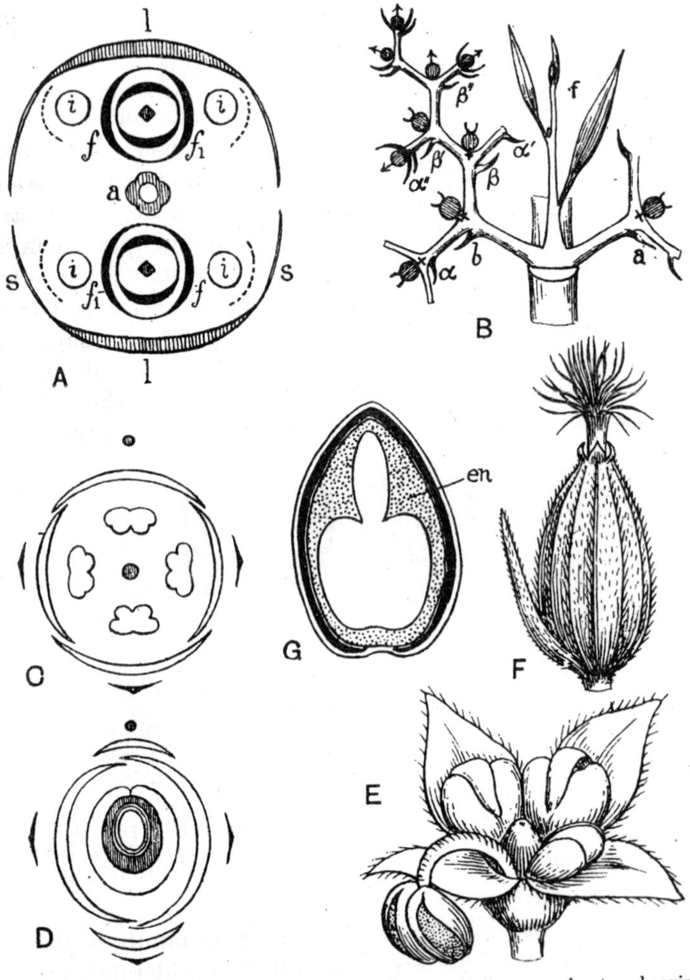

FIG. 14. A. Ground-plan of inflorescence of *Urtica dioica*; *a*, main stem bearing a pair of leaves, *l*, in the axils of which arise two foliage-shoots, the first pair of (unequal) leaves of which are represented by f, f_1; an inflorescence, *i*, is borne in the axils of the suppressed pair of fore-leaves of each shoot, indicated by a dotted line, and apparently springs from the stipule, *s*, of the main leaf. B. Plan of inflorescence of *Parietaria erecta*. The inflorescences spring in pairs from the base of the short foliage-shoot *f*; the fore-leaf is here carried up on the axis of the inflorescence at *a*, *b*. The pairs of bracteoles $\alpha, \beta, \alpha', \beta'$, etc. are similarly carried up on the axes which they subtend. The branch development of the left-hand side only is carried out in the figure. C. Diagram of male, D. Diagram of female flower of *Urtica*. E. Male flower of *Parietaria*, one stamen has rolled back and dehisced. F. Female flower of same. G. Fruit of *Urtica* in section; *en*, endosperm. E, F, G much enlarged. (A, B, C, D after Eichler.)

The inflorescences are cymose. In the more complete cases two lateral inflorescences are borne at the base of a much shortened shoot which stands in the axil of a foliage-leaf. The arrangement may be further complicated by the absence of the bract (fore-leaf) subtending each inflorescence, as in the Nettle (fig. 14, A), or by their being pushed up on to the axis of the inflorescence below the first bifurcation as in *Parietaria* (fig. 14, B).

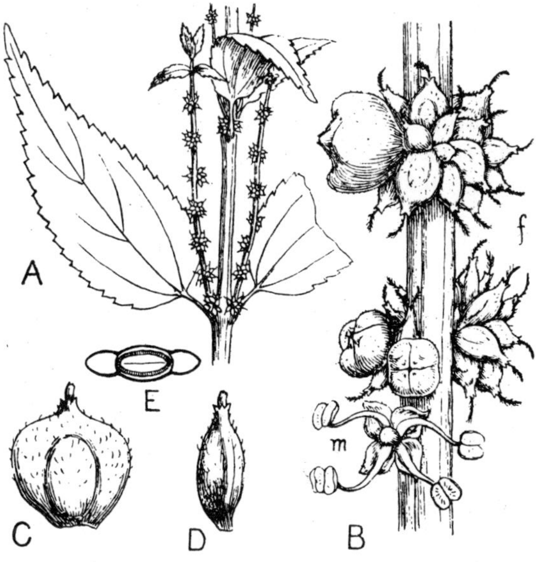

FIG. 15. *Boehmeria cylindrica*. A. Portion of shoot with inflorescence. B. Enlarged portion of inflorescence shewing male, *m*, and female, *f*, flowers. C, D. Fruit, front and side views, × 12. E. Fruit with seed in cross-section, × 12.

The short leafy shoot which bears the inflorescences is frequently suppressed so that a pair of inflorescences, or sometimes only one, are apparently borne in the axil of the foliage-leaf. In some cases the inflorescence is reduced to a single flower. The cymes are often much shortened, the flowers becoming crowded into head-like clusters, a number of which may be borne on an elongated axis as in *Boehmeria* (fig. 15), the whole appearing like a catkin.

The flowers are unisexual, and monoecious or dioecious. The occasional presence of bisexual flowers (as in *Parietaria*), the frequency of a pistil-rudiment in male and several cases of staminodes in female flowers point to their derivation from a bisexual flower.

The green perianth-leaves are free, or, especially in the female flower, more or less united. They are very often arranged in two alternating whorls each of two members, followed in the male by two dimerous whorls of stamens; 3- and 5-merous arrangements also occur. The pistil consists of one carpel containing a single erect ovule and bearing one style which ends in a brush-like stigma or bears along its length long collecting hairs. The flowers are wind-pollinated. The filaments are bent over inwards in the bud but when mature spring elastically backwards and outwards (fig. 14, E), thus scattering the pollen in a tiny cloud. The long stigmatic hairs are well adapted for reception of the pollen. The fruit is a small nut or drupe, often rendered more conspicuous by the persistent perianth which encloses it; or the fruits of a whole inflorescence may remain closely coherent. The seed contains an oily endosperm surrounding a straight embryo.

The family is essentially a tropical one, though several species, such as our common Nettle, are widely distributed and occur in large numbers in temperate climates. There are about 500 species (contained in 41 genera) of which about 33 per cent. occur in the New World, a similar proportion in Asia with the Indian Archipelago, about 14 per cent. in Africa, as many in oceanic islands, and only 3 to 4 per cent. in Europe. In the New World and in Asia Urticaceae form a much larger percentage of the vegetation of the islands than of the mainland, the proportion being 5 to 6 per cent. in the former but not more than 2 per cent. in the latter.

Parietaria debilis has a remarkably wide distribution, being found from Siberia to New Zealand and from North America to Argentina. *Urtica hyperborea* ascends to 15,000 ft. on the Himalayas and *U. andicola* to a similar height on the Andes.

Two genera are represented in the British Isles. *Urtica*, which has about 30 species distributed through the temperate regions of both hemispheres, is represented by *U. dioica* (Stinging Nettle), *U. urens* and *U. pilulifera* (Roman Nettle). The two former are general, the last is a rare plant and occurs only in the east of England, chiefly near the sea. *Parietaria* has seven species, chiefly temperate, rarer in the tropics. The British species, *P. officinalis* (Pellitory of the Wall), occurs on old walls, hedgebanks, etc.; it is found in Central Europe and throughout the whole Mediterranean region.

Family III. MORACEAE

The plants of this family are nearly all trees or shrubs (in *Dorstenia* herbs); it is distinguished from the last two families by the presence of latex, which is contained in long sacs, especially in the secondary cortex or the phloem.

As in Urticaceae, there is some variation in the form and position of the stipules. In the Mulberry and its allies they are small and lateral while in the Figs and allied genera they are intrapetiolar, each pair, as may well be seen in the common Indiarubber Plant (*Ficus elastica*), uniting to form a cap round the younger leaves. As the leaves unfold the stipules fall off, leaving a circular scar.

They are monoecious or dioecious, the unisexual flowers being borne in inflorescences built on the same plan as those of Urticaceae. In *Ficus* (the Figs) the cymes have coalesced to form a fleshy hollow axis bearing the flowers on the interior surface. The three scales which may often be observed at the base of the fruit are the bract, in the axil of which the main shoot is borne, and the bracteoles of its two lateral branches, the cymose inflorescences. In *Dorstenia* the coalesced axis is more or less flattened. There are generally four perianth-leaves, free or more or less united, and arranged in two dimerous whorls, as in the Nettle. In the male flower the stamens are generally equal in number and opposite to the perianth-leaves. The filaments are incurved in the bud in the Mulberry and its allies, or straight, as in *Ficus*. The number of stamens is sometimes reduced to two or one, as in *Ficus* (fig. 16), *Dorstenia* (fig. 17), *Artocarpus* (Bread-fruit).

In the female flower there are two medianly placed carpels, the hinder one of which shews various degrees of abortion. In *Artocarpus* it is sometimes developed, in *Morus* it is exceptionally represented by an empty ovary-chamber; in the same genus (fig. 18) and many others its presence is indicated only by a style similar to that of the anterior carpel. Occasionally the second style is represented only by a small protuberance, or it may be absent, and there is no longer any indication of the hinder carpel (*Chlorophora*). As in Ulmaceae, each chamber of the ovary contains a solitary pendulous, more or less curved, ovule. The fruit is generally a drupe; the perianth becomes fleshy and surrounds it. The coherence of the fruits of an inflorescence occurs in a much higher degree than in Urticaceae. In *Morus* (Mulberry) we can still distinguish the individual fruits with their enclosing perianth, but in *Artocarpus* (Bread-fruit (fig. 19) and Jack-fruit), fruits, perianth-

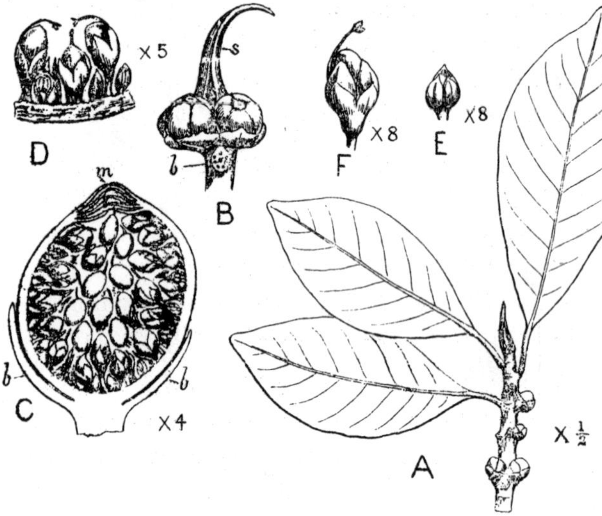

Fig. 16. *Ficus aurea.* A. Branch with young figs. B. Apex of shoot with two figs; *s*, stipule; *l*, leaf-scar. C. Vertical section of fig; *b*, basal bracts; *m*, mouth. D. Section of small portion of fig shewing flowers. E. Male flower. F. Female flower. (From *Flor. Jam.*)

leaves and axis become fleshy and united into one solid mass. The embryo is generally curved and surrounded by a fleshy endosperm.

The family is widely distributed throughout the warmer parts of the earth, containing about 1000 species very unequally divided among less than 60 genera. The genus *Ficus* contains 700 species, and is generally spread through tropical and sub-tropical regions; many species are epiphytic, sometimes forming so tight a network of roots round the stem of the host-plant as ultimately to strangle it.

The Fig of the Mediterranean region is *Ficus Carica.* In this and other edible species pollination is effected by a gall-wasp which lays its eggs in the ovaries of the female flowers and by its visits transfers pollen. In several species an interesting adaptation has arisen. There are two kinds of female flowers, long-styled and short-styled. The latter are termed gall-flowers, as in these the insect lays its egg; they are consequently barren and

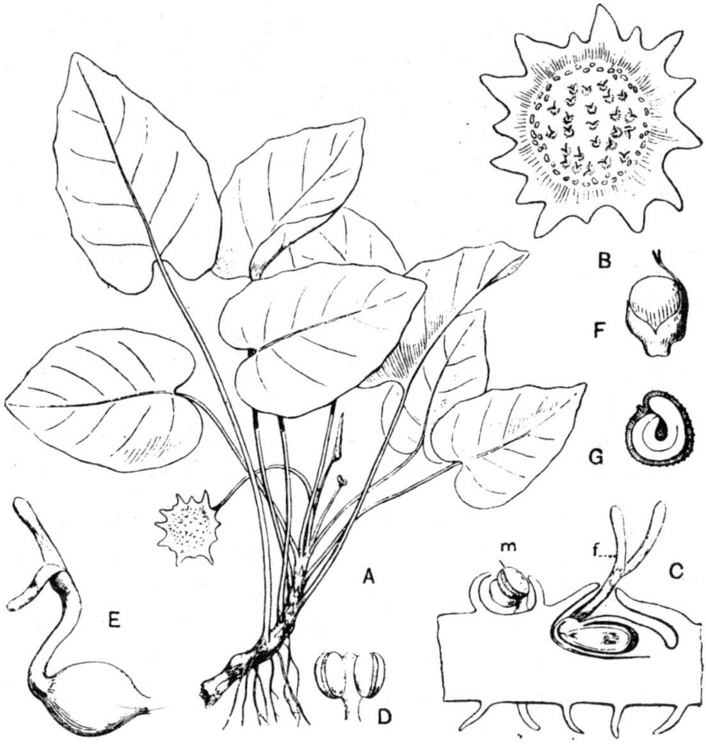

FIG. 17. *Dorstenia cordifolia.* A. Plant, ½ nat. size. B. Inflorescence seen from above, × 2. C. Section through portion of receptacle shewing male, *m*, and female, *f*, flowers, × 16. D. Stamen, × 16. E. Pistil, × 4. F. Endocarp escaping from ovary, × 2. G. Endocarp in section, shewing seed with embryo.

the style no longer bears stigmatic hairs. The long-styled bear well-developed stigmatic hairs; the insect cannot reach the ovaries, and these flowers are the seed-producers or "seed-flowers." *Ficus elastica* (Indiarubber plant), known in this country as a window-plant, becomes in the damp forests of India and Malaya a very large tree, covering the ground around its base with its much branched vertically flattened tortuous roots. *F. bengalensis*

(Banyan), like *F. elastica*, generally starts as an epiphyte, but soon destroys its host-tree and with its numerous descending aërial-roots covers a large area. A specimen in the Calcutta botanic garden has a main stem 45 ft. in girth and a crown of foliage 900 ft. in circumference.

Morus (Mulberry) contains 10 species of trees or bushes in the temperate parts of the northern hemisphere and in the mountains of the tropics. *M. alba*, a native of China, has been cultivated from

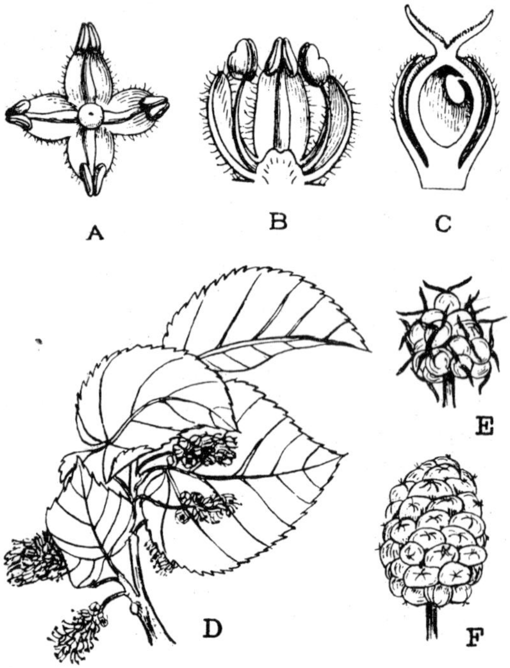

FIG. 18. A–C. *Morus nigra*. A. Male flower. B. Same in vertical section. C. Female flower in median vertical section. D. Twig of *M. alba* with male catkins. E. Female inflorescence. F. Fruit. A, B, C, E enlarged; D, F slightly reduced.

the earliest times in Asia and since the twelfth century in Europe, especially in the Mediterranean region.

The Black Mulberry (*Morus nigra*) has been cultivated in Asia and some parts of Europe from a remote period, chiefly for its leaves as providing food for silkworms. A better silk is obtained when the White Mulberry (*M. alba*) is the food-plant. *M. nigra* is said to have been introduced into England in 1548, the first trees being planted in the gardens of Syon House. It ripens its fruit well in the south of England.

Broussonetia has two or three species in Eastern Asia. The bark is used for paper-making in Japan. *Dorstenia* has about 150 species, one of which is East Indian, the rest chiefly tropical African and tropical American. *Artocarpus* has 40 species, native from Ceylon through the Indian Archipelago to China. *A. incisa* (Bread-fruit), native of the Sunda Islands, and *A. integrifolia* (Jack-tree), a native of India, are now generally cultivated in the tropics, especially in the islands of the Pacific. *Antiaris*, a tropical Asiatic genus of five to six species, includes the Upas tree (*A. toxicaria*), the latex of which is very poisonous. On the contrary, in *Brosimum galactodendron*, the Cow-tree of Venezuela, the latex is sweet and

FIG. 19. *Artocarpus incisa* (Bread-fruit), with male inflorescence, *m*, and fruit; much reduced.

nutritious. *Cecropia* is a tropical American genus with 30–40 species, which are rich in caoutchouc. Some species shew remarkable adaptations for housing and feeding ants, in the form of hollowed internodes and development of food-bodies on the leaf-stalks. The majority of the genera are monotypic or contain only a few species. Numerous fossil species of *Ficus* have been described, chiefly from leaves; many of these determinations are extremely doubtful, but it seems probable that the genus existed as far north as Greenland in the Cretaceous period, and was generally distributed in North America and Europe in the Tertiary period up to Miocene times.

Family IV. CANNABINACEAE

A very small family containing only two genera, *Humulus*, with two species, one of which, *H. Lupulus*, is the Hop, and *Cannabis sativa* (Hemp).

They are aromatic herbs with no latex. The leaves are palmate or palminerved, with persistent stipules. The flowers are normally dioecious, and borne in cymose inflorescences, which are loose and many-flowered in the male, few-flowered

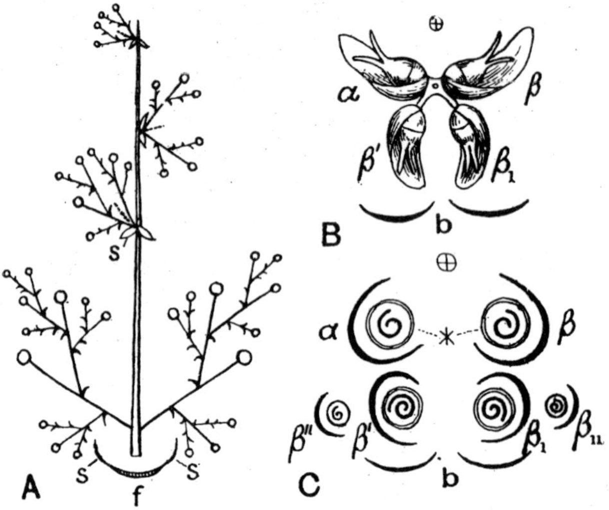

FIG. 20. *Humulus Lupulus*. A. Plan of male inflorescence; *f*, subtending leaf represented only by its pair of stipules, *s*, *s*. B. Bract (*b*) from female inflorescence, represented by its pair of stipules, subtending four flowers. C. Plan similar to B but subtending six flowers. In B the α', and in C the α' and α'' bracteoles are suppressed. (After Eichler.)

and closely packed in the female. In *Humulus* it is not unusual to find both sexes in the same plant, and Tournois[1] has shewn that the monoecious condition can be induced experimentally. The male flower has a five-leaved perianth, and five opposite stamens, with erect filaments. In the female a small cup-like perianth, with no indication of division into parts, surrounds the base of the central ovary, which is

[1] Tournois, J. *Ann. Sci. Nat.* (*Bot.*), Sér. 9, XIX, 49 (1914).

capped by a pair of styles or stigmas and contains a single pendulous upcurved ovule. The dry fruit contains a single seed with a coiled (*Humulus*) or curved (*Cannabis*) embryo and a small amount of fleshy endosperm.

The two species of *Humulus* are perennial twining plants, the twining following the course of the sun. The male inflorescence springs from the axil of a leaf which is represented only by its pair of stipules (fig. 20, A). The main axis is indefinite, bearing a number of lateral cymes which start as dichasia but become unilateral and scorpioid above. The

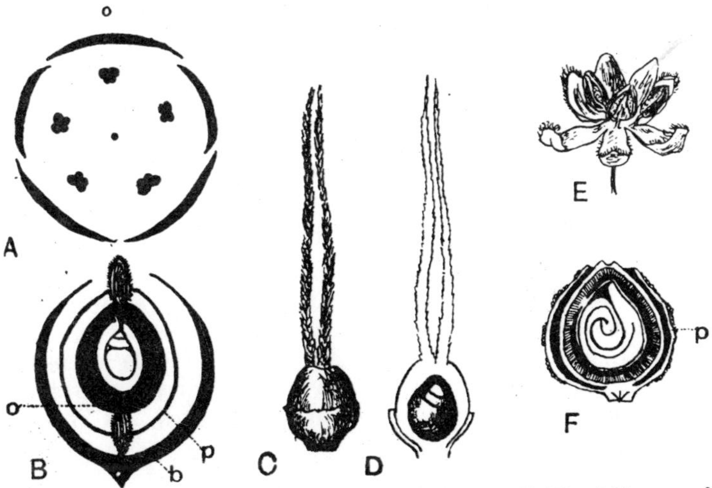

FIG. 21. A. Floral diagram of male flower of *Humulus*. B. Floral diagram of female flower of *Cannabis*; *b*, bract; *p*, perianth; *o*, ovary. C. Female flower of *Cannabis*. D. Same in vertical section. E. Male flower of *Cannabis*. F. Fruit of *Humulus* in vertical section enveloped in the gland-bearing perianth, *p*, and shewing the coiled embryo within the seed. C, D, E, F enlarged. (A, B after Eichler; C–F after Warming.)

female flowers form cone-like inflorescences consisting of a short axis bearing a number of bracts which are opposite below, alternate above. Each bract is represented only by a pair of stipules and subtends two, four or six flowers with a cymose arrangement. A central flower is absent (fig. 20, C) but its bracteoles α and β each subtend a flower. When more than two flowers are present a unilateral cyme is produced instead of a single flower (fig. 20, B, C). The bracteoles, like the stipules of the main bract, become large and membranous and in *H. Lupulus* (Hop) bear numerous yellow glandular hairs which secrete the lupulin to which the plant owes its economic value. Hop, which is perhaps wild in England, is spread over the temperate parts of the northern hemisphere. The second species, *H. japonicus*, is a native of China and Japan; it contains no lupulin-glands. Hemp is an annual plant, probably a native of Central Asia, but now generally cultivated in warm countries. The male inflorescence is somewhat similar to that of *Humulus*, but the main axis is suppressed while its two lateral branches produce each a many-flowered cyme. In the female a flower is borne on each side of the base of a leafy shoot. Hemp is useful for textile purposes on account of its long and strong bast-fibres. The seed contains an oil and a cultivated variety of the East contains in its leaves a narcotic resin for which it is used in the preparation of an intoxicating drink, haschisch.

The two following orders, Proteales and Santalales, are characterised by typically bisexual flowers, generally adapted for insect-pollination. In this connection the generally simple perianth is often petaloid and there is a tendency to zygomorphy; the stamens are also frequently united with the perianth-leaves. The Proteales shew the simpler type of flower, with free hypogynous petals; in the Santalales the petals are frequently united below into a cup and the flower is epigynous.

Both orders shew marked specialisation to a particular set of conditions. Proteales are a characteristic xerophytic group of the southern hemisphere, while Santalales are more or less adapted to a parasitic mode of life.

While undoubtedly allied to each other, it is difficult to associate these orders with other groups. The flowers, considered from a biological standpoint, hold an advanced position from the point of view of adaptation to entomophily; but the general plan of arrangement of the flower suggests its derivation from a type in which the differentiation of two distinct series in the perianth had not yet occurred.

Order 8. *PROTEALES*

Family. PROTEACEAE

Flowers bisexual or unisexual through abortion, cyclic, tetramerous, regular or sometimes zygomorphic; insect-pollinated. Perianth in one series, petaloid, hypogynous. Stamens opposite the segments and usually united with them, the anthers only being free. Carpel solitary, ovules 1–many, with two integuments, on the ventral suture. Fruit a follicle or indehiscent. Seeds without endosperm.

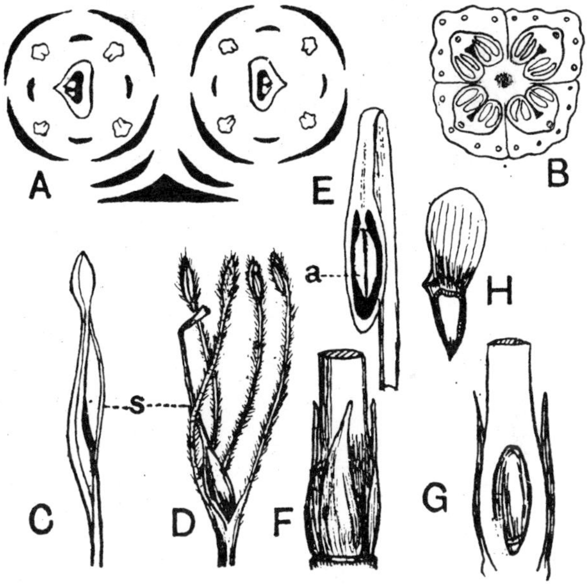

Fig. 22. *Banksia.* A. Diagram of pair of flowers. B. Transverse section of flower-bud. C, D. Opening bud and flower; *s*, style. E. Apex of sepal and anther, *a*. F. Base of ovary and hypogynous scales. G. Same in vertical section shewing ovule. H. Seed opened shewing embryo. C, D about twice nat. size; B, E, F, G several times enlarged. (A, B, E, F, G after Le Maout and Decaisne; C, D after Engler.)

Usually woody plants, with alternate entire or pinnately cut leaves without stipules. Flowers in spikes, racemes or heads, often showy.

One family, Proteaceae, with about 50 genera and 1100 species mainly in the drier regions of Australia and South Africa.

The plants are usually small trees or shrubs, rarely perennial herbs. The leaves are leathery and remarkably varied in form, their shape, position and structure shewing adaptation to xerophytic conditions. They are generally narrow, sometimes cylindrical as in species of *Hakea*, or often pinnately or bipinnately divided. The epidermis is more or less strongly cuticularised and often hairy-felted or silky; T-shaped hairs

FIG. 23. Branchlet of *Banksia serrata* with fruiting "cone," × $\frac{1}{3}$.

are characteristic of the large genera *Grevillea* and *Hakea*. The stomata are variously protected by developments of the cuticle of the guard-cells or by sinking in pits. The internal structure is bifacial or centric. The assimilatory tissue is protected from excessive insolation often by a vertical position of the blade, and also frequently by the presence of a hypoderm, or sclerotic layer, between the epidermis and the assimilatory tissue; isolated sclerotic cells of various shapes are often present to prevent collapsing of the delicate cells of the mesophyll from excessive loss of water.

The family was divided into two groups by Robert Brown (who made a careful study of the genera[1]) according to the nature of the fruit, either indehiscent (*Nucamentaceae*) or follicular (*Folliculares*). In the former group the flowers are solitary in the axils of the foliage-leaves or bracts, in the latter usually geminate with sometimes, as in *Banksia*, an additional smaller bracteole for each flower. The solitary flowers or pairs are very often associated in racemes, spikes or heads, more rarely cymose. In *Protea* the large heads of flowers are surrounded by large, often brightly coloured, involucral bracts.

The perianth consists of four leaves placed diagonally to the bract, at first more or less coherent but becoming ultimately more or less free; they are usually long and narrow and of equal, rarely unequal, breadth. The filaments of the stamens are generally coherent with the corresponding perianth-segment, the anther only being free. In association with a tendency to zygomorphy in the flower one anther is sometimes aborted. At the base of the floral axis are nectar-secreting swellings, very varied in form, sometimes as scales alternating with the perianth-segments.

The pistil is sessile, or borne on a more or less developed gynophore. The style is often long and thin and thickened only at the end where the stigma is borne; there is often a pollen-collecting apparatus below the thickened end. The flowers are generally proterandrous and entomophilous; a tendency to zygomorphy is frequent—in many cases (as in the *Embothrieae*) the perianth is curved to one side, and the end of the style swells on this side to an oblique or lateral disc in the middle of which is the small stigma on which the pollen brought from one flower will be deposited by insects when probing the nectary at the base of another.

The monocarpellary ovary is unilocular; the ovules, several, two or solitary, sometimes numerous, are borne on the ventral suture and are anatropous, or hang from the top of the chamber and are orthotropous.

[1] *Trans. Linn. Soc.* x, 15 (1811) and *Prodromus Florae Novae-Hollandiae* (1810)

The fruit is a follicle or nut, more rarely a drupe; the pericarp is generally thick and often, as in *Hakea* and *Xylomelum* (Wooden Pear), very hard and woody. The seeds have usually a thin coat and are often winged and adapted to wind-distribution. The fruiting inflorescence in *Banksia* forms woody cones, often very large, in which the dehiscent fruits are embedded (fig. 18).

The great majority of the species occur in those subtropical districts of the southern hemisphere in which a wet period alternates regularly with a dry period, as in South-west and Eastern Australia, and the south-west district of the Cape; the xerophytic structure already referred to is associated with this distribution. More than half the species are Australian, and more than one-fourth occur in the south-west Cape district. Two occur in New Zealand, one of which, *Knightia excelsa* (Rewa-Rewa), forms a large tree reaching 100 ft. in height. About 30 species occur in New Caledonia, and a similar number extend into tropical Eastern Asia; a few species are found on the mountains of tropical Africa as far north as Abyssinia (*Protea*), and two in Madagascar. The family is poorly represented in South America; there are seven species in Chile and 36 in tropical South America.

The more important genera are *Protea*, 80 species, mainly at the Cape, with a few on the mountains of tropical Africa; *Leucadendron* (Cape, 70 species); *L. argenteum* is the Silver-tree; *Grevillea* (Australia, 160 species), *Hakea* (Australia, 100 species), *Banksia*[1] (Australia, 50 species), *Dryandra*[2] (Australia, 50 species). A large number of fossil leaves, fruits and winged seeds from the Tertiary strata of Europe were assigned by Ettingshausen and Unger to this family, but the determination is very doubtful. The wood is generally hard, but trees large enough to yield timber are few; such are the Australian *Grevillea robusta* and the New Zealand *Knightia excelsa*. Many species are grown in northern latitudes as greenhouse plants (see the temperate house at Kew). *Telopea speciosissima* is the Waratah, the national flower of Australia.

[1] Named by the younger Linnaeus in honour of Sir Joseph Banks, the great patron of Science and botanical explorer who accompanied Captain Cook on his first voyage of exploration, 1768–1771.

[2] Named by Robert Brown in honour of Jonas Dryander, librarian and curator of Sir Joseph Banks's herbarium.

Order 9. *SANTALALES*

Flowers bisexual or unisexual, cyclic, regular, rarely with a tendency to zygomorphy. Perianth generally in one series with the stamens opposite its segments, sometimes petaloid. Carpels generally two or three, rarely only one; ovary inferior, generally unilocular, with a few naked ovules pendulous from the apex or from a central free placenta; or ovules not differentiated from the placenta. Fruit a one-seeded nut or drupe or a berry-like pseudocarp. Endosperm present, generally enveloping the straight embryo.

Woody plants or herbs often more or less parasitic, with alternate or opposite simple leaves; leafless in extreme forms (Balanophoraceae).

In addition to the four families here treated, all of which have a more or less parasitic habit, there are several allied to Santalaceae. These are the two small families Opiliaceae (tropics), and Grubbiaceae (South Africa), and the family Olacaceae (tropics); cases of hemiparasitism are recorded in the last named[1].

The order is allied to Proteales, differing mainly in the inferior ovary. It shews remarkable deviations from the normal in the development of the ovule, associated with the parasitic habit, which takes an extreme form in the Balanophoraceae in which the chlorophyll-containing assimilating organs have completely disappeared and the female flower has also become greatly reduced.

Family I. SANTALACEAE

The plants are hemiparasitic green herbs, shrubs or small trees, some of which live on branches of trees, as in the family Loranthaceae, while others are root-parasites. An example of the latter is found in the British representative, *Thesium humifusum* (Bastard Toad-flax). The seedling becomes attached to the root or rhizome of the host-plant by suckers formed on the branches of the primary root. *Thesium*, which contains about 250 species, chiefly in the temperate zones,

[1] See Barber, C. A. "Studies in root-parasitism." *Memoirs of the Depart. of Agriculture in India*, I, 1 (1906).

with its highest development in South Africa, has generally an herbaceous habit, the plants often becoming woody at the base, with alternate or opposite narrow one-nerved leaves. *Osyris* (South Europe, Africa, India) is a small genus of shrubs, with narrow or broader leaves. *Santalum* (Sandal-wood) with eight to nine species, extending from India through the Malay Archipelago to Australia and the Pacific Islands, contains small trees or shrubs with generally opposite broad penninerved leaves. Other genera, including several Australian, have a broom-like habit with much branched green stems bearing small scale-like leaves; of these, *Exocarpus*, a characteristic Australian genus, bears the hard fruit (a nut) on a swollen fleshy edible stalk (Australian Cherry). The leaves are exstipulate and generally glabrous.

The small regular flowers are bisexual or unisexual by reduction. They are solitary or in a spike, raceme, or head, the flowers standing singly in the axil of the bract, or forming dichasia by the production of a flower in the axil of each bracteole.

The flowers have a simple perianth, sepaloid or petaloid in character, with the parts more or less united below into a tube, and also adnate to a perigynous or epigynous nectar-secreting disc, which is scarcely developed in *Thesium*. The generally four to five lobes have a valvate aestivation. The stamens are equal in number and opposite to the perianth-leaves, to the base or on the tube of which they are attached; there is generally a little tuft of hairs on the perianth-leaf just behind each stamen, and the disc may be produced into short lobes alternating with them. The short filament bears a two-celled anther attached at or near the base. There are three to five carpels. The ovary is more or less inferior; rarely, as in *Santalum*, nearly superior. It is unilocular, having a terminal style with a capitate or lobed stigma, and containing a central placenta bearing one to five, generally three, naked pendulous ovules. Pollination by insects is favoured by the nectar-secreting disc, and often by the smell of the flowers and colour of the perianth; proterandry has been observed in some cases. Where there is more than one ovule, only one is fertile; its embryo-sac becomes much elongated before fertilisation, growing downwards into the tissue of the placenta and protruding above into the ovary-cavity. The fruit is indehiscent, forming a nut or drupe, containing a

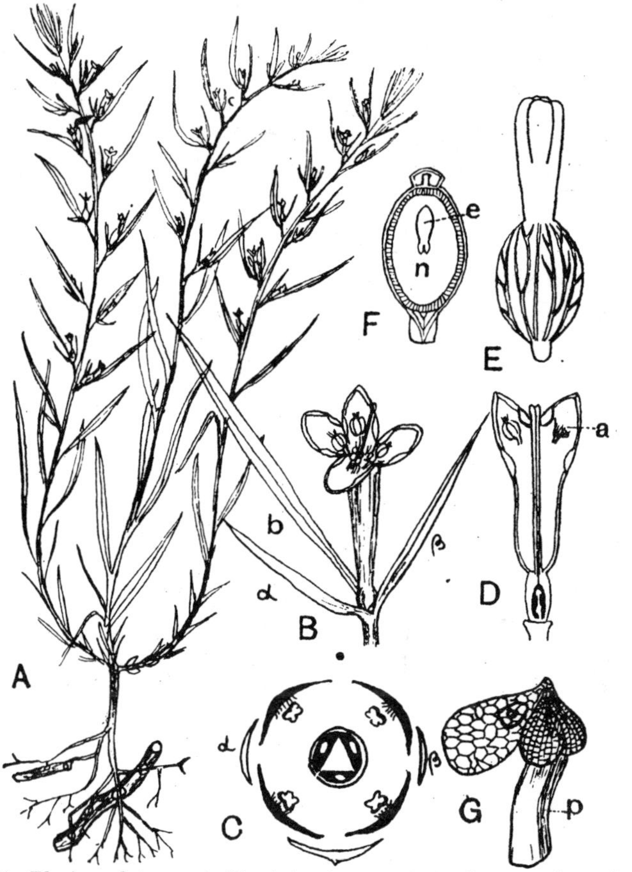

FIG. 24. *Thesium alpinum.* A. Plant shewing a root-attachment to host-plant. B. Flower with bract, *b*, attached to the pedicel and pair of bracteoles *a*, *β*. C. Diagram of flower. D. Flower in vertical section; the anther has been removed at *a*, revealing the tuft of hairs. E. Ripe fruit enveloped in the dried perianth. F. Fruit in longitudinal section; *n*, endosperm; *e*, embryo. G. Placenta, *p*, bearing three ovules, the left-hand one is developing to form the single seed. A slightly reduced; B, D, E, F × 5; G much enlarged. (After Hieronymus, except C, G after Le Maout and Decaisne.)

single seed which has no seed-coat but consists of a quantity of white fleshy endosperm, filling the ovary cavity, within which is embedded a straight embryo with an upwardly directed radicle.

The family contains 26 genera and about 400 species which are widely distributed in tropical and temperate zones. Besides *Thesium* with 16–18 species, two other genera have European representatives, *Comandra*, which has one species in Hungary (the three others are West American), and *Osyris*; *Osyris alba* is a shrub widely spread in the Mediterranean region.

Santalaceae are very closely allied to Loranthaceae, and the two families were united by Baillon; they are alike in their hemiparasitic habit, poorly differentiated ovule, and anomalous embryology. Santalaceae are distinguished by the simpler morphology of the fruit, the 4- to 5-merous flower, and the constant absence of a calyculus.

Santalum album is the Sandal-wood of Indo-Malaya; other species of this and other genera yield useful wood. The sweet pericarp of many species is edible.

Appendix to Santalaceae

The genus *Myzodendron* contains nine species of subshrubby green plants of temperate South America living hemiparasitically on the branches of the Antarctic Beeches. It may be regarded as a much reduced ally of Santalaceae. The flowers are very small, arranged without bracts or bracteoles in small heads or spikes which are again spicate. The male are naked, consisting of two or three stamens arranged round a small gland-like disc; the anthers are one-chambered. The female have a superior trimerous perianth, and a pistil of three carpels forming a unilocular ovary containing a central placenta from the apex of which hang three ovules. The fruit is one-seeded and indehiscent, with three wings or angles and three long feathery bristles alternating with the three stigmas by which the nut becomes attached to the branch of the future host. The single seed has a rudimentary seed-coat, and contains endosperm surrounding an embryo, the radicle of which forms a sucker, and is expelled from the fruit by the growth of the long negatively heliotropic hypocotyl. A full description with excellent figures will be found in Hooker's *Botany of the Antarctic Voyage*, I, Pt 2.

Family II. LORANTHACEAE

An order of hemiparasitic shrubs, having green leaves but deriving a considerable portion of their nourishment from the tissues of the host-tree to which they are attached by modified roots (suckers or haustoria). The stem-development is sympodial and may form a dichasium as in Mistletoe (*Viscum*) where each successive axis bears a pair of persistent green leaves, the growth being continued next season by the lateral buds, while the terminal bud produces the short inflorescence.

The leaves are simple and entire, frequently thick and leathery and long persistent; stipules are absent.

In germination the hypocotyl grows out from the end of the seed, which has become attached to the branch of its

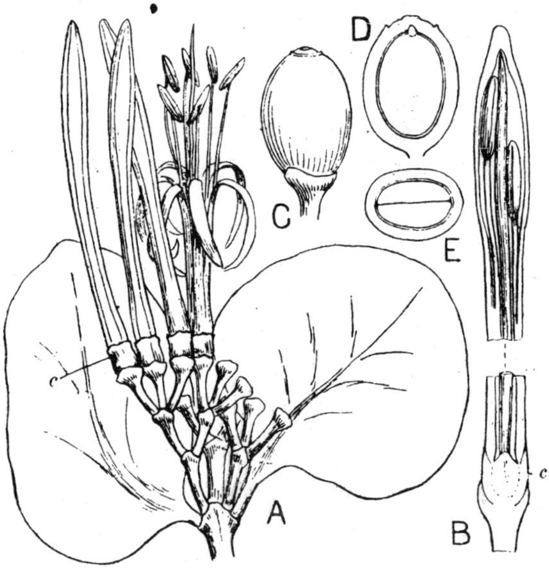

FIG. 25. *Psittacanthus claviceps.* A. Portion of shoot with inflorescence. B. Flower in longitudinal section. C. Fruit. D. Fruit and seed cut lengthwise shewing the embryo with its large cotyledons filling the seed. E. Fruit cut across. *c*, calyculus.

host by its viscid covering, and curving downwards fixes its swollen sucker-like ends to the branch (fig. 27, H). The rest of the embryo-axis bearing the small cotyledons, which have now absorbed the endosperm, is then drawn out from the seed-coat and the embryo stands erect; the first pairs of leaves of the plumule are small and transitory, the energy of the plant being devoted to producing the haustorium (root) which grows from the centre of the attaching sucker, penetrates the bark and reaches the vascular tissue of the host. Support and a supply of nourishment being assured, the first green foliage-leaves are developed.

The flowers, which are bisexual or unisexual, stand singly, or in dichasia, each in the axil of a bract; a number of bracts may be arranged in a racemose inflorescence; when the flowers are stalked, the bract is adnate to the stalk. In Mistletoe a dichasium of male or female flowers ends the leafy shoot.

In *Dendrophthora* and *Phoradendron* the spike consists of two or more nodes with decussating bracts, while the flowers

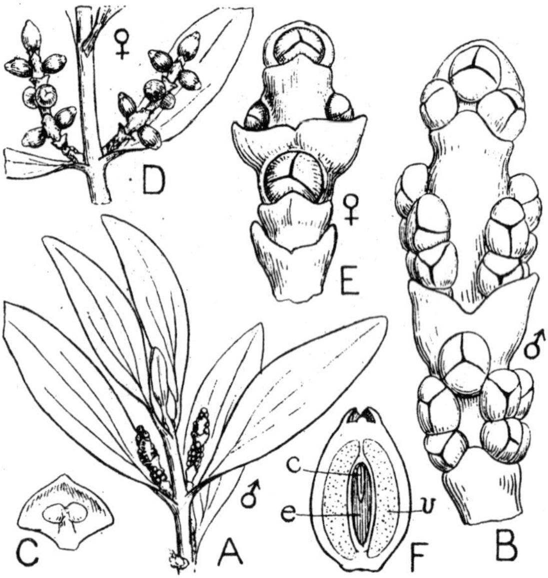

FIG. 26. *Phoradendron Wattii.* A. Portion of male shoot with axillary spikes. B. Male spike. C. Petal with stamen. D. Pair of female spikes in position. E. Female spike. F. Fruit cut lengthwise; *v*, viscid layer; *e*, endosperm; *c*, embryo.

spring either singly or in vertical rows from the internodes, in the tissue of which they are sunk in pits (fig. 26).

The flowers are regular or with a tendency to zygomorphy; the receptacle is cup-shaped and from the edge springs a perianth of generally two similar 2- to 3-merous whorls, the members of which are free or united, and small and sepal-like as in the subfamily Viscoideae (figs. 26, 27) or petaloid, often large and brilliantly coloured as in the Loranthoideae

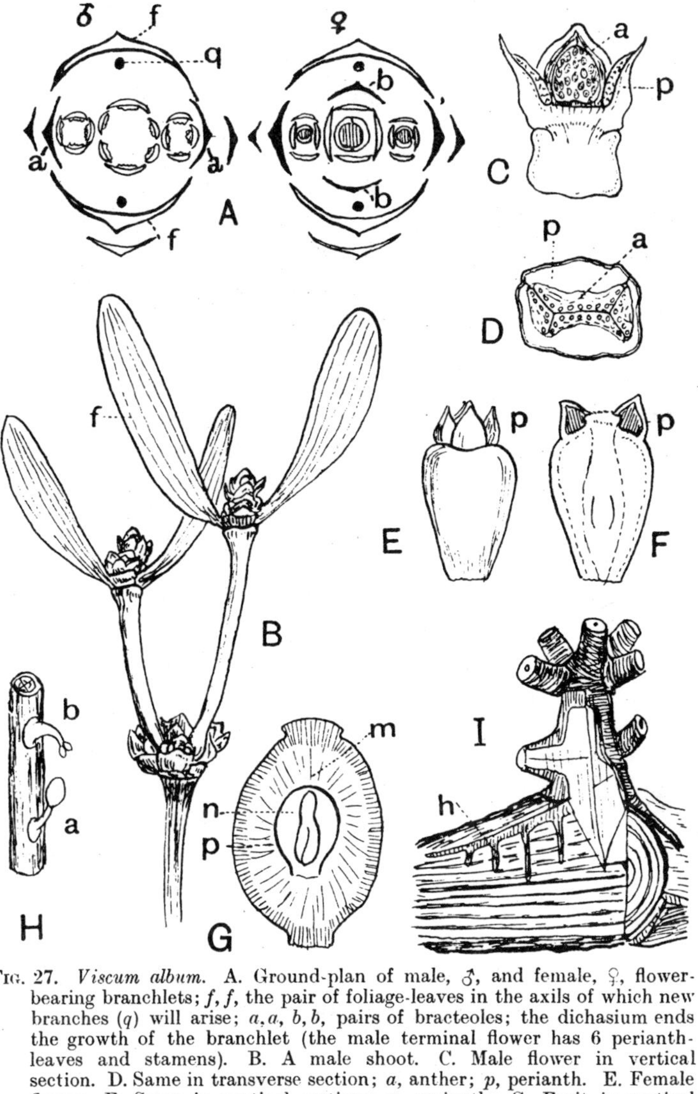

FIG. 27. *Viscum album*. A. Ground-plan of male, ♂, and female, ♀, flower-bearing branchlets; *f*, *f*, the pair of foliage-leaves in the axils of which new branches (*q*) will arise; *a*, *a*, *b*, *b*, pairs of bracteoles; the dichasium ends the growth of the branchlet (the male terminal flower has 6 perianth-leaves and stamens). B. A male shoot. C. Male flower in vertical section. D. Same in transverse section; *a*, anther; *p*, perianth. E. Female flower. F. Same in vertical section; *p*, perianth. G. Fruit in vertical section; *m*, viscid layer; *p*, pericarp; *n*, endosperm. H. Two stages in germination. At *a*, the plantlet is attached by a sucker-like disc and still bears the hard pericarp coat; at *b* the latter has been cast off and the pair of small cotyledons are seen. I. Attachment of a mature plant to the woody tissue of the branch of the host, seen in section, the course of a haustorium on the outside of the wood with branches penetrating the wood at intervals is shewn at *h*. C–G enlarged. (A, D after Engler; B, C, E, F, G after Le Maout and Decaisne.)

(fig. 25). In the latter a slightly toothed or irregular rim (calyculus) is present below the perianth, and is sometimes regarded as a calyx. The stamens are equal in number and opposite to the perianth-leaves with which they are more or less united. In *Viscum* the fusion is complete, the petal and stamen rising from the floral axis as one structure on the front of which a large number of pollen-sacs are produced (fig. 27, C, D). This separation of the pollen in numerous small chambers occurs elsewhere in the family. The ovary is unilocular and contains a large central placenta from which the ovules are not differentiated. Treub shewed in *Elytranthe globosa* that groups of sporogenous cells arise at three or four points, indicating the presence of an equal number of carpels; from one of these an embryo-sac is produced which becomes remarkably elongated upwards into the tissue of the ovary-wall and the style. In *Viscum* an ovary-cavity is scarcely distinguishable, a group of sporogenous cells is produced in the placenta, of which generally one only develops an embryo-sac.

Pollination may be effected in unisexual flowers by aid of the wind, but in the conspicuous brightly coloured flowers of many Loranths the effective agents are insects or small birds which come for the nectar secreted in the bottom of the perianth-tube at the base of the style. Explosive flowers occur, as in several species of *Loranthus*, the buds remaining closed till a tap on the apex causes a sudden separation of the perianth-segments and explosive dehiscence of the stamens. The small greenish flowers of the Mistletoe (*Viscum album*), which is dioecious, contain nectar and are visited by flies.

The fruit is a pseudocarp comparable to an apple, the ovary becoming united with the receptacular cup to form a berry-like, more rarely drupaceous fruit. At the limit between the fleshy receptacle and the true pericarp a layer of viscid substance is developed. The true pericarp is membranous or crustaceous and contains generally one seed, or sometimes, as in Mistletoe, two or three. The viscid substance causes the seed to cling to the beak of fruit-eating birds, to be subsequently rubbed off on the branch of a tree in the operation of cleaning.

The embryo generally lies in the axis of a fleshy endosperm, the cotyledons being appressed face to face; in *Psittacanthus* the endosperm is absorbed by the embryo which has two large fleshy cotyledons filling the seed (fig. 25, D).

The 27 genera, containing about 1000 species, fall into two sub-families LORANTHOIDEAE and VISCOIDEAE, distinguished by the presence or absence respectively of a calyculus. To the former belong 10 genera, including *Loranthus*, the largest genus of the family with about 500 species in the tropics of the Old World, with a few extratropical, such as *L. europaeus*, a parasite on Oak and Chestnut in Central and Southern Europe; and *Psittacanthus*, with about 50 species in tropical America. The largest genus of Viscoideae is *Viscum*, with about 60 species widely distributed in the Old World. *Viscum album* (Mistletoe) is the only British representative of the family; it is found on many different trees, rarely on the Oak. Here also belong *Dendrophthora*, *Phoradendron* and *Arceuthobium*; the last-mentioned is a genus of small leafless parasites growing on Conifers. The "wooden-roses" often shewn as curiosities are the swollen places of attachment of a Loranth to the woody tissue of its host.

Family III. BALANOPHORACEAE

A family of root-parasites, leafless and without chlorophyll, inhabiting the warmer parts of the world. The vegetative organs are reduced to fleshy yellowish or reddish, generally tuber-like, branched rhizomes which are attached by suckers to the woody root of a host-plant. The rhizome is generally naked; scale-leaves are seldom present. The inflorescence is borne on an erect cylindrical axis projecting above the ground; it often originates in the interior of the rhizome and after breaking through remains surrounded at the base by a sheathing outgrowth. When young it bears numerous scale-leaves. The small sessile or shortly-stalked flowers are generally crowded in a head or spike and rarely form a panicle as in the Cape genus *Sarcophyte*. They are unisexual, and monoecious or dioecious. The male generally have a simple perianth of three to four leaves, united below into a tube, with an equal number of stamens, one in front of each perianth-segment. The one- to many-chambered anthers open by pores or slits; a pistil-rudiment is sometimes present. The female are generally naked, rarely with a superior perianth; the pistil is formed of one, two or three carpels, with a unilocular ovary and one or two styles. The ovules, one to three in number, are naked, sometimes reduced to an embryo-sac, and often united with the ovary-wall. The fruit is a nut containing one seed which consists merely of a

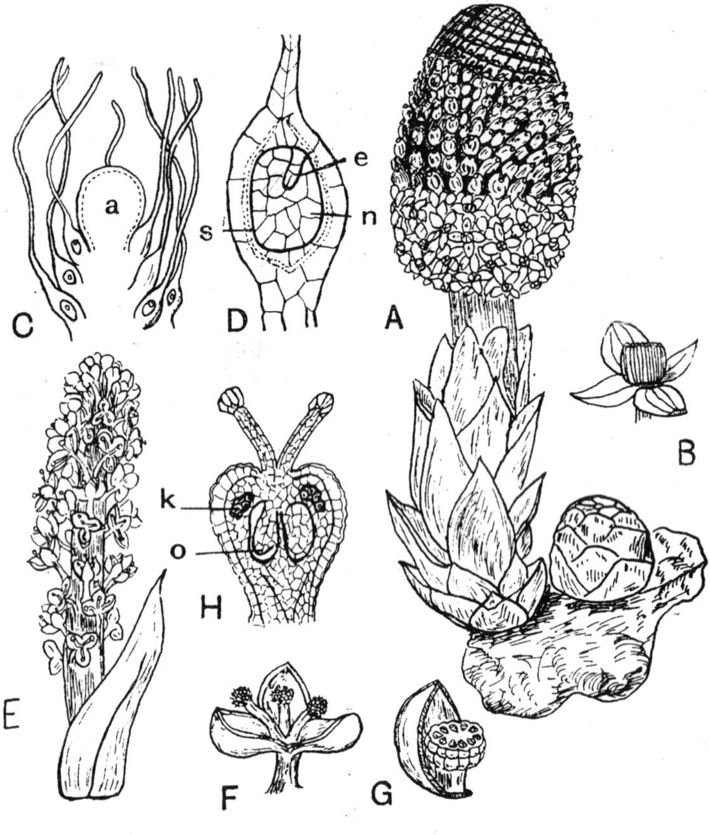

FIG. 28. A. Male plant of *Balanophora dioica*. B. Male flower of *B. elongata*. C. Portion of female spike cut vertically shewing sterile apex, *a*, and several flowers each consisting of an ovary with a single pendulous ovule and a long thread-like style. D. Fruit of *B. elongata* in vertical section; *s*, seed; *e*, embryo; *n*, endosperm. E. Branch of male inflorescence of *Sarcophyte sanguinea*. F. One male flower of same, consisting of three perianth-leaves and three stamens. G. One perianth-leaf with stamen, in bud, the anther cut across. H. Ovary of *Lophophytum mirabile* in vertical section; *o*, ovule; *k*, group of sclerenchyma. F × 3, G × 8; B, C, D, H variously enlarged. (After Baillon, Hooker, Eichler and Engler.)

fleshy oily endosperm enclosing at the apex a small undifferentiated embryo.

Little is known about the method of pollination; some species are visited by insects, probably attracted by the smell, that of *Sarcophyte sanguinea* resembling decaying fish; others are said to be wind-pollinated.

A. Ernst (1) has studied the formation of the embryo in different species of *Balanophora*, and finds that the embryo-sac is either derived directly from the embryo-sac-mother-cell, or arises, after a single division, from the upper daughter-cell: the nucleus has the somatic number of chromosomes. The endosperm is formed entirely from the upper polar nucleus; and after the first division of the latter into a small upper endosperm-cell and a large basal haustorial cell, the cavity of the embryo-sac develops. These results agree with those of Treub (2) and Lotsy (3), but their subsequent conclusions that the embryo develops apogamously from the endosperm appear to be incorrect, for Ernst found that while the endosperm-cell regularly develops into an eight-celled mass of endosperm, the egg-cell does not degenerate as supposed, but persists as a small shrunken cell, which after several divisions gives rise to a small undifferentiated embryo. The investigations of Van Tieghem and Hofmeister, taken in conjunction with the work of Ernst, shew that the embryo of the Balanophoraceae usually arises from the egg-cell, but that in exceptional cases another cell of the egg-apparatus may take part in its formation; in the majority of the genera embryo-formation takes place after fertilisation, but in *B. elongata* and *B. globosa*, and also in *Rhopalocnemis phalloides* and *Helosis guyanensis*, it is of a parthenogenetic character.

There are 14 genera with about 40 species which are found generally in tropical woods and savannahs where they grow attached to the roots of trees or shrubs. The genera fall into groups which are often confined to different geographical areas. The largest genus, *Balanophora*, has 11 species in the Indo-Malayan region and extending to tropical Australia and the Pacific Islands.

The family, which is of doubtful affinity, is here placed near Santalaceae on account of the antepetalous stamens, and the frequency of a 2- to 3-carpellary ovary containing a free placenta.

REFERENCES

(1) ERNST, A. "Embryobildung bei *Balanophora*." *Flora*, n.f. VI, 129 (1913).

(2) TREUB, M. "L'Organe femelle et l'Apogamie du *Balanophora elongata* Bl." *Annales Jard. Botan. Buitenzorg*, XV, 1 (1898).

(3) LOTSY, J. P. "*Balanophora globosa* Jungh." *Ibid.* XVI, 174 (1899).

Family IV. CYNOMORIACEAE

The genus *Cynomorium* has been included in Balanophoraceae but is now generally regarded as comprising a distinct monotypic family; it has been placed by Engler near Haloragaceae, owing to its resemblance in floral structure to *Hippuris*. It differs from Balanophoraceae in the presence of an integument round the single pendulous ovule, and of a seed-coat. The flowers are unisexual, or bisexual, with a simple perianth and one stamen. It is a monotypic genus found round the shores of the Mediterranean and on the Central Asiatic steppes, growing on the roots of shrubby or herbaceous salt-loving plants. The plant is of a brilliant red colour, whence its name *Cynomorium coccineum*.

Order 10. *ARISTOLOCHIALES*

Flowers bisexual or unisexual, cyclic, regular or zygomorphic. Perianth in one series, united or free, petaloid or fleshy. Stamens equal in number with the perianth-segments, or twice as many or indefinite, free or united with the style on a central column (gynostemium). Ovary inferior, four- to six-celled with axile placentas, or unilocular with parietal placentas; ovules numerous, generally with two integuments. Fruit a capsule or berry. Seeds with endosperm and sometimes perisperm, and a small embryo.

Herbs or woody climbers with simple scattered exstipulate leaves, or parasitic plants without chlorophyll and much reduced vegetative structure.

The position of Aristolochiales next to Santalales is that adopted by Engler and Warming. Others regard the order as allied to the Ranales, relying on the trimery of the flowers, the extrorse anthers and the presence of secretory cells in the leaves in Aristolochiaceae. Thus Wettstein places them immediately after Lauraceae, and Hallier derives them from the Berberidaceae through the Lardizabalaceae. It is generally agreed that Rafflesiaceae and Hydnoraceae should be placed in the same order with Aristolochiaceae, representing much reduced parasitic forms.

Family I. ARISTOLOCHIACEAE

A small family of six genera and about 400 species distributed through the hot and temperate parts of the world. The plants are herbs, or have a woody, generally a climbing, stem. The leaves are alternate, stalked, simple and entire, or sometimes three- to five-lobed, and often have a heart-shaped base; in most species oil-containing cells are found in the blade, and also in the tissue of the stem. The only British species is *Asarum europaeum* (Asarabacca), a rare perennial herb with a subterranean rhizome and a creeping aerial stem; the latter is a sympodium; each year's growth is represented by a short axis bearing a few scale-leaves and a pair of long-stalked kidney-shaped foliage-leaves; in the axil of the upper leaf arises the bud which will continue the growth next season while the main axis ends in a flower. Other members of the genus (60 species) have a similar habit.

The greater number of species (300) are included in *Aristolochia*, the species of which are woody climbers or herbs with a perennial rhizome as *A. clematitis*, a central and southern European species which has become established near old ruins in some parts of England. The flowers in *Aristolochia* stand singly or several together in a leaf-axil; in *A. clematitis* there are five to ten arranged in a zig-zag and below these two to five leaf-buds with a similar arrangement (fig. 29).

FIG. 29. Ground-plan of a group of six flowers with four foliage-buds beneath them in the leaf-axil of *Aristolochia clematitis*. (After Eichler.)

The flowers are bisexual and epigynous, with a simple, generally trimerous, petaloid perianth which is regular and bell-shaped in *Asarum*, and zygomorphic and tubular or pitcher-like in *Aristolochia*, having a constricted neck above a swollen base in which the essential organs are contained, and becoming dilated above with a more or less funnel-shaped mouth. In *Asarum* there are sometimes present three small teeth alternating with the perianth-leaves which may represent an inner perianth-whorl.

The stamens, which vary in number from 6 to 36, are free, or united with the style to form a gynostemium. The inferior ovary is generally four- or six-celled with several or numerous anatropous ovules borne on an axile placenta in each chamber.

Asarum europaeum has 12 free stamens, and a six-celled ovary, the chambers alternating with the inner whorl of six stamens. Six is the usual number of stamens in *Aristolochia*; they are arranged on the gynostemium at equal distances alternating with the six carpels, or are associated in pairs (fig. 30, B).

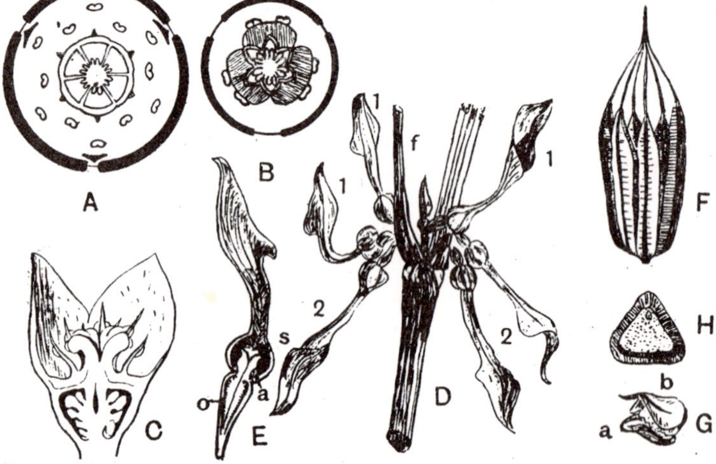

FIG. 30. A. Floral diagram of *Asarum europaeum*. B. Floral diagram of *Aristolochia clematitis*. C. Vertical section of flower of *Asarum*, $\times \frac{2}{3}$. D. Portion of stem of *Aristolochia clematitis* with leaf-stalk, *f*, in the axil of which are flowers of different ages; 1, flowers which have not been pollinated; 2, pollinated flowers; $\times \frac{2}{3}$. E. A single flower in longitudinal section, before pollination; *a*, anther; *s*, stigma; *o*, ovary; nat. size. F. Dehiscing capsule of *A. brasiliensis* in its normal pendulous position, $\times \frac{1}{3}$. G. Seed of *A. Sipho*; *a*, true seed; *b*, enveloping outgrowth from the hilum; nat. size. H. Seed of *Aristolochia* in section shewing small embryo and copious endosperm. (A and B after Eichler; D after Sachs; C after Le Maout and Decaisne; G after Schnizlein.)

The flowers are proterogynous and pollinated by small flies which are attracted by the lurid, spotted, and evil-smelling perianth. In *A. clematitis* the perianth somewhat resembles the bract of *Arum maculatum* on a much smaller scale and with an elongated narrow neck. This neck is lined with downward-pointed stiff hairs. Before pollination the flowers are more or less ascending with a widely open mouth; flies bringing pollen from another flower have easy access to the lower chamber, but their return is prevented by the downward-pointing bristles in the neck. The stigmas which crown the gynostemium project over the anthers, but after pollination they wither and become erect, thereby exposing the dehiscing anthers; the hairs in the neck of the perianth also wither, allowing the escape of the insect, and the flower gradually droops, the mouth of the funnel becoming closed.

The fruit is a capsule dehiscing septicidally; the seeds vary in shape; they are generally more or less flattened, and are often partially covered by a spongy development of the raphe which separates when the seed is ripe. They contain a copious endosperm and a small embryo.

Family II. RAFFLESIACEAE

A small family of parasitic herbs, mostly tropical, of which the vegetative organs are reduced to a mycelium-like tissue ramifying through the cambium and adjoining layers generally of the woody root of the host-plant. The flower-buds are produced within the host by a local growth of the tissue of the parasite, which breaks through and expands above ground, and are either solitary and terminal or, in *Cytinus*, arranged, each in the axil of a bract, in a dense spike. They are unisexual, and shew considerable variety in form and size; those of *Apodanthes*, a stem-parasite from tropical South America, being quite small, while in *Rafflesia* (Malay Archipelago) they are colossal and in *R. Arnoldi* are said to measure a yard across. They have a regular spreading superior perianth of four, five or six members, preceded, except in *Cytinus*, by an equal number of alternating scale-leaves, while in the centre rises a column, the upper surface of which spreads like a disc and in the female flower bears the stigmatic surface on the incurved edge; the inferior ovary is unilocular with numerous ovules borne on four to eight parietal placentas, or consists, as in *Rafflesia*, of an irregular complex of chambers bearing numerous ovules on their walls; the ovules are generally more or less erect and have a single integument. The male flowers have an indefinite number of anthers arranged in a ring on the lower side of the edge of the column or in *Cytinus* (fig. 31, H) around the swollen head; they vary in structure and in the number of the pollen-chambers.

The fleshy succulent fruit is crowned by the persistent column and contains numerous minute seeds with a hard testa, and a few-celled undifferentiated embryo surrounded by a layer of large oily endosperm-cells.

The eight genera, with about 50 species, are mostly tropical. They fall into three tribes, the *Rafflesieae* in the Himalayan region (*Sapria*), Siam (*Richthofenia*) and the Malay Archipelago (*Rafflesia*);

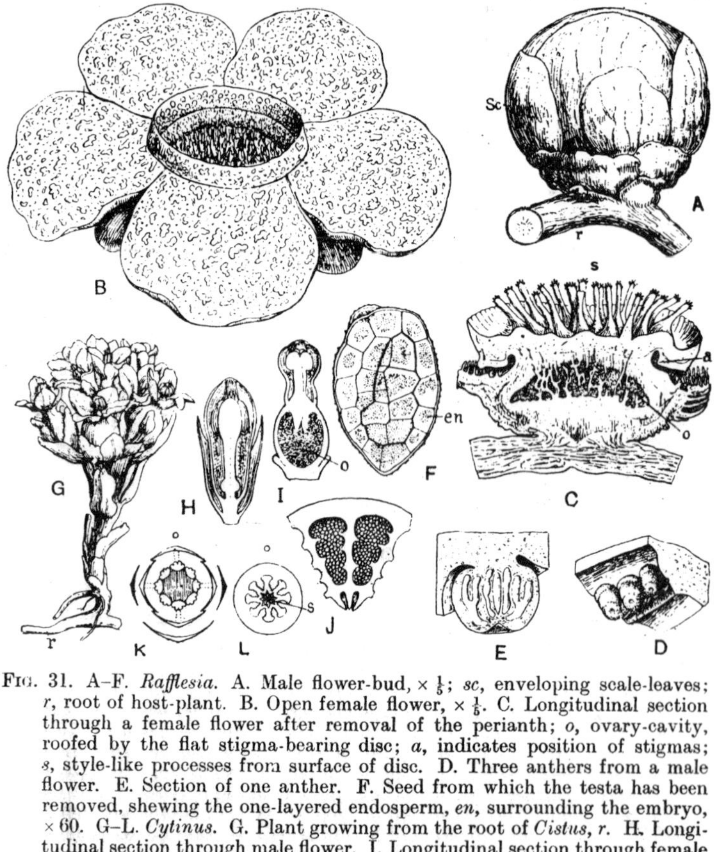

FIG. 31. A–F. *Rafflesia*. A. Male flower-bud, × $\frac{1}{5}$; *sc*, enveloping scale-leaves; *r*, root of host-plant. B. Open female flower, × $\frac{1}{5}$. C. Longitudinal section through a female flower after removal of the perianth; *o*, ovary-cavity, roofed by the flat stigma-bearing disc; *a*, indicates position of stigmas; *s*, style-like processes from surface of disc. D. Three anthers from a male flower. E. Section of one anther. F. Seed from which the testa has been removed, shewing the one-layered endosperm, *en*, surrounding the embryo, × 60. G–L. *Cytinus*. G. Plant growing from the root of *Cistus*, *r*. H. Longitudinal section through male flower. I. Longitudinal section through female flower; *o*, ovary-cavity. J. Portion of cross-section of ovary, shewing two chambers. K. Diagram of male flower. L. Diagrammatic section of ovary of female flower; *s*, position of stigmas. H, I, J enlarged. (B, E after R. Brown; F after Solms; G–J after Le Maout and Decaisne; K, L after Eichler.)

the *Apodantheae* with *Apodanthes* in South America, and *Pilostyles*, mainly South American, but with species in South California, Angola, and Western Asia; and the *Cytineae* with *Cytinus*, which has one species at the Cape of Good Hope while another, *C. Hypocistis*, parasitic on the roots of *Cistus*, accompanies its host throughout the Mediterranean region, extending northwards along the Atlantic coast of France, and *Bdallophyton* with two species in Mexico. The family is thus a widely distributed one.

Family III. HYDNORACEAE

Contains two genera only, *Hydnora* with about eight species in Africa and the Malagasy Islands, and *Prosopanche* with a single species from the pampas of Argentina. The vegetative structure consists of a branched creeping cylindrical or angular rhizome

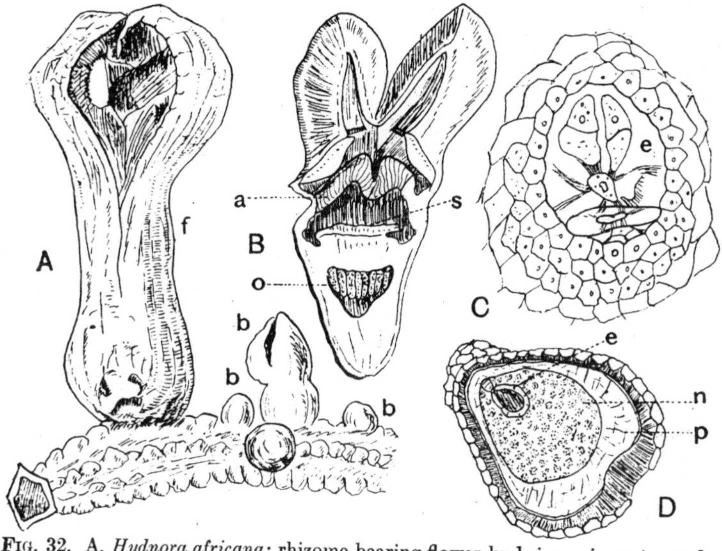

FIG. 32. A. *Hydnora africana*; rhizome bearing flower-buds in various stages, *b*, and an open flower, *f*. B. Flower in section shewing two perianth-leaves and corresponding stamens, *a*; *s*, stigma; *o*, placentas bearing ovules. C. *Prosopanche*; embryo-sac with its contents, *e*, buried in the tissue of the placenta. D. Ripe seed of same in longitudinal section, *e*, embryo; *n*, endosperm; *p*, perisperm. A and B $\frac{3}{4}$ nat. size; C and D much enlarged. (A after Sachs; B after R. Brown; C after De Bary; D after Solms.)

growing from the point of attachment to the woody root, generally of an Acacia or Euphorbia. From the rhizome spring large fleshy solitary flowers which project above the surface of the ground. The flowers are bisexual with a whorl of three or four thick fleshy perianth-leaves united below into a tube which springs from the top of the ovary; the lobes have a valvate aestivation. The stamens, which are equal in number to the perianth-lobes, form in *Hydnora* a sessile 3- to 4-rayed ring inside the perianth-tube, containing on the upper side a large number of pollen-sacs which open by longitudinal slits. In *Prosopanche*, which has a trimerous flower, the three stamens are united into a cap covering the stigmas and three alternating staminodes. The ovary is unilocular, and is roofed in by a large number of flat closely packed placentas, the upper surface of which forms the stigma while the lower grows down into the cavity of the ovary and bears, in *Hydnora*, on its surface numerous orthotropous, sessile ovules each with a thick integument; in *Prosopanche* the ovules are reduced to naked embryo-sacs buried in the tissue of the placenta (fig. 32, C). The fruit has a tough outer layer which splits obliquely and encloses a fleshy pulp in which are numerous spherical seeds, with a hard seed-coat. The small undifferentiated embryo is surrounded by both endosperm and perisperm, the cell-walls of which are much thickened and horny (fig. 32, D).

This family is often included in the last, from which it is distinguished by the structure of the androecium and of the ovary, and by the presence of perisperm in the seed.

The following three orders represent natural and well-defined groups which occupy at the present time a somewhat isolated position.

The flowers may be extremely reduced and naked, though usually bisexual, as in Piperales, but are generally provided with a perianth, which in Polygonales consists typically of two similar whorls. Centrospermae include a large series of forms ranging from a simple type of flower resembling that of Urticales to elaborate forms with well-differentiated calyx and corolla, a double series of stamens and a highly developed many-ovuled pistil, and shewing an advanced stage of adaptation to insect-pollination.

Order 11. *POLYGONALES*

Family POLYGONACEAE

Flowers generally bisexual, small, regular, usually borne in large numbers in compound inflorescences; trimerous (more rarely dimerous) and cyclic, or acyclic. Perianth hypogynous, generally uniform, more rarely the outer and inner whorls different. Stamens usually six to nine, sometimes fewer.

Carpels superior, generally three, rarely two; ovary unilocular containing a single erect orthotropous ovule. Flowers wind- or insect-pollinated. Fruit a three-sided or biconvex nut. Seed containing a more or less excentric or lateral embryo which is variously folded or straight, and a copious mealy endosperm.

Herbs, rarely shrubby or arborescent plants, with generally spirally arranged simple leaves with an ocreate stipule.

One family, Polygonaceae, containing about 32 genera and about 800 species, chiefly in the north temperate zone.

Allied on the one hand to Urticales, on the other to the earlier families of Centrospermae; compare the general structure of the flower and the unilocular ovary with a single ovule. The intrapetiolar stipule noticed in some members of Urticales is, as the ocrea, a constant character in Polygonales. An interesting feature is the multiplication of the stamens and the relation between the two-whorled trimerous and the acyclic perianth.

This is a very natural group, the members, though shewing considerable variation, having certain well-marked features in common, namely, the unilocular ovary with a solitary erect ovule, and an easily observed leaf-character, the almost constant presence of an ocrea. The ocrea is absent only in a group of genera which, with one exception, are practically confined to Western America, most of the species inhabiting either California or Chile or both. The exception is *Koenigia islandica*, a small low-growing annual, often not an inch high, which grows in arctic and subarctic regions, and in the Himalayas.

The stem is herbaceous or sometimes woody; and very often swollen at the nodes. In *Muehlenbeckia platyclada*, a native of the Solomon Islands, the stem and branches are flattened, forming ribbon-like cladodes, jointed at the nodes. Climbing species also occur, as for instance our native *Polygonum Convolvulus* (Black Bindweed).

The leaves are scattered, simple and generally entire; they are sometimes lobed as in Rhubarb (*Rheum*) and often in Sheep's Sorrel (*Rumex Acetosella*). They are mostly smooth, but sometimes, especially in mountain species, woolly or covered with a thick felt. The edges of the leaf are rolled back in the bud. Crystals of calcium oxalate are frequently present in the cells. The small flowers are borne

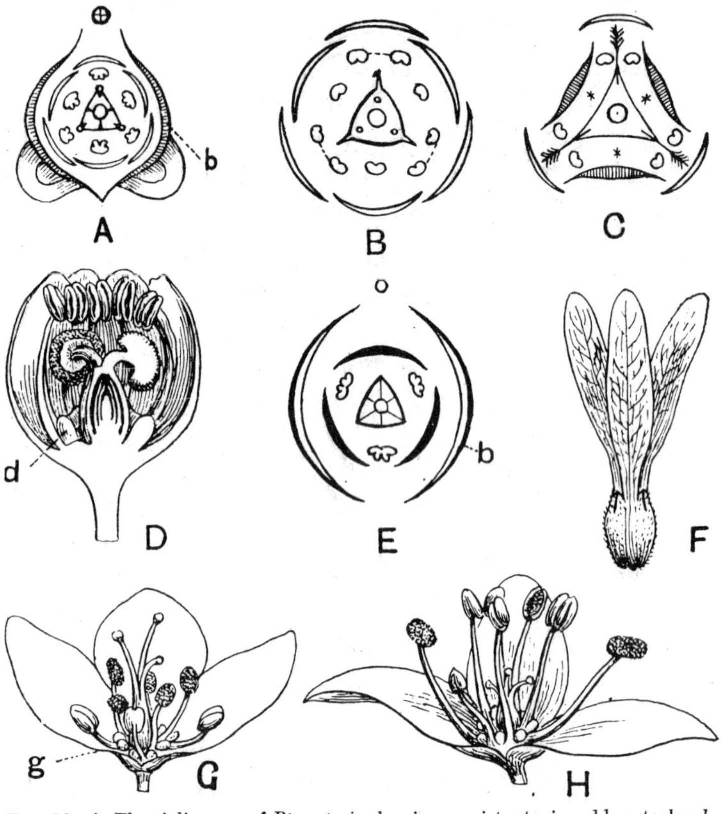

FIG. 33. A. Floral diagram of *Pterostegia* shewing persistent winged bracteoles, *b*. B. Floral diagram of *Rheum*. C. Floral diagram of *Rumex*. D. Flower of *Rheum officinale* in vertical section; *d*, nectar-secreting disc. E. Floral diagram of *Koenigia*; *b*, bracteoles. F. Fruit of *Triplaris*. G, H. Dimorphic flower of *Polygonum Fagopyrum*, shewing long- and short-styled forms; *g*, nectar glands. (A, B, C, E after Eichler; D after Luerssen; F after Dammer; G, H after H. Müller.)

in large numbers in compound inflorescences, the branches of which are cymose. They shew two forms of arrangement, cyclic, with regularly alternating isomerous whorls, and acyclic.

In *Pterostegia*, a monotypic Californian genus, which derives its name from the fact that the two bracteoles persist and form each a winged bladder enveloping the fruit, there are five regularly alternating whorls, each of three members (fig. 33, A). The two outer belong to the perianth, then follow two staminal whorls, while the triangular ovary occupies the centre. Other arrangements may be derived from this type by multiplication or suppression. Frequently doubling occurs in the outer whorl of stamens, as for instance in *Rheum* (fig. 33, B), or *Rumex* (Dock) (fig. 33, C); in the latter the inner staminal whorl is suppressed. In *Koenigia* often only one perianth-whorl and one staminal are developed (fig. 33, E). More rarely the cyclic flowers are dimerous, as in *Oxyria* (Mountain Sorrel), a monotypic arctic-alpine genus; the outer staminal whorl is doubled (fig. 34, C).

In acyclic flowers a 5-merous perianth is followed by a 5- to 8-merous androecium.

Karl Schumann[1] found that in *Polygonum* the order of development does not follow a spiral arrangement but depends on the growth of the floral axis and the mutual pressure of the members, organs being produced where there is most room. The order of succession of the perianth-leaves is indicated in the figure (fig. 34, A). The first staminal rudiment is posterior; this is quickly followed by a pair opposite sepals 1 and 2, which immediately split lengthwise, next arises one opposite sepal 4 and finally one opposite sepal 5. In *P. Bistorta* (fig. 34, B) the rudiment opposite sepal 4 also splits and there are eight stamens in the flower. Three of these, namely the single ones opposite sepals 3 and 5 and the one of the antero-lateral pair which is nearest to sepal 1, stand higher in the floral axis and form the inner whorl; the pistil originates in the triangular space left between them. In *P. orientale* (fig. 34, A) the staminal rudiment in front of sepal 4 does not split, and the development of the pistil is governed by the pressure of the two inner stamens and forms a transverse pair of carpels. In *P. tinctorium*, sepal 5 and the associated staminal rudiment may also disappear and a regular dimerous flower results which, according to Schumann, exactly resembles in arrangement that of *Oxyria digyna* (fig. 35, C). In *P. diospyrifolium* (fig. 34, D) the inner whorl of stamens is absent altogether. Outgrowths of the receptacle of the nature of glands (disc-developments) may occur within the staminal series (fig. 34, A, B).

[1] *Neue Untersuchungen über den Blüthenanschluss*, 1890, 327.

The perianth-members even when occurring in two distinct whorls are generally all alike, and green, white or red. They are free or more or less connate, and persist till the fruit is ripe, often playing an important part in its distribution. Thus in *Rumex* (Dock) the three outer leaves remain small but the three inner grow considerably and completely envelop the fruit which appears surrounded by three membranous wings (fig. 36, C). One or more of these wings may bear on the

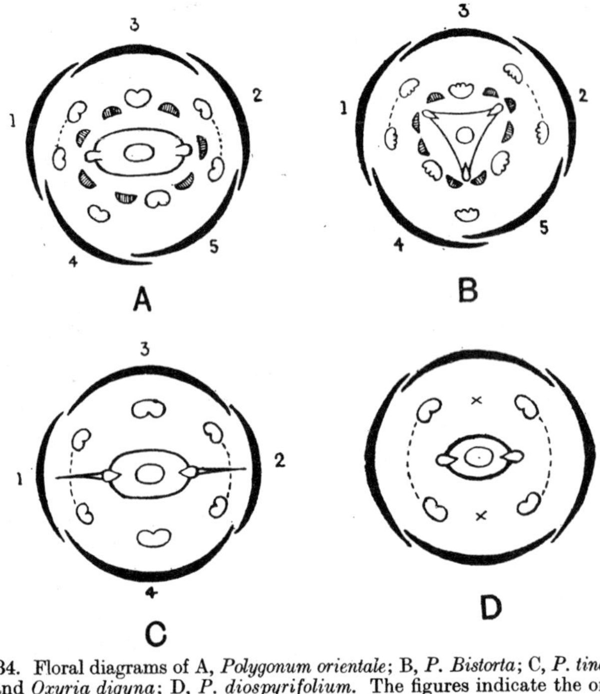

FIG. 34. Floral diagrams of A, *Polygonum orientale*; B, *P. Bistorta*; C, *P. tinctorium* and *Oxyria digyna*; D, *P. diospyrifolium*. The figures indicate the order of succession of the sepals. The shaded portions inside the staminal series represent the disc-development. (After K. Schumann.)

back a large fleshy wart (callus), which forms an attractive object. Frequently the wings are provided with bristles which are often hooked at the tip and will readily cling to a bird's feathers or anything soft or furry. Less often it is the outer perianth-whorl which functions as an agent of distribution as in the tropical South American genus *Triplaris*, where the three outer leaves develop into long flat membranous, almost erect wings, the whole appearing like a shuttlecock (fig. 33, F).

Pollination is effected by wind- or insect-agency or in some cases by incurving of the stamens on to the stigmas of the same flower. The Docks (*Rumex*) are wind-pollinated and have large, brush-like stigmas; the flowers are moreover pendulous on slender stalks (fig. 36).

Rheum and *Polygonum* are entomophilous and have accordingly capitate stigmas and nectaries near the base of the stamens (fig. 33, D, G, H). The flowers are rendered conspicuous chiefly by association in large numbers, or by crowding, as in *Polygonum Bistorta* (Bistort).

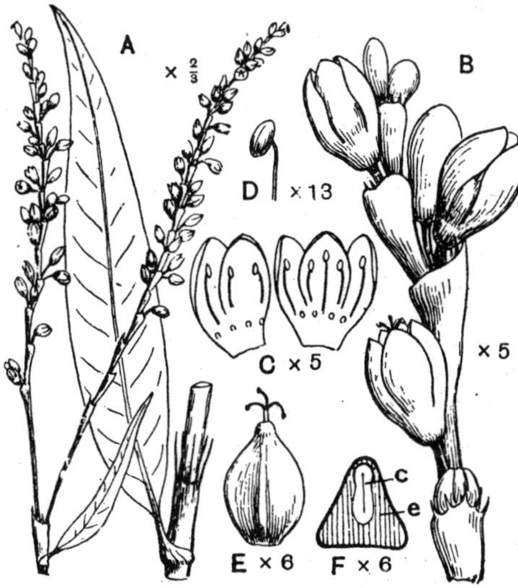

FIG. 35. *Polygonum punctatum*. A. Leaf and flower-spikes. B. Portion of flower-spike. C. Perianth cut open. D. Stamen. E. Nut. F. The same in transverse section; *c*, embryo; *e*, endosperm. (From *Flor. Jam.*)

Species of *Polygonum* afford an interesting series of adaptations for favouring cross-pollination, or failing this, self-pollination.

In Buckwheat (*P. Fagopyrum*) the flowers are dimorphic (fig. 33, G, H). They are made conspicuous by their white or red perianth, their number, and their perfume. Nectar, secreted by eight glands at the base of the stamens, lies at the bottom of the shallow outspread perianth and is accessible to short-lipped insects. Of the eight stamens the five outer dehisce introrsely, the three inner extrorsely, and as the nectary stands between the two series, insect-visitors will get dusted with pollen on both sides. The two forms of flower have respectively long styles and short stamens or short styles and long stamens, the anthers in one form being on a level with the stigmas in the other form. H. Müller records 41 species of insect-visitors, nearly all bees or flies. In *P. Bistorta* the crowding of the flowers in a spike and their coloured perianth make them conspicuous. Nectar is secreted by eight red swellings at the base of the stamens. Cross-pollination is ensured by dichogamy, the flowers being markedly proterandrous. In the first stage the anthers protrude from the flower and not until these have dehisced, withered and mostly fallen do the long styles pro-

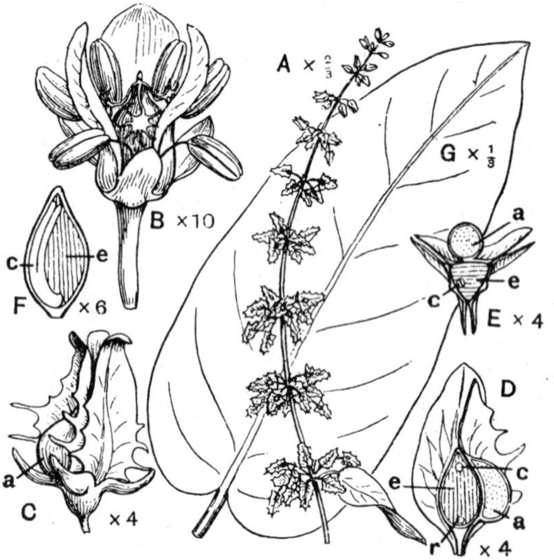

FIG. 36. *Rumex obtusifolius.* A. Portion of panicle in fruit. B. Flower. C. Persistent perianth enclosing fruit. D. Fruit with persistent perianth cut lengthwise. E. The same cut transversely. F. Fruit cut lengthwise at right angles to D. G. Leaf. *a*, callus; *c*, cotyledons; *e*, endosperm; *r*, radicle. (From *Flor. Jam.*)

trude their stigmas. In *P. Persicaria* the white or red flowers are crowded in a short spike. Owing to the smaller size of the flower and the spike, the absence of perfume and the smaller supply of nectar, insect-visits are much rarer than in the two preceding cases. "It is therefore," to quote Hermann Müller, "of more importance for self-fertilisation to be possible in default of insect-visits, than for cross-fertilisation to be absolutely insured when insect-visits do take place." The structure of the flower is therefore different from that of *P. Fagopyrum* and *P. Bistorta*. At the base of each of the five perianth-segments is a gland secreting a thin layer of nectar. Alternating with the perianth-segments are five stamens spreading outwards away from the stigmas which ripen at the same time and stand on the same level with the anthers. The other three stamens are reduced in size or have entirely disappeared. When present they bend towards the middle of the flower and come in contact with the stigmas, effecting self-pollination. In *P. minus* the flowers are about as large as those of *P. Persicaria* but are green and form looser spikes and are therefore less conspicuous and less visited by insects. Müller finds accordingly that the inner series of stamens which curve inwards persist much more frequently than in *P. Persicaria*. Finally in *P. aviculare* (Knot-grass) the flowers are solitary and very small. They are odourless, apparently without nectar, and are very rarely visited by insects. They pollinate themselves by the incurving of the three inner stamens. The efficiency of self-pollination in this case is demonstrated by the facts that every flower sets seeds and that Knot-grass is one of our commonest weeds.

The fruit is a dry one-seeded nut, three-sided when, as usual, three carpels were present in the flower, two-sided when, as in *Oxyria*, it was bicarpellary; both forms sometimes occur on the same plant, as in *Polygonum Persicaria*. The seed contains a nutritious mealy endosperm in which the straight or curved embryo is more or less embedded. In Buckwheat the cotyledons are folded on one another, appearing in transverse section to be twisted in the form of the letter S with endosperm occupying the sinus.

The north temperate zone is the principal home of the family. Of the 32 genera a few only are tropical; *Coccoloba* for instance, which is also remarkable for the perianth becoming fleshy in the fruit, has 125 species confined to tropical and subtropical America[1]. Relatively few species are found in the southern hemisphere, or, as *Koenigia islandica* and *Oxyria digyna*, in the colder parts of the northern. A few genera and species have a very wide distribution. *Polygonum*, for instance, occurs in all the five continents, spreading from the limits of vegetation in the northern hemisphere to the mountains of tropical Africa and the Cape, through the highlands of tropical Asia to Australia, and in the New World as far south as Chile. Most of the genera, on the other hand, have a limited distribution; four occur only in California, one is limited to Mexico, another to the island of San Domingo. Several genera (*Atraphaxis*, *Calligonum* and *Pteropyrum*) are Asiatic steppe-plants forming shrubs with slender woody stems and small leaves. *Rheum* (Rhubarb) is central Asiatic; the rootstocks of several species have

[1] *C. Uvifera* (Sea-side Grape) is a small tree common on the shores of the West Indies and eastern tropical America.

long been used in medicine. Three genera are represented in the flora of the British Isles: *Polygonum* by 14 species; *P. aviculare* and *P. Convolvulus* are common in fields and waste places, *P. Bistorta*, *P. amphibium*, *P. Persicaria* and *P. Hydropiper* occur in wet or damp situations, while *P. maritimum* is a sea-shore plant of the south-west of England: *Rumex*, by 14 species, several of which are common in waste places; *R. Acetosa* (Sorrel) contains a large amount of acid-potassium-oxalate: *Oxyria digyna* (Mountain Sorrel), taking its generic name from the acidity of its leaves, is an alpine plant found from North Wales and Westmorland to Orkney and also in the south-west of Ireland.

Order 12. *PIPERALES*

Flowers generally bisexual, but sometimes unisexual, minute, without a perianth, arranged in spikes, and each in the axil of a bract; bracteoles absent. Stamens (one to ten) and carpels (one to four) variable in number, but the flower may generally be derived from a trimerous type. Carpels free or united. Ovule with two or sometimes one integument, orthotropous, usually solitary. Seed relatively large, containing a small embryo at the apex of a copious endosperm, or endosperm and perisperm are present.

Generally herbs or shrubs with entire leaves, stipules present or absent; oil-containing cells are present in the leaves.

Mainly tropical or subtropical containing the important family Piperaceae and two small families.

Allied to the last order (Polygonales) in the trimerous relation of the flower and the orthotropous solitary ovule but regarded by Engler as representing one of the most primitive types of Dicotyledons and placed by him near the head of the group. A relationship has also been suggested between Piperales and the monocotyledonous family Araceae to which there is some resemblance, especially in the character of the inflorescence.

Family I. PIPERACEAE

Flowers generally in fleshy spikes. Ovary one-celled with a basal orthotropous ovule, which in *Peperomia* has only one integument. Seed with endosperm and perisperm.

Herbs or shrubs, sometimes lianes, rarely trees, with more than one ring of vascular bundles in the stem.

Widely distributed in the tropics of both hemispheres; a few species are extra-tropical. More than 1300 species, mainly included in the two large genera *Piper* and *Peperomia*; there are also seven small genera.

The species of *Piper* (over 700 in number) are generally shrubs, often climbing, rarely trees or herbs, with scattered conspicuously nerved stipulate leaves and terminal spikes of

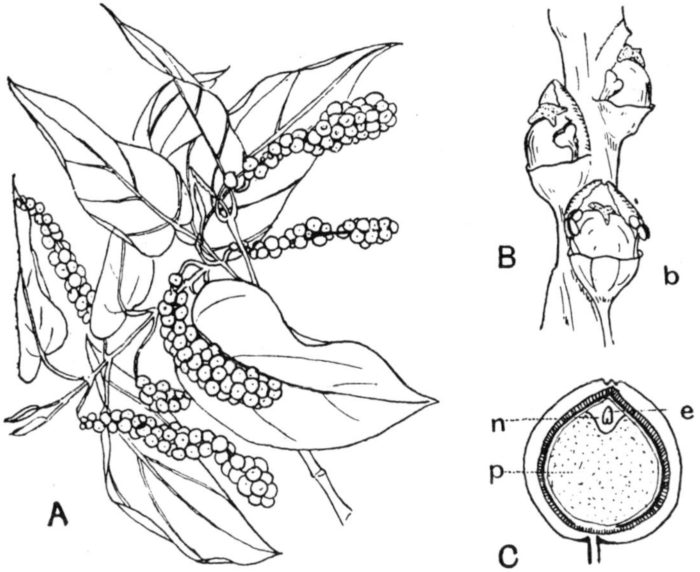

FIG. 37. *Piper nigrum*. A. Shoot bearing spikes of fruit, × ½. B. Small portion of flower-spike shewing three flowers, × 4; *b*, bract. C. Fruit in longitudinal section, × 4; *e*, embryo; *n*, endosperm; *p*, perisperm. (After Baillon.)

flowers which become pushed aside and leaf-opposed by the development of the axillary bud. The species of *Peperomia* (over 600) are annual or perennial herbs, often fleshy, with exstipulate leaves which are sometimes opposite or whorled; the spikes are terminal or axillary; sometimes, as in *Piper*, opposite the leaves. In Mexico, Central America and on the mountains of South America, reaching an altitude of 13,000 feet above sea-level in the Andes of Peru and Bolivia, are a number of geophilous species (1). These are often less than 2·5 cm. high and have a tuber formed from the swollen hypocotyl suggesting a miniature Cyclamen, or a short thickened rhizome.

The naked flowers are often more or less sunk in the fleshy axis of the spike on which they are generally closely arranged, each in the axil of a bract, the form of which is very variable.

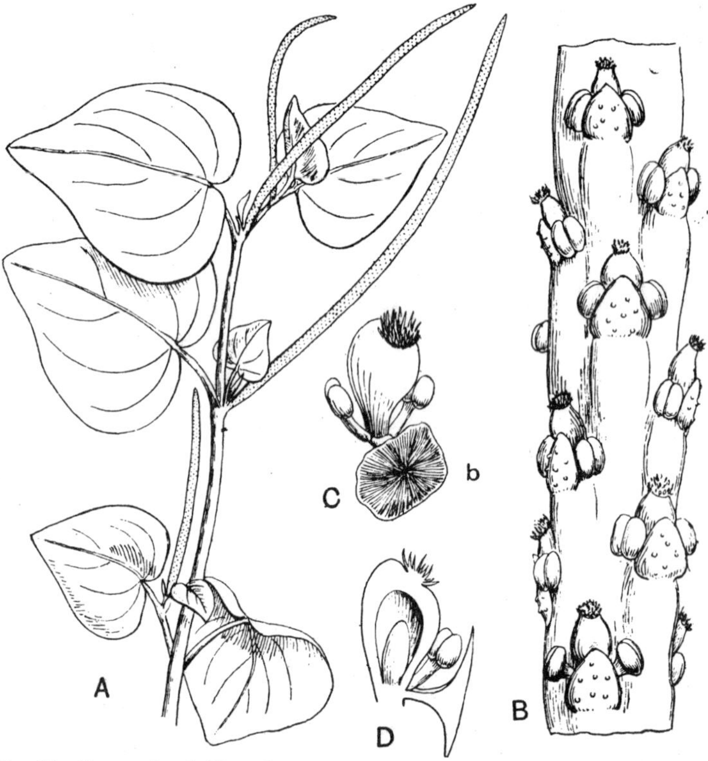

FIG. 38. *Peperomia.* A. Shoot bearing flower-spikes of *P. pellucida*, × $\frac{3}{4}$. B. Portion of flower-spike of the same, × 8. C. Bract, *b*, with flower, of *P. blanda* much enlarged. D. Same in longitudinal section. (C, D after Baillon.)

The number of stamens varies from one to ten, but most of the flowers may be derived from a trimerous type with two whorls, each of three stamens, and a tricarpellary pistil, as occurs in *Piper Amalago*; in other species of *Piper* the two posterior members of the inner whorl are absent, or the whole inner whorl. *P. nigrum* has two stamens only, the posterior one of the inner whorl having also aborted (fig. 37, B), and this arrangement is constant in *Peperomia*, which may be regarded as a constantly reduced form. In *Piper*, section *Ottonia*, the flower is usually tetramerous. The number of carpels is indicated in *Piper* by the stigmas, two to four (or rarely five), which are sessile or borne on a short style. In *Peperomia* the stigma is simple (often brush-like) and frequently lateral. The one-celled ovary contains a single erect ovule springing from the base of the chamber with one (*Peperomia*) or two integuments. The small berry contains a large seed, at the top of which is a small embryo enclosed in a fleshy endosperm below which is a copious mealy perisperm.

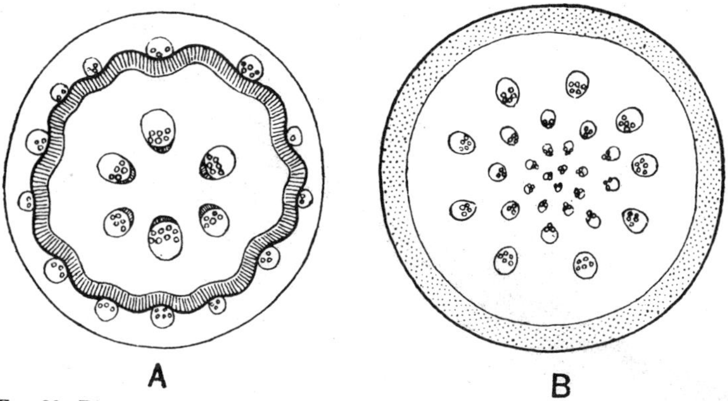

FIG. 39. Diagrammatic cross-sections of the stems of A, *Piper nigrum*; B, *Peperomia magnoliaefolia*, shewing position of the vascular bundles. The dark ring in A represents sclerenchyma, in B collenchyma. (After Engler.)

The internal anatomy is of exceptional interest. The stem contains, besides an outer ring of bundles, which may become united to form a continuous ring producing a sheath of secondary wood, also a number of internal bundles which are scattered or arranged in one or two rings as seen in transverse section. In *Peperomia* the outer ring of bundles does not become united and the general arrangement in transverse section recalls that of a monocotyledon (fig. 39, B)—the bundles are, however, open. In the fleshy leaves of *Peperomia* a well developed aqueous tissue is formed below the upper epidermis. Resin- or oil-containing secretory sacs are generally distributed, both in the epidermis and ground-tissue; to these is due the sharp aromatic taste of the fruits, leaves and roots of many species which are specially rich in these sacs.

Pepper is prepared from the dried berries of *Piper nigrum*; other species also yield similar condiments.

The genus *Peperomia* which, as already noted, differs from *Piper* in the constant reduction of the floral type, the presence of a single integument around the ovule, and an apparently reduced type of stem-structure, shews also a remarkable departure from the normal course in the development of the female gametophyte. The usual polarity in the embryo-sac is absent. The first four nuclei produced are large and arranged tetrahedrally; from these 16 parietal nuclei are formed, one of which at the micropylar end forms the egg, an adjoining one functions as a synergid, eight of the remaining nuclei unite after fertilisation to form the primary endosperm-nucleus, and the remaining six preserve their parietal position and are finally cut off by walls and do not in any way suggest antipodal cells. Campbell (2) regards this as a primitive state, and points out the striking similarity with *Gnetum* amongst Gymnosperms. Johnson (3), however, infers from the result of his investigation of the development in *Piper* and *Heckeria*, in which the embryo-sac contents are normal, that the course of events in *Peperomia* is specialised rather than primitive, a view which finds support in the indications of reduction in floral and stem-structure.

REFERENCES

(1) Hill, A. W. "A revision of the geophilous species of *Peperomia*." *Annals of Botany*, xxi, 139 (1907).

(2) Campbell, D. H. "Recent investigations upon the embryo-sac of Angiosperms." *American Naturalist*, xxxvi, 777 (1902).

(3) Johnson, D. S. "On the development of certain Piperaceae." *Botanical Gazette*, xxxiv, 321 (1902).

The small family Saururaceae differs from Piperaceae in the structure of the pistil which contains three or four free or united carpels; the two or more orthotropous ovules are borne on the edges of the carpels or, when these are united, on the parietal placentas. The stem-structure differs in having the bundles arranged in one ring. There are three genera: *Saururus*, with one species in Eastern Asia and one in Atlantic North America; and two monotypic genera, *Houttuynia* (Himalaya to Japan) and *Anemiopsis* (California).

Chloranthaceae is a small family of tropical and subtropical herbs, shrubs or trees with opposite simple stipulate leaves and spikes or cymes of small inconspicuous flowers, which are bisexual or unisexual and have sometimes a rudimentary perianth. The stamen is solitary or there are three which are united with each other and with the ovary. The ovary consists of a single carpel with one orthotropous ovule hanging from the apex of the chamber. The fruit is drupaceous; the seed contains a large amount of fleshy oily endosperm and a small embryo at the top. The vascular bundles are arranged in a single ring in the transverse section of the stem.

Order 13. *CENTROSPERMAE*

Flowers bisexual or unisexual by reduction, regular, generally pentamerous. Perianth simple or in two series, the inner of which is petaloid. Stamens in one or two series, rarely many, hypogynous or sometimes perigynous. Carpels one to five, rarely more, usually united and forming a one-celled, rarely several-celled, superior, rarely inferior, ovary, and containing $1-\infty$ campylotropous, more rarely anatropous, ovules with generally two integuments. Pollination anemophilous or entomophilous. Fruit generally a capsule or nut, rarely baccate. Embryo large, curved round the perisperm.

Generally herbs with simple, usually exstipulate, alternate or opposite leaves. Several families are characterised by abnormal stem-structure.

The order has also been called Curvembryeae from its characteristic curved embryo. The curved (campylotropous) ovule frequently becomes a kidney-shaped seed round the wall of which lies the embryo enclosing the perisperm; the endosperm is absorbed by the embryo during development. The seeds are solitary or numerous according as the ovary contains one or many ovules. When solitary the ovule springs from the base of the ovary; when numerous, the ovules are borne on a central axile placenta which is generally free, being formed from the inner (ventral) portions of the carpels which have become separated from the outer (dorsal) portions by the disappearance of the partition walls. Sometimes, as in *Viscaria* and *Silene*, the septa are present in the lower part of the ovary.

The families form a very natural association, combining with remarkably uniform characters of placenta, ovule and embryo, great diversity in the differentiation of the floral envelope and androecium. The monochlamydeous type of flower is represented by Chenopodiaceae and Amarantaceae; the stamens are in one series and the one-celled ovary contains generally a single basal ovule. Within these limits Chenopodiaceae shews a remarkable diversity in the form of the perianth and the degree of union of its parts, and in the form of the ovary and the means adopted for distribution of the one-seeded fruit.

The small family Phytolaccaceae shews great variety in the plan of structure of the flower, including genera (e.g. *Microtea*) in which the ground-plan is identical with that of Chenopodiaceae and Amarantaceae, but also forms in which a multiplication of stamens and carpels has taken place. This diversity has led to the suggestion[1] that the other families of the order may have originated from Phytolaccaceae by development along different lines. On this view Chenopodiaceae and Amarantaceae represent a variation on the simple type of flower. From the type with two staminal whorls and two whorls of carpels the families with more highly developed flowers have been derived (see diagram). Disappearance of the inner whorl of stamens and a whorl of carpels, accompanied by reduction in the pistil, give the ground-plan of the flower of Nyctaginaceae in which the perianth has become petaloid; a similar origin may be claimed for the simpler floral types of Aizoaceae (e.g. *Sesuvium*, *Mollugo*), while the more elaborate *Mesembryanthemum* type arises from a splitting of the staminal whorl to form series of petaloid structures and stamens.

The floral type of Portulacaceae may be derived from the same source by disappearance of the inner whorl of carpels and the approximation of a pair of bracteoles to the flower to form the dimerous calyx.

The typically 5-whorled flower of Caryophyllaceae represents a further elaboration and the alternative positions of the five carpels, noted in the description of the family, may be explained by the disappearance of one or other of the two whorls which existed in the *Phytolacca*-like ancestor.

The following diagram illustrates this view of the origin of the various families:

Caryophyllaceae
|
Amarantaceae, Chenopodiaceae — Phytolaccaceae — Aizoaceae, Nyctaginaceae
|
Portulacaceae

[1] See Pax in Engler and Prantl, *Pflanzenfamilien*, III, 1*b*, 68.

The order is of interest as indicating a passage from the monochlamydeous to the dichlamydeous type of flower, whether we take the view explained above that this passage had already occurred in the parent family and that the families as we now know them represent a segregation and further development of the types already laid down, or whether we assume that the families with the simpler type of flower represent the earlier forms. The simpler forms shew a similar plan of floral structure to the Urticales, while the more elaborate are typically dichlamydeous, reaching in the subfamily Silenoideae of Caryophyllaceae an advanced stage of differentiation in relation to the transference of pollen by means of insects.

Family I. CHENOPODIACEAE

Flowers monochlamydeous, bisexual or unisexual. Perianth 5- to 2-merous. Stamens equal to or fewer than, and opposite to, the perianth-leaves. Carpels generally two; ovary unilocular with one basal ovule. Fruit a utricle or nut surrounded by the persistent perianth.

Annual or perennial herbs, sometimes shrubs, with simple alternate leaves and small inconspicuous flowers.

Genera 101; species 1200. Cosmopolitan.

The plants are annual or perennial herbs, sometimes shrubs, very rarely forming small trees, as the Saxaul (*Haloxylon Ammodendron*), which, with its thick stunted trunk, and tufts of whip-like apparently leafless branches, forms a striking feature of the Central Asiatic steppes, sometimes reaching 20 feet in height. They are salt-loving plants (halophytes), the greater number of species occurring near the sea as Salt-wort (*Salsola Kali*), in salt marshes as Samphire (*Salicornia herbacea*), or in steppes and deserts which, once covered by the sea, have now a soil strongly impregnated with salt. For the same reason our common species of *Chenopodium* and *Atriplex* grow near dwellings and in cultivated soil, quickly covering soil which has become enriched in mineral salts by the decomposition of refuse or manure-heaps. The various forms of growth, as well as several characteristic peculiarities in structure, bear a close relation to the halophilous habit, and shew a striking similarity to desert forms (xerophytes), the functions of water-storage and checking of transpiration giving the key to the characteristic habit and structure. Coverings of hairs are frequent and their different forms may help to distinguish genera and smaller groups. Such are the mealy hairs of many species of *Chenopodium*, *Atriplex* and *Salsola*, which consist of a stalk bearing large thin-walled end-cells full of clear watery sap. The terminal bladders easily collapse and form the peculiar mealy covering. They are a means of storing water and lose their contents when the organ on which they are borne reaches a certain age. In some species of *Atriplex* which have to endure prolonged periods of drought, the closely packed hairs with their end-cells tensely filled with water form, during the rainy period, a glass-like covering. In the succeeding dry period their watery contents are gradually used up, the turgid cells collapse and the hairs become matted into a dense felt which forms an excellent check to loss of water from the young assimilating portions of the plant.

Our native species of *Chenopodium* and *Atriplex* have more or less hastate-triangular or rhomboid leaf-blades, but the great majority of members of the family have a much reduced leaf-surface, narrowly linear and cylindrical or semicylindrical in section. Examples are found in Britain in *Suaeda* and *Salsola*, while in *Salicornia* (fig. 42) the succulent jointed stems are apparently leafless, each internode ending in a narrow cup-like ring which embraces the base of the one above.

Later work by E. de Fraine (1) confirms the theory put forward by Duval-Jouve in 1868 that the succulent outer cortex of the stem-joints is foliar in origin, and derived from a basal development of the cup-like leaf-sheath of the pair of leaves at the node above. This view is supported by the study of the development of the shoot behind the apex, of the anatomy and the course of the vascular bundles, and by the fact that the cortex separates at the end of the season in a manner analogous to that of leaf-fall.

The most general means for ensuring a water-supply to the green assimilating cells is a system of water-storing tissue, consisting of large colourless cells filled with water holding sodium chloride and other mineral salts in solution, and also frequently containing crystals of calcium oxalate. In dorsiventral leaves the water-storing cells occupy the upper and lower faces of the leaf, the green palisade layers coming between, or the latter are associated with the veins, the aqueous tissue forming the remainder of the mesophyll. In concentric leaves the great bulk consists of aqueous tissue, the green assimilating cells forming an outer ring beneath the epidermis.

A. F. W. Schimper shewed the reason for this similarity between halophytes and xerophytes. Plants absorb water with difficulty when solutions of high concentration are supplied to the roots, and concentrated solutions also hinder assimilation in the green cells. Halophytes, therefore, take up as little as possible of the solutions presented to them in the soil, in order that nutrition may not be prejudiced by an excess of salt in the tissues. Hence the necessity of checking transpiration, for which similar means have been adopted to those in xerophytic plants.

The small flowers are usually united into small dense cymose inflorescences, which, beginning as dichasia, pass into unilateral cymes (fig. 40, A), or are uniparous from the first. The flowers are bisexual or unisexual, and, except in *Beta*, hypogynous. The most complete type of flower conforms to the formula P5, A5, G2–3; other forms may be regarded as derived from this. There is no indication of a second perianth-whorl, or of antepetalous stamens, the five stamens being opposite the perianth-leaves, which are more or less united with one another; the degree of union varies from an almost free condition to complete union, forming a pitcher- or funnel-shaped perianth. The number of perianth-leaves and stamens is very variable, even in the same genus, and even on the same plant. Thus, of our eleven British species of *Chenopodium*, seven have a pentamerous perianth and androecium, in three species the lateral flowers of each cluster are usually 2- to 4-merous and the terminal 5-merous, while the flowers of *C. Bonus-Henricus* are described as all 5-merous or the lateral 2- to 3-androus. *Atriplex* has unisexual flowers (bisexual are also sometimes present), of which the male are 3- to 5-merous, while the female have only two perianth-leaves or are naked but have a pair of lateral bracteoles. *Salicornia* (fig. 42), with bisexual flowers, has a 3- to 4-lobed or truncate perianth sunk in the spike, and one rarely two stamens.

The perianth-leaves are usually similar, but sometimes differ in size and structure, as for instance in *Halocnemum* where there are three, two opposite and hooded, the odd one slightly concave; in *Corispermum* the hinder leaf is much larger and toothed, in *Alexandra* (Central Asia) the two lateral leaves are boat-shaped with a broad wing-like keel, the remaining three flat and without an appendage. In con-

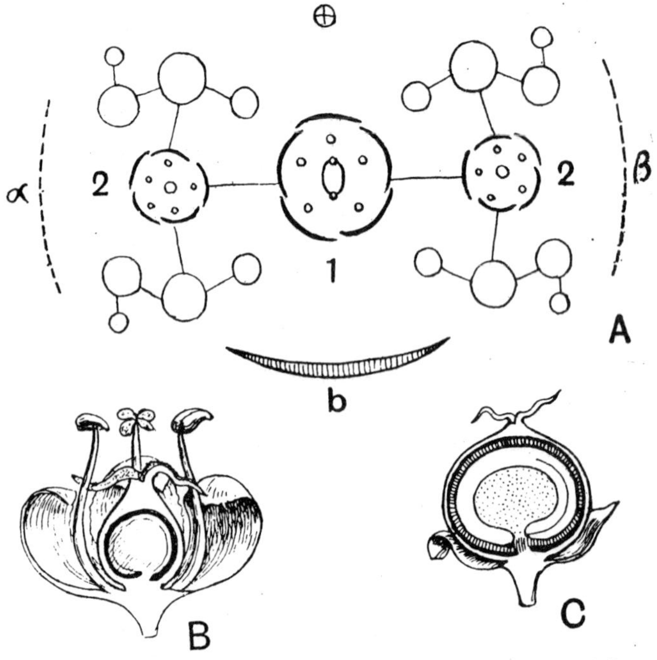

FIG. 40. A. Plan of inflorescence of *Chenopodium album*; 1, central flower of dichasium springing from axil of main bract, *b*; 2, lateral flowers in axils of bracteoles *α* and *β*, the branches of the secondary axes pass into unilateral cymes. B. Longitudinal section of flower of *C. Bonus-Henricus*. C. Seed of same in vertical section. B, C enlarged. (A after Eichler; B, C after Baillon.)

sistence they are either green and herbaceous as in *Beta*, *Chenopodium*, *Atriplex*, *Suaeda* and genera allied to these, or hyaline, forming a thin delicate membrane, as in *Corispermum*, *Salicornia*, *Salsola* (fig. 43) and allied genera. They are persistent in the fruiting stage, sometimes developing to form means of distribution for the enclosed fruit, such as wings, spines or a fleshy covering simulating a berry.

The stamens are free or united at the base, and attached to the floral axis, to the base of the perianth-leaves, or to a disc which sometimes forms small outgrowths between the stamens (fig. 41, C); these outgrowths have been erroneously regarded as representing an inner whorl of staminodes. The narrow subulate filament is attached to the back of the anther which is four-celled and opens by a pair of longitudinal slits which are lateral or turn inwards. Before the flower opens the anthers are bent inwards. The pollen is spherical with

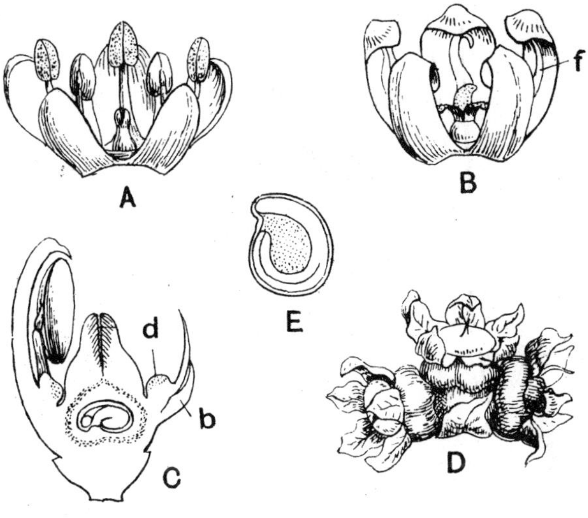

FIG. 41. *Beta vulgaris*. A. Flower in the first or staminate stage. B. Flower in the later or pistillate stage; *f*, the withering filaments from which the anthers have fallen. C. Longitudinal section through a flower-bud, shewing the arrangement of the parts; *b*, subtending bract; *d*, glandular disc. D. A cluster of fruits. E. Seed in vertical section. All enlarged. (D after Baillon; the rest after Volkens.)

numerous roundish pores. The unilocular ovary varies widely in form and is usually drawn out at the apex into a style which bears two, sometimes three or four, very rarely five, filiform or sometimes broad and flat stigmas. The position of the campylotropous ovule varies greatly according to the length and form of the funicle, from pendulous to transverse; the funicle springs from the base of the ovary except in *Beta* where it is attached laterally (fig. 41, C).

But little is known as to pollination. Volkens (2) points out various facts which indicate that wind can only be of subordinate significance. Such are the comparatively heavy pollen, the absence of slender flexible stamens, and the gradual opening of the individual flowers, as well as of the flowers of the inflorescence taken collectively, one of the characteristics of anemophily being the simultaneous opening of a large number of flowers. Cross-pollination is favoured by the frequence of dicliny, and, in bisexual flowers, of dichogamy, both proterandry, as in *Salicornia herbacea* and *Beta vulgaris* (fig. 41, A, B), and proterogyny as in many species of *Chenopodium*.

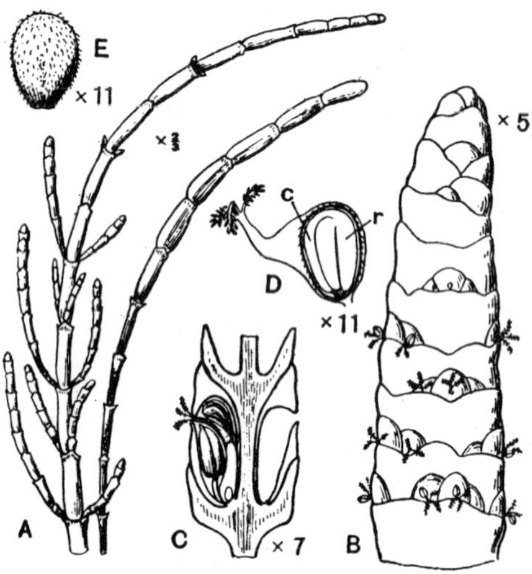

FIG. 42. *Salicornia ambigua.* A. Portion of stem and branch. B. Portion of stem in flower. C. Small portion of B cut lengthwise, shewing a flower enclosed in the perianth and another perianth empty. D. Fruit cut lengthwise; *c*, cotyledons; *r*, radicle. E. Seed. (From *Flor. Jam.*)

The one-seeded fruit is indehiscent or opens before or during germination by a transverse lid, as in *Beta* (fig. 41, D). The pericarp is membranous, crustaceous, or leathery, rarely stony or fleshy, and occasionally adnate to the seed-coat. It is generally more or less completely enveloped by the persistent perianth. The means of distribution of the fruit are very varied. Berry-like fruits are formed by the pericarp or floral envelope becoming fleshy. Distribution by clinging to the coats of animals is favoured by development of thorns on one or more of the perianth-leaves, or the enveloping bracteoles become simple or branched thorns as in *Spinacia oleracea* (Spinach). Adaptations for distribution by wind occur in the development of hairs on bracteoles or perianth; or wings of very varied form are developed on the fruit itself or on the perianth (fig. 43, C) or bracteoles; or the pericarp, floral envelope, or bracteoles become light and spongy. In *Beta* the fruits, which are sunk in the fleshy floral axis, become coherent by the enlarged hardened bases of the sepals.

The form of the seed is governed by that of the ovule, and is lens- or kidney-shaped, horizontal or vertical, with a superior, lateral or inferior micropyle; the testa is horny, leathery or membranous, and smooth or granulate.

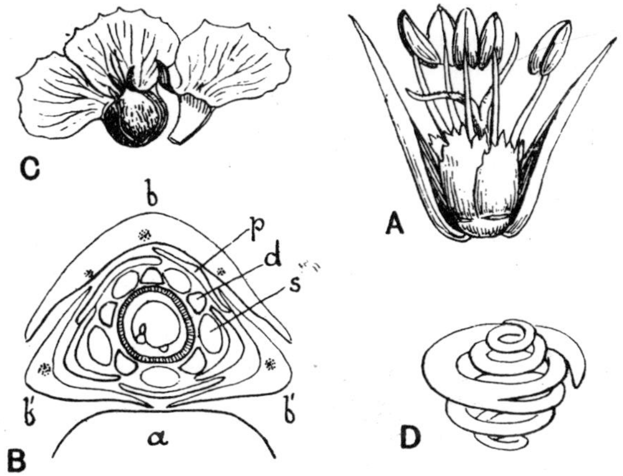

FIG. 43. *Salsola Kali.* A. Flower with the pair of bracteoles, × 6. B. Cross-section through base of flower shewing arrangement of parts in a horizontal plane; *a*, axis on which the flower stands in the axil of the bract *b*; *b′*, bracteoles; *p*, perianth; *s*, stamens; *d*, lobe of disc, alternating with the stamens; in the centre is the ovary containing the ovule; × 5. C. Fruit surrounded by the persistent perianth-segments (three are shewn) from the back of which a membranous wing has developed, × 5. D. Embryo, × 8. (After Volkens.)

The character of the embryo has been used for the separation of the family into two sections. In the larger, *Cyclolobeae*, the seed contains a curved embryo surrounding a floury perisperm; this includes *Chenopodium* (fig. 40), *Beta* (fig. 41), *Atriplex* and others, also *Salicornia* (fig. 42, D) in which there is little or no perisperm. *Salsola*, *Suaeda* and allied genera constitute a smaller section (*Spirolobeae*) in which the embryo is rolled up spirally and almost completely fills the seed (fig. 43, D).

Bunge recognised the following main areas of distribution: the Australian low-lying salt plains; the South American pampas; the North American prairies; the coast-land of the Mediterranean; the South African Karroo; the Red Sea basin; the south-west shores of the Caspian; the Central Asiatic basin; and the salt-steppes of Eastern Asia.

The most important economic plant is the Beet (*Beta vulgaris*), owing to its tendency, common to several plants of the family with herbaceous roots, to swell under cultivation by increase of the parenchymatous elements. The Beet also supplies a good illustration of an anatomical peculiarity which is very general in the family, namely the formation of short-lived concentric rings of cambium, the new one appearing in the cortex just outside the circle of bundles produced by the last. The varieties include the Sugar-beet, and the Mangold. *Spinacia oleracea* is Spinach. In steppe and desert country the members of the family serve as fodder-plants, as for instance the Australian Salt-bushes (species of *Atriplex*, *Kochia* and *Rhagodia*). Species of *Chenopodium* are cultivated as pot-herbs (e.g. *C. Bonus-Henricus* and others), or as grain on account of the mealy perisperm, as the South American *C. Quinoa*.

REFERENCES

(1) Fraine, E. de. "The anatomy of the genus *Salicornia*." *Journ. Linn. Soc.* (*Botany*), xli, 317 (1913).

(2) Volkens, G. in Engler and Prantl, *Pflanzenfamilien*, iii, 1*a*, 47.

See also Ulbrich, E. in *Pflanzenfam.* Ed. 2 (Harms), xvi, c (1934).

Family II. AMARANTACEAE

This family is scarcely separable from Chenopodiaceae with which it was united by the French botanist Baillon. It differs in having the perianth dry, membranous and often white or coloured, not green and herbaceous as generally in Chenopodiaceae. The plants are herbs or shrubs with opposite or alternate simple generally entire exstipulate leaves which are often more or less thickly covered with hairs. As in Chenopodiaceae, the vascular bundles in the stem are arranged in several concentric rings or more or less irregularly. The small flowers form inflorescences which are often dense and showy as in Foxtail (*Amarantus*), and Cockscomb (*Celosia cristata*) where it is fasciated. Each flower has a pair of large membranous persistent bracteoles which may be sterile, the inflorescence being a simple or branched spike or raceme; or branching occurs in the axil of the bracteoles forming dichasia, which are borne on a simple or racemosely or cymosely branched main axis. Frequently lateral flowers are aborted and their place is occupied by simple or branched prickles, or tufts of hair, which persist till the fruit is ripe and serve as a means of distribution, the partial inflorescences becoming detached from the main axes, as in *Cyathula*.

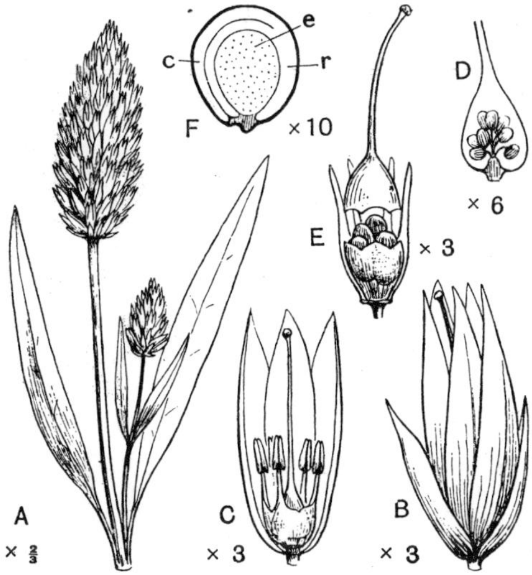

FIG. 44. *Celosia argentea*. A. Upper portion of plant with inflorescence. B. Flower with bract and one bracteole. C. Flower with perianth partly removed. D. Ovary cut lengthwise. E. Fruit shewing dehiscence. F. Seed cut lengthwise; *c*, cotyledons; *r*, radicle; *e*, perisperm. (From *Flor. Jam.*)

The structure of the flower corresponds with that of Chenopodiaceae. As a rule, a simple generally regular 5-merous perianth is followed by an opposite whorl of five stamens, while a unilocular superior ovary with one, two or three styles occupies the centre of the flower. The membranous perianth-leaves are free or more or less united; they may become woody at the base enclosing the fruit. They are often hairy; in species of the Australian genus *Ptilotus* the long silky hairs form wind-floats. The stamens are generally united below into a membranous tube, which may bear simple, lobed or fringed, petaloid outgrowths between each stamen (fig. 45, D). In the greater number of the genera each half of the anther is 2-celled, in a smaller number ultimately one-celled. The ovules, which are solitary except in *Celosia* (fig. 44, D) and a few allied genera, are campylotropous and borne, either pendulous or erect, on long or short funicles springing from the base of the ovary-chamber.

The fruit is generally dry, either with a hard pericarp forming a one-seeded nutlet, or with a thin wall forming a utricle containing one (as in *Achyranthes*) or more seeds, in

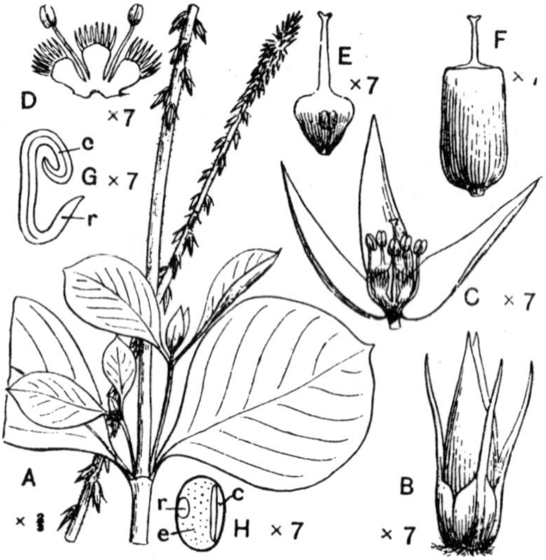

FIG. 45. *Achyranthes indica.* A. Upper part of plant with inflorescence. B. Flower with pungent bract and bracteoles. C. Flower with two perianth-segments removed. D. Portion of androecium shewing two stamens and three fimbriated outgrowths. E. Ovary shewing the ovule. F. Utricle. G. Embryo. H. Seed in section. *c*, cotyledons; *r*, radicle; *e*, perisperm. (From *Flor. Jam.*)

the latter case dehiscing transversely (as in *Celosia*) or irregularly. In a few genera only it forms a one- or several-seeded berry.

The seeds, which are generally lenticular, have a rough or polished testa; the embryo lies close to the seed-coat surrounding a central mass of mealy perisperm.

The family contains 64 genera with about 800 species distributed throughout the warmer parts of the earth. The chief centres of distribution are tropical America and India.

Several species are cultivated as ornamental plants, as *Celosia cristata* (Cockscomb), *Amarantus* (Foxtail), and *Gomphrena globosa*.

The family is subdivided into three tribes as follows:

Celosieae. Ovary with two to many ovules; anther-halves 2-celled. Chief genus *Celosia* with 60 species.

Amaranteae. Ovary with a single ovule; anther-halves 2-celled. Chief genera *Amarantus* (50 species), *Aerva* (10 species), *Ptilotus* (100 species).

Gomphreneae. Ovary with a single ovule; anther-halves ultimately 1-celled. Chief genera *Alternanthera* (170 species), *Gomphrena* (90 species), *Iresine* (70 species).

Family III. NYCTAGINACEAE

This family is well characterised by the petaloid tubular to funnel-shaped perianth, the lower portion of which persists, enveloping the fruit and simulating a pericarp (the whole is termed an anthocarp); the true fruit is an achene, the single erect seed containing a large, straight, curved or folded embryo which surrounds a mealy perisperm. The plants are herbs, shrubs as in *Mirabilis Jalapa* (Marvel of Peru, a well-known ornamental shrub), or trees. *Bougainvillea spectabilis*, a native of Brazil, is a handsome climber often seen in greenhouses. Needle-like crystals of calcium oxalate (raphides) are very generally present, occurring even in the flower. The stem-structure is anomalous; cambium is developed outside the original ring of bundles from which new closed bundles and intermediate tissue are produced; the bast in the new bundles is extremely rudimentary[1].

The leaves are generally opposite, those of each pair being often very unequal; they are simple and exstipulate. The inflorescence is cymose and an important feature is the mode of development of the uppermost bracts. In *Mirabilis* (fig. 46, A) the dichasium has a tendency to monochasial development in the higher branches; each flower is surrounded at the base by an involucre of five sepal-like bracts. In *Oxybaphus* (fig. 46, B) two of the five involucral bracts are fertile so that there are three flowers, a terminal and two lateral in each involucre. In *Bougainvillea* (fig. 46, C) there are three involucral bracts, each subtending a flower with which it has become adnate below. Owing to absence of the terminal

[1] See De Bary. "Comparative Anatomy of Phanerogams and Ferns." (English Edition), 590 (1884).

flower there are again three flowers within each involucre, which is here large and petaloid. In *Boerhaavia* and others the involucral leaves are reduced to teeth or scales. The perianth is folded in

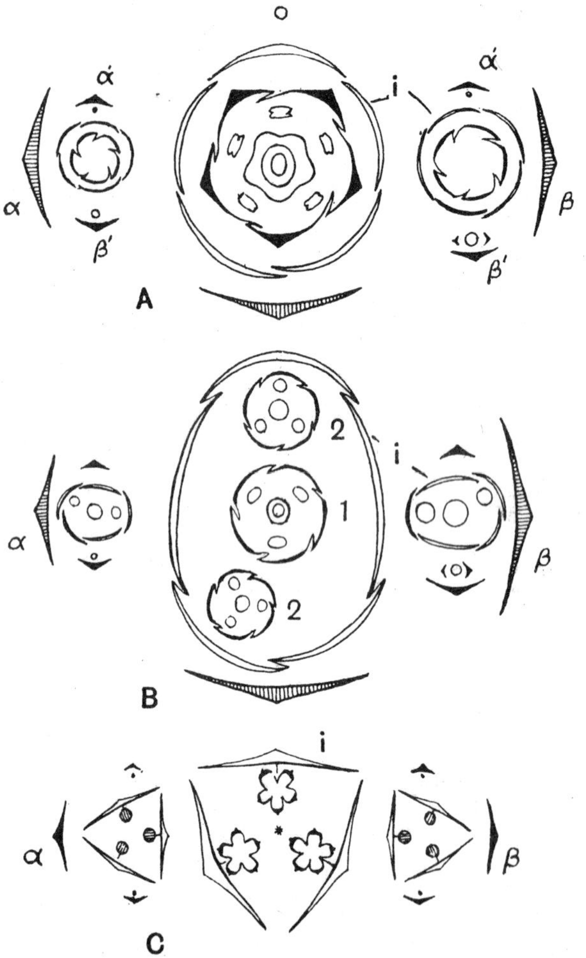

FIG. 46. Diagrams of inflorescence and flowers in A, *Mirabilis*; B, *Oxybaphus*; C, *Bougainvillea*; *i*, involucre of bracts; *α*, *β*, *α′*, *β′*, pairs of bracteoles. (After Eichler.)

the bud; in *Mirabilis* it is also contorted as in the corolla of Convolvulaceae. The number of stamens is very variable; in *Mirabilis Jalapa* they are equal in number to the petals, but there may be fewer or more (1–30). The filaments are often unequal.

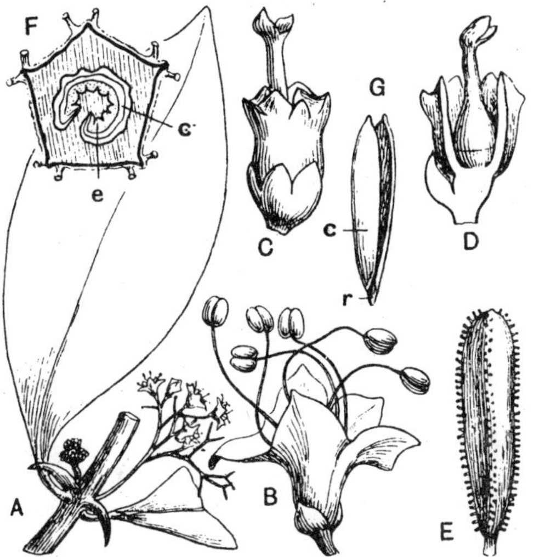

FIG. 47. *Pisonia aculeata.* A. Portion of male flowering branch, × ⅔. The thorns in the leaf-axils are reduced branches. B. Male flower, × 5. C. Female flower, × 5. D. Female flower with involucre and perianth cut lengthwise. E. Anthocarp, × 2. F. Anthocarp cut across, × 5. G. Embryo. *c*, cotyledons; *e*, perisperm; *r*, radicle. (From *Flor. Jam.*)

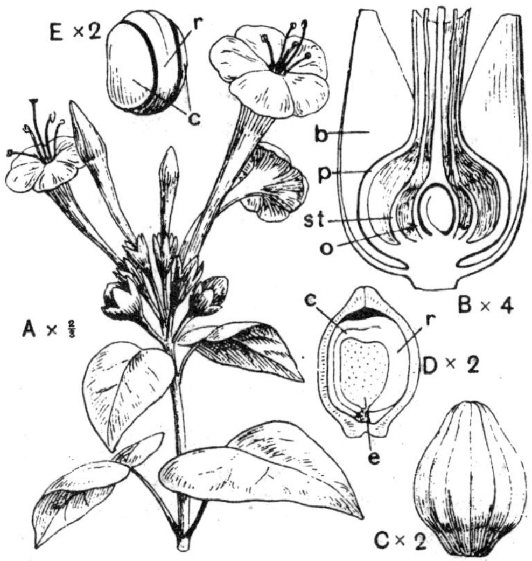

FIG. 48. *Mirabilis Jalapa.* A. Flowering shoot. B. Lower part of flower in vertical section; *b*, bract; *p*, perianth; *st*, stamen; *o*, ovary. C. Anthocarp. D. Same in vertical section; within the crustaceous wall formed from the persistent perianth-base and pericarp is seen the seed; *c*, cotyledons; *r*, radicle; *e*, perisperm. E. Embryo; *c*, outer cotyledon; *r*, radicle. (From *Flor. Jam.*)

The flowers are bisexual, or sometimes, as in *Pisonia*, unisexual by suppression of stamens or pistil (fig. 47). The pistil consists of a single carpel; the free ovary bears a long simple style and contains a solitary basal ovule, the micropyle of which points to the bottom of the ovary (micropyle inferior). The persistent portion of the perianth enveloping the fruit becomes leathery, hardened or fleshy, and striated or ribbed, and is often a means of distribution by the development of mucilage or by glandular hairs, as in some species of *Pisonia* (fig. 47, E), or by hooks as in the Australian and Polynesian *Pisonia grandis*. In *Bougainvillea* the persistent bracteoles form wind-floats.

The seed has a thin hyaline testa or the seed-coat is completely adherent to the pericarp. The cotyledons are unequal, the inner being smaller than the outer. In *Abronia* the inner is almost aborted in the seed and remains small for some time after germination, but in *A. umbellata* ultimately becomes the larger and leaf-like in appearance.

There are 30 genera, containing about 300 species, in the warmer parts of both hemispheres, but especially in America.

Family IV. PHYTOLACCACEAE

A small family of about 115 species in 17 genera, comprising herbs, shrubs (a few climbing) and trees, widely distributed in tropical and temperate climates but found mainly in the warmer parts of America. The leaves are simple, entire, generally exstipulate, and glabrous. The inconspicuous flowers are arranged in racemose or cymose inflorescences. The flowers, which are bisexual or sometimes unisexual by reduction, have a simple regular perianth of four to five members which are free or united at the base, and persistent. The stamens and carpels vary widely in number, and variations occur in one and the same species. In the simplest case, the stamens form a single whorl the members of which alternate with (*Rivina laevis*) or are opposite to (*Microtea*) the perianth-leaves (fig. 49, A, B), while a single carpel occupies the centre of the flower. Often there are two alternating whorls of stamens the number of which may be increased by doubling in one or both whorls (see diagrams of *Phytolacca decandra* and *P. icosandra*, fig. 49). In *Stegnosperma* (a genus with one species in Central America and the West Indies) the outer whorl of stamens is replaced by a whorl of five petals.

A similar variation occurs in the pistil; there may be as many as 10 carpels, and these are free or united; the ovary is superior, except in the monotypic Mexican genus *Agdestis* where it is inferior. Each carpel bears a single style and contains a solitary ovule, in the position of which there is again considerable variation. The form of the fruit varies widely. Where one carpel only is present it is indehiscent either forming a berry (*Rivina*, fig. 50) or having a leathery or dry pericarp. Where two carpels are united in the flower the fruit is a schizocarp, the two one-seeded mericarps being often winged or thorny. When the carpels are numerous the

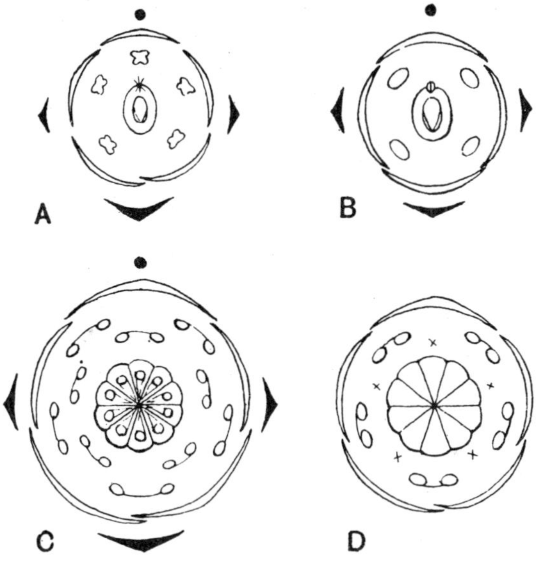

FIG. 49. Floral diagrams of A, *Microtea*; B, *Rivina laevis*; C, *Phytolacca icosandra*; D, *P. decandra*. Bracts and bracteoles are indicated in A, B and C. (After Eichler.)

fruit consists of an equal number of separate indehiscent nutlets, or forms a multilocular capsule each chamber splitting when ripe to set free the solitary seed. The small laterally compressed seeds are lenticular or kidney-shaped and contain a large embryo curved round the mealy or fatty perisperm. The testa is rarely smooth, generally wrinkled or warty or sometimes hairy; a membranous aril is often present. In *Phytolacca* (fig. 51) the many-seeded fruit is at first berry-like, the pericarp subsequently becoming drier. *P. decandra* is often cultivated as an ornamental plant for its showy spike of deep-purple berries, which yield a deep red dye used for colouring wines, etc.

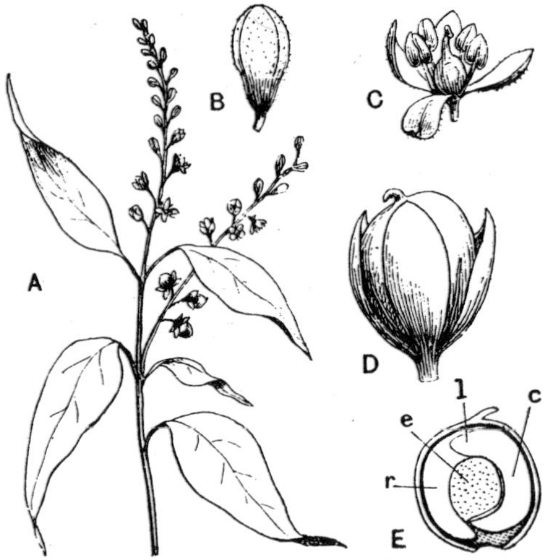

FIG. 50. *Rivina humilis.* A. Upper portion of branch, × ⅔. B. Flower-bud, × 7. C. Flower, × 7. D. Fruit, × 5. E. Section of seed, × 5; *e*, perisperm; *c*, cotyledons; *l*, lobed base of cotyledon; *r*, radicle. (From *Flor. Jam.*)

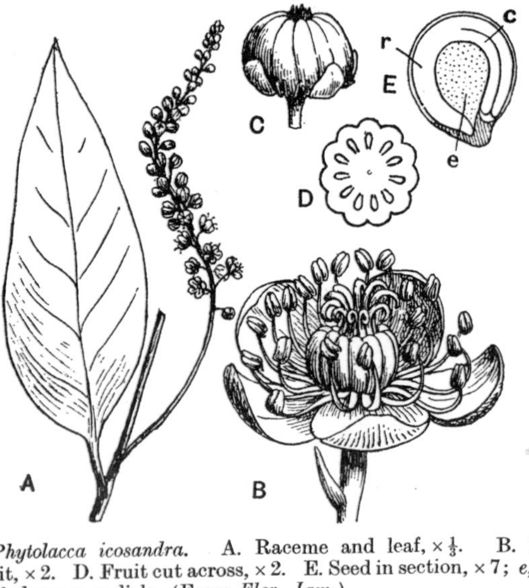

FIG. 51. *Phytolacca icosandra.* A. Raceme and leaf, × ⅓. B. Flower, × 5. C. Fruit, × 2. D. Fruit cut across, × 2. E. Seed in section, × 7; *e*, perisperm; *c*, cotyledons; *r*, radicle. (From *Flor. Jam.*)

Family V. AIZOACEAE

(FICOIDEAE of Bentham and Hooker)

Flowers bisexual, typically monochlamydeous, cyclic. Perianth 4- to 5-merous. Stamens 5 (3) or ∞, the outer members often forming several series of petaloid structures. Carpels 2 to ∞ ; ovary superior or inferior, 2- to multi-locular; ovules ∞, anatropous or campylotropous. Capsule various.

Herbs or undershrubs with filiform or fleshy leaves and cymose flowers.

Genera 23; species 1100.

The members of this family are found in dry climates in the warmer parts of the earth. They are herbs or low shrubs with simple opposite or alternate generally exstipulate leaves. The leaves are often succulent, and narrow, spathulate, or rhombic in shape, containing layers of water-storing tissue between the peripheral green assimilating layers. In many species of *Mesembryanthemum* the epidermis has a granular coating of wax; coverings of hairs of various forms are also frequently present, giving the plant a greyish-green appearance. The familiar Ice-plant (*M. crystallinum*) owes its crystalline look to a covering of bladder-like hairs which serve as water-reservoirs enabling it to thrive in dry, sandy soil at the Cape, in South and West Australia, California, and throughout the Mediterranean region. In others the leaves are narrow and filiform, or sometimes reduced to scales, as in the Australian genus *Macarthuria*, recalling in their Broom-like habit Mediterranean species of *Cytisus*. The perennial species shew abnormal thickening, depending on the short-lived activity of the cambium of the original ring of bundles and the successive production of new rings outside the old.

Flowers are terminal (many species of *Mesembryanthemum*) or axillary (*Tetragonia*), or form small crowded dichasia passing in the higher branches into monochasia. In *Tetragonia* two flower-buds may be arranged serially in one leaf-axil and a leaf-bud may also be present.

The arrangement of the parts of the flowers shews considerable variation but may be derived from a simple type containing three whorls, a perianth, staminal and carpellary, as seen in *Sesuvium pentandrum* (fig. 52, A). The number of stamens may be reduced as in *Mollugo verticillata*, a very variable South American species (fig. 52, C). More often, however, splitting of the original five staminal rudiments occurs, producing a larger number of stamens (fig. 52, B) which are free or united in bundles, or monadelphous at the base. In many cases, as

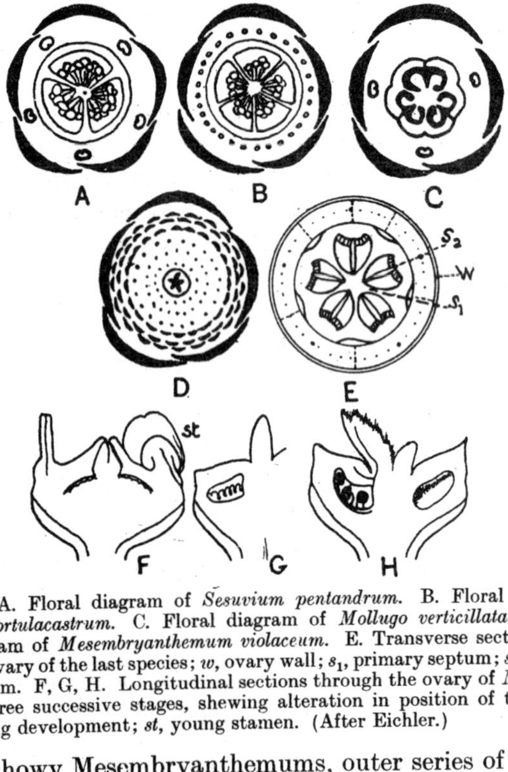

FIG. 52. A. Floral diagram of *Sesuvium pentandrum*. B. Floral diagram of *S. Portulacastrum*. C. Floral diagram of *Mollugo verticillata*. D. Floral diagram of *Mesembryanthemum violaceum*. E. Transverse section through the ovary of the last species; *w*, ovary wall; s_1, primary septum; s_2, secondary septum. F, G, H. Longitudinal sections through the ovary of *M. violaceum* in three successive stages, shewing alteration in position of the placenta during development; *st*, young stamen. (After Eichler.)

in the showy Mesembryanthemums, outer series of members resulting from this splitting become petaloid, forming a showy corolla-like series (fig. 52, D). The superior or inferior pistil consists generally of three to five united carpels, with a corresponding number of ovary-chambers, in the inner angle of which are borne rows of anatropous to campylotropous ovules. In some species of *Mesembryanthemum* the placentas, originally axile, become, through unequal growth of the ovary, first basal and finally parietal (fig. 52, F, G, H), a more or less complete secondary or "false" partition being also formed between the parietal placenta and the central axis (fig. 52, E, s_2). The fruit is generally a leathery capsule with loculicidal dehiscence, but septicidal and transverse dehiscence also occur. The capsule of *Mesembryanthemum* is hygroscopic, opening when wetted and spreading its valves star-wise; the valves close again in the dry (fig. 53, C, D). In *M. edule* the

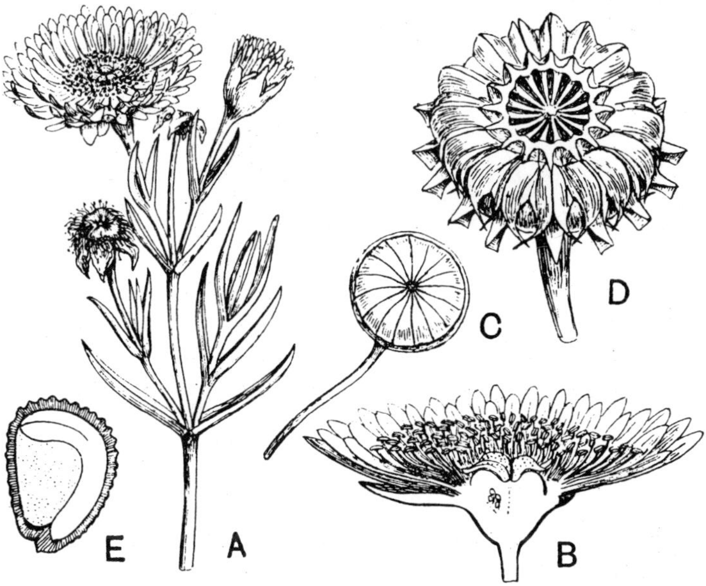

FIG. 53. *Mesembryanthemum.* A. Flowering shoot of *M. blandum.* B. Flower of same in vertical section. C. A closed fruit. D. Fruit which has opened on exposure to moisture. E. Seed cut lengthwise, enlarged. (A, B after Baillon; E after Le Maout and Decaisne.)

fruit is fleshy and edible, the numerous seeds being embedded in a sweet, slightly acidulous pulp. *Tetragonia* has a nut-like or drupaceous indehiscent fruit containing several seeds. The seed contains a large embryo curved round a mealy perisperm.

The greater number, more than 800, of the species belong to *Mesembryanthemum*, and are found chiefly in the dry sandy districts of South Africa but also in widely separated similar regions from Australia to California. South Africa is the chief centre of distribution of the family, members of which extend the range through tropical Africa to Arabia and the Mediterranean region. There is also a small centre in West and South Australia, and a few genera (e.g. *Mollugo*, *Tetragonia*) spread through the tropics of both hemispheres. *Tetragonia expansa*, a widely distributed species (Japan, Australasia, Polynesia, and extratropical S. America), is cultivated as a pot-herb and known as New Zealand Spinach.

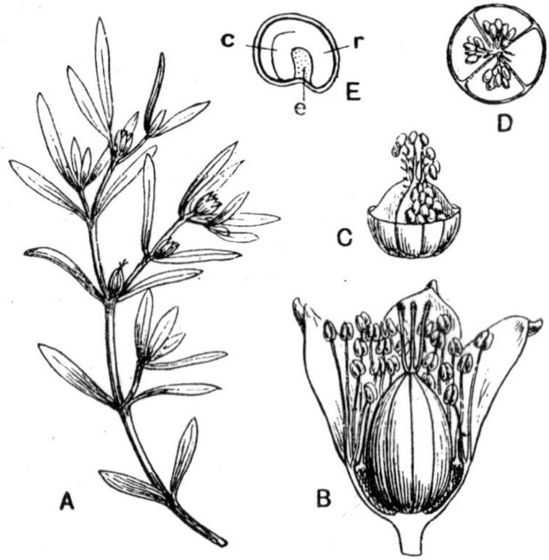

FIG. 54. *Sesuvium Portulacastrum.* A. Portion of plant in flower, $\times \frac{2}{3}$. B. Flower with part of perianth and some stamens removed, $\times 3$. C. Capsule after dehiscence (circumsciss), $\times 3$. D. Transverse section of lower portion of capsule, $\times 3$. E. Seed in section, $\times 10$; *e*, perisperm; *c*, cotyledons; *r*, radicle. (From *Flor. Jam.*)

The family is subdivided according to the character of the perianth and the relative position of the ovary.

Subfamily MOLLUGINOIDEAE. Perianth-segments almost free. Petaloid staminodes generally absent. Ovary superior. About 7 genera with 60 species in the warmer parts of the world.

Subfamily FICOIDEAE. Perianth-segments uniting to form a longer or shorter tube. Includes the following tribes.

Tribe *Sesuvieae*. Ovary superior. Capsule opening by a lid. Staminodes absent. *Sesuvium* (fig. 54) (8 species), sea-shore plants on tropical and subtropical coasts. *Trianthema*, about 15 species, widely distributed, chiefly in the Old World tropics.

Tribe *Aizoeae*. Ovary inferior. Capsule loculicidal. Staminodes absent. Dry sandy or stony places mainly in South Africa. *Aizoon hispanicum* is a characteristic species of the Mediterranean region. Contains 6 genera with 49 species.

Tribe *Mesembryanthemeae*. Ovary inferior. Petaloid staminodes absent in *Tetragonia*, present and numerous in *Mesembryanthemum* forming showy white, yellow or rose-coloured flowers suggesting Compositae. Species of *Mesembryanthemum* are grown as ornamental plants; the fleshy rhombic leaves of some South African species borne in pairs on the shoot resemble stones in appearance[1]. The stamens of some species are irritable, bending inwards when touched by bees or other insects. The fruits of *M. acinaciforme* and *M. edule* are eaten in South Africa under the name Hottentot-figs. *M. edule* has become naturalised near the sea in the Isle of Wight, Cornwall and the Channel Islands.

Aizoaceae resemble Phytolaccaceae in their morphologically apetalous flowers, and, like them, have an anomalous stem-structure. They are distinguished by the typically syncarpous ovary with numerous ovules, the frequently tubular perianth and often opposite leaves.

Family VI. PORTULACACEAE

A small family of about 500 species in 19 genera, generally distributed in warm and temperate climates, especially on the Pacific side of the American continent. It comprises herbs or small shrubs with succulent leaves and scarious stipules (*Claytonia* is exstipulate). The flowers are generally small but sometimes showy. The most general type contains two sepals, situated anteroposteriorly, four to five petals, an equal number of ante-petalous stamens, and three carpels. Such for instance is *Claytonia perfoliata*, a small succulent annual, a native of western North America, which is naturalised in many places in England. *Montia*, a cosmopolitan monotypic genus, and the only true British representative of the family, differs in the suppression of the two anterior stamens. It is a small glabrous annual with spathulate subopposite leaves and minute white flowers, growing in wet and damp places. The number of stamens may be increased, as in *Portulaca oleracea* (Purslane). The superior ovary (half inferior in *Portulaca*) is unilocular, bears a simple or three- or more-branched

[1] See Marloth, R. "Mimicry among plants." *Trans. S.A. Phil. Soc.* XV (1904) and XVI (1905). See also Marloth, *Flora of South Africa*, I, 194, for an excellent illustrated account of the family in South Africa.

style and contains two to many campylotropous ovules on a central basal placenta. The fruit is a capsule dehiscing transversely, or by two or three valves. The compressed seeds contain an embryo coiled round a mealy perisperm. *Portulaca grandiflora* is a favourite garden plant.

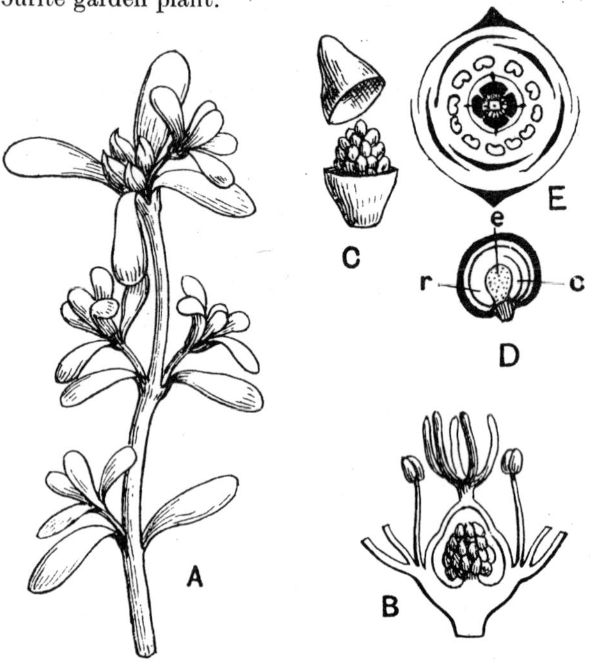

FIG. 55. *Portulaca oleracea.* A. Upper part of branch bearing flower-buds at apex, × ⅘. B. Flower in section with base only of sepals and petals indicated, × 5. C. Fruit shewing transverse dehiscence, × 3. D. Seed in section, × 12; *c*, cotyledons; *r*, radicle; *e*, perisperm. E. Floral diagram. (E after Eichler.)

Family VII. BASELLACEAE. See Appendix

Family VIII. CARYOPHYLLACEAE

Flowers generally bisexual, rarely unisexual by reduction, 5- or 4-merous, with calyx and corolla. Sepals free or united, sometimes scarious, persistent. Corolla rarely absent, often clawed. Stamens 8–10 in two whorls, or fewer, hypogynous. Receptacle sometimes elongated between the corolla and stamens. Carpels five, often fewer, united to form a one-celled or incompletely chambered ovary, with two to five styles free or united below; ovules generally numerous, more rarely few to solitary, on a basal or central placenta formed from the ventral portion of the carpels. Fruit a capsule opening apically by long or short valves, equal or double in number to the carpels; rarely a berry.

Herbs or undershrubs with entire, generally narrow, opposite, rarely alternate leaves; stipules sometimes present. Inflorescence cymose, generally many-flowered; more rarely flowers solitary.

Genera 80; species about 2000; mostly in the temperate regions of the northern hemisphere, but also in the southern hemisphere and a few in the mountains of the tropics.

A family common in Britain where the members are easily distinguished by the herbaceous stem with swollen nodes and opposite leaves, the cymose inflorescence, the white or pink flowers with their parts in fives (excepting the reduction to three or two which often occurs in the pistil and rarely in the androecium), and the characters of the pistil, seed and embryo already indicated for the order.

The family as a whole is one of annual or perennial herbs, sometimes woody below, or in the warmer parts of the world with even a bushy growth. The leaves are simple, entire, sessile and generally exstipulate; the pairs decussate. One leaf of a pair is formed earlier than the other and bears in its axil a more vigorous bud, which is often the only one to develop a branch.

The inflorescence is always definite, the main axis ending in a flower. Occasionally the flower is solitary as in Corn Cockle (*Githago segetum*). Generally, however, a dichasium is formed, a branch being borne in the axil of each of the bracteoles of the main axis. In the ultimate branches the dichasia pass into scorpioid cymes.

In the most complete type of flower there are five whorls with five members in each, namely five sepals, five petals, two pentamerous whorls of stamens and, crowning the floral axis, five carpels forming a compound ovary in which the number of members is indicated by the five styles, the five double rows of ovules on the central placenta, and sometimes also by the five partition walls remaining in the lower part of the ovary (fig. 56, J).

The flower is apparently obdiplostemonous; when viewed from above the anthers of the outermost whorl are seen to be opposite the petals. not alternate with them. Study of the floral development, however, shews that the antepetalous stamens arise in regular order, after the antesepalous. Also that the insertion of the filaments is normal and that the vascular bundles which pass into the antesepalous stamens from the floral axis stand in the latter outside those passing into the antepetalous. Obdiplostemony is therefore not real but only apparent and due to the mechanical pushing outwards of the upper portions of the stamens of the actually inner whorl.

Another difficulty which is not so readily explained arises from the fact that the carpels (as judged from the position of the styles which are in the same line as the midrib of the carpellary leaf) are sometimes opposite the sepals as in *Viscaria* or *Spergula* (fig. 56, G, J), sometimes opposite the petals as in *Githago*. It has been assumed that this variation arises from the fact that in the typically complete flower, the pistil, like the androecium, consists of two whorls, but one is always absent, in *Viscaria* the antepetalous whorl, in *Githago* the antesepalous.

The ovules are borne on a central placenta which is well developed in the Silenoideae where its relation to the carpels is indicated by the occasional presence of septa in the lower part of the ovary corresponding in number to the carpels; and also by the presence of corresponding lines between the double rows of ovules. In the Alsinoideae the placenta is often much shortened, becoming basilar; where the ovule is solitary it is attached to a longer or shorter funicle springing from the base of the ovary-chamber (fig. 56, A).

The flowers shew a considerable range of complexity, and may be derived either by elaboration or reduction from a type such as occurs in *Cerastium* or *Spergula*—a simple open flower with five regular alternating whorls represented by the formula S5, P5, A5 + 5, G5 (fig. 56, G). In this connection the family falls into two subfamilies, Alsinoideae, with the simpler or a reduced type of flower, and Silenoideae with a higher type elaborated in association with a more specialised entomophily and characterised by a tubular development of the calyx, stalked, often coloured petals, and frequently by presence of an internode above the calyx.

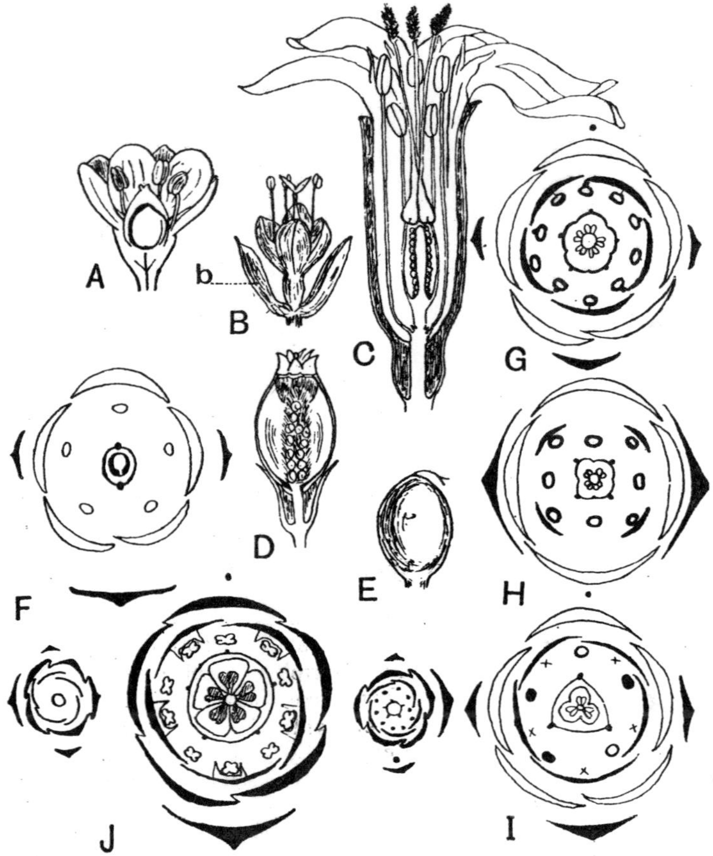

FIG. 56. A. Flower of *Corrigiola littoralis* in vertical section. B. Flower of *Lyallia*; *b*, bracteoles. C. Flower of *Silene nutans* in vertical section. D. Fruit of same in vertical section. E. Fruit of *Cucubalus*. F. Floral diagram of *Paronychia*. G. Floral diagram of *Spergula arvensis*. H. Floral diagram of *Sagina procumbens* (terminal flower). I. Floral diagram of *Stellaria media*. J. Dichasium of *Viscaria vulgaris*. The position of the styles is indicated by the dots on the outer wall of the ovary in the floral diagrams. (A after Baillon; B after Hooker; C, D after Wettstein; F–J after Eichler.)

Subfamily ALSINOIDEAE. Flowers with free sepals, petals of simple form or absent, stamens sometimes perigynous, styles free or united. An example of the most complete type is the flower of *Spergula arvensis* (fig. 56, G), a small open flower with simple white petals succeeded by two whorls of stamens (of which sometimes only one is developed) and five carpels. *Sagina* shews a similar structure but the parts of the flower are often in fours (fig. 56, H). In *Sagina apetala* the petals are minute or absent. *Cerastium* has notched petals and in *C. semidecandrum* the inner whorl of stamens is usually suppressed. In *Stellaria* (figs. 56, I, 57) the flower is usually 5-, rarely 4-merous, but the carpels are reduced to three. Considerable variation occurs in one and the same species. The common Chickweed (*Stellaria media*, fig. 57) is a most variable

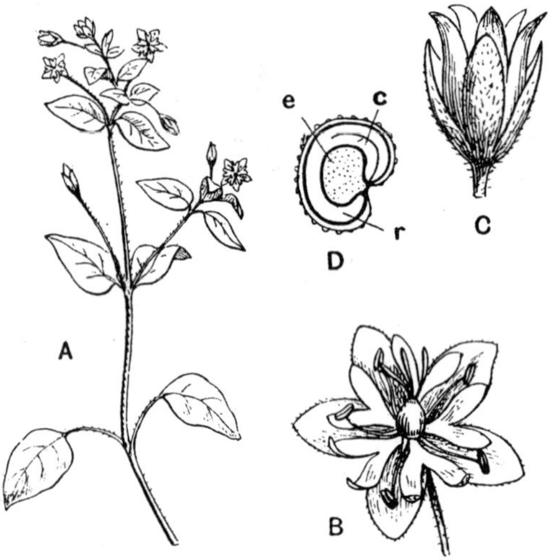

FIG. 57. *Stellaria media*. A. Branch with flowers. B. Flower, ×4. C. Capsule open, ×4; five of the six teeth are seen. D. Seed cut lengthwise, ×20; *c*, cotyledons; *e*, perisperm; *r*, radicle. (From *Flor. Jam.*)

plant. The flower may have the formula S5, P5, A5 + 5, G3, or the antepetalous stamens may be absent (S5, P5, A5, G3) and also some (rarely all) of the antesepalous stamens, a common formula being S5, P5, A3, G3. The petals, which are more or less deeply divided, are shorter than the sepals and sometimes altogether absent. In these genera the stamens are often slightly perigynous. The many-ovuled one-celled ovary becomes a capsular many-seeded fruit opening by teeth which are generally twice the number of the carpels (fig. 57, C). Other genera are *Holosteum*, *Minuartia* (*Alsine*), *Arenaria* and others; these differ in the number of the styles (three or four), the entire or notched form of the petals, and the number of valves in the capsule. *Spergula* (Spurrey) and *Spergularia*, which differ in having small scarious stipules at the base of the leaves, are separated as the tribe *Sperguleae* from the typical *Alsineae* (to which the other genera above-mentioned belong).

Polycarpon and allied genera differ in having the styles united below, and form a distinct tribe, *Polycarpeae*. Many of them are tropical and subtropical species and natives of the southern hemisphere. *Pycnophyllum*, a native of the South American Andes, growing near the snow-line, has a caespitose habit with short slender stems closely covered with narrow scale-like leaves suggesting *Lycopodium*. *Lyallia*, a genus with a single species in Kerguelen Island, forms a dense cushion of stout dichotomously branched stems closely covered with small leaves. The flowers in the *Polycarpeae* are reduced; the petals are often absent and the stamens reduced in number—*Lyallia* has an apetalous flower with three stamens (fig. 56, B); the ovary contains a few ovules and the fruit is a one-seeded capsule.

These reduced forms suggest the passage to the tribes *Paronychieae* and *Sclerantheae*, small sand- and steppe-plants with inconspicuous flowers, generally arranged in small dichasia. The corolla is small or absent, the antepetalous stamens are wanting or represented by small scales, the receptacle is often strongly perigynous, the ovary contains only one ovule (or very few) and the fruit is a one-seeded nut. Membranous stipules are generally present as in *Corrigiola*, *Paronychia*, *Herniaria* and *Illecebrum* forming the tribe *Paronychieae*, or absent as in *Scleranthus* (tribe *Sclerantheae*).

Subfamily SILENOIDEAE. The sepals are united into a tube, the petals are often coloured (red) and are differentiated into a stalk and limb; at the junction of stalk and limb, or the throat of the corolla, there is often an outgrowth, or ligule, forming the corona of the flower; it is present in *Dianthus* (Pink), many species of *Silene* (Catchfly), and *Lychnis* (Campion), absent in *Githago* and *Gypsophila*. An internode is frequently developed above the calyx on which petals, stamens and pistil are carried up (fig. 56, C). The flower is 5-merous, with two whorls of stamens. The styles are free and indicate the number of the carpels, generally five in *Lychnis*, three in *Silene*, and two in *Dianthus* and *Saponaria*

(Soapwort). The ovary is sometimes chambered at the base by septa (fig. 56, J) representing the lateral walls of the carpels; these disappear above. The fruit is a capsule opening by teeth the number of which is generally double that of the carpels. *Cucubalus* has berry-like fruits which do not open (fig. 56, E); this genus contains a single species, *C. baccifer*, the range of which extends from central Europe to Japan and which formerly occurred as an introduced plant in the Isle of Dogs.

Associated with the different degrees of elaboration in the flower are differences in their relation to insect-visitors and the manner of pollination. The simpler, more open flowers of Alsinoideae with easily accessible nectaries at the base of the outer stamens invite short-lipped flies or bees, and such visits are efficacious in the transport of pollen from one flower to another. On the other hand, when the nectar is concealed at the bottom of a deep tube, as in the Silenoideae, only the visits of insects with a long proboscis, such as the larger bees, moths or butterflies, are effectual in cross-pollination. Such visitors are more especially catered for by the strong sweet scent, often, as when night-flying moths are invited, emitted only at night, by the larger white or brightly coloured corolla, and by the fine markings (honey-guides) round the entrance of the tube; characters which are generally associated with the visits of Lepidoptera. The inconspicuous flowers of the *Paronychieae*, in which nectar when present is easily accessible, attract only minute short-lipped insects.

The chances of cross-pollination are frequently increased by dichogamy in the form of proterandry, rarely of proterogyny. For instance, in *Cerastium arvense*, there are three stages in the history of the mature flower. In the first the anthers of the outer whorl of stamens dehisce and are covered with pollen, while those of the inner whorl are not yet full-grown, and the styles are curled inwards, hiding their stigmatic surfaces. In the second stage the outer whorl of anthers becomes bent out of the way, and the now fully-developed and dehiscing inner whorl occupies its place. In the third the stigmas, which in the second stage still kept their papillar surfaces turned inward towards each other, are spread out above the withered stamens, turning their receptive surfaces upward. As, however, in this movement the ends of the stigmas often come in contact with the inner anthers still bearing their pollen, self-pollination is ensured if cross-pollination fails.

In *Stellaria media* the flowers, appearing as they do at all times of the year except in severe frost, are for a great period shut off from insect-visits and depend largely on self-pollination; the number of abortive stamens is greater the colder the time of year. The fertile stamens, especially when the number has been reduced to three, mature in slow succession; the stigmas follow so closely that they expand fully while the second and third stamens dehisce, and in absence of insects regularly pollinate themselves by contact with the anthers.

It has been observed that most of the distinctly proterandrous Alsinoideae are also gynodioecious, plants with small pistillate flowers being found in bloom chiefly at the beginning of the flowering period of the larger-flowered bisexual plants. The staminate condition of the bisexual plants is thus counterbalanced by the presence of the subsidiary female plants.

The seeds are generally more or less flattened, sometimes laterally and then attached by the edge, sometimes, as in *Dianthus*, dorsally, when the hilum is on the ventral surface. *Dianthus* is exceptional in having the embryo straight and surrounded by the mealy perisperm, and not, as is usual, curved round the outside of the latter.

The family is a very widely distributed one, reaching from the arctic through the temperate and tropical zones to the antarctic. Alsinoideae have a more general distribution than the more specialised Silenoideae, which are, for instance, absent from Australia; Silenoideae are most frequent in the north-temperate zone, where the Mediterranean region may be regarded as their centre of distribution, containing a large number of species and several characteristic endemic genera. *Silene* and *Dianthus*, both British genera, are widely distributed. By far the greater number of species are Mediterranean, but the former occurs on the mountains of tropical Africa and in Mexico and the latter at the Cape. Several species of *Silene* (*S. Cucubalus*, *S. maritima*, etc.), and of *Lychnis* (*L. Floscuculi*, *L. diurna*) are found in arctic regions. Many genera of Alsinoideae are very widely distributed, for instance the cosmopolitan *Stellaria*, *Cerastium*, *Minuartia* (*Alsine*), *Arenaria* and *Spergula*. *Sagina* extends from the north temperate zone to Abyssinia and Mexico. The Mediterranean is one of the richest areas, and contains several small or monotypic genera. The tribe *Paronychieae* contains about 90 species in warm dry regions.

The family is of little economic value. The roots of *Saponaria* (Soapwort) contain saponin, which yields a lather with water. Several genera are well known in gardens, chiefly *Dianthus*, which includes the pinks (derived from *D. plumarius*), carnations and picotees (from *D. Caryophyllus*) and Sweet William (from *D. barbatus*); others are *Lychnis*, *Silene*, and *Gypsophila*.

Grade B. DIALYPETALAE

Flowers generally bisexual, unisexual only in derived forms. Perianth (except in derived forms) generally in two series, the outer forming a calyx of sepals, the inner a corolla of free petals. Pollination generally entomophilous, or in derived forms anemophilous.

Order 1. Ranales.
- Family I. Magnoliaceae.
- ,, II. Annonaceae.
- ,, III. Myristicaceae.
- ,, IV. Calycanthaceae.
- ,, V. Monimiaceae.
- ,, VI. Lauraceae.
- ,, VII. Ranunculaceae.
- ,, VIII. Berberidaceae.
- ,, IX. Lardizabalaceae.
- ,, X. Menispermaceae.
- ,, XI. Nymphaeaceae.
- ,, XII. Ceratophyllaceae.

Order 2. Rhoeadales.
- Family I. Papaveraceae.
- ,, II. Capparidaceae.
- ,, III. Cruciferae.
- ,, IV. Resedaceae.

Order 3. Sarraceniales.
- Family I. Sarraceniaceae.
- ,, II. Nepenthaceae.
- ,, III. Droseraceae.

Order 4. Parietales.
- Family I. Cistaceae.
- ,, II. Bixaceae.
- ,, III. Tamaricaceae.
- ,, IV. Frankeniaceae.
- ,, V. Elatinaceae.
- ,, VI. Violaceae.
- ,, VII. Flacourtiaceae.
- ,, VIII. Passifloraceae.
- ,, IX. Caricaceae.
- ,, X. Loasaceae.

Order 5. Peponiferae.
- Family I. Cucurbitaceae.
- ,, II. Begoniaceae.
- ,, III. Datiscaceae.

Order 6. Guttiferales.
- Family I. Dilleniaceae.
- ,, II. Ochnaceae.
- ,, III. Marcgraviaceae.
- ,, IV. Ternstroemiaceae. (Theaceae.)
- ,, V. Guttiferae.
- ,, VI. Dipterocarpaceae.

Order 7. Malvales.
- Family I. Tiliaceae.
- ,, II. Malvaceae.
- ,, III. Bombacaceae.
- ,, IV. Sterculiaceae.

Order 8. Tricoccae.
- Family I. Euphorbiaceae.
- ,, II. Buxaceae.
- ,, III. Callitrichaceae.

Order 9. Geraniales.
- Family I. Geraniaceae.
- ,, II. Oxalidaceae.
- ,, III. Balsaminaceae.
- ,, IV. Tropaeolaceae.
- ,, V. Linaceae.
- ,, VI. Zygophyllaceae.
- ,, VII. Malpighiaceae.

Order 10. Rutales.
- Family I. Rutaceae.
- ,, II. Simarubaceae.
- ,, III. Burseraceae.
- ,, IV. Meliaceae.

Order 11. Sapindales.
- Family I. Anacardiaceae.
- ,, II. Sapindaceae.
- ,, III. Aceraceae.
- ,, IV. Hippocastanaceae.

Anomalous Family. Polygalaceae.

Order 12. Celastrales.
Family I. Celastraceae.
,, II. Staphyleaceae.
,, III. Aquifoliaceae.
,, IV. Empetraceae.

Order 13. Rhamnales.
Family I. Rhamnaceae.
,, II. Vitaceae.

Order 14. Rosales.
Family I. Crassulaceae.
,, II. Saxifragaceae.
,, III. Cunoniaceae.
,, IV. Pittosporaceae.
,, V. Podostemaceae.
,, VI. Hydrostachyaceae.
,, VII. Hamamelidaceae.
,, VIII. Platanaceae.
,, IX. Rosaceae.
,, X. Connaraceae.
,, XI. Leguminosae.

Order 15. Myrtiflorae.
Family I. Thymelaeaceae.
,, II. Elaeagnaceae.
,, III. Lythraceae.
,, IV. Punicaceae.
,, V. Sonneratiaceae.
,, VI. Rhizophoraceae.
,, VII. Combretaceae.
,, VIII. Lecythidaceae.
,, IX. Myrtaceae.
,, X. Melastomaceae.
,, XI. Onagraceae.
,, XII. Haloragaceae.
,, XIII. Hippuridaceae.

Order 16. Opuntiales.
Family Cactaceae.

Order 17. Umbelliflorae.
Family I. Araliaceae.
,, II. Umbelliferae.
,, III. Cornaceae.

Order 1. *RANALES*

(POLYCARPICAE of Eichler)

Flowers with parts arranged spirally or spirocyclic or cyclic, regular, or in higher forms zygomorphic. Floral axis convex, sometimes much elongated, the flower being hypogynous, sometimes in derived forms concave and more or less united with the pistil, the flower being perigynous or epigynous. Perianth generally either petaloid or differentiated into a calyx and corolla. Stamens generally numerous. Carpels numerous to solitary, generally free. Seeds generally with copious endosperm, sometimes also with perisperm.

Herbs or woody plants.

The earlier families of the order include woody plants with oil-containing cells in the parenchymatous tissues; in the later families oil-containing cells are absent and the plants are generally herbaceous.

A. *Woody plants. Oil-cells present in the parenchymatous tissue.*

Family I. MAGNOLIACEAE

Flowers bisexual or sometimes unisexual, regular, hypogynous, spiral or spirocyclic; floral axis convex or elongated. Perianth whorled, and sometimes more or less distinguished into calyx and corolla. Stamens usually numerous. Carpels

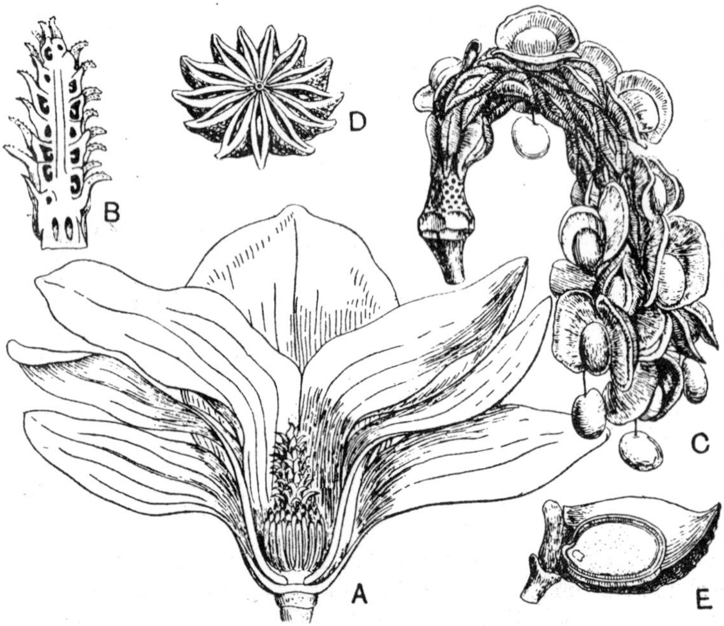

FIG. 58. A. Flower of *Magnolia heptapeta* cut open to shew androecium and gynoecium. B. Gynoecium of same cut lengthwise. C. Fruit of same. D. Fruit of *Illicium anisatum*. E. Portion of axis of same with one follicle and seed cut lengthwise, × 2. A, B, C reduced; D slightly reduced. (A, B after Wettstein; C after Baillon; D, E after Engler.)

usually numerous and free, sometimes whorled, rarely united. Ovules parietal, anatropous with two integuments. Fruit a follicle, a winged nut or a berry. Seeds containing a copious endosperm with a small embryo.

Shrubs or trees with alternate entire, rarely lobed, often stipulate leaves, and generally solitary, terminal or axillary flowers. Oil-containing cells are present in the parenchyma of stem and leaves.

Genera 18; species about 300. Tropics of Asia and America extending into the north temperate zone in the Himalayas, Eastern Asia, and Atlantic North America (in Tertiary times the distribution was circumpolar), and in the genus *Drimys* and allies into the south temperate zone in America and Australasia.

The genera fall into three subfamilies characterised by the degree of development of the floral axis, presence or absence of stipules and habit.

~~Subfamily~~ I. MAGNOLIOIDEAE. The leaves bear sheathing stipules which enclose the next younger leaf in the bud-stage. The flowers are usually bisexual with an elongated floral axis. There are ten genera. *Magnolia* includes about 70 species of shrubs and trees with entire leaves and terminal flowers, which sometimes appear in the spring before the leaves (Asiatic species), or with the leaves, as in the North American species; *M. grandiflora* of the southern United States, and well known in cultivation, is evergreen. The leaves of the floral envelope are usually in three whorls of typically three members each; all are petaloid or the outer whorl is sepaloid. An indefinite number of stamens arranged spirally is followed by numerous free carpels, also spirally arranged on the much elongated axis. The fruits open along the dorsal suture and the one or two seeds, which have a fleshy coat, hang suspended by the unrolled spiral vessels of the funicle (fig. 58, C). The 70 species occur in tropical Asia, Eastern Asia and Atlantic North America; fossil species indicate that in Cretaceous and Tertiary times the area of distribution extended right across the north temperate zone and as far north as Greenland and Spitsbergen and also included Australia. *Liriodendron*, represented by *L. Tulipifera* (Tulip-tree), a native of Atlantic North America, and a closely allied species in China, differs in having blunt commonly four-lobed leaves, while the flower has three sepals succeeded by two trimerous whorls of yellow petals; the numerous fruits are one-seeded and winged. The present anomalous distribution is explained by the existence of nearly related fossil species in the Tertiary beds of Europe and Greenland. There are eight other genera in Eastern Asia and tropical America.

Subfamily 2. ILLICIOIDEAE. Evergreen trees or shrubs with exstipulate leaves and a short convex floral axis. There are six genera and all except *Illicium* are remarkable for the absence of vessels in the secondary wood, which resembles that of Conifers. The flowers are solitary or in terminal or axillary inflorescences, bisexual, or sometimes diclinous in *Drimys*. *Illicium* has numerous perianth-leaves (the outer shorter and sepaloid, the inner becoming petaloid) spirally arranged, as are the stamens, and a whorl of 8–20 free carpels, which in the fruit form a radiating star of one-seeded follicles. The 20 species occur in Atlantic North America and in Eastern Asia. Fossil species occur in the Tertiary beds of North America and England, and in the Pliocene of Australia. In *Drimys* the six or more petals are enclosed when young by a membranous sac which splits into two or four sepals; the carpels form a whorl, and contain each many ovules. *D. Winteri* (Winter's Bark) occurs in the highlands from Mexico to the Straits of Magellan; other species occur in Australia, New Guinea, the Philippines and Borneo. *Zygogynum*, a small genus from New Caledonia, has a cup-like calyx, and a few unequal petals; the many-ovuled carpels are united and the fruit is a berry.

Subfamily 3. SCHIZANDROIDEAE. Climbing shrubs with deciduous exstipulate leaves and solitary axillary flowers; the flowers are unisexual with the parts arranged spirally on a convex axis which may become elongated in the fruit. The numerous carpels are free, contain two ovules, and remain closed in fruit forming a berry. There are two genera, the species of which occur in tropical and Eastern Asia with one in Atlantic North America.

The flowers in this family contain no nectar and are pollinated by insects which find shelter in them.

The species of *Magnolia* are cultivated for their foliage and large showy flowers; the Tulip-tree is also grown as an ornamental tree and affords useful timber. Star-anise, used medicinally and in the manufacture of liqueurs, is the fruit of *Illicium verum* (China);

the bark of *Drimys Winteri* is known medicinally as Winter's Bark.

Some stress from the point of view of phylogeny, as indicating primitiveness, has been laid on the Conifer-like character of the wood of *Drimys* and its allies. The nature of this is, however, discounted by the fact that the structure of the bast is characteristically dicotyledonous, having sieve-tubes with companion-cells. Further, Strasburger has shewn that the course of events in the embryo-sac is typically dicotyledonous. The structure of the wood of the genus *Illicium* is normal. More recently Jeffrey and Cole (4) have found that, as a result of injury, peculiar tracheary structures are developed in the root of several species, which may be regarded as a reversionary return of the vessels formerly occurring in the genus. Magnoliaceae have played an important part in recent attempts to trace the origin of Angiosperms. In his original system, Hans Hallier (2) regarded the family as representing the most primitive of existing Angiosperms, and compared the elongated floral axis of *Magnolia*, bearing numerous spirally arranged free sporophylls, with the sporophyll-bearing axis of *Bennettites*. A hypothetical group, *Drymyto-magnolieae*, combining the presumed primitive stem-structure of *Drimys* with the acyclic arrangement of the sporophylls of *Magnolia*, was regarded as connecting Magnoliaceae and the other polycarpic families of Dicotyledons with a *Bennettites*-like ancestor. This view has been developed by Arber and Parkin (1), and Lotsy (5) in his system of Angiosperms also starts with the Magnoliaceae. More recently, however, Hallier (3) has revised his system and now regards Berberidaceae as the earliest existing representative of the Angiosperms, and derives both Dicotyledons and Monocotyledons from a primitive type of this family, *Proberberideae*.

REFERENCES

DANDY, J. E. "The genera of Magnoliaceae." *Kew Bulletin*, 1927, 257.

(1) ARBER, E. A. N. and PARKIN, J. "On the Origin of Angiosperms." *Journ. Linnean Soc.* (*Botany*), XXXVIII, 29 (1907).

(2) HALLIER, HANS. For a sketch of this system and references to other papers by the same author, see "Provisional Scheme of the Natural (Phylogenetic) System of Flowering Plants." *New Phytologist*, IV, 151 (1905).

(3) —— "L'Origine et le Système Phylétique des Angiospermes exposés à l'aide de leur arbre généalogique." *Archives Néerlandaises d. Sciences Exactes et Naturelles*, sér. III, B. I, 146 (1912).

(4) JEFFREY, E. C. and COLE, R. D. "Anatomy of *Drimys*." *Annals of Botany*, XXX, 359 (1916).

(5) LOTSY, J. P. *Vorträge über Botanische Stammesgeschichte*, III (1911).

Family II. ANNONACEAE

A family widely spread in the tropics of both worlds containing about 800 species very unequally distributed among 62 genera. They are woody plants, sometimes climbing, with alternate entire exstipulate leaves, and large regular hypogynous bisexual, rarely unisexual, flowers which are often solitary. In some cases the flowers are borne on the old wood (cauliflorous). The perianth (absent in the Australian genus *Eupomatia*, which has also a perigynous flower) consists of three whorls with three, rarely two, members in each whorl. The outer whorl forms a calyx and has a valvate aestivation, the sepals are free or united below; the two inner whorls are valvate, or slightly imbricate and are petaloid; they are alike or, as in *Monodora*, strikingly different in size and form.

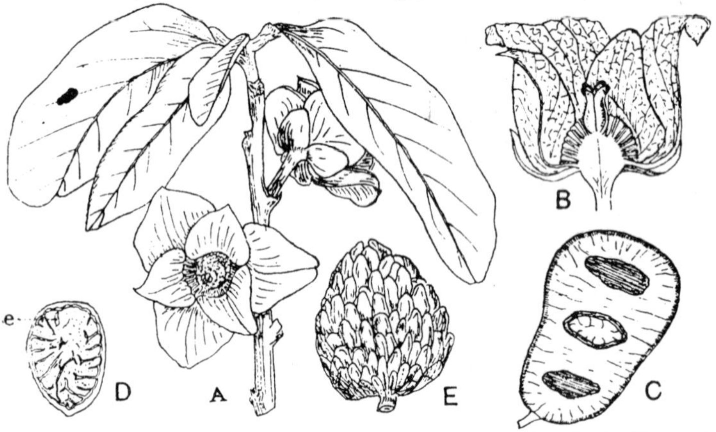

Fig. 59. A–D. *Asimina triloba*. A. Flowering shoot, reduced. B. Flower in vertical section, about nat. size. C. Fruit cut lengthwise shewing three seeds, $\times \frac{1}{2}$. D. Seed in section shewing embryo, *e*, in the ruminate endosperm, nat. size. E. Fruit of *Annona squamosa*, reduced. (A after Le Maout and Decaisne; B–E after Baillon.)

Above the perianth the axis forms a large convex receptacle on which are spirally arranged an indefinite number of stamens succeeded by numerous carpels. The stamens are peculiar and consist of a short thick filament bearing a pair of extrorse anthers above which is an outgrowth of the connective varying in form in different genera. The anatropous ovules are generally numerous in a double row on the ventral suture of the carpel. The fruits are fleshy and form separate berries which, when many-seeded, shew constrictions or fleshy ingrowths of the pericarp between individual seeds; or the ovaries become united with each other and the floral axis to form a fleshy mass in which the seeds are embedded as in *Annona squamosa* (Custard Apple) (fig. 59, E).

The seeds are large and occupied almost entirely by a deeply ruminated endosperm in the upper portion of which is embedded a small embryo. Oil-sacs occur throughout in the parenchyma, even in the parts of the flower. The large fleshy fruits of many species are eaten, for instance, the Custard Apples (*Annona*); other species are strongly aromatic or acrid.

Annonaceae are most nearly allied to the Magnoliaceae; a constant distinction is the ruminated endosperm.

Family III. MYRISTICACEAE

A small tropical family of aromatic plants allied to Annonaceae in the trimerous perianth, and the structure of the seed. Like Annonaceae also they are woody plants with alternate exstipulate

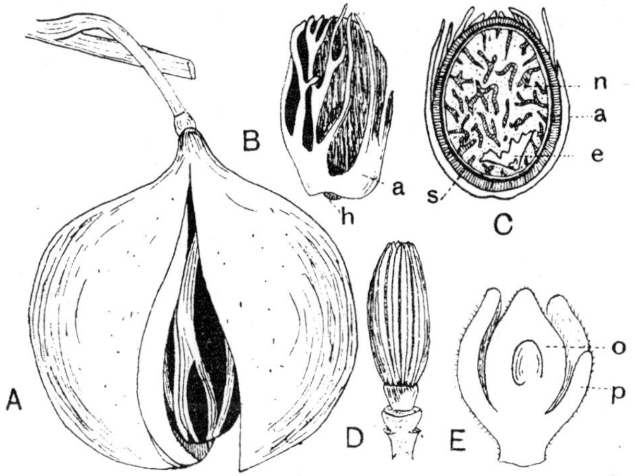

FIG. 60. *Myristica fragrans*. A. Fruit, the pericarp is splitting, exposing the seed. B. Seed surrounded by the aril, *a*; *h*, hilum. C. Seed in section; *a*, aril; *s*, seed-coat; *n*, ruminate endosperm; *e*, embryo. D. Androecium of male flower. E. Female flower in longitudinal section; *p*, perianth; *o*, ovary. A, B, C reduced; D, E enlarged. (A after Le Maout and Decaisne; B and C after Luerssen; D and E after Baillon.)

coriaceous simple entire leaves, and oil-sacs are generally distributed in the parenchyma. The inconspicuous flowers are racemose, dioecious and regular, with acyclic arrangement. The simple perianth is tubular or bell-shaped below dividing above into generally three limbs. The male flowers have a varying number of stamens (3–18) united by their filaments (monadelphous) into a central column which may become expanded above into a disc; the anthers are extrorse. The female flowers have a solitary free carpel with a short or obsolete style and containing a solitary basal anatropous ovule. The fruit is fleshy, the pericarp splitting along both the ventral and dorsal sutures and exposing the large seed which is more or less completely enveloped by a branching fleshy aril springing from the base of the short funicle. The hard testa encloses a deeply ruminate aromatic oily endosperm, in the base of which is a small embryo with a short radicle and leafy spreading cotyledons.

The species are spread through the tropics of both Worlds but are most numerous in tropical Asia. *Myristica fragrans* (Nutmeg) (fig. 60), a native of the Moluccas, is now widely cultivated; both the seed, nutmeg, and the aril, mace, are used as spices.

The family was regarded by Alphonse De Candolle as representing a single genus containing about 80 species. O. Warburg[1] in a more recent monograph subdivides it into 11 genera, which form geographical groups, and contain about 250 species.

Family IV. CALYCANTHACEAE

A small family containing the single genus *Calycanthus* with a remarkable distribution; two or three species in China (sometimes separated as a distinct genus, *Chimonanthus*), three species in the Southern United States, one in California, and one in Queensland. They are shrubby plants with opposite entire leaves, bisexual acyclic flowers with numerous petaloid perianth-leaves, 10–20 stamens, and about 20 free carpels situated on the inside of a cup-shaped floral axis and containing each two anatropous ovules with two integuments. The one-seeded indehiscent fruits are enclosed in the enlarged floral axis. There is hardly any endosperm, the seed being nearly filled with the spirally rolled cotyledons of the large embryo. Oil-sacs occur generally distributed in the parenchyma of all parts of the plant. Outside the normal ring of bundles in the stem are four bundles with inverted orientation, the xylem being towards the outside.

They resemble Magnoliaceae and Annonaceae in the acyclic flowers and the presence of oil-sacs, but are distinguished by the opposite leaves, and the large embryo which nearly fills the seed; also by the perigynous flower which, however, occurs in the anomalous genus *Eupomatia* of Annonaceae. *Calycanthus florida* (Carolina Allspice) (fig. 61, A), native in the southern United States, with fragrant deep red flowers, is a garden-plant.

[1] "Monographie der Myristicaceen." *Nova Acta Acad. Leop. Carol. Bot.* LXVIII (1897).

Family V. MONIMIACEAE

A tropical and subtropical family containing 32 genera of trees or shrubs, rarely climbing, with evergreen leathery, generally opposite, exstipulate leaves, and containing aromatic oil-cells. The flowers are solitary or cymose and usually unisexual. The floral stıucture is remarkably varied, but generally characterised by a more or less cup-shaped, sometimes very deeply hollowed receptacle on the inside of which the stamens and carpels are borne (fig. 61, B, C). The perianth-leaves (sometimes wanting) are small and four to many in number; in the latter case the inner are petaloid. The stamens are generally numerous, but sometimes few; the dehiscence of the anthers is longitudinal, transverse or sometimes valvular and the short filaments are sometimes pro-

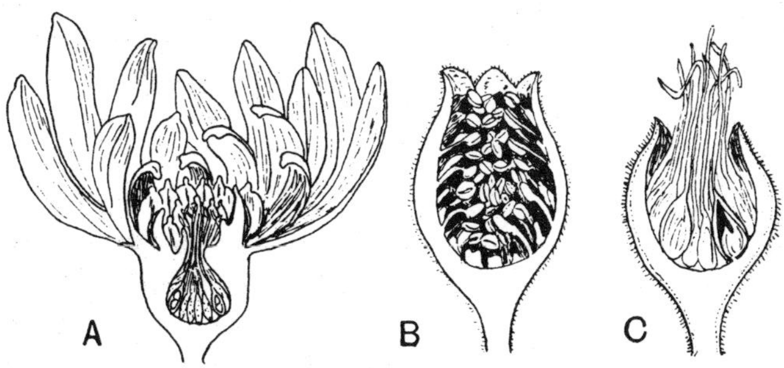

FIG. 61. A. Flower of *Calycanthus florida*, in vertical section, $\frac{3}{4}$ nat. size. B, young male, and C, female flower of *Monimia* in vertical section, enlarged. (After Baillon.)

vided with a pair of lateral appendages (compare Lauraceae). The carpels vary similarly in number but are generally numerous; they are free and contain each a single ovule, the position of which varies. The one-seeded indehiscent fruits are generally enclosed in the enlarged floral axis. The small embryo is enclosed in a copious, not ruminate, endosperm.

The family is of considerable interest from its relation to Magnoliaceae on the one hand and Lauraceae on the other. The structure of the flowers suggests Magnoliaceae with a concave floral axis, in which the numerous stamens and free carpels are depressed. The small embryo embedded in a copious endosperm also recalls Magnoliaceae, but the biological character of the flowers—inconspicuous and generally asexual—is very different. In the perigyny, and the occasional valvate dehiscence of the anthers and appendaged filaments, an affinity is indicated with the Lauraceae. There are about 350 species.

Family VI. LAURACEAE

A family of woody plants, excepting the parasitic *Cassytha*, with coriaceous, generally smooth, simple and entire leaves, which are exstipulate and arranged alternately, though sometimes, by shortening of the internodes, apparently opposite or whorled. A character is the presence in the leaves

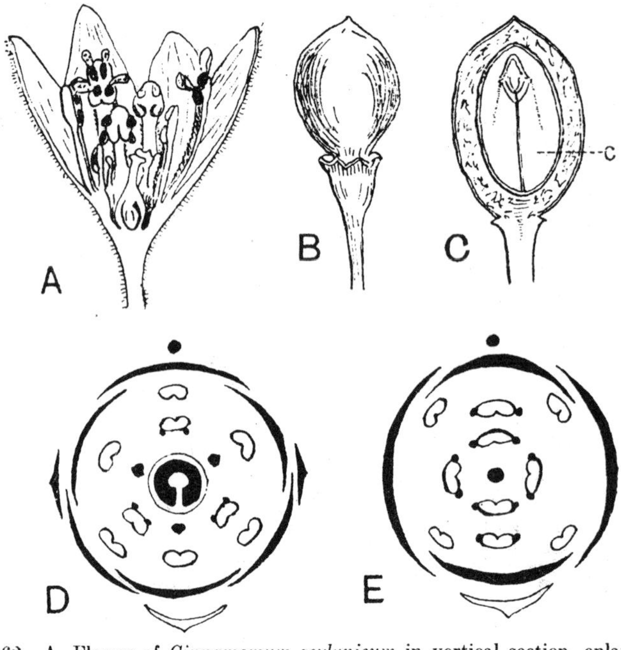

FIG. 62. A. Flower of *Cinnamomum zeylanicum* in vertical section, enlarged. B. Fruit of *Sassafras officinale*, nat. size. C. Berry, enclosing seed, of *Laurus nobilis*, in vertical section; *c*, cotyledon; enlarged. D. Floral diagram of *Cinnamomum zeylanicum*. E. Floral diagram of male flower of *Laurus nobilis*. (A, B, C after Baillon; D, E after Eichler.)

and cortex of cells containing ethereal oil; mucilage-sacs may also be present. The flowers are generally in many-flowered terminal or axillary inflorescences of a racemose type, which, however, may pass in the ultimate branchings into dichasia. They are bisexual, or sometimes unisexual by reduction, and regular, with the parts arranged in generally trimerous whorls. The two outer whorls form a perianth, the members of which are generally similar, rarely, as in *Cassytha* (fig. 64), the outer are smaller than the inner; they are more or less united at the base and are situated with the stamens on the edge of the concave receptacle from the bottom of which springs the pistil. The typical flower contains four whorls of stamens which are rarely all fertile, generally the innermost is reduced to a whorl of staminodes, as in *Persea* (fig. 63) and Cinnamon (*Cinnamomum zeylanicum*, fig. 62, D), or it is absent. Further reduction occurs in some genera, only two whorls, or even a single whorl, being functional. These variations in the androecium together with the number, two or four, of pollen-sacs, and the introrse or extrorse dehiscence of the anthers, supply useful generic characters. A constant character is the mode of dehiscence of the anthers by a valve opening from below upwards. The anthers of the third whorl have often an extrorse dehiscence, those of the other whorls being introrse, and their filaments frequently bear a pair of lateral glandular outgrowths, as do also sometimes those of the outer whorls.

The pistil consists of a one-celled ovary containing a single pendulous anatropous ovule, and bearing a style ending in a stigma which is often three-cleft. The three-cleft stigma suggests the presence of three carpels, which Eichler considered to be represented in the pistil; Payer, Baillon and others, following the development of the pistil, regarded it as monocarpellary, but more recently Mirande[1] concludes, especially from his investigation of *Cassytha*, that the pistil is composed of several, usually three carpels, one posterior, which is prolonged into a style and stigma, and two lateroanterior which abort. The fruit is a berry, rarely a drupe as in *Persea* (fig. 63), and more or less enveloped by the cup-shaped receptacle, and sometimes also by the persistent base of the perianth, as in *Cassytha* (fig. 64). The single seed is filled by the large embryo consisting of a pair of large fleshy straight cotyledons protecting the small plumule, and a short superior radicle.

Laurus (Laurel) differs from the usual case, which we

[1] *Comptes Rendus*, Paris, CXLV, 570 (1907).

have considered, in having dioecious or polygamous flowers with the parts arranged in fours. In the male a four-leaved perianth is followed by a varying number of stamens, generally 8–12, in two or three whorls, the innermost of which bear a pair of lateral glands (fig. 62, E). The female flowers have generally a whorl of staminodes alternating with the perianth-leaves.

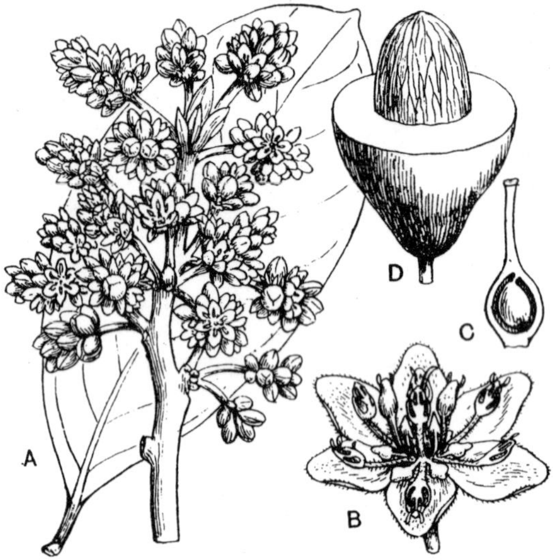

FIG. 63. *Persea americana.* A. Leaf and inflorescence, × $\frac{2}{3}$. B. Flower, × 3. C. Pistil cut open to shew ovule, × 3. D. Drupe, the upper part cut away to shew the stone, × $\frac{1}{2}$. (From *Flor. Jam.*)

Cassytha consists of herbs resembling Dodder in habit (fig. 64), having slender cylindrical stems attached by suckers to their host-plant; leaves are absent or scale-like. The small flowers are generally collected in heads or spicate, having three sepals, three petals, 3 + 3 + 3 stamens and an inner whorl of three staminodes, as usual in the family. The petals are much larger than the sepals. The perianth-tube is little developed in the flower but increases considerably in the fruit forming a succulent outer covering.

The 40 genera contain about 1000 species; they are common forest-trees in tropical Asia and America. There seem to be two main centres of distribution: one is in tropical south-eastern

Asia whence the family ranges northwards to South China and Japan and southwards to Australia; a second is in Brazil from which it extends southwards to the Island of Chiloe and is represented northwards by a few genera in the United States, *Sassafras*, and *Lindera* reaching Southern Canada. Africa is poor in species. The only European representative at the present day is the Bay or true Laurel

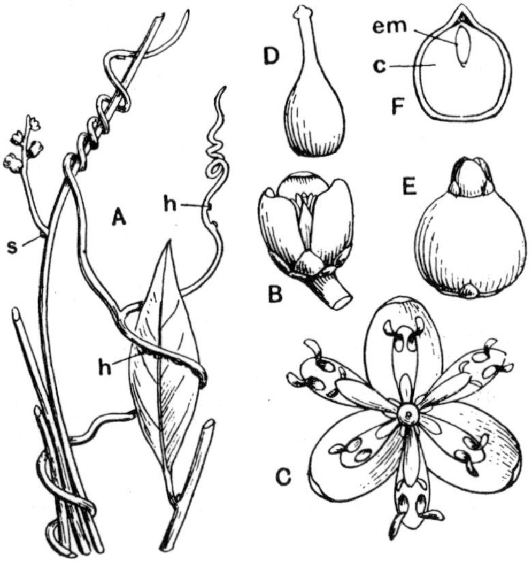

FIG. 64. *Cassytha filiformis.* A. Portion of stem with flowers, shewing attachment to leaf of host-plant by haustoria (*h*), × $\frac{2}{3}$; *s*, scale-leaf. B. Flower, × 6. C. Flower spread out to shew arrangement of parts, × 7. D. Pistil. E. Fruit, × 2; it is enclosed by the succulent perianth-tube which has become much increased in the fruit and bears the persistent segments. F. Seed in section shewing one cotyledon, *c*, and the axis of the embryo, *em*. (From *Flor. Jam.*)

(*Laurus nobilis*) in the Mediterranean region, probably originally a native of Western Asia. Fossil leaves, and also fruits and flowers of several genera found in Miocene strata in Europe, North Asia, Greenland and elsewhere, indicate a much more northerly extension of the genus in Tertiary times.

Many of the species are of economic or medicinal value on account of the volatile oil which they contain. *Cinnamomum zeylanicum* and *C. Camphora*, both Asiatic trees, yield respectively cinnamon and camphor. The North American Sassafras (*S. officinale*), a deciduous-leaved tree, yields a scented wood; a species of *Nectandra* yields the Greenheart wood of Demerara, and there are other valuable timber-trees. *Persea americana*, a native of tropical America, now widely cultivated, has a delicious pear-shaped fruit (a drupe), the Alligator or Avocado Pear.

B. *Generally herbaceous plants. Oil-cells absent. There is evidence from their arrangement that the sepals are derived from bracts and, on the other hand, that the petals are generally stamens which have become nectar-secreting structures or merely coloured attractive organs.*

Family VII. RANUNCULACEAE

Flowers bisexual, spiral or spirocyclic, rarely cyclic, regular, more rarely medianly zygomorphic, hypogynous. Perianth simple and petaloid, and generally succeeded by nectar-secreting structures of various forms, or differentiated into a distinct calyx and corolla. Stamens generally indefinite and free. Carpels indefinite to few or solitary, rarely united, with numerous to few ovules with one or two integuments. Pollination generally by aid of insects. Fruit a follicle or achene, rarely a berry. Seeds containing a copious oily endosperm with a small embryo.

Generally perennial herbs with usually alternate, more or less divided leaves, and solitary or cymose, often showy, flowers.

Genera 30; species about 1200. Chiefly in the temperate and colder parts of the earth.

The cotyledons are generally epigeal in the form of stalked or sessile small ovate green leaves; the stalks are sometimes united, either partially or completely, forming a tube through which the epicotyl breaks laterally as in Monocotyledons; this occurs in species of *Anemone*, *Delphinium* and others. The union may extend also to the blade as in *Eranthis* and *Ranunculus Ficaria*. In some cases, as in *Clematis recta*, *Paeonia* and others, the cotyledons are hypogeal.

The plants are generally annual or perennial herbs, persisting by means of a rhizome, or by tuberous roots as in *Paeonia*, species of *Ranunculus*, *Aconitum* and others.

The leaf-base is usually broadened into a sheath which is

sometimes elongated into a pair of lateral stipular lobes as in *Thalictrum*, the *Batrachium* section of *Ranunculus* (Water Crowfoot) and others. The blade is sometimes entire; it is then narrow as in *Myosurus* and some species of *Ranunculus* (as *R. Lingua* and *Flammula*) or cordate as in *R. Ficaria* and *Caltha*. Generally, however, the blade is palmately lobed, divided or compound, rarely pinnately as in *Xanthorrhiza* and species of *Clematis*. Special forms are the dissected submerged leaves of the aquatic species of *Ranunculus* and the climbing leaves of *Clematis*; in the latter the petiole, rachis and sometimes even the blade are sensitive to contact. In *C. aphylla* the whole of the leaf becomes a tendril and the work of carbon-assimilation is done by the green cortex of the stem.

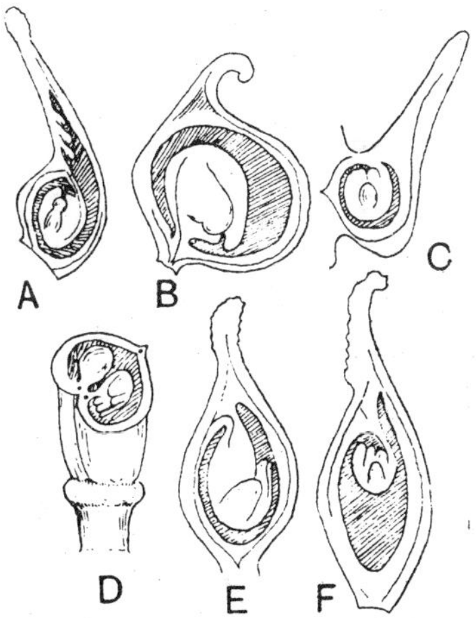

FIG. 65. Longitudinal (A, B, C, E, F) and transverse (D) sections of carpels of A, *Anemone nemorosa*; B, *Ranunculus acris*; C, *Myosurus minimus*; D, *Delphinium Consolida*; E, *Callianthemum rutifolium*; F, *Thalictrum minus*. All enlarged. (After Prantl.)

The course of the vascular bundles in the stem is in some genera (*Actaea*, *Cimicifuga* and *Thalictrum*) suggestive of the Monocotyledons.

The family may be divided into three tribes:

Tribe 1. *Helleboreae*, in which the ovules are arranged in two rows along the ventral suture of the carpels (fig. 65, D) and the fruit is a follicle or rarely a berry.

Tribe 2. *Anemoneae*, in which a solitary ovule arises at the base of the ventral suture (fig. 65, B, C, F) and the fruit is an achene.

Tribe 3. *Paeonieae*, which is distinguished by the large development of the outer ovule-integument; the ovules are arranged as in *Helleboreae* and the fruit is a follicle with fleshy walls.

Great reduction in the number of ovules occurs in some

of the *Hellebcreae* as in *Xanthorrhiza*, where there are two, while the fruit is a one-seeded pod, and in *Callianthemum*, where only one ovule comes to maturity (fig. 65, E) and the fruit is one-seeded and indehiscent. On the other hand, rudimentary ovules may occur above the solitary ovule in *Anemoneae* as in *Anemone nemorosa* (fig. 65, A); and the ovule may become pendulous by elongation of the basal portion of the suture (fig. 65, F).

The most constant feature of the family is the internal arrangement of the seed, which contains a small embryo in the apex of a copious oily endosperm.

Tribe 1. *Helleboreae*. A comparative study of the genera of this tribe by Schrödinger (1) is here summarised. In the more primitive forms (*Caltha, Helleborus*, etc.) the perianth is spiral and is continued from the spiral arrangement of the foliage-leaves; in the later forms it becomes cyclic. The perianth is therefore here regarded as having originated from the foliage-leaves. In most genera "honey-leaves" are present; these are shewn to be the transformed lowermost members of the rows of stamens, and the origin of the corolla. The primitive form of flower contained three kinds of organs: (1) perianth or calyx, (2) stamens, which may in part be changed into honey-leaves, and (3) carpels. Two developmental series are distinguished, namely:

(*a*) *Isopyroideae*, in which the members of the corolla, when present, alternate with the sepals, the petals (honey-leaves) and staminal rows being opposite the spaces between the sepals, and

(*b*) *Trollioideae*, where the corolla when present is episepalous.

The *Isopyroideae* have all actinomorphic flowers and are hence less advanced than the *Trollioideae*, the higher members of which (*Aconitum* and *Delphinium*) are markedly zygomorphic.

(*a*) *Isopyroideae*. In *Helleborus* the flower has a green, reddish or white persistent perianth of five leaves, followed by a varying number of short tubular honey-leaves. The very numerous stamens (generally more than 100) are in 13 rows and usually the first member of each row is converted into a honey-leaf (nectary). In some species, as *H. foetidus*, only eight or fewer of the outermost stamens form nectaries, and cases occur where there are only five such which alternate with the sepals. Such cases supply the transition to the arrangement in *Isopyrum*, where the honey-leaves are in a whorl of five alternating with the five sepals, and thus may be regarded as a definite corolla. As in *Helleborus*, there are about 13 rows of stamens, but the total number is greatly reduced.

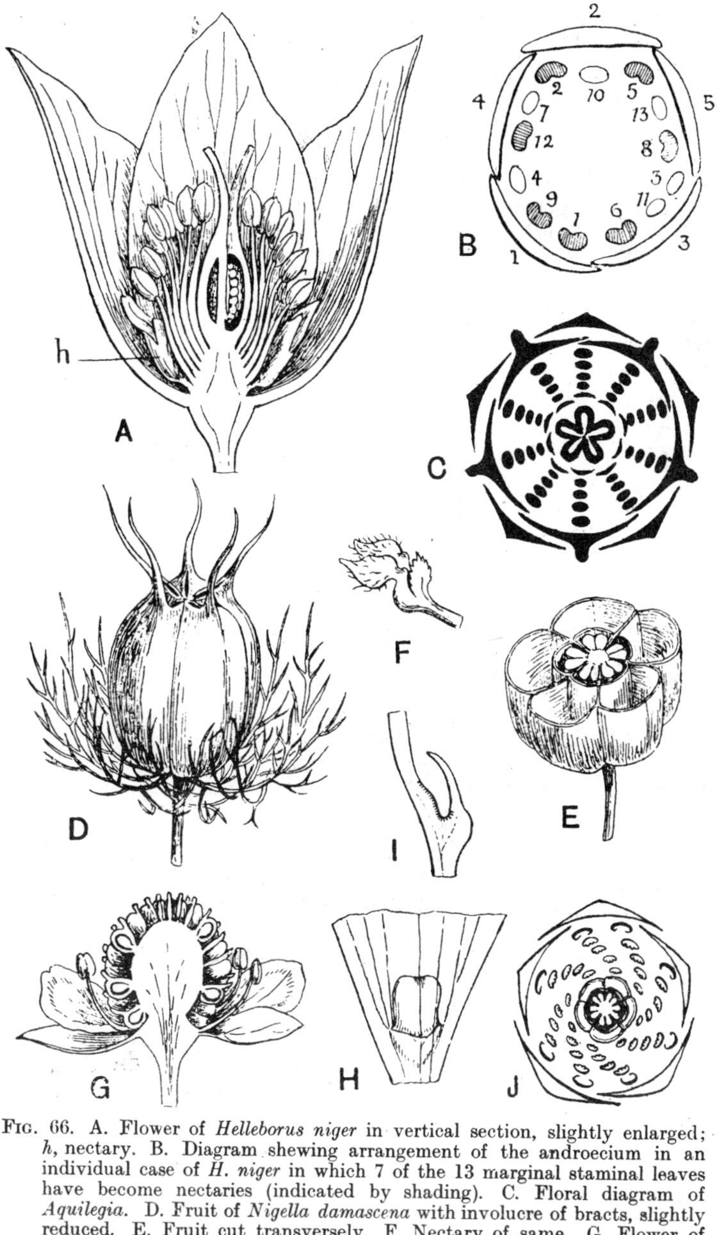

FIG. 66. A. Flower of *Helleborus niger* in vertical section, slightly enlarged; *h*, nectary. B. Diagram shewing arrangement of the androecium in an individual case of *H. niger* in which 7 of the 13 marginal staminal leaves have become nectaries (indicated by shading). C. Floral diagram of *Aquilegia*. D. Fruit of *Nigella damascena* with involucre of bracts, slightly reduced. E. Fruit cut transversely. F. Nectary of same. G. Flower of *Ranunculus sceleratus* in vertical section. H. Base of petal of *R. acris*. I. The same in median longitudinal section. J. Floral diagram of *Nigella*. F, G, H, I enlarged. (A after Berg and Schmidt; B after Schrödinger; C, E, F after Warming; D, G, J after Baillon; H, I after Prantl.)

Isopyrum again supplies the transition to *Aquilegia* where the flower consists of 5-merous whorls throughout and is constructed on the plan K5, C5, A5 + 5 + 5 (+ 5 + 5), G5. The foliage-leaf of *Helleborus* is palmately cut while that of *Isopyrum* is ternately divided and resembles in this *Aquilegia*. The tubular honey-leaf of *Isopyrum* is in some species very obliquely cut and approaches in form the spurred honey-leaf characteristic of *Aquilegia*.

The other genera of the tribe, *Anemopsis*, *Cimicifuga*, *Actaea*, *Coptis* and *Xanthorrhiza*, resemble *Isopyrum* and *Aquilegia* in the generally ternately divided foliage-leaf and the tubular honey-leaf, but differ in the toothed margin of the leaf-segments and the more open character of the nectary. In *Actaea* and *Cimicifuga*, which are sometimes united as a single genus, there are generally four sepals, which soon fall, and four or fewer petal-like "honey-leaves" which are sometimes devoid of nectar, generally numerous stamens and one (*Actaea*) or one to several (*Cimicifuga*) carpels. Flowers with trimerous calyx and corolla also occur. In *Coptis* there are generally five sepals, alternating with which are five smaller honey-leaves, some of which are, however, often incompletely developed and without nectar; the numerous stamens are often arranged in 13 rows, and the carpels (10–1) are often stalked. The flowers of *Xanthorrhiza* (a monotypic genus of the woods of Atlantic North America) are, like those of *Aquilegia*, cyclic and pentamerous, but small and inconspicuous, with fewer staminal whorls. They have the formula K5, C5, A5 (or 5 + 5), G5. *Eranthis* (Winter Aconite) resembles *Helleborus* in floral structure but the sepals, which vary in number from five to nine (generally six) (2), are smaller, yellow and deciduous; the radical leaves are palmate, the cauline form an involucre just below the flower.

(*b*) *Trollioideae*. The lowest floral stage is represented by *Caltha*, where there are no honey-leaves, but a nectary is present on the carpel; the plant closely resembles *Trollius* in the form and development of its leaf. There are from 5–15 deciduous white or yellow perianth-leaves, 80–150 stamens in 21 rows, and 5–10 carpels. The flower of *Trollius* is similar but a varying number of the first members of the staminal rows become transformed into small narrow honey-leaves which have a short claw and a small flat blade with a nectary at its base. Schrödinger explains the variation in the number of the staminal rows by exigencies of space on the floral axis.

In *Nigella* the five petaloid perianth-leaves are generally followed by eight well-developed honey-leaves (nectaries) and an eight-rowed superposed androecium; the carpels (5–12) are more or less united to form a compound ovary.

The zygomorphic genera (*Delphinium* and *Aconitum*) are derived from a *Nigella*-like flower. The stamens and carpels are spirally arranged and shew no trace of the median symmetry which is determined by the structure of the calyx and corolla. The calyx is arranged in the quincuncial whorl, so frequent in Dicotyledons, that is the second-developed sepal is median and posterior; in *Delphinium* this is spurred. There are originally, as in *Nigella*,

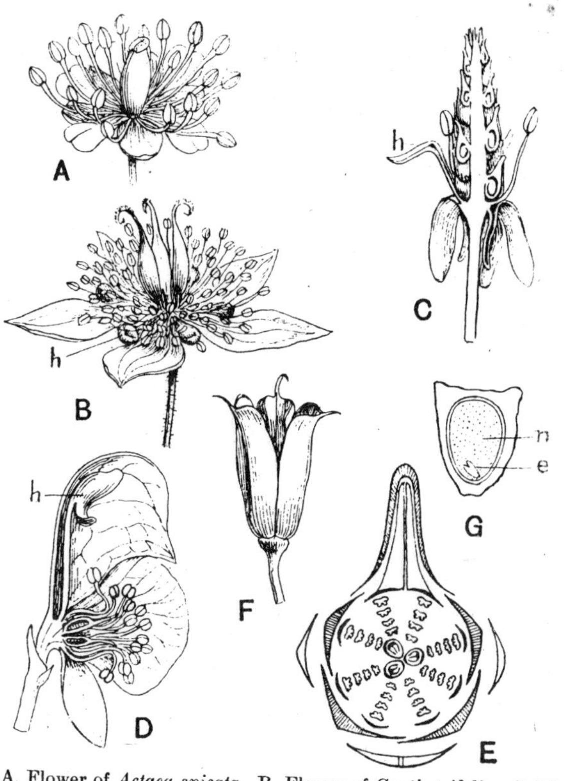

FIG. 67. A. Flower of *Actaea spicata*. B. Flower of *Coptis trifolia*. C. Flower of *Myosurus minimus* in vertical section. D. Flower of *Aconitum Napellus* in vertical section. E. Floral diagram of *Delphinium*. F. Fruit of *Aconitum*. G. Seed of same in vertical section. *e*, embryo; *h*, honey-leaf; *n*, endosperm. A, B, C, F and G variously enlarged. (A, B, C after Baillon; D after Warming.)

eight honey-leaves, a pair opposite each of the three older sepals and one opposite each of the two younger sepals. Of these the four in the lower part of the flower are either suppressed or form slender functionless structures; the median lateral form attractive petal-like structures which have lost the nectar-secreting function but are closely associated in the insect-attracting function with the median posterior pair which have a projecting blade and a nectar-secreting spur sunk in the spur of the back sepal. In the section *Consolida* (which includes *D. Ajacis*, a central and southern European species formerly found in Cambridgeshire) the two median petals have become united into a single structure with one spur. In *Aconitum* the completely free median petals are less conspicuous and are borne on long slender stalks and hidden in the large hooded dorsal sepal. In both genera there are usually three carpels which are free, but the *Consolida* section of *Delphinium* has only one.

The fruit in the *Helleboreae* is generally a follicle containing few to many seeds; the single carpel of *Actaea* forms a many-seeded

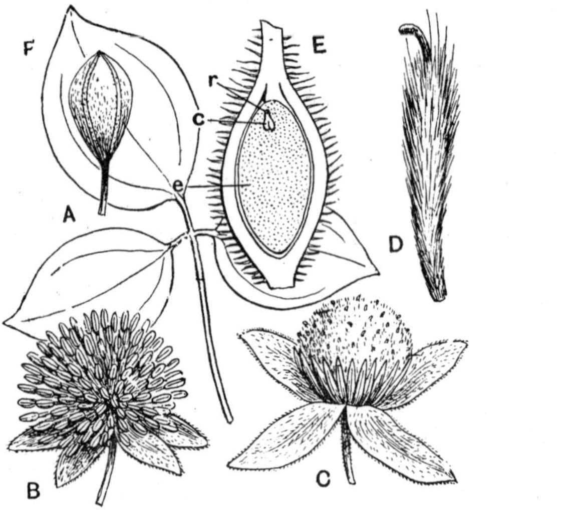

FIG. 68. *Clematis dioica.* A. Bud of male flower, ×2. B. Male flower open, ×2. C. Female flower, ×2. D. Carpel, ×5. E. Achene cut lengthwise, ×7; *e*, endosperm; *c*, cotyledons; *r*, radicle. F. Leaf. (A, B after Eichler.) (From *Flor. Jam.*)

indehiscent berry. In *Xanthorrhiza* the carpels contain a pair of ovules, and the fruit is a one-seeded pod. In *Nigella* the carpels are more or less united and in *N. damascena* (Love-in-a-Mist) the fruit forms a septicidal capsule which has large hollow spaces between the inner and outer layers of the pericarp.

Tribe 2. *Anemoneae.* The flowers in this tribe are all actinomorphic and suggest a more primitive type than the *Helleboreae.* They have not, however, been so exhaustively studied as has the latter tribe. We may distinguish several groups. The first comprises *Anemone*, herbaceous plants, rarely shrubby, with palmately or rarely pinnately cut radical leaves, and a whorl of three stem-leaves forming an involucre which sometimes, as in *A. Hepatica*, stands close beneath the flower. The flowers are usually solitary. The five, six or more perianth-leaves are petaloid; true petals are absent and there are numerous stamens (according to Schrödinger in 13 rows) and carpels. The single ovule is pendulous, and in the subgenus *Pulsatilla* the achene is crowned by a long hairy style. Warming associates with *Anemone* the genus *Thalictrum* (Meadow-rue) with alternate ternately highly compound leaves and numerous small flowers in corymbs or panicles. There is no involucre, and the perianth consists of four to five small greenish leaves which soon fall. The numerous stamens are longer than the perianth, the ovule is pendulous (fig. 65, F) and the fruit is a small head of achenes sometimes reduced to a single achene.

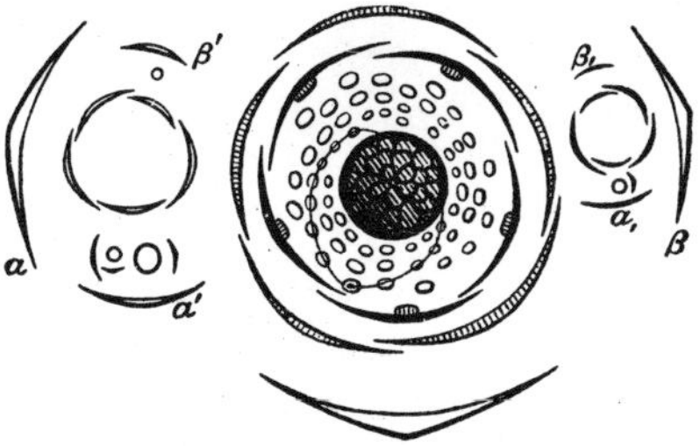

FIG. 69. Inflorescence (axillary dichasial cyme) of *Ranunculus acris* with details of central flower shewing spiral arrangement. Stamens in $\frac{8}{21}$ phyllotaxy. (After Eichler.) α, β, bracteoles of central flower in the axil of each of which arises a lateral flower with bracteoles α', β' and α_1, β_1; branching is repeated in the axils of α', α_1.

A second group is represented by *Clematis*, which differs from the other genera of the tribe in the usually valvate aestivation of the four or more petaloid sepals, and in the opposite simple or compound leaves. The plants are generally shrubby climbers, clinging by means of the sensitive leaf-stalks. The flowers are sometimes dioecious (fig. 68). There are numerous stamens and carpels, the ovule is pendulous and the achene generally bears a long hairy persistent style, whence one of the popular names, Old Man's Beard, of our British species, *C. Vitalba*.

The third group includes *Myosurus* and *Adonis* in which the ovule is pendulous (fig. 65, C) and *Ranunculus*, where it is erect (fig. 65, B). The flowers have generally five green sepals followed by five or more coloured petals generally bearing a nectary, and numerous spirally arranged stamens and carpels. *Myosurus* (Mouse-tail) includes a few species of small annual herbs with narrow radical leaves and a slender scape ending in a flower. The generally five pale yellow narrow sepals have a small basal spur;

the petals, of the same number and alternating with the sepals, have a stalk and narrow limb bearing a shallow nectar-pit; there are few stamens and a large number of spirally arranged carpels on an elongated axis (fig. 67, C). The few species of *Adonis* (Pheasant's eye) are herbs with much divided leaves, and solitary terminal flowers in which five sepals are succeeded by 8–16 yellow or red flat nectar-less petals and numerous stamens (according to Schrödinger, 21 rows) and carpels on a convex axis. *Ranunculus* is a large genus of annual or perennial, usual acrid, herbs with generally palmately divided leaves and solitary or cymose yellow or white flowers. The generally five sepals are green and fall

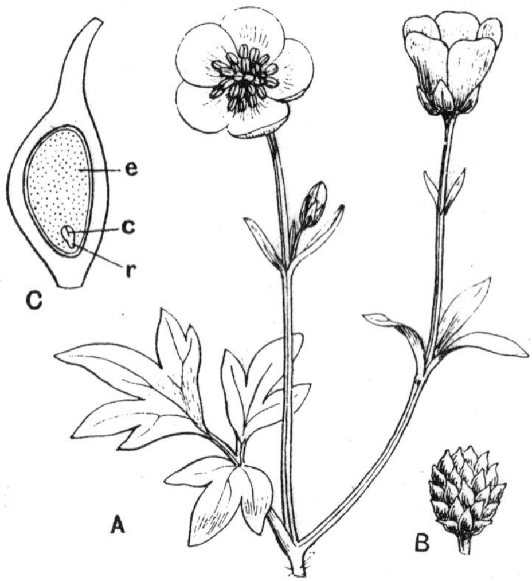

FIG. 70. *Ranunculus repens.* A. Upper portion of stem, × ⅔. B. Fruit, × 1½. C. Achene cut lengthwise, × 10; *e*, endosperm; *c*, cotyledons; *r*, radicle. (From *Flor. Jam.*)

early, the five or more petals are generally larger than the sepals and bear a nectary above the base. The numerous stamens are arranged in 13 rows. The section *Ficaria*, represented in the British Isles by *R. Ficaria* (Lesser Celandine), has entire heart-shaped leaves, and generally three sepals and seven to eight (4–11) petals.[1] The section *Batrachium* contains water- or marsh-plants

[1] Salisbury, *Annals of Botany*, XXXIII, 47 (1919), has recently studied the variation in number in the parts of the flower in this and other genera. He concludes that the flower of Ranunculaceae has been derived from a trimerous type, which has in many cases become obscured by multiplication of parts and consequent changes in phyllotaxy, or by fusion and abortion. He also supports the view that the perianth or calyx is derived from the bracts and the petals from the staminal series.

with leaves often submerged and much divided, sometimes also with entire floating leaves; the petals are white with a naked yellow basal gland and the achenes are wrinkled. In the Spearworts, *R. Lingua* and *R. Flammula*, found in wet places, the leaves are undivided. The presence or absence of a scale covering the nectary is a useful character for distinguishing species as are also the characters of the achene, which is variously veined, smooth, tubercled, hispid or, as in our cornfield species, *R. arvensis*, large and spine-bearing.

Tribe 3. *Paeonieae.* This comprises the genus *Paeonia* (Peony), with 15 species, natives of temperate Asia, Europe and the Mediterranean region, with one species in California. *P. mascula*, a native of southern Europe, has become established on Steep Holm at the mouth of the Severn. They are mostly herbs, sometimes shrubby, with deeply cut leaves and large solitary showy flowers. The flowers are acyclic or hemicyclic; in the former the spiral is often obviously continued from the upper foliage-leaves, which by shortening of the internodes and a gradual change in form pass into the persistent sepals, generally five in number, and these in turn into the large coloured petals. The indefinite stamens occupy several turns of the spiral. The two to five free carpels have a fleshy wall and bear along the ventral suture a double row of ovules in which the well-developed outer integument exceeds the inner—a character which distinguishes them from the rest of the family. Definite nectar-secreting structures are absent but there is a disc, or ring-like swelling of the receptacle, round the base of the carpels; the Japanese species, *P. Moutan*, is said to be pollinated by beetles which lick the disc, which is remarkably developed in this species. In the well-developed outer integument of the ovule *Paeonia* resembles Berberidaceae with which it is classed by Hallier. It has also been regarded as a distinct family (3).

The flowers of Ranunculaceae have in their conspicuous corolla, or petaloid calyx, or bright coloured stamens (*Thalictrum*), a ready means of attraction for insects which visit them for nectar, or where this is absent, as in *Anemone* (e.g. *A. nemorosa*), *Clematis* and *Thalictrum*, for pollen. As regards their pollination by insects, the flowers fall under two heads. The common regular open type with easily accessible nectar or pollen, such as the Buttercup, invites, and can make use of, very various insects. Generally the anthers of the outer stamens dehisce first, the immature stigmas being meanwhile covered by the inner stamens. The stigmas, however, mature while the innermost anthers still contain pollen, so that both self- and cross-pollination are rendered possible. On the other hand, in the more elaborately constructed flowers of *Aquilegia*, *Delphinium* and *Aconitum*, the shape of the flower and the position of the nectar, which in the two former is concealed at the bottom of a long spur, exclude the smaller insects, and the flowers are pollinated only by a few species of bees which have a sufficiently long proboscis. In other cases cross-pollination is favoured by well-marked proterandry.

The long feathery styles on the achenes of most species of *Clematis* and of the *Pulsatilla* group of *Anemone* aid in the distribution of the fruit by means of wind. A hook-shaped persistent style, or the hooked spines or tubercles on the surface of the achene in species of *Ranunculus* (as *R. arvensis*), favour dissemination by animals or birds, to the fur or feathers of which the fruits are able to cling. Distribution by ants (myrmecochory) occurs in species of *Helleborus*; the ants are attracted by an oil-containing swelling (elaiosome) on the raphe. A similar function is suggested for a swelling on the base of the achene in *Anemone Hepatica*, *Ranunculus Ficaria* and *Adonis vernalis*.

The 30 genera contain about 1200 species which are distributed through the temperate and cold regions of the world, but more especially beyond the tropics in the northern hemisphere. The presence of identical or closely allied species in North America and Central and Eastern Asia points to a former more intimate relation than now holds between the floras of the two hemispheres.

Helleboreae includes 15 genera and is almost exclusively north temperate or subarctic; exceptions are found in some species of *Delphinium* on the Himalayas and mountains of Abyssinia, and in a section of *Caltha*, with species on the Andes, in the extreme south of the American continent, in New Zealand and in Australia.

The eight genera of *Anemoneae* are, on the contrary, very widely distributed, occurring in the tropics and beyond to southern antarctic regions. Thus while the great majority of the 90 species of *Anemone* are temperate European, Asiatic or North American, a small section is confined to tropical East Africa and the Cape, another to the Cape alone, another to the Cordilleras of Chile, while individual species have a wide range, such as *A. multifida* in the mountains of North America, in Chile and at the Straits of Magellan. Species also occur on the Himalayas, in further India and in Tasmania. The 250 species of *Ranunculus* are chiefly temperate, arctic, and alpine plants of the northern hemisphere. *Clematis* (170 species) has a wide distribution but is rarer in the tropics than in temperate regions.

A certain interest attaches to this family from its relation to the development of a natural system of classification. It was by a careful study of the genera that Antoine Laurent de Jussieu was led to appreciate the relative value of the characters contained in the number, position, arrangement and structure of the organs of the flower, fruit and seed. He realised that plants shewing such wide differences in the form and structure of the floral envelopes as do *Aconitum*, *Ranunculus* and *Clematis* may yet be sufficiently similar in other points to justify their inclusion in the same family. The clue to a natural system, that is, one which brings together the most closely allied plants, lies in the discrimination between important and subordinate characters. Thus in Ranunculaceae we find, associated with considerable variation in other features, free hypogynous perianth-leaves, numerous stamens, a superior ovary and seeds with copious endosperm in which is embedded a minute straight embryo. With rare exceptions the carpels also are free.

REFERENCES

(1) SCHRÖDINGER, R. "Der Blütenbau der zygomorphen Ranunculaceen etc." *Abhandl. k. k. zool.-botan. Gesellsch. Wien*, IV, 5 (1909).

(2) SALISBURY, E. J. "Variation in *Eranthis hiemalis, Ficaria verna* etc." *Annals of Botany*, XXXIII, 47 (1919).

(3) WORSDELL, W. C. "The affinities of *Pæonia*." *Journ. of Botany*, XLVI, 114 (1908). See also *Annals of Botany*, XXII, 663 (1908).

Family VIII. BERBERIDACEAE

Flowers bisexual, generally cyclic, regular, hypogynous, 3- or sometimes 2-merous. Perianth of two to four whorls, generally differentiated into a calyx and corolla, often succeeded by two whorls of "honey-leaves." Stamens generally in two whorls, rarely more; anthers with generally valvate dehiscence, sometimes longitudinal. Carpel generally solitary; ovules numerous to few (rarely one) on the ventral suture or solitary at its base; integuments two, the outer strongly developed and longer than the inner. Fruit a berry or variously dehiscing, rarely an achene. Seed containing a copious endosperm and a generally small, straight embryo; an aril of various form is sometimes present.

Perennial herbs or shrubs with simple or compound exstipulate leaves and flowers (often yellow) solitary or in raceme-like cymose inflorescences.

Genera 10; species about 250. Chiefly in the north temperate zone.

There are two well-marked subfamilies, Podophylloideae and Berberidoideae.

Subfamily Podophylloideae. Characterised by flowers without honey-leaves, and foliage-leaves which are not pinnately divided. *Hydrastis* has two species, natives respectively of Atlantic North America and Japan. The thick yellow rhizome of *H. canadensis* contains the alkaloid hydrastin. The plant is a low perennial herb and sends up each spring a single long-stalked palmately lobed radical leaf and a simple stem bearing two leaves below the single terminal greenish-white flower. The flower has three petal-like caducous sepals, no petals, and numerous stamens and carpels; there are two ovules on the ventral suture of each carpel, and the fruit is a head of crimson 1–2-seeded berries. The genus is of special interest as suggesting a link between this family and Ranunculaceae in which it has also been included (in the tribe *Paeonieae*). In habit it resembles *Podophyllum*, as also in the trimerous perianth, while the numerous stamens (with longitudinal dehiscence) and carpels indicate its affinity with Ranunculaceae. The sympodial rhizome and manner of growth are characteristic of the subfamily. Thus in *Podophyllum peltatum*, the May Apple of the woods of Atlantic North America, the stout rhizome bears each season a large round seven- to nine-lobed umbrella-like leaf, and flowering stems bearing two smaller leaves, from the fork of which springs a nodding white flower. Other species occur in the Himalayas and Eastern Asia, and have solitary or cymose flowers. The flower is trimerous, the two outer whorls are sepaline and fugacious, the two inner (rarely more) large and petaloid and there are six to nine stamens with valvate dehiscence of the anthers. As in the rest of the family, there is a single carpel which bears numerous ovules on a well-developed placenta. The fruit is a many-seeded berry; each seed is enclosed in a pulpy aril-like development of the placenta. The allied genera *Jeffersonia* and *Diphylleia* have each one species in Atlantic North America and another in Manchuria or Japan. *Achlys*, with one species in Pacific North America and another closely allied to it in Japan, has no perianth, nine stamens, and carpels with a solitary basal ovule forming an achene in the fruit. The rhizome of *Podophyllum peltatum* yields the drug podophyllin, a combination of a resin and a glucoside.

Subfamily BERBERIDOIDEAE. Honey-leaves are present in the flower, and the foliage-leaves are pinnately compound or simple. *Berberis* and *Mahonia* (often united in one genus, *Berberis*) are shrubs bearing raceme-like inflorescences of yellow flowers on short lateral shoots. In *Berberis* the leaves are simple and on the elongated shoots are often reduced to simple to five-partite thorns; *Mahonia* has pinnately compound leaves. The flowers are trimerous; one or two outer whorls are sepal-like, the next is larger and petal-like; then follow two whorls of petal-like leaves with a pair of nectaries above the short claw. The six stamens have anthers with valvate dehiscence. The fruit is a one- to few-seeded berry. The numerous species (together about 225) are widely distributed in the north temperate zone, and spread from North America through Mexico along the chain of the Andes to Tierra-del-Fuego. *Berberis vulgaris* (Barberry) is native in England in copses and hedges; other species are well known in cultivation.

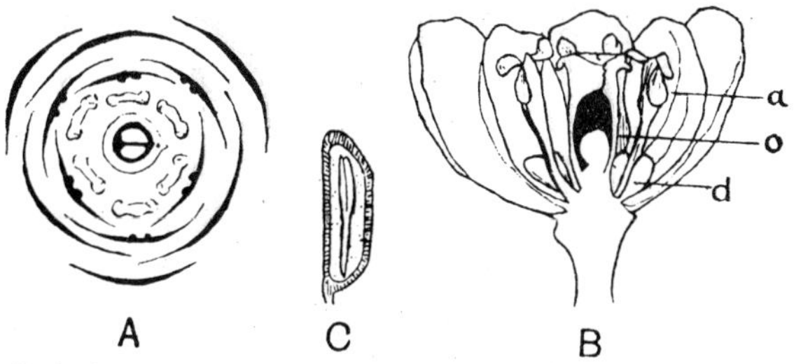

FIG. 71. *Berberis vulgaris*. A. Floral diagram. B. Vertical section of flower; *a*, anther dehiscing; *d*, nectary; *o*, ovary shewing one ovule. C. Vertical section of seed, shewing embryo surrounded by endosperm. B and C enlarged. (After Le Maout and Decaisne.)

In the remaining genera the inflorescence is terminal. *Nandina domestica* is a Chinese-Japanese shrub with large bi- to tri-pinnately compound leaves, compound inflorescences and trimerous many-whorled flowers. The carpels contain two ovules on the ventral suture. The remaining genera are perennial herbs.

Epimedium has ternate or highly compound leaves and simple or compound inflorescences of generally dimerous flowers; one to three whorls are sepal-like, two larger and petal-like, then follow two whorls with nectaries in the form of a pit or a spur. The carpel contains numerous ovules on the ventral suture, and the fruit is a two-valved capsule. The seeds bear a membranous aril. There are about twelve species which occur in southern Europe, temperate Asia, and western North America. *Leontice*, with about twelve species in the north temperate zone, has raceme-like inflorescences of yellow trimerous flowers, and frequently a large tuber-like rhizome; the fruit is dry, dehiscing more or less irregularly.

In *Berberis* and *Mahonia* the filaments of the stamens are sensitive to contact, springing inwards when touched.

The common Barberry was one of the flowers figured and described by Karl Sprengel in his work on the mechanisms of flowers and their relation to insects. Sprengel thought the irritability of the stamens, causing them to spring inwards when touched, was a device to ensure self-pollination. Hermann Müller, however, shewed that although self-pollination may ultimately ensue, we probably have here a means for promoting cross-pollination. The flowers are made conspicuous by the three large bright yellow petals. The six honey-leaves have each a pair of large nectaries just above their base. As the flower opens the six anther-filaments bend outwards and come to lie each opposite the middle line of a honey-leaf. The nectar collects between the stamens and ovary and a bee in search of it will touch with its proboscis the adjacent bases of two filaments. The anthers open as soon as the petals expand, the little valve moving up and rotating at the same time so that the masses of pollen which it carries with it come to face the centre of the flower. When the stamens are stimulated by the touch of the bee, they spring inwards and the pollen-bearing valves strike the insect's head. As the stigma is represented by the papillose sticky edge of the disc surmounting the ovary, the bee, on visiting the next flower, will probably deposit on the stigma some of the pollen derived from the last. It is interesting to note that in the position assumed by the stamens as a result of irritation, the pollen-bearing valves are above, not on a level with, the stigmatic ring. Müller observes that, failing insect-visits, self-pollination is possible, the masses of pollen coming into contact with the stigma by the bending inwards of the anthers as the flowers wither. He also states that hive-bees after being once struck by the springing inward of the stamens do not probe the same flower again but fly immediately to another. Humble-bees, on the contrary, were observed to thrust the proboscis again and again into the same flower, in which case doubtless some pollen was placed on the stigma of its own flower.

The cotyledons are carried up in germination and become green; in *Podophyllum* and *Leontice* the long stalks are united and the plumule breaks through laterally, as in several species of Ranunculaceae. In the stem of *Podophyllum, Diphylleia* and *Leontice,* the arrangement of the vascular bundles is very irregular in transverse section. These monocotyledonous characters associated with the trimery of the flower have suggested a phylogenetic relationship with the Monocotyledons as a group. Hallier finds in the Berberidaceae the origin of the Angiosperms as a whole, both Monocotyledons and Dicotyledons, the former having a distinct line of descent.

Apart from these larger questions, there is general agreement as to the affinity of Berberidaceae with Ranunculaceae, in which trimery of the flower is often seen in *Eranthis,* and it may further be pointed out that a trimerous arrangement may easily be derived from the $\frac{2}{5}$ arrangement of the perianth-leaves which is a characteristic of the Ranunculaceae.

Family IX. LARDIZABALACEAE

A small family with 20 species in 8 genera, in the Himalayan region, Eastern Asia and Chile. With the exception of *Decaisnea* (Himalayas and W. China)—an erect tree-like shrub with very long pinnate leaves—the plants are lianes with palmately compound leaves. The flowers are racemose, rarely solitary, regular, trimerous, and often unisexual by reduction of the stamens or carpels. *Akebia quinata,* a Japanese species frequently cultivated in gardens, has three brown-purple perianth-leaves, no "honey-leaves," and two whorls of free stamens or carpels. In *Lardizabala* (Chile) the plants are dioecious; the flowers have six coloured perianth-leaves, six smaller honey-leaves, six stamens united by their filaments (rudimentary in the female flower) and three free carpels (rudimentary in the male flower).

The stamens dehisce extrorsely by longitudinal slits. The carpels, which are always free, contain numerous anatropous ovules sessile in longitudinal rows on the side walls of the ovary; the ovules have two integuments. The fruit is a berry, sometimes opening on the inside; the seeds contain a copious endosperm and a small straight embryo. The juicy berry is often edible, as in *Akebia quinata.*

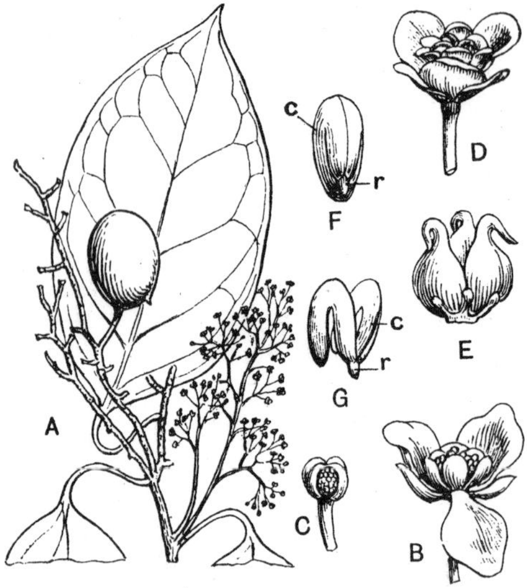

FIG. 72. *Hyperbaena domingensis.* A. Portion of stem with flowers and ripe fruit, × $\frac{2}{3}$. B. Male flower with one of the interior sepals bent back, × 10. C. Stamen, × 30. D. Female flower, × 10. E. Pistil from female flower, × 60. F, G. Embryo, × $\frac{2}{3}$; *c*, cotyledons; *r*, radicle. (A, F, G after Miers; B–E after Eichler.) (From *Flor. Jam.*)

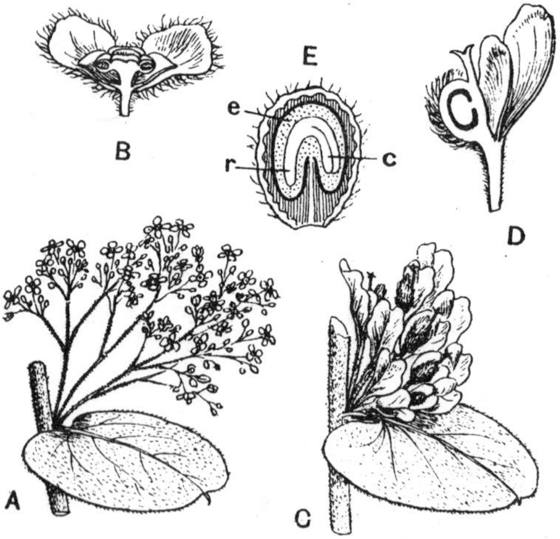

FIG. 73. *Cissampelos Pareira.* A. Male inflorescence, × $\frac{2}{3}$. B. Male flower in section, × 10. C. Female inflorescence, × 6. D. Female flower in section, × 12. E. Drupe cut lengthwise, × 4; *c*, cotyledons; *e*, endosperm; *r*, radicle. (After Baillon.) (From *Flor. Jam.*)

Family X. MENISPERMACEAE

An important tropical family with about 70 genera and 300 species. A few genera pass into the north temperate zone in North America, the eastern Mediterranean region and Eastern Asia, and also into the south temperate zone. Fossil species from Tertiary beds in Europe and North America indicate a former more extended northerly range.

The plants are generally lianes with simple lobed palmately veined leaves and small dioecious flowers in axillary racemes. Rudimentary stamens or pistils are generally present in the female or male flowers respectively. The flowers are cyclic, 3- or sometimes 2-merous,and generally regular, with two whorls each of sepals, petals, and stamens (fig. 72); the petals, which may be compared with the honey-leaves of the previous families, are smaller than the sepals and sometimes absent. Increase or diminution in the number of the sepals and stamens occurs, and occasionally union of the parts in calyx and corolla, more often in the androecium; in *Cissampelos* and allied genera the stamens unite completely, forming a central column (fig. 73, B). The carpels are free and three or rarely more, sometimes solitary, with a single hemianatropous pendulous ovule, with two integuments, on the ventral suture. The fruit is a drupe, the apex of which through strong dorsal growth often approaches the base. Endosperm is present, sometimes ruminate, or absent; the embryo is often bent. In *Cissampelos* (fig. 73, D) the female flower is remarkably zygomorphic, having a single sepal, two often united petals, and a single carpel.

As frequently occurs in climbing woody plants, abnormal secondary growth is characteristic of many genera. After the first or second year's growth the original cambium ceases to function, and its place is taken by a ring of meristem in the cortex in which new collateral bundles are formed. Repetition of this process leads to the formation of concentric rings of bundles.

Family XI. NYMPHAEACEAE

Flowers bisexual, regular, hypogynous to perigynous, acyclic, hemicyclic or cyclic. Perianth-leaves free, six to indefinite, generally differentiated into a calyx and corolla, the petals often passing into the stamens (6–∞). Carpels 3–∞, free or united; ovules 1–∞ on the inner walls of the carpel, each with two integuments. Seeds often arillate, generally with both perisperm and endosperm; embryo well-developed with fleshy cotyledons.

Water- or marsh-plants with generally submerged or floating leaves and large showy solitary flowers.

Genera 8; species about 100. Cosmopolitan.

The cotyledons are hypogeal; in germination the plumule elongates and bears first an awl-shaped leaf (except in *Nelumbo*, where the adult peltate form is at once assumed); this is generally followed by lanceolate, sagittate, or ovate submerged leaves which precede the development of the adult floating form. In *Cabomba* a pair of awl-shaped first leaves is succeeded by a decussating pair of the characteristic much divided submerged leaves. The stem is a rhizome which is short, thick and erect, as in *Victoria*, where it lives for a few years, and in *Euryale* for one year; or a long-lived branched rhizome creeping in the mud, as in our native *Nuphar* and *Nymphaea*. The vascular bundles have the woody elements much reduced; in the absence of cambium and in their scattered arrangement in the parenchymatous ground-tissue they recall the arrangement characteristic of Monocotyledons. Laticiferous vessels are present. So-called stellate hairs, or radiating supporting cells, frequently occur in the angles of the walls of the large intercellular spaces, especially of the leaf-stalks.

The family falls naturally into three subfamilies, two of which, Cabomboideae and Nelumbonoideae, have free carpels, whilst in the third and largest, Nymphaeoideae, the carpels are united. Nelumbonoideae, which contains the single genus *Nelumbo*, is distinguished by the fact that except for a thin membrane of endosperm the embryo completely fills the seed; in the other two subfamilies both endosperm and perisperm are present.

Subfamily 1. CABOMBOIDEAE. Contains two genera, *Cabomba* and *Brasenia*. *Cabomba* has six species in the warmer parts of America. The erect sympodial rhizome bears closely arranged scale-leaves in the axils of which arise long floating stems which bear decussating deeply cut submerged leaves, and, at the flowering period, spirally arranged peltate round floating leaves. The flowers originate laterally, and Raciborski (1) compares the relation of leaf and flower with that obtaining between the leaf and tendril in the Vitaceae. At the growing point of the stem, leaf-primordium and flower-primordium originate alternately in spiral succession; later the internode between each pair of primordia elongates so that leaf and flower stand at the same level. Vegetative

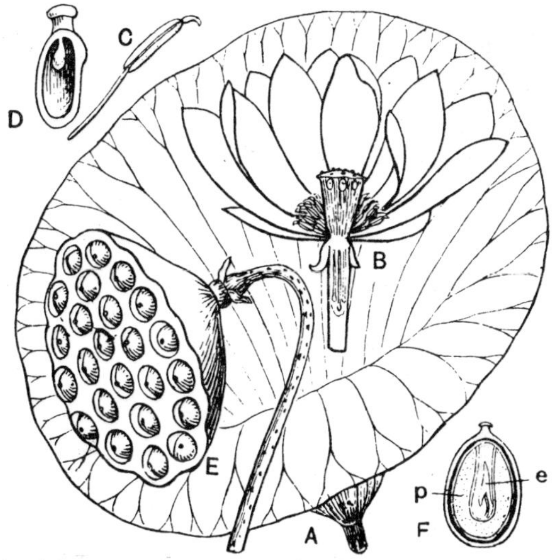

FIG. 74. *Nelumbo lutea.* A. Leaf, × 1/8. B. Flower in section, × 1/4. C. Anther, nat. size. D. Carpel detached, with ovary cut lengthwise, × 2/3. E. Receptacle in ripe fruit, × 1/3. F. Fruit cut lengthwise, nat. size; *e*, plumule; *p*, cotyledon. (From *Flor. Jam.*)

buds are developed in the leaf-axils. The cyclic flower has three sepals, three petals alternating with these, three to six stamens and usually (two or) three free carpels. Sepals and petals are yellow in the widely distributed *C. aquatica* (Mexico to South Brazil) which has proterogynous flowers, white in *C. caroliniana* (south-eastern United States) and red-violet in the other species. A few ovules are borne on the sides of the carpels. The fruits are indehiscent pods which contain generally three pendulous seeds. The embryo is surrounded by a small layer of endosperm, and there is a copious perisperm.

Brasenia contains one species, *B. purpurea*, which is cosmopolitan outside Europe, and is represented by fossil species in the Tertiary strata of Europe. In habit it resembles *Cabomba*, but the stems bear only the peltate floating leaves. The trimerous calyx and corolla are followed by numerous stamens and six or more carpels. The young shoots are covered with a remarkably thick coating of mucilage which protects the stem and lower leaf-surface from contact with the water until the cuticle is developed.

This subfamily recalls the Ranunculaceae in its small flowers with free, hypogynous parts and in its fruit; and in the dimorphic leaves of *Cabomba*, the *Batrachium* section of *Ranunculus*.

Subfamily 2. NELUMBONOIDEAE. Contains the single genus *Nelumbo* (fig. 74) with two species—*N. lutea*, with yellow flowers in Atlantic North America and southwards to Columbia, and *N. nucifera*, with rose-coloured flowers, in the warmer parts of Asia and north-east Australia—the sacred lotus of the Hindu. The long-stalked concave peltate leaves and large showy flowers are carried high above the water. The numerous free perianth-leaves and stamens are arranged spirally beneath a large obconical receptacle, in the flat upper surface of which the numerous carpels are buried, each in a round pit. Each carpel contains a pendulous anatropous ovule on the side towards the axis, and ripens to an indehiscent nut with a very hard pericarp, completely filled by the embryo. The embryo (Lyon (2)) consists of two large white fleshy hemispherical cotyledons between which lies a well-developed green plumule enclosed by a thin membrane of endosperm. Opposite the plumule is the radicle which is functionless, its place being taken by adventitious roots developed on the plumule. The two cotyledons originate round the base of the plumule in the form of a ring which later becomes two-lobed. This mode of origin has given rise to the suggestion that *Nelumbo* has only one cotyledon.

Subfamily 3. NYMPHAEOIDEAE. Contains five genera characterised by the union of the carpels into a single many-chambered ovary bearing above a flat radiating stigma; the flowers are hypogynous, perigynous or epigynous. *Nuphar* has seven species in the north temperate zone; *N. luteum* is our Yellow Water-lily. The rootstock creeps in the mud and bears large thin crumpled submerged leaves, without stomata, and roundish floating leaves with a deeply two-lobed base, bearing stomata on the upper face. The yellow flowers arise in the axils of small bracts and are few in number compared with the leaves. The five large yellow sepals have a $\frac{2}{5}$ arrangement, the small scale-like petals, which bear a nectary on the back, are numerous, arranged in a spiral, and pass into the numerous stamens, above which rises the large syncarpous pistil formed of numerous (10–16) carpels with as many chambers, and ray-like stigmas corresponding with the middle line of each carpel. The numerous ovules cover the side-walls. In the ripe fruit the internal parenchymatous tissue contains intercellular spaces filled with air. The fruit ultimately bursts by the swelling of mucilage, the outer rind is pushed off and the parenchymatous tissue of each carpel protrudes and is set free; it floats in the water and is carried by currents. The small seeds contain a small embryo at one end embedded in a scanty endosperm below which is a copious perisperm.

Nymphaea contains 32 species in warm and temperate climates; *N. alba* is our White Water-lily. *N. Lotus*, another white-flowered species, is the lotus-flower of ancient Egypt. Other species have red or blue flowers. The short stout horizontal rhizome bears the large floating leaves and extra-axillary flowers which take the place of a leaf in the spiral development at the growing-point. The first developed leaf in the seedling is awl-shaped, succeeded by lanceolate to elliptical leaves, and later by the characteristic adult orbicular or oval form. In some tropical species the lower end of the rhizome becomes covered with cork and forms a resting tuber which persists during periods of drought. The four sepals are followed by numerous spirally arranged large spreading petals which pass gradually into the numerous stamens, and, like these, are perigynous, apparently springing from the sides of the ovary. The number of carpels is very variable (5 to 35). They are sunk in the receptacle with the upper half of the ventral suture, forming the roof of the ovary-cavity, exposed and bearing the stigmatic papillae, and the lower half fused with the central axis. The back of each carpel is more or less prolonged into a stylar process (fig. 75, E). An indication of an apocarpous ancestry is found in the varying degree of lateral union of the carpels; in some species the walls between the cells of the ovary consist of distinct lamellae which easily separate but generally the tissue of the

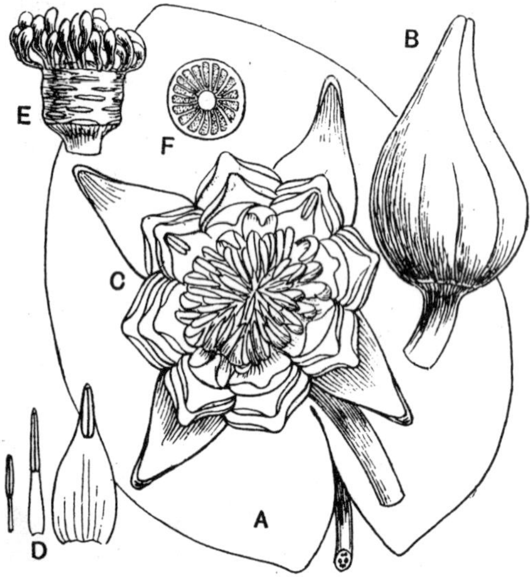

FIG. 75. *Nymphaea amazonum*. A. Leaf, × ½. B. Unopened flower, × ½. C. Flower, × ½. D. Stamens shewing variation in form from the innermost to the outermost, × ⅔. E. Ovary shewing scars of petals and stamens and bearing the stylar processes, × ⅔. F. Ovary cut across, × ⅔. (After Conard.) (From *Flor. Jam.*)

partition is continuous. The ovules are numerous, as in *Nuphar*. The fruit, which forms a spongy berry, ripens beneath the surface of the water, where it is drawn by the spiral twisting of the flower-stalk. It dehisces regularly by swelling of the mucilage which surrounds the numerous seeds; the seeds rise to the surface, rendered buoyant by air which is enclosed within a fleshy aril. The structure of the seed resembles that of *Nuphar*.

Two genera, *Euryale* (one species in Eastern Asia) and *Victoria* (two species in the Amazon district), are distinguished by their completely epigynous flowers and the copious development of prickles on the surface of the organs. The best known species, *Victoria regia*, has leaves six to seven feet in diameter and flowers to a foot or more across. It grows in the backwaters of the Amazon, often covering the surface for miles. The seeds are eaten and known as *mais del aqua*. The flowers are extra-axillary but enveloped when in bud by the strongly developed stipule of the adjacent leaf.

Both self-pollination and cross-pollination are known to occur. In some cases, as in a small group of species of the genus *Nymphaea* inhabiting tropical America, the anthers burst and pollination takes place before the flowers open. The efficacy of self-pollination in these plants is evident from the fact that ten to thirty thousand or more perfect seeds are formed in a single fruit. The flowers open at night, and in one case, *Nymphaea amazonum*, remain open only for 20 to 30 minutes. *Nymphaea alba* is also self-pollinated, the innermost stamens opening first, while the stigma is still in a receptive condition. Other species of the same genus, in which the outermost stamens are the first to dehisce, are adapted for insect-pollination, and the same applies to *Nuphar* and *Victoria*.

Cook (3) has shewn that the development of the endosperm is peculiar. A transverse wall is formed between the two nuclei resulting from the first division of the endosperm-nucleus, dividing the embryo-sac into an upper and a lower portion. Endosperm is formed only in the upper portion, the lower portion forming a haustorium which penetrates the nucellus. In the development of the embryo the character of the suspensor varies widely in the different genera and it is sometimes absent.

The Cabomboideae, as already indicated, suggest a close affinity between this family and the Ranunculaceae. The scattered distribution of the conducting bundles recalls *Podophyllum* and allied genera in the Berberidaceae, with which family also the frequent presence of an aril in the seed suggests an affinity. On the other hand, the presence of latex in the tissues, the superficial placentation of the ovules and the large peltate-rayed stigmas may be compared with the Papaveraceae, a member of the next order, Rhoeadales.

REFERENCES

(1) RACIBORSKI, M. "Die Morphologie der Cabombeen und Nymphaeaceen." *Flora*, LXXVIII, 244 (1894).

(2) Lyon, H. L. "Observations on the Embryogeny of *Nelumbo*." *Minnesota Botanical Studies*, II, 643 (1901).

(3) Cook, M. T. *Bull. Torrey Botan. Club*, XXIX, 211 (1902).

See also Conard, H. S. "The Waterlilies. A Monograph of the genus *Nymphaea*." *Carnegie Institution, Washington*, 1905.

Family XII. CERATOPHYLLACEAE

A small family containing one genus, *Ceratophyllum*, with three species of rootless submerged water-plants with whorls of simple much dissected leaves and unisexual flowers. The hypogynous perianth, numerous stamens spirally arranged on a convex receptacle, and solitary carpel indicate its affinity with the Ranales. It grows in stagnant water in all parts of the world.

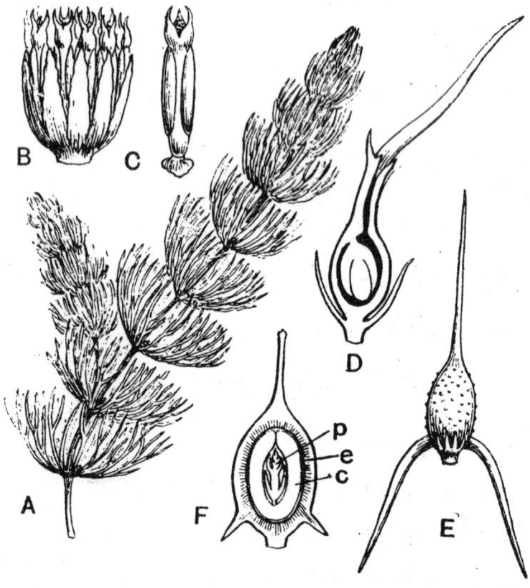

Fig. 76. *Ceratophyllum demersum.* A. Portion of plant, × ⅔. B. Male flower. C. One stamen. D. Female flower cut lengthwise shewing the ovule. E. Fruit. F. Fruit cut lengthwise; *c*, cotyledons; *e*, endosperm; *p*, plumule. B–F enlarged. (B–E after K. Schumann; F after Baillon.) (From *Flor. Jam.*)

The plants are slender fragile herbs, with the sessile exstipulate leaves dichotomously cut into narrow linear segments which are shallowly serrate on the outer edge. A branch appears only in the axil of one leaf of a whorl. The flowers are solitary and axillary. The male consists of a perianth of about 12 subulate lobes united at the base, and 12–16 stamens with a very short filament, large extrorse anther-cells, and a fleshy connective produced above the anther-cells and bearing two or three sharp points. The roundish or elongated pollen-grains have only one coat. The similar perianth of the female flower persists round the medianly placed carpel which bears a long slender pointed style and contains a single ovule pendulous from the top of the cell. The fruit is a nut crowned by the persistent style and bearing a pair of spurs or other outgrowths at the base. The seed contains a large embryo, the two thick cotyledons surrounding a well-developed many-leaved plumule, around which is a scanty large-celled endosperm; the radicle is very short. The embryology resembles that of the Nymphaeaceae in the separation of the lower part of the embryo-sac as an haustorium.

Order 2. *RHOEADALES*

Allied to Ranales in the hypogynous flowers with free sepals, petals, and stamens, the last being also frequently indefinite. It differs in the general arrangement of parts in twos and fours and in the union of the two or more carpels into a one-chambered ovary with generally numerous ovules on parietal placentas; or the ovary becomes bilocular by development of a septum between the placentas; as this dividing wall is not produced by the ingrowth of the ventral edges of each carpel but is an additional structure arising late in the history of development of the ovary, it is known as a false septum. The stigmas sometimes occupy the normal position above the midrib of the carpel, but are more often commissural, that is, placed immediately above the placentas. Flowers cyclic, the members of successive whorls alternating regularly. Seeds generally small, often curved; embryo sometimes small in a copious endosperm, more often large and filling the seed.

Herbs with scattered entire or more or less divided leaves without stipules; occurring in temperate and cold regions.

The arrangement of the parts of the flower in this order has provoked much discussion, and there is an extensive literature on the subject. The most satisfactory view seems to be that adopted by Čelakovský of a gradual reduction, especially in the androecium; this is exemplified in Papaveraceae and also in Capparidaceae; the subfamilies Hypecoideae and Fumarioideae of Papaveraceae and the family Cruciferae represent extreme forms of reduction which have become constant in each group. Resedaceae stands somewhat apart from the other families.

Family I. PAPAVERACEAE

Flowers bisexual, regular or zygomorphic. Sepals two or sometimes three, imbricate, generally falling very soon. Petals four or sometimes six, rarely more, regularly alternating in two whorls, imbricate and often crumpled in bud, deciduous, usually brightly coloured (absent in *Macleaya* and *Bocconia*). Stamens generally indefinite in 2- or 3-merous regularly alternating whorls, more rarely 3- or 6-merous, or four (*Hypecoum*), or two and tripartite (Fumarioideae); anthers bilocular (but see Fumarioideae) dehiscing longitudinally. Carpels two to numerous, ovary generally unilocular with the same number of many-ovuled parietal placentas often more or less produced into the ovary-cavity, and rarely meeting in the centre forming a multilocular ovary, sometimes spuriously bilocular by a false septum joining the placentas; ovules anatropous or campylotropous, ascending or horizontal, rarely few, solitary from the base of the cavity in *Bocconia*; stigmas equal in number to the placentas, sometimes distinct and corresponding with the midrib of the carpel (alternating with the placentas), sometimes forming a radiating lobed structure on the top of the ovary or short style, the radii corresponding with the placentas (commissural). Fruit generally a capsule opening by pores or valves; rarely indehiscent. Seeds globose or ovoid, smooth or minutely warted or reticulate. Embryo small in an oily endosperm.

Generally herbs sometimes becoming shrubby below, more rarely shrubs, very rarely a tree (*Bocconia*), with alternate, entire or more often lobed or cut leaves; latex-containing sacs or vessels often present. Flowers often showy, solitary at the end of the main and lateral shoots, sometimes forming dichasia or unilateral cymes.

Genera 28; species about 700; mostly northern extra-tropical.

The genera fall into distinct subfamilies, Papaveroideae and Fumarioideae. The latter has been regarded as a distinct family, Fumariaceae, but the two groups have

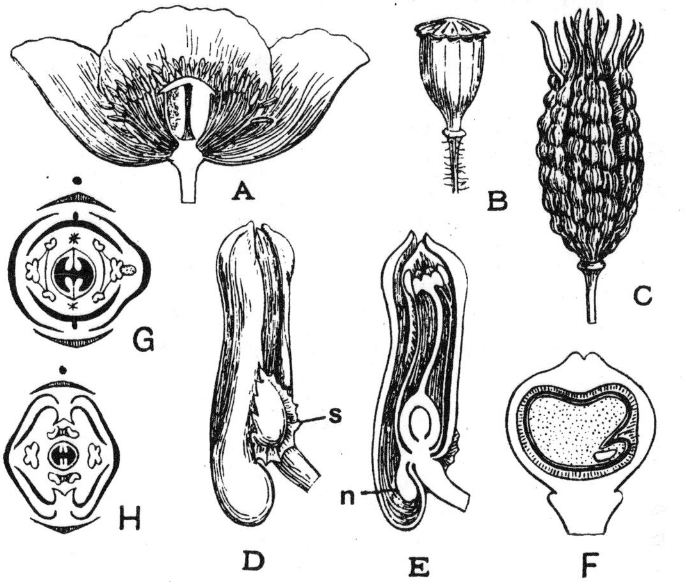

FIG. 77. A. Flower of *Papaver Rhoeas* in vertic[illegible] section, slightly reduced. B. Fruit of same, slightly reduced. C. Fruit of *Platystemon californicus*, slightly enlarged. D. Flower of *Fumaria officinalis*; *s*, sepal, ×4. E. Same in vertical section; *n*, nectar-secreting spur of stamen. F. Fruit of same, ×6. G. Floral diagram of *Fumaria*. H. Floral diagram of *Hypecoum*. (A–F after Baillon; G, H after Eichler.)

important characters in common and are, moreover, connected by an intermediate genus, *Hypecoum*, which may be regarded as representing a third subfamily.

Subfamily I. PAPAVEROIDEAE. Flowers regular, petals not spurred, rarely absent; stamens generally numerous, free; carpels two or many. Herbs, shrubs or rarely trees, with a milky or hyaline juice, and entire or pinnately cut leaves. The androecium shews remarkable variation both in the number and the arrangement of the members in the whorls, and much has been written on the subject of the floral formula and very diverse views have been maintained[1]. A study of the development shews that the staminal whorls contain two or a multiple of two members (or in the trimerous flowers three or a multiple), the numbers varying greatly in different genera or species. De Candolle regarded the flower as typically tetramerous, a view which involves the assumption of a reduction in the calyx and corolla. Čelakovský, on the other hand, insists on the typical dimery of the flower, recognising at the same time a frequent increase, often progressive in the inner whorls, in the number of members in the whorls of the androecium. He regards the larger number of members as indicating an earlier condition and an affinity with the Ranales; genera with fewer stamens have been derived by reduction from the older forms, and an extreme stage of reduction is represented by *Hypecoum* and the Fumarioideae. A similar reduction has occurred in the Capparidaceae, the most reduced form of which, *Cleome*, closely resembles the form which has become constant in Cruciferae.

The largest genus, *Papaver*, contains about 100 species, chiefly from Central and Southern Europe and temperate Asia. *Papaver nudicaule* (Iceland Poppy) grows in the Arctic regions of both hemispheres and also in the mountains of central Asia and Colorado. Two closely allied species characterised by a thick covering of thorns, the one in Australia the other at the Cape, are outlying representatives of this essentially northern genus. The poppies are annual or sometimes perennial herbs, growing erect, with lobed or cut, generally long-stalked leaves and nodding flower-buds. They contain a milky juice (latex) which is found in a network of vessels running through the whole plant and often accompanying the sieve-tubes of the vascular bundles. *P. somniferum* is extensively cultivated in India and elsewhere for the preparation of opium, which is the dried latex of the unripe fruit.

The long-stalked regular flowers are dimerous (sometimes

[1] For a *résumé* see Fedde, *Pflanzenreich*, IV, 104 (1909), in which a discussion of the subject with references to literature will be found.

trimerous), the two sepals are merely protective and fall before the flower opens, the bright-coloured crumpled petals occupy the next two whorls, and are followed by numerous whorls of stamens. The large central ovary is formed of 4 to 16 carpels, and roofed by a flat or conical disc formed by the union of the radiating stigmas. Below each stigma the placentas, bearing numerous ovules, project into the ovary-cavity, almost meeting in the centre. The fruit is a capsule which opens by means of small valves beneath the persistent stigma-lobes. When the capsule is shaken with sufficient force the small seeds are jerked out through the pores. The flowers contain no nectar and are visited for the sake of the pollen by bees and other insects, which find the broad central stigma a convenient resting-place. There is thus a good chance of pollen from another flower being deposited on the stigmas, but, as in many cases the lower parts of the stigmas get covered with the pollen from adjoining anthers, the chance of self-pollination is not excluded.

Meconopsis has 41 species, mainly in south and east Central Asia and China, two occur in Pacific North America and one, *M. cambrica*, in Western Europe, which is found in the west of England, Yorkshire, Wales and Ireland. They are perennial herbs with a yellow juice, and yellow, blue or purple flowers, solitary or in cymes or racemes. A distinction from *Papaver* is the presence of a distinct style on which the capitate four- to six-rayed stigma is borne. The ovoid or elongated capsule opens by a corresponding number of valves below the persistent style.

The closely allied genus *Argemone* (nine species, America) has sometimes trimerous flowers. The plants are herbs with prickly leaves and fruits, and white or yellow flowers. *A. mexicana* (Mexican Thistle) is widely naturalised in the Old World. The capsule opens by four to six valves at the apex (fig. 78).

Glaucium luteum (Yellow Horned Poppy), found on our sandy sea-shores, is the British representative of a genus with about 21 species, occurring chiefly in the Mediterranean region, but having a single species in China. The generic name is derived from the greyish coating of wax which gives a glaucous appearance to the stem and leaves of our native and some other species. The long curved pod is a distinctive feature; it is formed from two carpels, and becomes two-chambered by the formation of a septum uniting the two parietal placentas. It dehisces when ripe by the separation, from above downwards, of the walls of the carpels from the placentas, forming two long narrow valves. *Chelidonium* has a single species, a native of the eastern northern hemisphere from Europe to Japan; in Britain it is a doubtful native. It is an erect branching perennial herb with much divided leaves and a yellow juice. The rather small yellow flowers are umbellate. The fruit as in *Glaucium* is the product of two carpels, and is a long narrow capsule, with no septum, splitting by two thin valves from below upwards. The seeds are myrmecochorous, having an appendage on the raphe, which functions as an elaiosome and is attractive to ants.

Eschscholtzia, a genus of the Pacific states of North America, is characterised by its perigynous flower, sepals, petals and stamens being elevated on a cup-like projection of the floral axis, the bicarpellary ovary springing from the bottom of the cup. The two sepals are coherent and are pushed off in the form of a cap by the growth of the petals.

Platystemon, a genus confined to Pacific North America, is of special interest as indicating a passage from the last order, Ranales. The floral axis is somewhat depressed and bears on its edge a trimerous calyx followed by a corolla of two alternating trimerous whorls. The numerous stamens are in several whorls and the pistil is formed of many carpels united laterally to form a multilocular ovary but free in the upper style-bearing portion; the style is a continuation of the middle line of the carpel and bears the stigma. In the ripe fruit the carpels become separated (fig. 77, C). The trimery of the perianth, which occurs also in other genera of the subfamily, and especially the incomplete union of the carpels, suggest an affinity with the Ranales.

In two closely allied genera, *Macleaya* (China and Japan) and *Bocconia* (fig. 79) (Central and South America, and West Indies), the corolla is absent and the small flowers are arranged in many-flowered compound racemes. As in *Chelidonium*, the two valves of the fruit separate completely (fig. 79, E, F);

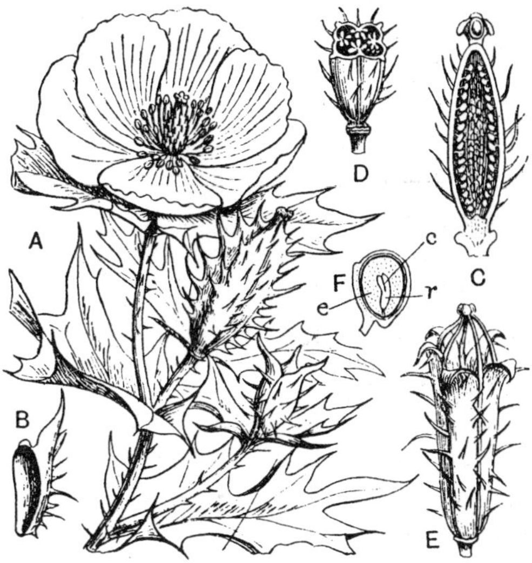

FIG. 78. *Argemone mexicana*. A. Upper portion of plant with bud, fruit and flower, × ⅔. B. Sepal, × ⅔. C. Pistil cut lengthwise, × 2½. D. Pistil cut across, × 2½. E. Ripe capsule, × ⅔. F. Seed cut lengthwise, × 4; *e*, endosperm; *c*, cotyledons; *r*, radicle. (After A. Gray.) (From *Flor. Jam.*)

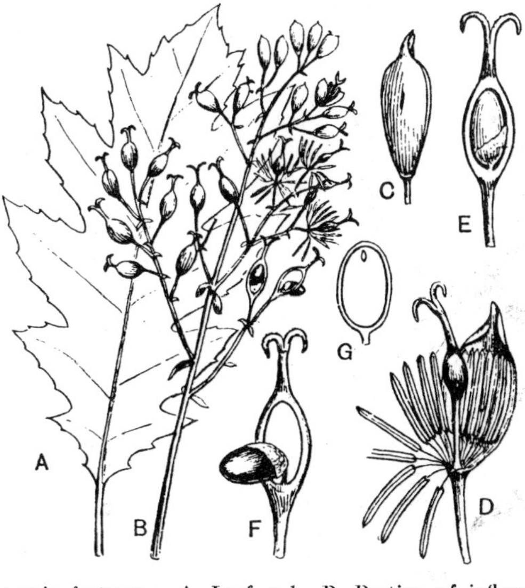

FIG. 79. *Bocconia frutescens*. A. Leaf, × ¼. B. Portion of inflorescence, × ⅔. C. Flower-bud, × 2. D. Flower just opening with one sepal removed, × 2. E. Fruit with one valve gone; note oblique aril enveloping base of seed, × 2. F. Fruit with both valves gone, × 2. G. Seed cut lengthwise, × 2. (From *Flor. Jam.*)

the seed (solitary in *Bocconia*) has a basal arillary appendage. The five species of *Bocconia* include great diversity of habit; one is an annual herb, others are more or less shrubby, and *B. arborea* is a small tree.

Subfamily II. HYPECOIDEAE. The flower is dimerous and symmetrical (fig. 77, H) with four stamens, one opposite each petal; the outer petals are generally trilobed, the inner tripartite (except in *Pteridophyllum*). The bicarpellary ovary forms a long pod which generally becomes divided by late developing transverse septa into one-seeded joints which separate when ripe. Annual or perennial herbs with pinnately cut leaves. Latex-containing vessels are absent. The genus *Hypecoum* contains 15 species in the Mediterranean region, Central Asia and China. *Pteridophyllum*, with a single species in Japan, differs in having entire petals, the inner median pair slightly narrower than the outer.

Subfamily III. FUMARIOIDEAE. Distinguished from Papaveroideae by the marked dissimilarity between the two whorls of petals, by the saccate or spur-like development at the base of one or occasionally both of the outer petals, and by the two tripartite stamens in the transverse plane (fig. 77, G). They also differ in the absence of latex, but have oil-containing sacs. The plants are herbs with scattered pinnately much divided leaves, by means of which they often climb. *Fumaria capreolata* and *F. officinalis*, for instance, climb by means of the petioles, while the British species of *Corydalis* (*C. claviculata*) has branched tendrils terminating the leaf-stalks.

The relatively small flowers are borne in racemes which are terminal on the stem and branches; sometimes through the stronger growth of the axillary shoot they appear to spring from opposite the leaves. The bracts are sometimes suppressed, an occurrence which becomes a constant character in the next family.

The closed flowers are in striking contrast with the open ones of Papaveroideae and are associated with a difference in the method of pollination, which is here effected by nectar-seeking insects, while the much smaller amount of pollen is placed in a suitable position. The two sepals are reduced to small scales and very soon fall (fig. 77, D, *s*). The outer pair of petals are large and more or less completely enclose the rest of the flower; in *Fumaria* they form a tube, and one of them is developed at the base into a sac-like spur (fig. 77, D, E). In *Dicentra* they are in the form of a lyre, and, both having a broad saccate base, the flower is symmetrical. The inner pair of petals are smaller; they closely envelop the stamens and with their adhering hood-like upper portions form a cap over the anthers and the central stigma. The middle division of each stamen bears a perfect anther with two chambers, the two lateral divisions each a half-anther with one chamber. At the base of the common filament is a nectary, on each side of the flower in *Dicentra*, on that side only which bears the spur in *Fumaria* (fig. 77, E, *n*). The nectar-seeking bee, in thrusting its proboscis down into the nectar-containing sac or spur, displaces the hood-like apex of the inner petals and exposes the pollen-covered anthers. It thus gets dusted with pollen, some of which in a subsequent visit to another older flower will be transferred to the stigma.

In *Corydalis cava*, cross-pollination is essential, since, as Hildebrand has shewn by experiment, the flowers are sterile to their own pollen, and but very slightly fertile to pollen from another flower of the same plant, and only thoroughly fertile when impregnated with pollen from a different plant. In connection with these facts it is of interest to note that the bees habitually go from below upwards on each plant, and thus tend to bring pollen to the lower and older flowers of one plant from the upper younger ones of another. The small flowers of *Fumaria*, on the contrary, are fertile with their own pollen, and the inconspicuous-flowered species, at any rate, are very sparingly visited by insects and generally self-pollinated.

Two explanations have been offered of the unusual character of the androecium. De Candolle suggested that the two half-anthers belonged to a pair of median stamens, such as are present in *Hypecoum*, each of which had become split, the halves being displaced right and left to become attached to the lateral stamens. The theory suggested by Asa Gray is, however, more probable, as it is borne out by the history of development of the androecium. He regarded each lateral group as a single stamen which has become tripartite, and this view is supported by the fact that the whole structure proceeds from a single protuberance on the floral axis; an analogy is afforded by the tripartite petals of *Hypecoum*.

The flowers of *Corydalis* and *Fumaria*, with their lateral spur, are of interest as an example of that rare form of zygomorphy in which the plane of symmetry passes through the transverse and not, as usual, through the median plane.

The ovary is one-celled with parietal placentas and develops into a two-valved many-seeded capsule; in *Fumaria*, however, there are only two ovules, and the fruit is an indehiscent one-seeded nut (fig. 77, F). Many species of *Corydalis* are myrmecochorous, the seeds bearing an appendage of the raphe which functions as an elaiosome and is attractive to ants.

Several species of *Corydalis* and *Dicentra* have only one cotyledon, and associated with this is the formation of a subterranean tuber or a rhizome bearing fleshy scales. In *Corydalis cava* the tuber originates from the swollen hypocotyl, in *C. solida* it is the thickened root.

The five genera contain about 200 species spread over the north temperate zone; a large number occur in the Mediterranean region.

Family II. CAPPARIDACEAE

There are about 450 species in 40 genera distributed through the warmer parts of the world and including herbs and shrubs without latex, with alternate simple or palmately compound leaves with or without stipules, and solitary or racemose flowers subtended by a bract but without bracteoles.

As the structure of the flower indicates, the family occupies an intermediate position between Papaveraceae and Cruciferae. The most reduced type of flower is illustrated by *Cleome tetrandra* (fig. 80, A); there are two outer median and two inner transverse sepals, four petals arranged diagonally and four stamens alternating with them; the ovary is bicarpellary. This may be compared with the plan of flower typical for Cruciferae. Numerous variations from this plan arise by multiplication of the median stamens only or of both median and lateral, and by conversion of one or more stamens into petaloid staminodes; or suppression may occur. Thus in *Cleome spinosa* (fig. 80, B) the horizontal plan of flower resembles that of Cruciferae, but the stamens are never tetradynamous in Capparidaceae. The numerous stamens of *Capparis* arise by serial as well as lateral splitting of the four rudiments. This type of flower is regarded by Čelakovský as the more primitive, a similar reduction having taken place in this family to what has occurred in Papaveraceae.

Another characteristic feature is the development of an internode between petals and stamens (androgynophore) or especially between stamens and pistil (gynophore) (fig. 81, B), the pistil often projecting considerably beyond

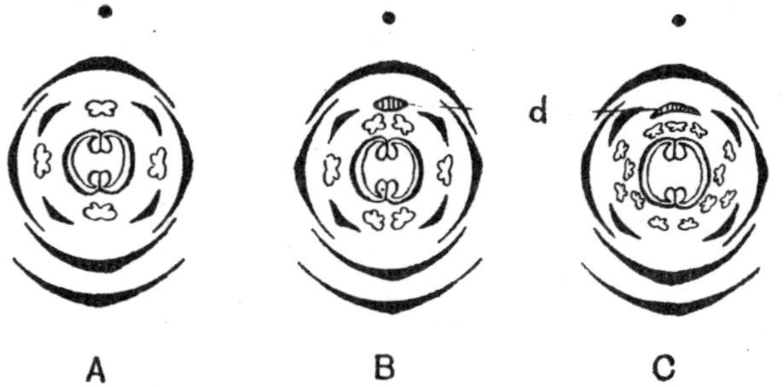

FIG. 80. Floral diagrams of A, *Cleome tetrandra*; B, *C. spinosa*; C, *Polanisia graveolens*; *d*, disc. (After Eichler and Pax.)

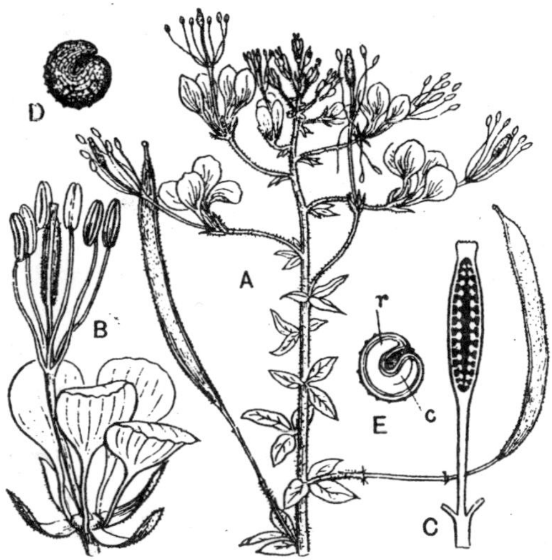

FIG. 81. *Gynandropsis pentaphylla*. A. Portion of flowering branch, × ⅔. B. Flower, × 3. C. Ovary cut lengthwise, × 4. D. Seed, × 7. E. Seed cut lengthwise, × 7; *c*, cotyledons; *r*, radicle. (After A. Gray.) (From *Flor. Jam.*)

the flower. Lateral outgrowths of the axis are also frequent in the form of swellings, discs or tubular structures, generally in a median position, either inside or outside the corolla (fig. 80, B, C); these disc-formations suggest affinity with the Resedaceae. The various deviations from the tetramerous ground plan result generally in the formation of a medianly zygomorphic flower; the zygomorphy is sometimes emphasised by unequal development of members as in *Pteropetalum*, where the posterior petals are much larger than the anterior, or in many species of *Capparis* where the hinder sepal forms a hood-like structure; in *Emblingia* (West Australia) the sepals are united below forming a four-lobed calyx, while the two posterior petals form a large hood protecting the stamens and pistil. The sepals are generally free but sometimes more or less united below; in a few small genera the sepals do not separate when the flower opens but the "capsular" calyx splits lengthwise along one or both sides or transversely like a pyxidium[1]. Except in the anomalous genus *Emblingia* the petals are free. In several small genera the petals are wanting.

The pistil is generally carried up on the long gynophore, and consists of two, sometimes several, carpels forming a one-chambered ovary which may become incompletely multilocular by the ingrowth of the parietal placentas. The ovules are numerous and campylotropous. The fruit in *Cleome*, *Polanisia* (fig. 82) and allied genera is siliquose, dehiscing as in Cruciferae by two valves, the seeds remaining attached to a replum. In *Capparis* and allies it is a berry (fig. 83) either more or less spherical or elongated. Nut-like and drupaceous fruits also occur. The kidney-shaped seeds are exalbuminous and contain a large variously folded embryo.

Pollination is effected by insect-aid; secretion of nectar has been observed on the disc-formations, and dichogamy has been noted in many cases, the anthers dehiscing before the stigmas are receptive.

The plants occur mostly in dry districts; associated with this habit is the frequent hairy or scaly covering of the leaves. Several shrubby species of *Capparis* occur in the Mediterranean region; in *C. spinosa* the stipules are represented by spines; the flower-buds of this species are the "capers" of economic use.

[1] Briquet, J. "La déhiscence des calices capsulaires chez les Capparidacées." *Arch. d. Sciences phys. et naturelles*, Genève, sér. 4, XXXVI, 534 (1913).

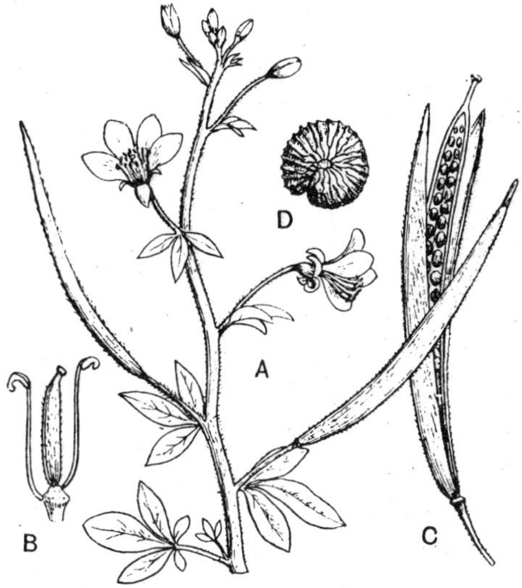

FIG. 82. *Polanisia viscosa*. A. Portion of flowering branch, nat. size. B. Flower with sepals, petals and most of the stamens removed, × 2. C. Ripe fruit, nat. size. D. Seed, enlarged. (From *Flor. Jam.*)

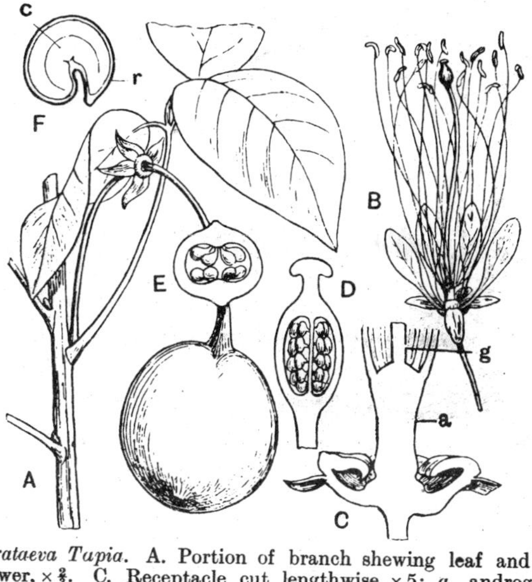

FIG. 83. *Crataeva Tapia*. A. Portion of branch shewing leaf and fruit, × ⅔. B. Flower, × ⅔. C. Receptacle cut lengthwise, × 5; *a*, androgynophore; *g*, gynophore. D. Ovary cut lengthwise, × 5. E. Ovary cut across, × 6. F. Seed cut lengthwise, × 1⅓; *c*, cotyledons; *r*, radicle. (From *Flor. Jam.*)

Family III. CRUCIFERAE

Flowers bisexual, regular. Sepals in two alternating dimerous whorls; petals four, diagonally placed. Stamens, tetradynamous—an outer pair in the transverse plane alternating with two longer pairs in the median plane produced by branching from a single origin. Carpels two, transverse. Ovary bilocular by development of a septum uniting the two parietal placentas; ovules generally numerous, anatropous or campylotropous. Stigmas generally commissural. Fruit usually a siliqua. Embryo filling the seed.

Generally annual or perennial herbs, rarely woody, with alternate leaves, and usually one-celled, simple or branched hairs. Flowers in racemes, generally without bracts or bracteoles.

Genera about 200; species about 2000. Mainly in the temperate and frigid zones of the northern hemisphere.

Though a world-wide family, Cruciferae resemble Papaveraceae in having their chief development in the northern hemisphere and especially in the Mediterranean region; also in being generally herbaceous plants, with scattered exstipulate leaves. Many are annuals; such for instance are some of the commonest weeds of cultivation, Charlock (*Brassica Sinapis*), Black Mustard (*B. nigra*) and Shepherd's Purse (*Capsella Bursa-pastoris*). Others are biennials, producing a rosette of radical-leaves in the first year, and in the second sending up a flowering shoot at the expense of the nourishment stored in the tap-root. Under cultivation this root may be induced to swell, as in the Turnip, Rape, Swede and others. Wallflower (*Cheiranthus Cheiri*) and Scurvy-grass (*Cochlearia*) are examples of perennials.

Special means of vegetative propagation occur by bulbils formed in the axils of the upper leaves in *Dentaria bulbifera*, Coral-root, so called from the form of the creeping scale-bearing root-stock, and on the leaves of *Cardamine pratensis*.

The inflorescence is a raceme which is indefinite (that is, does not bear a terminal flower), and on which both bracts and bracteoles are almost invariably suppressed. Bracts are occasionally present and the presence of bracteoles may sometimes be traced in young flowers, but evidence for the existence of the latter at some time in the history of the family is found in the position of the outer whorl of sepals which are median and not lateral as we should expect them

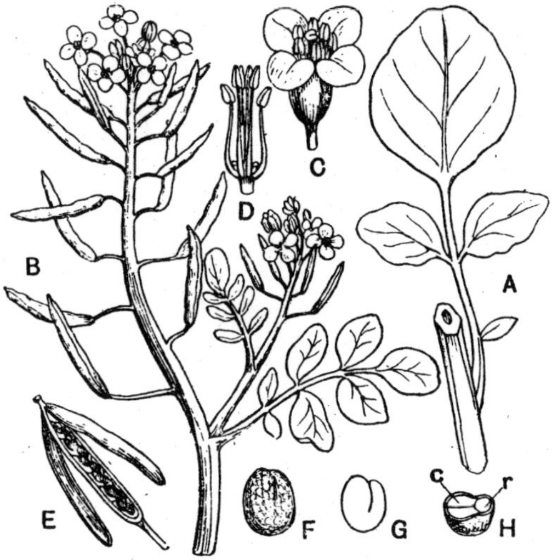

FIG. 84. *Nasturtium fontanum.* A. Leaf from base of plant, × ⅔. B. Upper portion of flowering branch, × ⅔. C. Flower, × 4. D. Stamens and receptacle, × 4. E. Siliqua slightly enlarged. F. Seed, × 8. G. Embryo, × 8. H. Seed cut across, × 8; *c*, cotyledons; *r*, radicle. (From *Flor. Jam.*)

to be in the absence of bracteoles. Recently E. R. Saunders[1], working with the Stock (*Matthiola incana*), has produced evidence for the view that it is only the free exserted part of the bract which has been suppressed, the basal extension being still formed and clothing the axis with a "leaf-skin" in the same manner as if the region above the exsertion level had attained full development. Another characteristic of the Crucifer raceme is its corymbose form in the flowering stage.

FIG. 85. Floral diagram of a Crucifer. (After Eichler.)

The flower bears a close relation to that of the preceding family, as a comparison of the floral diagrams indicates.

[1] *New Phytologist*, XXII, 150 (1923).

The regular flowers are bisexual and hypogynous, with four free sepals in two whorls, median and transverse, and four free diagonally placed petals. Whereas in Capparidaceae the stamens shewed great variety in number and arrangement, in Cruciferae one phase has become remarkably constant, namely a pair of lateral stamens with shorter filaments and four median stamens with longer filaments, which, though generally distinct, may occasionally be seen to arise from a single protuberance and may even remain more or less united in the adult flower. This tetradynamous character of the androecium was selected by Linnaeus (it had been previously noted by John Ray[1]) as the diagnostic feature of one of his 24 classes corresponding with our family Cruciferae. The pistil consists of two transversely placed carpels, bearing a double row of pendulous campylotropous ovules on the parietal placenta formed at the two sutures; the ovary becomes bilocular by the subsequent production of a septum between the two sutures: The stigma is knob-like or two-lobed, the lobes generally standing in the median plane, that is, over the line of union of the carpels, but in *Matthiola* (Stock) and *Moricandia* situated above the midrib. The sepals are sometimes saccate at the base, especially the two inner; the sacs correspond with the nectaries developed on the floral axis and serve as nectar-containing pouches. The sepals in the mature flower are relatively short and spreading, or relatively long and narrow, overlapping at the edges and forming a tube in which the slender claws of the petals are supported; the petals have a longer or shorter stalk or claw according as the sepals are closed or open. At the base of the stamens are nectar-secreting outgrowths of the floral axis which vary in form and number in different genera. They occur constantly on each side of the lateral stamens either as distinct outgrowths or joining and more or less completely enveloping the base of the filament. In addition smaller glands may occur between the median stamens and these may extend and unite with the glands surrounding the lateral stamens.

Departures from the normal type of flower occur. In

[1] *Historia Plantarum*, 1686–1704.

Iberis (Candytuft) and *Teesdalia* the flower becomes zygomorphic by the enlargement of the two outer petals; not infrequently the petals are very small or may be altogether absent, as in species of *Lepidium*, *Coronopus* (fig. 87, B), *Nasturtium* and others; in *Capsella Bursa-pastoris* the petals are sometimes replaced by four stamens. Reduction in the number of stamens occurs; thus in *Cardamine hirsuta* (fig. 86) the lateral stamens are generally absent, in species of *Lepidium* and *Coronopus* (fig. 87) doubling of the median stamens may also fail to take place, and the flower becomes

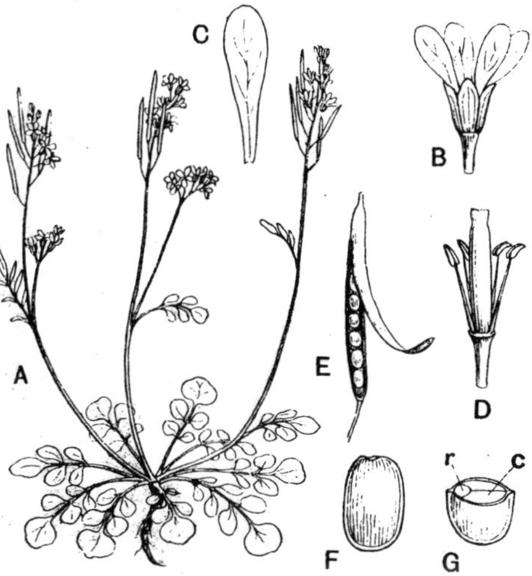

FIG. 86. *Cardamine hirsuta*. A. Plant, × ⅔. B. Flower, × 4. C. Petal, × 6. D. Pistil and stamens, × 5. E. Siliqua slightly enlarged. F. Seed, × 10. G. Seed cut across, × 10; *c*, cotyledons; *r*, radicle. (From *Flor. Jam.*)

diandrous. On the other hand, in *Megacarpaea* (fig. 90, F) the stamens are numerous—up to 16. Three or four fertile carpels occur abnormally in a considerable number of genera. The genera *Tetrapoma* and *Holargidium*, characterised by a four- (or more-) carpelled ovary, have since been recognised as abnormal forms of *Nasturtium* and *Draba* respectively. A tricarpellary form of *Lepidium sativum* (Cress) was found to transmit the abnormality by seed. Multicarpellary pistils have also been recorded in abnormal flowers in a considerable number of genera.

Several points call for comment in the flower of Cruciferae as described above, and there is an extensive literature on the subject. The flower is remarkably uniform, although, as pointed out, certain deviations occur, some of which are constant while others occur in abnormal flowers. The most obvious feature in the flower as described above is the corolla with its four diagonally placed petals which disturb the apparent dimerous character (S 2 + 2, C × 4, A 2 + 2, G (2)). The simplest explanation would be to regard the four petals as derived by splitting from two median members, in the same way as the two pairs of longer stamens are regarded as representing the division of two simple primordia. But there is no evidence for this in the floral development, and if this view be accepted we must regard it as a change in the arrangement of the parts of the flower all trace of which has been lost in the development of the flower at the present day. The arrangement in various abnormal flowers has been adduced in support of this view. Thus Velenovsky has described a flower of *Arabis alpina* with a dimerous corolla of two median petals, and Benecke an abnormal flower of *Eschscholtzia* (in Papaveraceae) in which the pair of median petals have become divided and transferred to the diagonal position, resembling precisely the normal corolla of Cruciferae. On the other hand, it has been maintained, especially by Wretschko, that the inner whorl of stamens represents four distinct diagonally placed members, thus corresponding in position with the four petals. Kunth, who accepted Lindley's interpretation of the ovary, regarded the flower as tetramerous throughout; the median members of the outer whorl of stamens having been suppressed.

The pistil also presents difficulties in the almost universal position of the stigmas above the commissure, or the vertical line of tissue at the juncture of the two carpels, and in the presence of the false septum. In *Matthiola* (Stock) and *Moricandia* the stigmas occupy the usual position above the midrib of the carpel. Commissural stigmas are common in Papaveraceae and occur also in Ericales and elsewhere. Lindley regarded the commissure as representing a true carpel on which the ovule-bearing placentas and stigmas were borne; the pistil on this view consists of two pairs of carpels, a transverse barren pair which ultimately separate as the pair of valves in the fruit, and a median pair forming the replum and false septum and bearing the seeds. As has been pointed out, four carpellary pistils are sometimes produced in abnormal flowers. Recently this view has been revived by E. R. Saunders in explanation of a series of abnormal developments of the pistil of the Stock (*Matthiola incana*). This author regards the commissures as representing consolidated carpels, and maintains that this view explains the development of the false septum which is due to a rearrangement of the tissues and an alteration of the direction of growth of the consolidated carpel; but the position of the ovules at the outer edges of the septum presents a serious difficulty which is not explained by the author. Similarly in the other families of the order the commissure is regarded as representing a solid carpel on which the placentas are borne while the apparently normal carpels, which alternate with the solid ones, are barren.

Both cross- and self-pollination of the flowers occur. The petals, which are generally white or yellow, more rarely lilac or otherwise

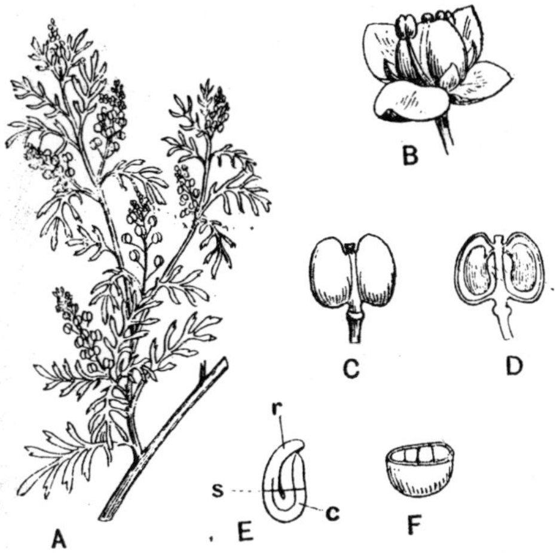

FIG. 87. *Coronopus didymus*. A. Portion of branch, × $\frac{2}{3}$. B. Flower (with four small petals and two stamens), × 16. C. Silicula, × 16. D. Silicula cut lengthwise, × 16. E. Embryo, × 30; *c*, cotyledons; *r*, radicle. F. Seed cut across where the line *s* is drawn in E, × 30. (From *Flor. Jam.*)

coloured, do not attract a great number or variety of insects. All or some of the anthers become twisted so that nectar-seeking insects will touch them with one side of their head and the stigma with the other. The capitate stigma is, however, in such close proximity to the anthers that very slight irregularity in the movements of the visitor will cause self-pollination, and this may also often occur by the actual dropping of pollen from the anthers of the larger stamens on to the stigma. Cleistogamic flowers occur in *Cardamine chenopodifolia*, and in *Subularia aquatica* when the water is unusually high, the flowers remaining submerged and closed. *Pringlea* (Kerguelen Island Cabbage) has become anemophilous in the absence of insects suitable for its pollination.

Hermann Müller, in reviewing the pollination-methods of the family, says:

"On the whole, Crucifers are far behind Umbellifers in the number and variety of their insect-visitors, both on account of their less conspicuous flowers and their less accessible nectar; and not rarely plants remain altogether unvisited. The possibility of self-fertilisation is useful if not necessary for the preservation of all the Crucifers that we have considered; in many we find that self-fertilisation takes place to a very considerable extent; and in several we have experimental evidence that it is productive of seed. Under these circumstances it would be better for the plant to forego attaining in its anthers the most favourable position possible for dusting nectar-seeking insects with pollen, if by doing so, while retaining the chance of cross-fertilisation if insects did come, it could fully insure self-fertilisation if they did not."

"The way in which conspicuousness is attained throughout the Umbelliferae by association of many flowers in one surface, and by asymmetrical development of florets for the common good, is exemplified only in isolated genera of Cruciferae (*Teesdalia*, *Iberis*); and, in *Teesdalia* at least, it does not so far insure cross-fertilisation that self-fertilisation may be dispensed with."

The fruit is a pod dehiscing in the great majority of cases by two valves separating from below upwards and leaving the placentas with the seeds attached to the replum, or framework of the septum. The pod is termed a *siliqua*, or when scarcely longer than broad, as in Shepherd's Purse, a *silicula*. There is some variation in the form of the pod and the arrangement of the seeds. In the silicula the seeds are generally in two rows in each chamber, and this holds also in many siliquas, as of *Diplotaxis*; but generally the seeds of the opposite placentas alternate, forming a single row in each chamber, and alternation may also occur between the seeds in each chamber so that the whole pod contains only one row, as in *Raphanus* (Radish). In the last case constrictions may be formed between the seeds, or transverse walls forming one-seeded segments, and the fruit may separate when ripe into one-seeded joints as in Wild Radish (*R. Raphanistrum*) (fig. 90, D). In *Cakile* the siliqua separates when ripe into two one-seeded indehiscent joints (fig. 89, D). In *Crambe* the pod has only two joints, the lower of which is barren, forming a short thick stalk, while the upper is inde-

FIG. 88. *Capsella Bursa-pastoris*. A. Plant, × ⅓. B. Bud, × 6. C. Flower, × 6. D. Pistil and stamens, × 12. E. Silicula, × 2. F. Silicula with one valve removed, × 2. G. Seed, × 8. H. Seed cut across, × 8; *c*, cotyledon; *r*, radicle. (After Sturm.) (From *Flor. Jam.*)

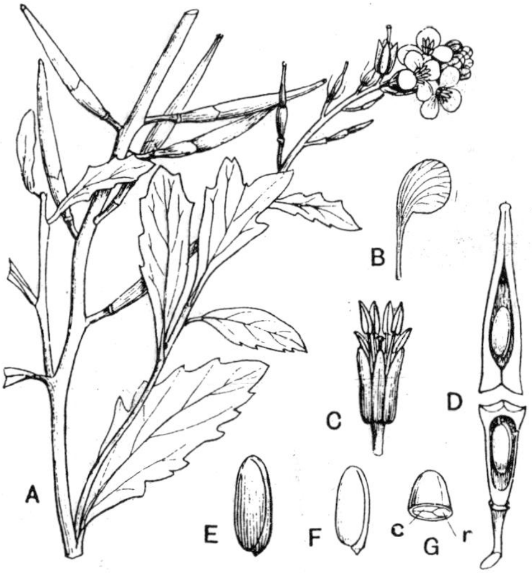

FIG. 89. *Cakile lanceolata*. A. Portion of plant, × ⅔. B. Petal, × 2. C. Flower without the petals, × 2. D. Fruit with the joints separated, slightly enlarged. The seed in the upper joint is erect from the base, in the lower pendulous. E. Seed, × 2. F. Embryo, × 2. G. Seed cut across, × 2; *c*, cotyledons; *r*, radicle. (A–D after Delessert.) (From *Flor. Jam.*)

hiscent, globose, and contains a single seed (fig. 90, C); *C. maritima* (Sea-kale) is a rare plant in Britain on sandy and shingly sea-coasts. *Isatis* has also indehiscent one-seeded pods; *I. tinctoria* (Woad) grows wild on cliffs by the Severn. Another variation in the structure of the pod is governed by the amount or direction of compression. It may for instance be flattened laterally, in which case the septum is very broad, as in *Lunaria* (Honesty) (fig. 90, E), or from front to back when

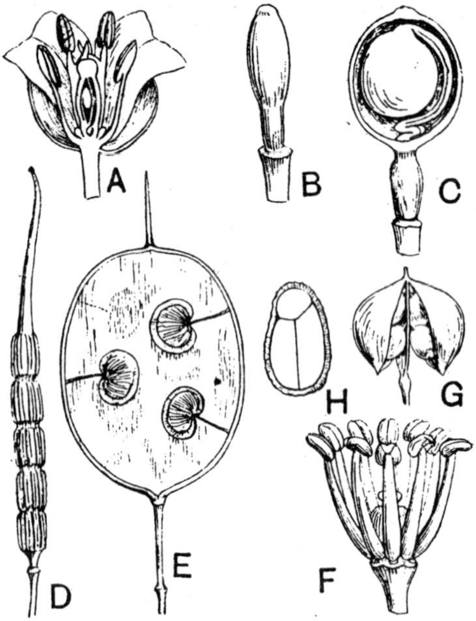

FIG. 90. *Crambe maritima*. A. Flower in vertical section, × 2. B. Young fruit. C. Ripe fruit, the upper joint cut open, exposing the seed. D. Fruit of *Raphanus Raphanistrum*. E. Fruit of *Lunaria*, the valve removed, shewing the seeds attached to the replum. F. Flower of *Megacarpaea*, after removal of sepals and petals, enlarged. G. *Cochlearia*, opening silicule, enlarged. H. Seed of same cut across, much enlarged. (A, F, G, H after Le Maout and Decaisne; D, E after Baillon.)

the septum becomes very narrow, as in Shepherd's Purse (fig. 88, F). On dehiscence of the fruit the seeds remain lightly attached to the replum and lie exposed on the thinly membranous septum; they are readily detached when the long dry stalk of the inflorescence is shaken by the wind. The seeds themselves are sometimes flat or even winged.

In *Morisia*, a small acaulescent herb, native of Sardinia and Corsica, the peduncle bends down after flowering and buries the closed pod in the ground. In *Anastatica* (Rose of Jericho), a native of the eastern Mediterranean region, the small plant, which is an annual, dries up at the end of the season and becomes detached from the soil; the branches bend together, protecting the numerous small fruits, and the whole plant is rolled along the ground by the wind.

The embryo is large, completely filling the seed. The relative position of radicle and cotyledons and the absence or degree of folding of the latter in the seed supply characters which have been used for distinguishing genera and groups of genera. The radicle with the hypocotyl may be bent upwards and lie along the edge of the cotyledons, o|| (radicle accumbent, embryo pleurorhizal) as in *Cheiranthus*, *Nasturtium*, *Matthiola*, or parallel to the face of the cotyledons, o= (radicle incumbent, embryo notorhizal) as in *Capsella* or *Sisymbrium*. In the cases above mentioned the cotyledons are flat, but they may be folded along the midrib as in *Brassica*, *Raphanus*, *Crambe*, o>> (embryo orthoplocal), or rolled so that a transverse section of the seed cuts them twice, o|| ||, as in *Coronopus* or *Bunias* (embryo spirolobal), or so folded as to be cut several times as in *Subularia*, o|| || || (diplocolobal).

In germination the cotyledons are epigeal, appearing above ground as the first green leaves of the plant; they are often notched at the apex as in Mustard.

Cruciferae contains more than 200 genera and 2000 or more species. Its distribution is world-wide, but the greater number are confined to the north temperate zone, and the family is specially developed in the Mediterranean region. Though remarkably distinct and readily characterised as a family, the subdivision into smaller groups is very difficult, and very various arrangements have been suggested.

Linnaeus seized on the marked difference between the pods of different genera to divide his class Tetradynamia into two orders, *Siliculosae* (fruit a silicula), and *Siliquosae* (fruit a siliqua). This was extended by De Candolle, and the subdivision of the family adopted in Hooker's *Student's Flora* is based upon this system. The proportions and mode of dehiscence of the pods supply characters for five large groups, and the relative positions of the radicle and cotyledons characters for further subdivision. The British genera are arranged as follows:

Group I. SILIQUOSAE. Has a siliqua and includes three tribes:

Tribe I. *Arabideae*. Includes *Cheiranthus* (*C. Cheiri*, Wall-flower), *Nasturtium* (*N. officinale*, Water-cress), *Barbarea* (Winter-cress), *Arabis* (Rock-cress), *Cardamine* (Bitter-cress) and *Matthiola* (Stock). In these the radicle is accumbent.

Tribe II. *Sisymbrieae*. Includes *Sisymbrium* (Hedge-mustard) and *Erysimum*; the radicle is incumbent.

Tribe III. *Brassiceae*. Includes *Brassica* (Cabbage, Mustard) and *Diplotaxis*. The arrangement of the embryo differs from that in Tribe II in having the cotyledons not flat, but folded along the midrib.

Group II. SILICULOSAE LATISEPTAE. Has a silicula with a broad septum.

Tribe IV. *Alyssineae*. Includes *Draba*, *Erophila*, *Alyssum* and *Cochlearia* (Scurvy-grass); also *Lunaria* (Honesty) with a very broad flat pod. The radicle is accumbent, and the cotyledons are flat

Tribe V. *Camelineae*. Radicle incumbent. Includes *Subularia aquatica* (Awl-wort), a submerged water-plant which is exceptional in the family in having, like *Eschscholtzia* in Papaveraceae, a perigynous flower; and *Camelina* (Gold-of-Pleasure).

Group III. SILICULOSAE ANGUSTISEPTAE. Has the silicula flattened at right angles to the plane of the septum, so that the latter becomes very narrow.

Tribe VI. *Lepidieae*. Includes *Capsella*, *Coronopus* and *Lepidium* (Cress). Radicle generally incumbent.

Tribe VII. *Thlaspideae*. Includes *Thlaspi* (Penny-cress), *Iberis* (Candytuft) and *Teesdalia*. Radicle accumbent.

Group IV. NUCUMENTACEAE. Is characterised by the reduction of the pod to a short, one-celled, one-seeded indehiscent fruit or nut.

Tribe VIII. *Isatideae*. Includes *Isatis* (Woad) and *Bunias*.

Group V. LOMENTACEAE. Has the pod constricted between the seeds or divided by transverse walls into one-seeded segments.

Tribe IX. *Cakileae*. Pod indehiscent, two-jointed. *Cakile* (Sea Rocket) (fig. 89), and *Crambe* (Sea-kale) (fig. 90, A–C).

Tribe X. *Raphaneae*. *Raphanus*. Pod constricted into one-seeded joints (fig. 90, D), sometimes few or, as in *R. sativus*, only one, which is very large.

The arrangement suggested by Pomel includes three main groups:

A. PLATYLOBEAE. Cotyledons flat; radicle accumbent or incumbent.

B. ORTHOPLOCEAE. Cotyledons longitudinally folded; radicle in the channel formed by the cotyledon.

C. PLEUROPLOCEAE. Cotyledons rolled or transversely folded. A subdivision of these main groups into tribes is based on the character of the fruit.

The arrangement adopted by Prantl in the *Pflanzenfamilien* depends primarily upon the presence or absence and character when present of the hairs; the family falls into two great groups:

A. Hairs unbranched or absent; no glandular hairs.

B. Hairs more or less branched, rarely unbranched or absent; glandular hairs also sometimes present.

Each group is divided into two tribes according as the stigma is equally developed all round or more strongly developed above the placentas; differences in the development of the style are also considered. For the smaller subdivisions the position of the cotyledons, the number of nectaries in the flower, the character of the fruit and the character of the surface-cells of its septum are taken into account.

In other arrangements special stress is laid on the form and distribution of the nectaries in the flower, as by Hayek, or, as by Schweidler, on the distribution of myrosin-cells in the tissues.

REFERENCES

The following are some of the more important works in which the floral structure of the family is discussed.

LINDLEY, J. *Botanical Register*, XIV, under Plate 1168 (*Eschscholtzia californica*) (1828).

KUNTH, C. S. "Ueber die Blüthen und Fruchtbildung der Cruciferen." *Physikal. Abhandl. d. K. Akad. Wissensch.* Berlin, 1832, 33 (1834).

EICHLER, A. W. "Ueber den Blütenbau der Fumariaceen und Cruciferen." *Flora*, 497 (1865); and *Blütendiagramme*, II, 192 (1878).

BENECKE, F. "Zur Kenntniss des Diagramms der Papaveraceae u. Rhoeadineae." Engler's *Botan. Jahrb.* II, 373 (1882).

POMEL, A. "Contribution à la classification méthodique des Crucifères." *Thèse, Faculté des Sciences*, Paris, 1883.

ČELAKOVSKÝ, L. "Das Reductionsgesetz der Blüthen, etc." *Sitzungsber. K. Böhm. Gesell. Wissensch.* no. 3 (1894).

SCHWEIDLER, J. H. *Bericht. Deutsch. Bot. Gesell.* XXIII, 274 (1905).

HAYEK, A. von. "Entwurf eines Cruciferen-Systems auf phylogenetischer Grundlage." *Botan. Centralbl. Beiheft.* XXVII, 127 (1911).

SAUNDERS, E. R. *Annals of Botany*, XXXVIII, 451 (1923).

Family IV. RESEDACEAE

A small family containing only 6 genera with about 70 species, of which between 50 and 60 belong to *Reseda*. The Mediterranean region is the great centre from which the family spreads eastwards through Persia to India, and southwards to the mountains of Abyssinia and Somaliland, reappearing at the Cape where it is represented by four endemic species of one genus, *Oligomeris*, a fifth species of which has a widely extended range from the Canary Islands through Northern Africa to Northern India and occurs also in California and New Mexico. A few species of *Reseda* range northwards, and the limit in this direction is reached in our two native species, *R. Luteola* and *R. lutea*, in Central Russia, North Germany and Scotland. Resedaceae are mostly herbs with alternate leaves provided with small glandular stipules. Their habit varies greatly according to the habitat. In the dry, hot districts of the Mediterranean region the leaves are small and caducous; in a damper climate they are large and well developed. The inflorescence is a raceme or spike, often, as in Cruciferae, ebracteate. The flowers are irregular, being zygomorphic in the median plane owing to a development of the axis, comparable to the gynophore of Capparidaceae, between the corolla and stamens, which becomes very pronounced on the posterior side of the flower, forming a characteristic disc. Associated with this, the posterior petals are larger and of a more complicated structure, while the stamens, which are inserted on the disc, become crowded towards the front, the few posterior ones being often shorter. Pentamery occurs only exceptionally, the number of sepals and petals generally varying between four and eight, and the stamens from 3 to 40. There are two to six carpels, rarely free, generally united into a lobed one-chambered ovary which does not become closed in at the top. The numerous ovules, which are more or less bent, are borne on two to six parietal placentas. The fruit is a capsule open at the top, and contains numerous small kidney-shaped seeds which are completely filled by the curved embryo. In *Ochradenus* the flowers are sometimes unisexual by abortion of stamens or pistil, and are apetalous; the fruit is also closed and berry-like; it contains a few species in the dry country of north-eastern Africa and Western Asia.

The cultivated Mignonette (*Reseda odorata*) will illustrate the functions played by the different parts in the process of pollination. Nectar is secreted by the smooth under-surface of the large posterior development of the disc, which from its yellow colour and conspicuous position serves as a guide to the nectar. The

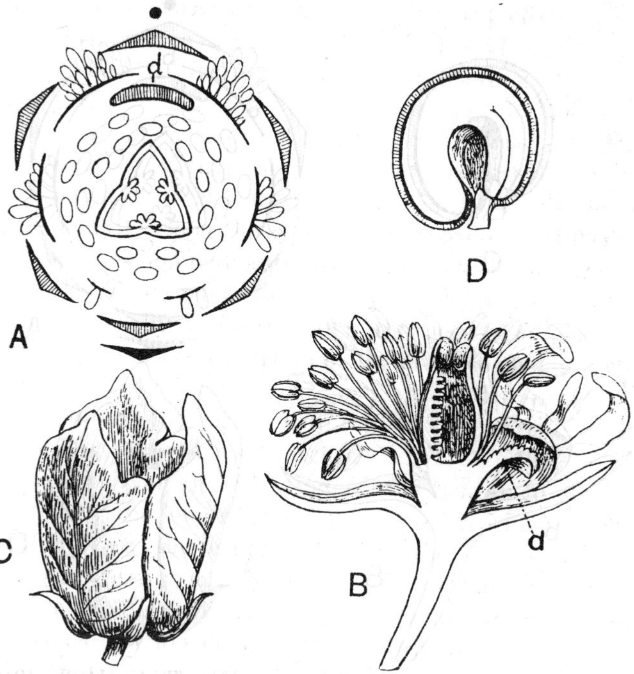

FIG. 91. *Reseda*. A. Floral diagram of *R. odorata*. B. Flower of *R. lutea* in vertical section. C. Fruit, D. Seed of *R. odorata*. *d*, disc. B, C, D enlarged. (A after Eichler; B, C after Baillon; D after Schnizlein.)

large expanded claws of the upper and median pairs of petals surround and protect the nectar-secreting surface, while their white fimbriated laminae, the red anthers and the strong scent serve to attract visitors. The stamens are originally bent downwards towards the front of the flower, but rise up towards the nectar-secreting disc as the anthers dehisce. The pistil projects considerably from the middle of the flower and the carpels terminate in papillose stigmas. The latter become readily covered with pollen by nectar-seeking insects which find the ovary a convenient resting-place. If insect-visits fail, self-pollination occurs, as the stigmas lie directly below the dehiscing anthers. Müller, who describes the process, states that plants which were kept protected from insects yielded capsules filled with good seed.

The family is most nearly related to the Capparidaceae, as seen by the parietal placentation, and the development of a gynophore and an excentric disc leading to median zygomorphy of the flower; the dry-country habit is also characteristic of both families. The marked tetramery of Capparidaceae is, however, absent in Resedaceae.

Order 3. *SARRACENIALES*

Flowers bisexual (unisexual in *Nepenthes*), regular, hypogynous; perianth generally distinguished into calyx and corolla. Pistil of three to five syncarpous carpels with parietal or axile placentas and indefinite ovules. Seeds small, endospermic. An order of damp-loving herbaceous plants with generally alternate entire leaves which are more or less modified for the capture of insects.

Family I. SARRACENIACEAE

A family of marsh-loving perennial herbs with pitcher-like leaves including three genera only, *Sarracenia* with seven species in Atlantic North America, *Darlingtonia*, a monotypic genus from California, and *Heliamphora*, with four species in British Guiana and Venezuela.

The leaves are radical and a central scape bears a single terminal flower or a few in a loose raceme. The flowers are large, spirocyclic, bisexual and regular, and shew considerable variation in the number of parts. A frequent arrangement is a calyx of five sepals followed by a corolla of five regularly alternating petals; the stamens are numerous and placed below the three- to five-celled ovary, which contains a large number of anatropous ovules borne on large axile placentas. A distinguishing feature of the genera is the great development of the style which in *Sarracenia* is dilated above into a large umbrella-like structure spread over the stamens and bearing the small stigmas below the apex of each of the five lobes. The fruit is a capsule splitting along the dorsal sutures into as many valves as there are carpels. The numerous seeds are small, containing a copious fleshy endosperm and a small embryo near the hilum. The membranous seed-coat generally forms a wing, which in *Sarracenia* is developed on one side of the seed. The leaves are effective insect-traps. The brilliant colouring of the upper part of the pitcher and the nectar secreted about the mouth serve to attract insects. In the lower part of the interior an area bearing reflexed hairs prevents the escape of insects which

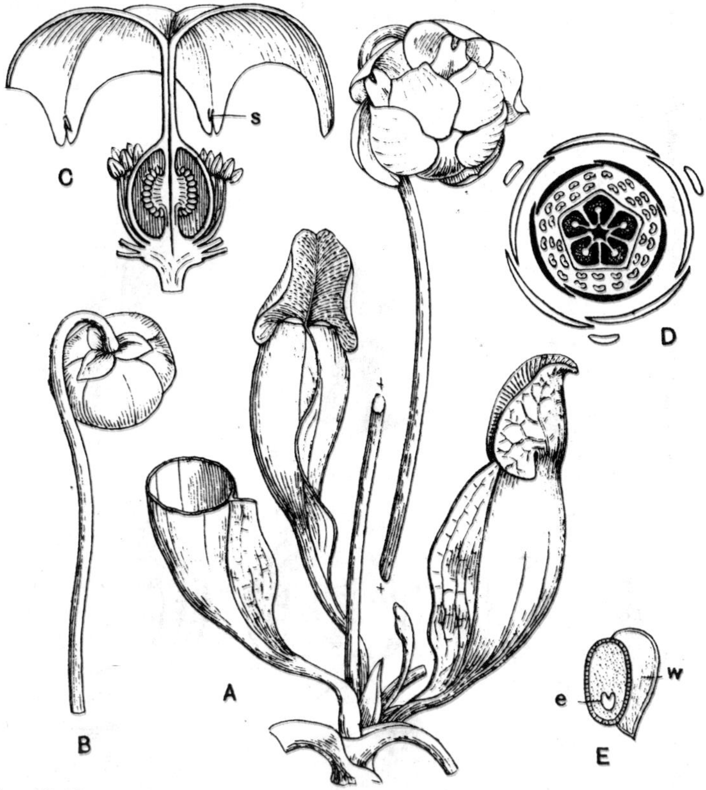

FIG. 92. *Sarracenia purpurea*. A. Portion of plant; the lower leaves have been removed and the left-hand one has been cut across, reduced. B. Flower-bud, reduced. C. Flower cut lengthwise after removal of the sepals and petals, nat. size; *s*, stigma. D. Floral diagram. E. Seed in vertical section, ×4; *e*, embryo; *w*, wing. (A, B after Asa Gray; C, E after Wunschmann.)

have fallen into the pitcher and which ultimately become drowned in the fluid therein secreted. The fluid contains a proteolytic enzyme by which the proteids of the insect are rendered soluble and then absorbed by glands on the pitcher-wall.

REFERENCE

MACFARLANE, J. M. "Sarraceniaceae." *Das Pflanzenreich* (1908).

Family II. NEPENTHACEAE

Contains one genus only, *Nepenthes* (Pitcher-plant), with about 60 species mostly inhabiting the islands of eastern tropical Asia, with its greatest development in Borneo, but extending west to

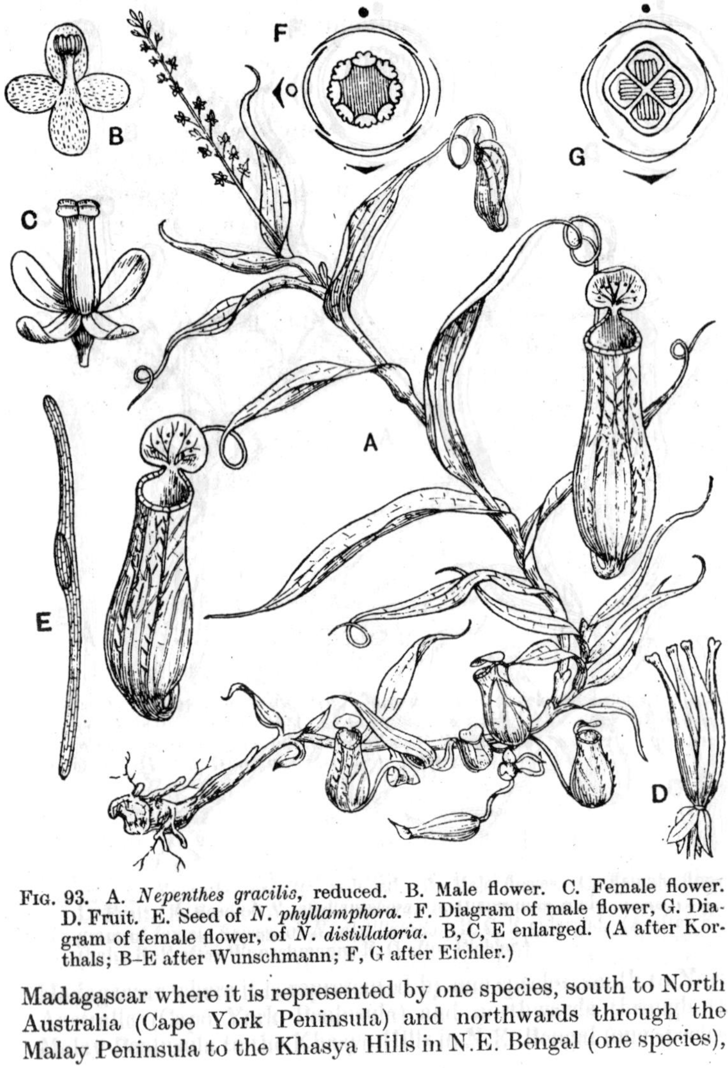

FIG. 93. A. *Nepenthes gracilis,* reduced. B. Male flower. C. Female flower. D. Fruit. E. Seed of *N. phyllamphora.* F. Diagram of male flower, G. Diagram of female flower, of *N. distillatoria.* B, C, E enlarged. (A after Korthals; B–E after Wunschmann; F, G after Eichler.)

Madagascar where it is represented by one species, south to North Australia (Cape York Peninsula) and northwards through the Malay Peninsula to the Khasya Hills in N.E. Bengal (one species), and Southern China. They are shrubby plants, generally climbing by means of the leaves which are alternate, entire and remarkably modified. The most complete have a blade divisible into three parts, the lowest flat and performing the ordinary functions of a foliage-leaf, a median portion slender and tendril-like becoming coiled round suitable objects for support, and a terminal portion which is developed into the familiar pitcher with its lid. The pitcher contains a fluid with digestive properties(1), due to the presence of a proteolytic enzyme secreted by glands lining the lower portion of the wall; the brilliant colouring of the upper portion of the structure and the nectar secreted near and about the rim attract various insects which on falling into the pitcher become drowned and are digested in the fluid, the soluble proteid being ultimately absorbed by the walls of the pitcher.

The flowers are small, regular and dioecious and borne in simple or compound racemes. The perianth consists of two dimerous whorls, succeeded in the male by a varying number of stamens with filaments united into a tube and in the female by a superior four-chambered ovary, containing numerous anatropous ovules on axile placentas. The fruit is an elongated leathery capsule separating into four valves and containing numerous sawdust-like long narrow winged seeds. The fleshy endosperm contains a straight embryo. Many species and hybrids are cultivated in greenhouses.

REFERENCES

1. HEPBURN, J. S. "Biochemical Studies of Insectivorous plants." *Contributions from the Botanical Laboratory, Univ. of Pennsylvania*, IV, 419 (1919).

See also MACFARLANE, J. M. "Nepenthaceae." *Das Pflanzenreich* (1908).

Family III. DROSERACEAE

A small family of insectivorous plants containing four genera with 87 species, 84 of which belong to *Drosera* (Sundew), a genus widely distributed through the temperate and tropical regions of both hemispheres. The other genera have a limited distribution—*Drosophyllum*, with a single species from Morocco to Portugal and South Spain, *Dionaea* (Venus's fly-trap), another monotypic genus, confined to the south-eastern United States and *Aldrovanda* with a single species in Central and Southern Europe, North and East Asia, India (Bengal), and Australia (Queensland).

The great majority are small herbs inhabiting sphagnum-bogs like our British Sundews, or localities which are relatively damp during the growing season. *Aldrovanda* is a rootless swimming water-plant recalling *Utricularia* in habit.

The germination of those species of *Drosera* which have been studied shews absence of a primary root, its place being taken by a protocorm-like development of the hypocotyl bearing long attaching hairs; in *Drosophyllum* and *Dionaea* a primary root is developed. The protocorm is a temporary structure and is replaced by adventitious roots developed from the stem, which, though few in number, are often large and serve as organs of storage as well as of absorption. The method of growth and life-history varies widely, and may be either photophilous (epigeal) or geophilous. In the epigeal, the axis ends in a bud capable of indefinite growth. The species are sometimes short-lived, as in *Drosera indica*, a native of the tropical monsoon region, which grows in a saturated soil and runs through its life-cycle from germination to seeding in one wet season; more generally they are perennial, as in our native *D. rotundifolia*, and adapted to climates which shew a marked periodicity. In this species the stem bears long internodes until it reaches the surface of the sphagnum-layer in which it is growing, the internodes then remain short and a rosette of leaves spreading on the surface is formed. Above the rosette the terminal bud remains enveloped by the stipules. During winter the bud with the decaying leaf-rosette becomes buried by the growing Sphagnum; in spring the terminal bud renews its growth, its small lower leaves becoming separated by long internodes until the surface of the moss is again reached when a new leaf-rosette is formed. The inflorescence springs from the axil of a surface-leaf (fig. 94, A).

The geophilous species are adapted mainly to conditions where a damp winter alternates with a dry season. In a few cases, as in the two Cape species *D. cistiflora* and *D. pauciflora*, the plant persists by means of a swollen root lying close beneath the soil-surface, but generally the persistent organ is a bulb which is buried deeply in the soil (fig. 94, B). This mode of life characterises the large Australian subgenus *Ergaleium*. The portion of the stem below ground bears leaf-structures with a much reduced blade which function as rhizoids; when the surface of the soil is reached either a leaf-rosette is formed, or the stem continues to elongate and the lower leaves are reduced to small scales with no blade-development. In the axil of the upper scale-leaves is formed a pair of functional leaves, while higher on the stem the subtending leaf itself develops a blade. The higher and younger leaves shew a gradual suppression of the two accessory leaves. In species inhabiting cold hard clay soils and similar unfavourable localities, development proceeds no further and the main axis ends in an inflorescence (as in the West Australian *D. microphylla* and *D. Menziesii*), but where soil and climatic conditions are favourable to a more vigorous development a climbing or branching habit is developed. In climbing species the leaf of the first order has a much elongated petiole and the tentacles of its blade form the organs of attachment (fig. 95, B). In branching species the secondary axes may also, like the primary, end in an inflorescence.

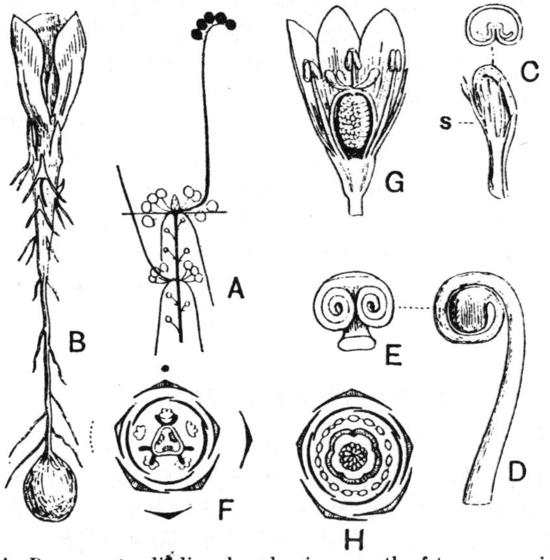

FIG. 94. A. *Drosera rotundifolia*, plan shewing growth of two successive years. The horizontal line indicates the surface of the bog. B. *D. erythrorhiza*, hypogeal portion of stem growing from a bulb and bearing leaves modified to form rhizoids, at the apex a rosette of leaves is formed, × 1½. C. *D. rotundifolia*, young leaf, the blade is bent over against the petiole and is enveloped by the fimbriated stipule, *s*; above is a sectional view of the blade; × 6. D. *D. binata*, lateral view of young leaf, × 6. E. Inside view of same, shewing the coiled halves of the blade. F. Floral diagram of *D. rotundifolia*. G. Vertical section of flower of same. H. Floral diagram of *Dionaea*. (A–E after Diels; F, H after Eichler; G after Le Maout and Decaisne.)

The bulb consists of very closely united leaves, of which only the tips are free and form a peristome-like crown surrounding the base of the hypogeal caulome. The bulb is renewed each year by a lateral bud, developed at the base of the caulome, which grows into the older bulb, gradually absorbs its store of nourishment and finally replaces it, except for the external scales which form a protecting outer coat. When the plant has died down the hypogeal caulome does not completely perish but its dead remains are found clinging round the new hypogeal stem which is developed next season. They form a kind of velamen consisting of long strips of tissue the cell-walls of which have numerous oblique pores, the whole forming a capillary system which holds moisture for the supply of the rhizoids of the new caulome. In some species, as *D. auriculata*, the bud which forms the new bulb develops externally and is carried deeper into the soil at the end of a stout stolon.

Drosophyllum has a short woody, sometimes branched, stem; the long narrow crowded leaves are borne on the upper part of the stem and branches, which below are densely clothed with the persistent remains of earlier leaves; the stem passes above into the branched cymose inflorescence. *Dionaea* has a perennial rhizome; the short stem bears a rosette of leaves and a long scape bearing an umbel-like cyme of flowers.

The degree of differentiation of the leaf is remarkably varied. In some sections of *Drosera* and in *Drosophyllum* the long narrow leaf shews little or no distinction between petiole and blade, but in the great majority a broad blade is plainly distinguished from the petiole. In *Dionaea* and *Aldrovanda* the two halves rise upward on stimulation, the midrib acting as the hinge (fig. 95). Stipules are absent or developed in a less or greater degree. In some cases they are indicated merely by fimbriated outgrowths on the side of the leaf-base. In *Drosera longifolia* and *D. rotundifolia* they form a narrow fimbriated ligule at the base of the petiole analogous to the intravaginal scale in the leaves of Monocotyledons. In a series of South-West Australian xerophilous species which persist through a long dry summer on sandy soil, the stipules form an important intrapetiolar growth by which the young blade is protected in the bud. The degree of development of the petiole and its function also vary widely. On the one hand, as in our British species with leaf-rosettes, or the Australian climbing forms, it serves to place the leaf-blade in a suitable position, while in other cases it assumes, more or less, the function of assimilation, becoming broadly winged, as in a very marked degree in *Dionaea* (fig. 95, A). In *Dionaea* the lower part of the petiole is swollen, containing storage-tissue in which starch is deposited, and the same occurs in *Drosera binata* (Eastern Australia) where the lower part of the petiole persists after the decay of the upper part and the forked blade.

The mode of development of the leaf-blade also varies. In a small southern group of *Drosera* the young blade is folded lengthwise; in others it is folded across the middle and attains its full size by intercalary growth; in others again, as in

D. rotundifolia, the young blade is bent over to lie close against the petiole (fig. 94, C) with margins inrolled. In *Drosophyllum* and those species of *Drosera* with leaves in which long-continued apical growth occurs, the blade is folded circinately (fig. 94, D, E). In the mature leaf, where there is a distinct blade, this is often more or less spathulate or sometimes round; in long-stalked cauline leaves the blade is often peltately inserted on the petiole.

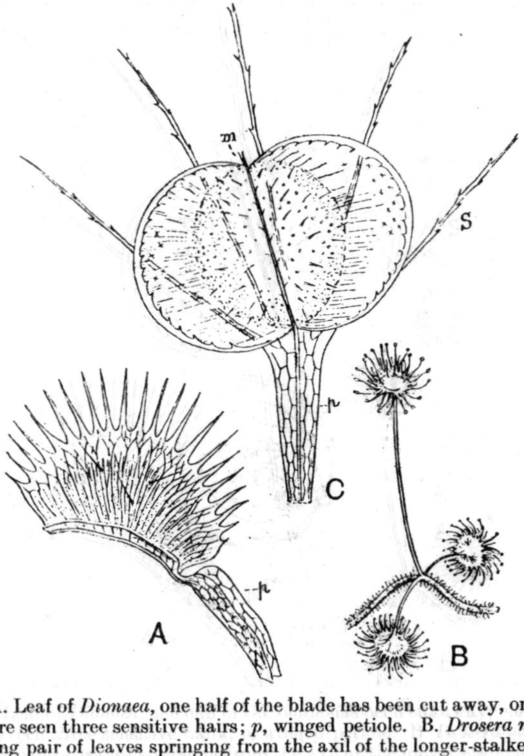

FIG. 95. A. Leaf of *Dionaea*, one half of the blade has been cut away, on the other half are seen three sensitive hairs; *p*, winged petiole. B. *Drosera macrantha*, shewing pair of leaves springing from the axil of the longer-stalked primary leaf. C. Leaf of *Aldrovanda vesiculosa*, enlarged; *p*, petiole; *m*, midrib of leaf; *s*, bristle-like outgrowths below the blade. (A after Sachs; B after Diels; C after Fenner.)

Outgrowths of the leaf in the form of hairs, glands or tentacles are developed in connection with modification for the attraction, capture and digestion of insects. Capitate glandular hairs which function as digestive glands secreting a proteolytic ferment occur on the leaf-surface in all genera. The tentacles which occur on the surface and edge of the leaf of *Drosera* and *Drosophyllum* are complicated outgrowths with a swollen head which bears a secreting layer, exuding a viscid fluid, and contains a mass of tracheids communicating by a cord of tracheids with the vascular tissue of the leaf. Sensitive hairs of a complicated structure occur on the leaf of *Aldrovanda* and *Dionaea* (fig. 95).

In response to mechanical or chemical stimuli the leaves execute movements, consisting in *Drosera* of an incurving of the tentacles towards the stimulated area, and in *Dionaea* and *Aldrovanda* of a closing together of the halves of the lamina. In *Aldrovanda* the outer portions of each half become appressed, the inner portions forming an inflated bladder. For details of these remarkable movements, and subsequent digestion and absorption of nitrogenous food-stuffs, works on physiology should be consulted.

In *Aldrovanda* the flowers are few in number and stand solitary in a leaf-axil; a few species of *Drosera* have a one-flowered scape but generally the flowers are borne in cymose inflorescences on terminal or axillary peduncles. They are bisexual and regular, generally pentamerous or sometimes tetramerous, with often increase in the number of the stamens and reduction in that of the carpels. Fig. 94, F (*Drosera rotundifolia*) represents the most general type of floral structure. An increase in the number of stamens (*Drosophyllum* and *Dionaea* have 10 to 20) is due, according to Payer, to the formation of a second whorl of five, and to doubling, especially in this second, epipetalous, whorl. The pollen-grains generally remain united in tetrads. The most important distinctions occur in the pistil; in place of the three parietal placentas of the *Drosera* type, there is in *Dionaea* (fig. 94, H) a large many-ovuled basal placenta; the form of the styles and stigmas also varies. The ovules are anatropous. The fruit is a many-seeded capsule in *Drosera* which dehisces loculicidally into three to five valves, according to the number of carpels. In *Dionaea* dehiscence is irregular, and in *Aldrovanda* the fruit remains closed, the seeds being set free by decay of the pericarp. The testa often loosely envelops the seed; the short straight embryo lies at the base of the copious oily endosperm.

There is some difference of opinion as to the position of this family. The grouping in one order, Sarraceniales, of the three families Sarraceniaceae, Nepenthaceae and Droseraceae is adopted by Engler, and also by Hallier, who, however, includes the first two in one family (together with Cephalotaceae). Hallier derives the order from the *Helleboreae* in Ranales. Diels, the monographer of the family in the *Pflanzenreich*, relying mainly on the parietal placentation of *Drosera*, regards the family as a member of the next order, Parietales, most nearly allied to Violaceae, as also does Wettstein.

Order 4. *PARIETALES*

Flowers regular or zygomorphic, generally bisexual with distinct pentamerous calyx and corolla. Stamens as many as the petals or more. Pistil most frequently of three united carpels, superior or more or less sunk in the floral axis, containing numerous ovules on parietal placentas; ovule with two integuments (one in Loasaceae). Seeds generally with endosperm. Herbs or woody plants with opposite or alternate generally stipulate leaves.

The families form several groups, the relationship between which is not clear. The parietal placentation is a constant character. The first five, Cistaceae, Bixaceae, Tamaricaceae, Frankeniaceae and Elatinaceae, have a starchy endosperm, and regular flowers with free petals and a superior ovary; in the first two the stamens are indefinite, in the other three they are whorled and definite, or if indefinite are arranged in bundles. The remaining families, Violaceae, Flacourtiaceae, Passifloraceae, Caricaceae and Loasaceae, have an oil- and proteid-containing endosperm. Violaceae are characterised by the pentamerous regular or medianly zygomorphic flowers with reduction to three in the gynoecium. The flowers of Flacourtiaceae shew great diversity of structure; the sepals are sometimes united below, the petals are free, the stamens usually indefinite and the ovary generally superior; disc-like developments of the floral axis occur in great variety. In the allied family Passifloraceae these developments take the form of the characteristic corona and the perianth is perigynous. The relationship of the last two families is doubtful; in Loasaceae the ovary is more or less inferior; in Caricaceae the petals unite to form a longer or shorter tube.

Family I. CISTACEAE

A small family (160 species in 7 genera) of herbaceous or shrubby plants with generally opposite stipulate or exstipulate leaves and regular flowers with indefinite stamens. The stamens are borne on an elongated and often disc-like growth of the floral axis beneath the pistil. Their development is basipetal, the oldest being found just below the pistil (fig. 96, A). This is explained as the result of an intercalary growth—a zone of growth is intercalated on the floral axis below the pistil and the stamens are formed on it in basipetal order. There are five sepals, the two outer generally smaller (fig. 96, B) and sometimes suppressed, and five, sometimes three, showy white or coloured caducous petals; reduction to three may take place in the gynoecium, as in *Helianthemum*, or there are five or even ten carpels as in *Cistus*.

The parietal placentas bear two to numerous more or less orthotropous, rarely anatropous, ovules on well-developed funicles; the placentas may be carried inwards by the growth of the edges of the carpels, the ovary becoming more or less completely three- or five-celled. The flowers are solitary or in cymose raceme-like inflorescences. Pollination is effected by aid of insects which collect the pollen; self-pollination also occurs. Cleistogamic flowers are formed in several species of *Helianthemum* and *Cistus*.

The fruit is a leathery or woody capsule, which splits into valves along the middle line of the carpels (loculicidally). The seeds are small, often angular by compression and frequently have a rough surface. The embryo is usually curved or coiled in a mealy or cartilaginous endosperm (fig. 96, E).

R. Gaume[1] has studied the germination, development and anatomy in several species of the family and finds that they shew great uniformity of structure. All the seeds, especially of perennial species, germinate quickly and easily. There are two types of seedlings, the *Helianthemum* type having oval stalked cotyledons and the *Cistus-Fumana* type with long linear sessile cotyledons. The perennial species have usually no underground stem, but

[1] *Revue générale de Botanique*, XXIV, 273 (1912).

Cistus umbellatus and *Helianthemum Chamaecistus* have well-developed stems capable of vegetative reproduction.

The plants are especially characteristic of the Mediterranean region. They love open dry sunny places with a chalky or sandy subsoil and, especially in the Spanish peninsula and North Africa, cover wide areas and are a characteristic feature of the evergreen bush vegetation of the Maqui. There is also a very much less

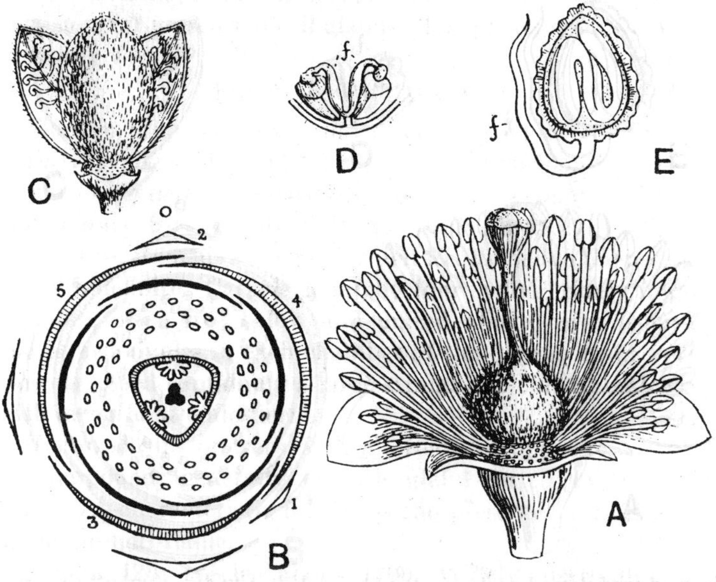

FIG. 96. A. Flower of *Cistus polymorphus*; the sepals and petals have been cut away to shew the basipetal development of the stamens; ×6. B–E. *Helianthemum Chamaecistus*. B. Floral diagram, the numerals indicate the order of development of the sepals. C. Opening capsule, ×3. D. Horizontal section through one of the valves of a capsule shewing the placenta bearing two seeds. E. Seed cut lengthwise, ×8. *f*, funicle. (A, C, D, E after Willkomm; B after Eichler.)

important centre of distribution in America, chiefly in the Eastern United States. The family is represented in Britain by four species of *Helianthemum* (Rock-rose), of which *H. Chamaecistus* is common on dry soils and is one of the most widely distributed species of the genus, occurring throughout Europe, in North Africa and in Western Asia; the three other British species are rare and of limited distribution. Associated with their dry habitat is the development of hairs. These take the form of long one-celled hairs, which are often associated in bundles or united to form flat scales, and of longer or shorter glandular hairs.

Family II. BIXACEAE

A small family represented by a single species, *Bixa Orellana*, a tree native of tropical America but cultivated throughout the tropics. The leaves are alternate, simple and palmately nerved, and the showy flowers are borne in panicles. There are five sepals, five petals, indefinite stamens, and two carpels with numerous anatropous ovules on parietal placentas. The fruit is a capsule, splitting loculicidally; the numerous seeds have a fleshy red seed-coat containing the colouring matter known in commerce as Anatto or Orlean, which is used as a dye-material.

Family III. TAMARICACEAE

A small family of four genera and about 100 species; chiefly steppe, desert and sea-shore plants of the Mediterranean region and Central Asia. The plants are shrubs or perennial herbs with alternate, exstipulate leaves, which are often small and narrow, giving the plant a heath-like habit. The flowers are solitary or in racemes, and ebracteolate; regular, bisexual, hypogynous, and pentamerous or tetramerous, with often increase in the number of stamens and reduction in the pistil. The ovary is unilocular with usually free styles and few or indefinite ascending anatropous ovules on basal (*Tamarix*) or parietal placentas. The fruit is a capsule, the seeds are hairy (*Tamarix*) with or without endosperm. The family falls into two distinct tribes:

Tribe 1. *Tamariceae*. Comprises *Tamarix*, the largest genus, with 64 species, one of which, *T. gallica* (Tamarisk) (fig. 97), is a doubtful native on our southern and eastern coasts, and *Myricaria*. It has small racemose flowers, free petals, and hairy seeds without endosperm.

Tribe 2. *Reaumurieae*. With solitary flowers, free petals and hairy seeds with endosperm. Two genera, *Reaumuria* and *Hololachne*, in the eastern Mediterranean region and Central Asia.

Family IV. FRANKENIACEAE

A small family of generally perennial herbs or undershrubs with stems jointed at the nodes and small opposite-decussate exstipulate leaves. The small regular hypogynous bisexual flowers are arranged in terminal or axillary cymes; beneath each flower is a pair of medianly placed sterile bracteoles alternating with the two lower normal (and fertile) bracteoles. The four to seven sepals are united into a tube for the greater part of their length;

alternating with them are an equal number of petals, each with a claw, to which is attached a ligular scale, and a spreading blade. There are six stamens in two trimerous whorls, the filaments being united for a short distance at the base; the versatile extrorse anthers dehisce longitudinally. The pistil consists of generally three carpels united to form a unilocular ovary bearing a filiform style which divides at the apex into three stigmas; the three

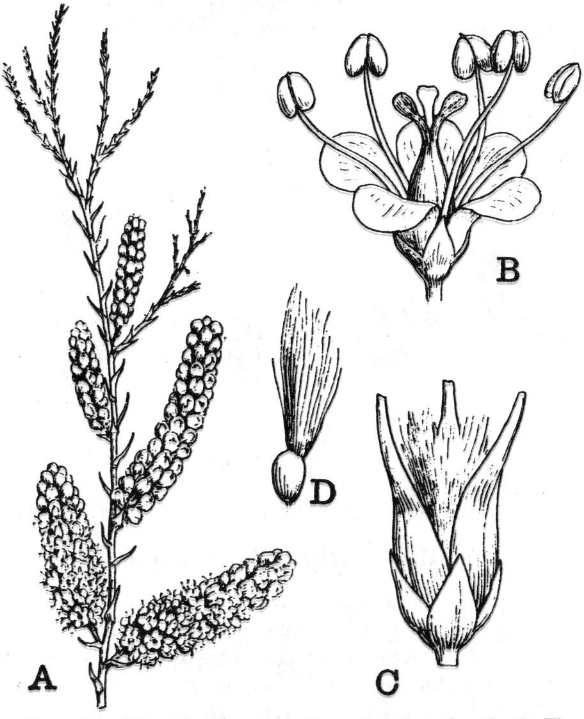

FIG. 97. *Tamarix gallica.* A. Flowering shoot slightly reduced. B. Flower, × 7. C. Capsule dehiscing, × 5. D. Seed, × 7.

parietal placentas are fertile only in the lower half, bearing indefinite anatropous ovules on long ascending funicles. Pollination is effected by aid of pollen-collecting insects. The fruit is a capsule included in the persistent calyx-tube, and dehisces loculicidally. The seeds have a crustaceous testa and contain a mealy endosperm surrounding the straight axial embryo.

There are four genera with 64 species, 60 of which belong to *Frankenia*. They are salt-loving maritime plants or inhabitants of dry localities, such as rocks, steppes and deserts. The halophytic and xerophytic habit finds expression in a hairy covering and revolute leaves giving a heath-like appearance; the structure of the leaf is intimately associated with the checking of transpiration and storage of water. The family is widely distributed in the temperate and warmer parts of the earth, the Mediterranean area being the chief centre of distribution; it is represented in Britain by *Frankenia laevis* (Sea-heath), a small procumbent plant with wiry branches and small rose-coloured flowers found by salt-marshes on the south-east coasts of England. The family is most nearly allied to the Tamaricaceae.

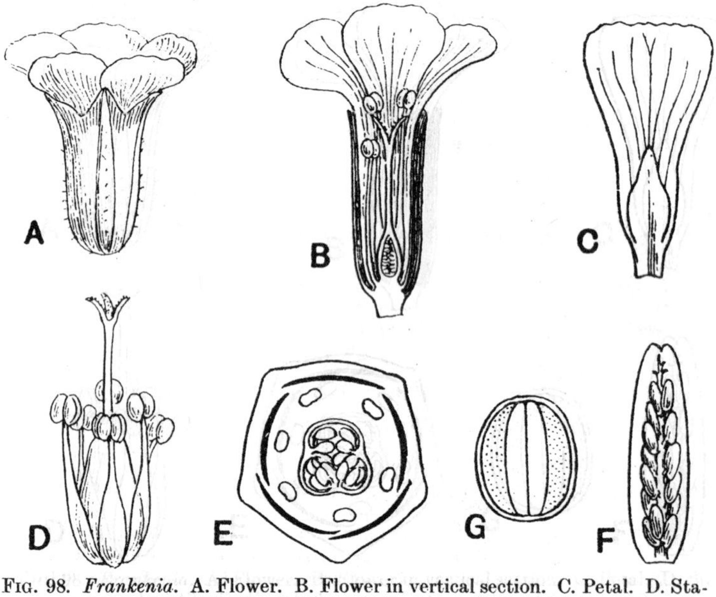

FIG. 98. *Frankenia*. A. Flower. B. Flower in vertical section. C. Petal. D. Stamens and pistil. E. Floral diagram. F. One valve of fruit bearing seeds. G. Seed cut transversely. All magnified. (After Le Maout and Decaisne.)

Family V. ELATINACEAE

A small family containing only about 40 species in two genera, but very widely distributed through the temperate and subtropical regions of both hemispheres. They are generally small annual glabrous herbs living in water or on mud, with creeping stems rooting at the nodes, and opposite or whorled, simple, entire or serrate, stipulate leaves. The cortex of the stem and root is interrupted by a ring of large longitudinal intercellular spaces. The flowers are solitary or form small dichasia in the leaf-axils. They are small, regular, bisexual, hypogynous, and conform to the formula Sn, Pn, An + n, Gn, n being 2, 3, 4 or 5. Sometimes the inner whorl of stamens is absent. The sepals are free or united at the base, the petals and stamens are free; sepals, petals and stamens persist till the fruit is ripe. The pistil is syncarpous; the number of chambers in the ovary and of the short free styles indicate the number of carpels present. The numerous anatropous ovules are arranged in two or more rows on a central column. Self-pollination has been observed in many cases, and in *Elatine* cleistogamy is known to occur. The fruit is a capsule

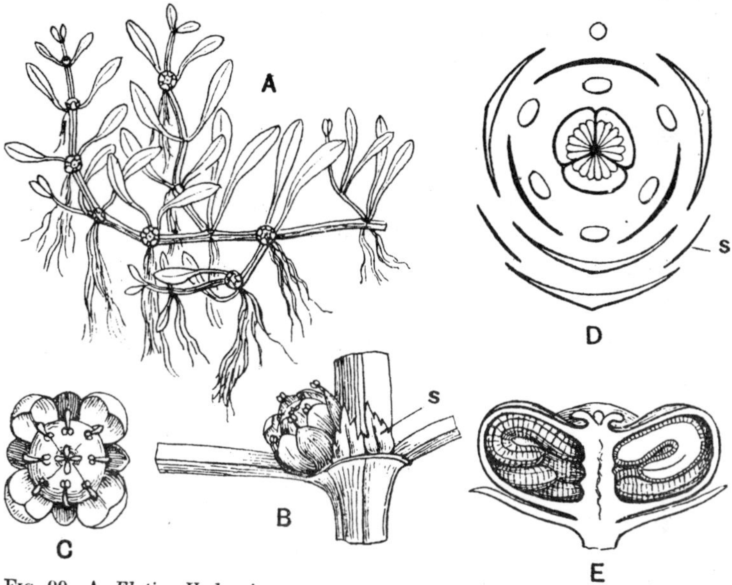

FIG. 99. A. *Elatine Hydropiper*, portion of plant. B. *E. Alsinastrum*, flower in leaf-axil, × 10. C. Flower of same seen from front, × 10. D. Floral diagram of *E. hexandra*. E. Fruit of *E. Hydropiper* in vertical section, × 12. *s*, stipules. (A, E after Seubert; D after Eichler.)

with septifragal dehiscence. The seeds are straight or curved, with a hard or membranous, smooth or rugose coat enclosing a straight or bent embryo, which consists of a large radicle and hypocotyl and two small cotyledons. Endosperm is absent.

The family is represented in our flora by two species of *Elatine*, both rare. *E. hexandra* is found on margins of ponds and lakes in the south of Scotland and in England (not in the eastern counties), and in north and west Ireland. It occurs also in Western and Central Europe and in the Azores. *E. Hydropiper*, another European plant, is very rare and in England and Ireland is recorded only from muddy ponds in a few counties.

The second genus, *Bergia*, is tropical and subtropical.

Family VI. VIOLACEAE

A small family, containing about 800 species in 18 genera, which are widely distributed in temperate and tropical regions but are found chiefly in the warmer parts of the earth. One genus, *Viola*, to which the British representatives of the family belong, contains about half of the species, the majority of which are confined to the temperate zones and the mountains of the northern hemisphere. A small group of species, characterised by an acaulescent habit with leaves arranged in a rosette, is endemic in the Cordilleras of South America, and another closely allied group is confined to the Chilian Andes. A few occur in subtropical Brazil, at the Cape, and on the mountains of tropical Africa respectively, eight are found in Australia and New Zealand and five are endemic in the Sandwich Islands. The British flora contains about sixteen species, of which *V. palustris*, *V. canina* (Dog-violet) and *V. tricolor* (Pansy) extend beyond the Arctic circle.

The plants are annual or perennial herbs, as in *Viola*, some species of which are shrubby below; in tropical and subtropical genera usually shrubs or small trees, rarely shrubby climbers.

The leaves are scattered, rarely opposite as in species of *Hybanthus*, simple and stipulate. In *Viola* the stipules may become large and leaf-like. The flowers spring from the axils of bracts, and the pedicels bear two bracteoles. They are solitary, as usually in *Viola*, or form axillary or terminal spicate, racemose, or panicled inflorescences. Half the genera have quite or almost regular flowers; of these the genus *Rinorea* (*Alsodeia*) (fig. 100, B, C) contains 260 species of trees or shrubs in the tropics of both Old and New Worlds. The remainder, as in *Viola*, are medianly zygomorphic owing to the larger size of the anterior petal which is spurred or gibbous.

Pentamery prevails below the gynoecium, five sepals, five petals and five stamens following in regular alternating whorls. The members of each whorl are generally free but sometimes connate at the base. The sepals are green in colour and generally uniform in size. In *Viola* they are appendiculate, that is, produced below their point of insertion. The petals are larger and coloured. The stamens have a very short filament. In zygomorphic flowers the connective of the anterior pair of anthers is spurred, as in *Viola* (fig. 100, D), or otherwise appendaged.

The one-chambered ovary is formed by the union of three carpels, of which the odd one is anterior. Three parietal placentas bear one to many anatropous ovules. The terminal style ends in a stigma which is extremely variable in shape even in the same genus.

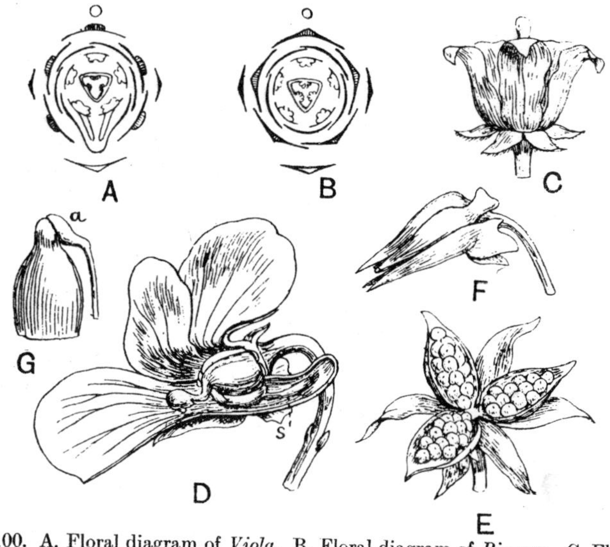

FIG. 100. A. Floral diagram of *Viola*. B. Floral diagram of *Rinorea*. C. Flower of *Rinorea macrocarpa*, × 3. D. Flower of *Viola tricolor*, cut open to shew relation between petals, stamens and ovary; *s*, spur of one of the two anterior stamens. E. Open fruit of same surrounded by persistent sepals. F. Cleistogamic flower of *Viola*, × 3. G. Stamen and pistil of same, the anther, *a*, closely appressed to the stigma. (A, B after Eichler; C after Martius, *Flor. Brasil.*; D, E after Oliver.)

These differences in form are associated with variations in the method of pollination which is effected by insect-visits. Sprengel, more than a hundred years ago, shewed the connection between the shape of the flower and the visits of bees, together with the importance of the latter for pollination. The spurs of the anterior anthers secrete nectar which is stored in the hollow spur of the front petal. In the Pansy (*V. tricolor*) the receptive stigmatic surface is situated in a groove on the front of the globular stigma, below it is a projecting valve which when pressed back covers the receptive surface. The stigma lies in a groove lined with hairs, formed by the anterior petal at the entrance to the spur. The anthers, which form a closed cone round the ovary, dehisce introrsely, and the pollen gets shaken into this hairy groove. To reach the nectar the bee thrusts its proboscis into the flower just below the stigma; in passing down the groove into the spur the proboscis becomes coated with pollen, and on being withdrawn presses back the valve on to the stigma which is therefore protected from contact with the pollen of its own flower. On entering the next flower the insect comes in contact with and deposits pollen on the upper surface of the valve; and cross-pollination is thus effected when the valve is again pressed back. Many species of *Viola,* such as our British *V. canina* and *V. odorata,* have, besides the large conspicuous flowers, small inconspicuous cleistogamic flowers (fig. 100, F, G). These appear later in the year, and are apetalous or have small equal petals; the anthers, which may be reduced to two, are closely applied to the stigma, and the style is much shortened. The pollen-grains germinate in the pollen-sacs and the tubes grow into the stigma through a specially prepared part of the upper wall of the anther[1]. Self-pollination is thus ensured.

The fruit is a capsule splitting elastically and loculicidally when ripe into three boat-shaped valves (fig. 100, E); in *Viola* these, on drying, may close along the central line and in so doing eject the smooth seeds one by one with considerable force. This ensures their distribution over a certain area, as they may be thrown a distance of several yards. The Sweet Violet (*Viola odorata*) and *V. hirta* bury their seed-capsules in the ground. In a few genera the fruit is a berry. The small seeds are obovoid or subglobose in shape, sometimes compressed, and are attached by very short funicles. The testa is generally hard and shiny. Many species, including both those with and without an explosive capsule mechanism, have an elaiosome (oil-body) on the seed, in the form of an appendage of the raphe, and their dissemination is aided by ants (myrmecochory). In the climbing genera *Anchieta* (tropical South America) and *Agatea* (Fiji and New Caledonia) the seeds are winged. The embryo is straight, lying in the long axis of the seed, which it nearly equals; it is surrounded by a copious fleshy endosperm.

[1] See Sablon, L. du. *Revue générale de Botanique*, XII, 305 (1900).

Numerous hybrids occur in the genus *Viola*. The garden Pansy is the result of hybridisation between forms of *V. tricolor*, a common cornfield weed, and other wild species, including *V. altaica*.

Family VII. FLACOURTIACEAE

A widely distributed tropical family of woody plants often forming tall trees, with generally alternate leaves in two rows; the leaves are generally thick, leathery, and evergreen, and are

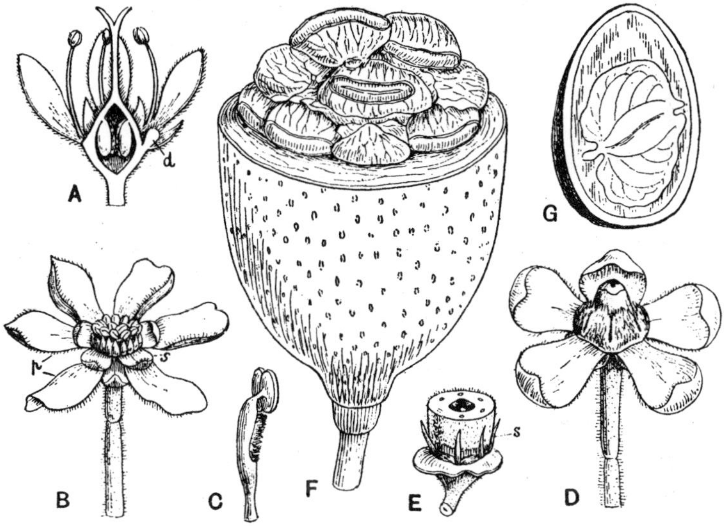

FIG. 101. A. Flower of *Homalium* in vertical section, × 6; *d*, disc. B–G. *Pangium edule*. B. Male flower, nat. size; *p*, petal; *s*, scale. C. Side view of stamen, enlarged. D. Female flower, nat. size. E. Lower part of the ovary, shewing outgrowths of the floral axis in the form of scales, *s*; enlarged. F. Fruit after removal of the upper portion of pericarp, revealing the seeds; reduced. G. Seed cut lengthwise shewing the embryo, ⅔ nat. size. (A after Warburg; the rest after Blume.)

provided with stipules which usually fall early. The flowers are generally small and arranged in lateral or terminal cymose inflorescences; in *Oncoba* (chiefly tropical Africa) they are often very large and sweet-scented and generally axillary. They are regular and generally bisexual but sometimes unisexual as in *Pangium* (fig. 101) (monoecious or dioecious). The structure is very varied, generally cyclic but sometimes spirocyclic. The number of sepals varies from 2–15, they are usually free, imbricate in the bud and equal in size; they are sometimes united below to form a short tube which is usually united with the ovary, the ovary becoming half inferior (fig. 101, A); in one genus, *Bembicia*, they are superior. The petals are sometimes wanting, as in *Casearia* (fig. 102, B); when present they are usually equal in number to the sepals (fig. 101, A); in *Oncoba* and allied genera more numerous; in *Dissomeria* there is an inner whorl of petals (fig. 102, A); they are free, usually imbricate in bud and larger than the sepals. In *Erythrospermum* and allied genera the spirally arranged sepals and petals pass gradually from one to the other, and may be petaloid or scale-like. The stamens are usually indefinite, sometimes arranged in bundles alternating with the sepals or opposite the petals; they are arranged in one whorl, or in two, or apparently irregularly; they are usually free. The

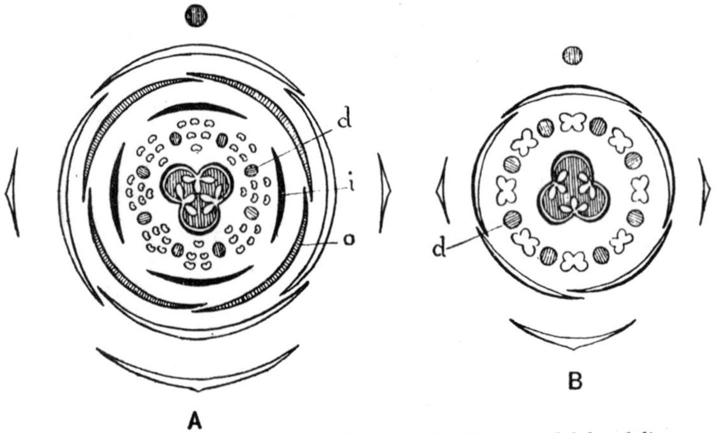

FIG. 102. Floral diagram of A, *Dissomeria*; B, *Casearia*; *d*, lobe of disc; *i*, inner petal; *o*, outer petal. (After Warburg.)

anthers are two-celled with longitudinal dehiscence; they are variously shaped and often drawn out into a point (*Oncoba*) or variously appendaged. Outgrowths of the floral axis are present in very great variety; in some cases, as in *Pangium*, in the form of scales opposite and often united with the base of the petals; in others in the form of glands occupying very various positions with relation to sepals, petals and stamens. *Flacourtia* has a ring-like disc; in *Casearia* there are developments of the disc between the stamens; occasionally a cup-like disc surrounds the ovary. In other cases a conspicuous variously appendaged corona is formed between petals and stamens. The pistil consists of 2–10 united carpels and is generally superior, but sometimes half-inferior as in *Homalium* (fig. 101, A) and other genera; in *Bembicia* it is quite inferior. It is usually unilocular, with three to five parietal placentas bearing numerous ovules the positions of which are very various. In *Flacourtia* and others it is incompletely chambered, rarely completely chambered. There is a single style, or several are present equal in number and alternating with the placentas.

That pollination is effected by insect-agency is indicated by the large size and bright colour of the flowers in some genera, their strong scent and especially by the great variety of intrafloral nectaries. Extrafloral nectaries occur on the leaves in many genera.

The fruit is a capsule or a berry, or a one- to many-seeded indehiscent structure. The seeds are generally small and numerous. Endosperm is present, usually in quantity; the straight embryo has a small root and usually large flat cotyledons. The hard seed-coat often bears a conspicuously coloured aril.

The fleshy acid fruits are sometimes edible, as in *Flacourtia*. *Pangium edule*, a large tree which occurs throughout the Malay Archipelago, has a very large egg-shaped indehiscent capsule, containing many large seeds with a copious oily endosperm (fig. 101, F, G); the seeds are eaten after long soaking in water to remove a poisonous constituent. In *Phyllobotryum* and other allied west tropical African genera the axis of the inflorescence has become united with the midrib of the large elongated leaf. In another West African genus, *Barteria*, the large internodes are hollow and afford shelters for ants.

There are about 70 genera and 800 species.

Family VIII. PASSIFLORACEAE

Contains 12 genera and over 500 species spread over the warmer parts of the world but largely American. It consists chiefly of herbs and shrubs with alternate, generally stalked, lobed and stipulate leaves, climbing by means of tendrils borne in the leaf-axil. The tendril may often be seen to correspond to the central flower of a dichasium or the first flower of a monochasium. The frequently large showy flowers are regular and bisexual or sometimes unisexual.

The large receptacle is often hollowed out like a cup or basin, as seen in the Passion-flower, and bears numerous filamentous or annular appendages between the corolla and stamens, which may be brightly coloured and form a conspicuous corona of great diversity in form. There are generally five sepals (more rarely three to eight), five petals (more rarely three to eight, sometimes absent), five stamens (more rarely four to eight or indefinite) and three to five carpels. The calyx and corolla are perigynous; the petals are almost invariably free. The stamens and pistil are raised on an internode (androgynophore) in the centre of the flower. The ovary is one-celled with parietal placentation, and several to numerous anatropous ovules. The styles, corresponding in number with the carpels, are free or united at the base; each bears a capitate stigma.

Pollination is effected by aid of insects, in the attraction of which the remarkable corona-developments doubtless play an important part; the flowers are often strongly scented and nectar is secreted on the receptacle. Humming birds also visit the flowers. Extrafloral nectaries occur on the leaf-stalks.

The fruit is a capsule, or, as in *Passiflora*, a berry with a leathery, fleshy or sometimes membranous wall enclosing a pulp in which the seeds are embedded; it generally opens either irregularly or loculicidally, and is rarely indehiscent. The generally numerous seeds have a sac-like aril, which is often red in colour and probably assists in the distribution of the seeds by birds. The embryo is enveloped by a fleshy endosperm.

About three-fourths of the species (more than 400) are included in the genus *Passiflora* (including *Tacsonia*), chiefly found in the warmer parts of America with a few species in Asia and Australia and one in Madagascar. Many species are known in cultivation as Passion-flowers and Tacsonias. Several are cultivated in the tropics for their edible fruit; *P. quadrangularis* is the Granadilla, a tropical American species. *Adenia* contains about 80 species in tropical Africa and Asia, one of which, *A. globosa*, a native of the desert country of Tanganyika territory, is an exception to the usual climbing habit, having a very thick globular fleshy stem with very small leaves. Another xerophyte, *Echinothamnus*, is a monotypic genus from Damaraland and forms a thick fleshy cushion-like growth fixed in rock-crevices by a strong tap-root. *Tetrapathaea* is a monotypic genus from New Zealand.

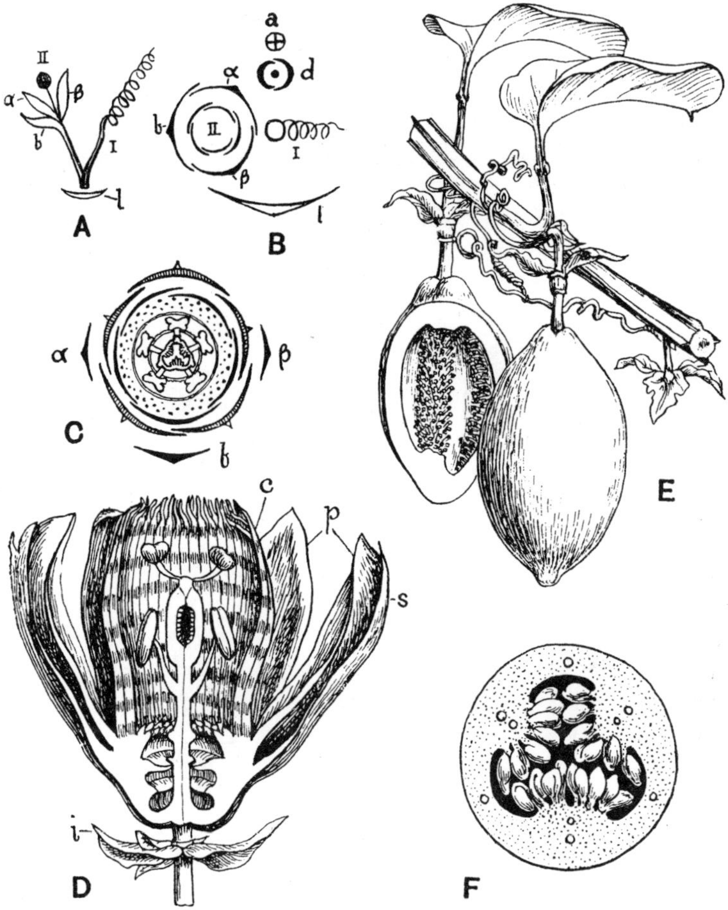

FIG. 103. *Passiflora*. A, B. Diagram and ground-plan of arrangement of flower of *P. coerulea*; *a*, main axis; *l*, foliage-leaf bearing in its axil a tendril, I, which corresponds to the central flower of a dichasium; II, lateral flower of the dichasium (the opposite lateral flower is undeveloped); *b*, its bract which has been raised on the flower-stalk and forms with the bracteoles (*α*, *β*) an involucre; *d*, a secondary bud borne in the axil of the leaf, *l*. C. Floral diagram. D, E, F. *P. alata*. D. Flower in longitudinal section, $\times \frac{3}{5}$; *i*, involucre; *s*, sepal; *p*, petals; *c*, corona. E. Branch bearing fruits, one of which is cut open shewing the seeds, $\times \frac{1}{3}$. F. Transverse section of ovary, enlarged. (A, B, C after Eichler; D, E, F from Martius, *Flor. Brasil.*)

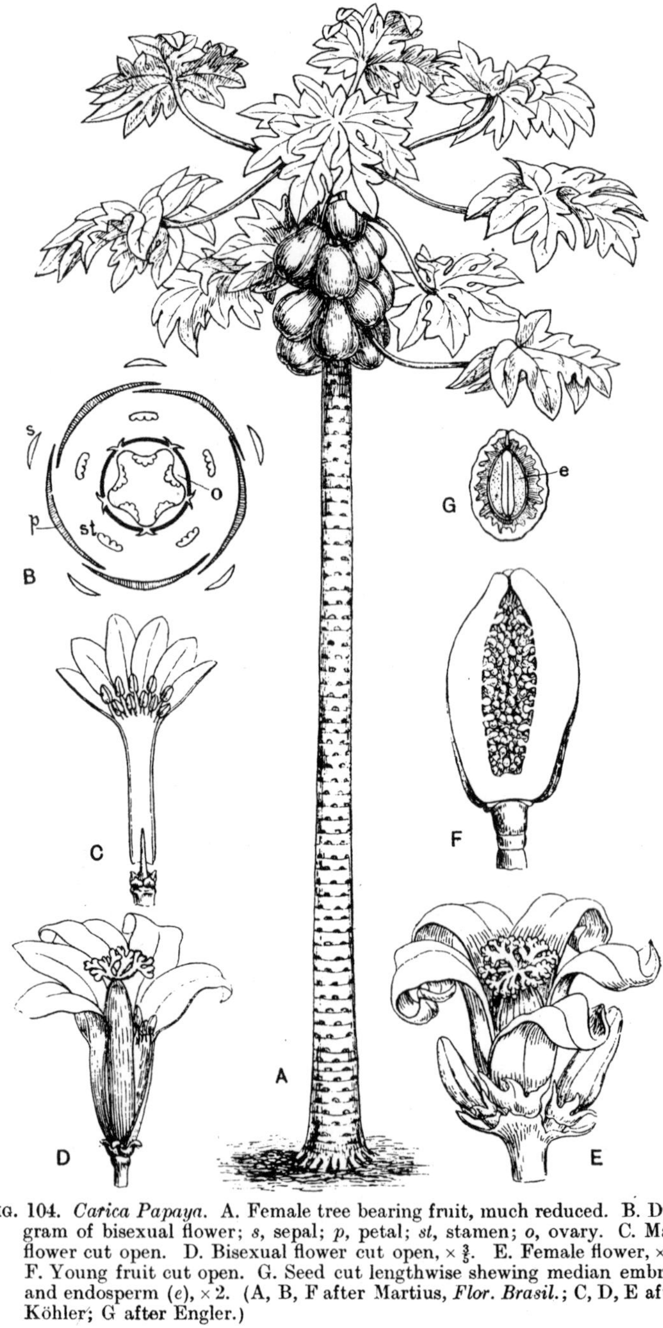

FIG. 104. *Carica Papaya*. A. Female tree bearing fruit, much reduced. B. Diagram of bisexual flower; *s*, sepal; *p*, petal; *st*, stamen; *o*, ovary. C. Male flower cut open. D. Bisexual flower cut open, $\times \frac{3}{5}$. E. Female flower, $\times \frac{3}{5}$. F. Young fruit cut open. G. Seed cut lengthwise shewing median embryo and endosperm (*e*), $\times 2$. (A, B, F after Martius, *Flor. Brasil.*; C, D, E after Köhler; G after Engler.)

Family IX. CARICACEAE

A small family of doubtful affinity represented mainly by the genus *Carica*, with 40 species in tropical and subtropical America. The best known is *C. Papaya* (Papaw), widely cultivated throughout the tropics for its large edible fruit and unknown in the wild state. The plants are small trees with succulent stems, and spirally arranged exstipulate leaves which are generally long-stalked and palmately or pinnately compound. The tissues are permeated with a network of laticiferous vessels, which in *C. Papaya* contain an active peptonising ferment, papain. The inflorescences are axillary with dichasial branching. The flowers are unisexual, the plants being monoecious or dioecious as in *C. Papaya*, where the female inflorescences are one- to three-flowered, while the male form richly branched pendulous panicles in which the terminal flowers are sometimes bisexual or female. The flowers are pentamerous and regular. The petals in the male flower are united below into a long tube on which the stamens are attached in two whorls at different levels. The corolla-tube in the female is short; the three to five carpels form a superior one- or five-chambered ovary with numerous anatropous ovules attached to parietal placentas, and bearing free styles. The fruit is a large berry containing numerous seeds. In *C. Papaya* it is one-chambered, the hollow cavity being lined by the seeds; where it is several-chambered the seeds are enveloped by a soft pulp derived from the partition walls. The testa consists of a soft fleshy outer coat and a woody inner coat. The embryo has two large flat cotyledons and lies in a soft oily endosperm. The small genus *Cylicomorpha* contains one species in the Cameroons and another in the mountains of east tropical Africa. Otherwise the family is tropical American.

Caricaceae has generally been placed near Passifloraceae, which it resembles in the structure of the ovary and fruit, but differs in the structure of the flowers—union of petals and double row of stamens—and the form of the vegetative organs. Van Tieghem regards it as allied to the Cucurbitaceae, on account of the structure of the ovule, which has a thick long-persistent nucellus, and two thick integuments; the vascular bundle from the raphe enters the inner integument at the chalaza and branches widely.

Family X. LOASACEAE

Herbaceous or shrubby plants, sometimes climbing, with alternate or opposite, entire, lobed or pinnately cut, exstipulate leaves, and generally more or less covered with hooked or sometimes stinging hairs. The flowers are bisexual. The receptacle is united with the ovary and forms a variously shaped tube. The flowers are very various in form, but generally have five sepals, as many alternating petals, which are usually free, and numerous stamens arranged in groups opposite the often concave petals; staminodes or nectar-secreting scales are frequently present opposite the sepals. The ovary is entirely or partly inferior; the three to seven carpels have each one to numerous ovules usually

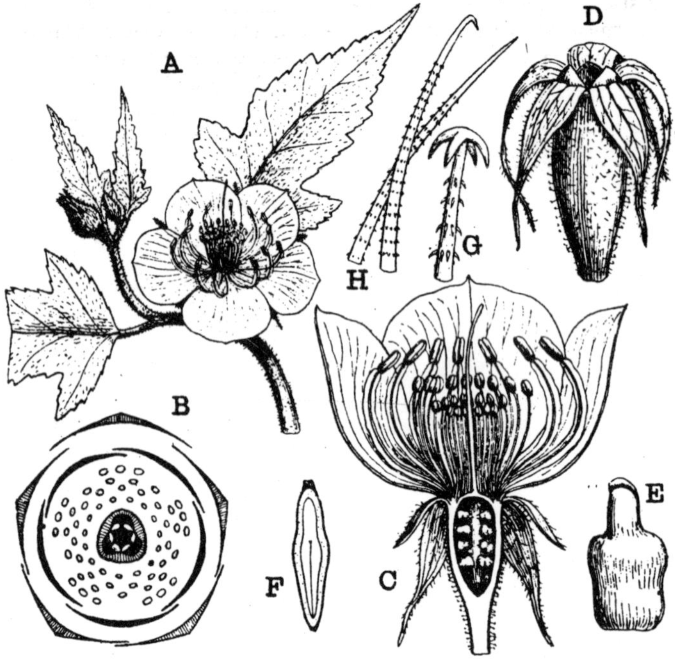

FIG. 105. *Mentzelia aspera*. A. Tip of shoot with flower, × $\frac{1}{5}$. B. Floral diagram. C. Flower in vertical section, × $2\frac{1}{2}$. D. Fruit shewing apical dehiscence, × 5. E. Seed, × 7. F. Seed cut lengthwise, × 7. G, H. Bristly hairs, × 60.

on parietal placentas; the ovule has only one integument. The fruit is a straight or spirally twisted capsule, usually with five to seven valves. The seeds contain endosperm. (Fig. 105.)

There are 13 genera and about 250 species. The chief centre of distribution is Chile, but the family is well represented in tropical South America, spreading southwards to Argentina and northwards through Mexico to California with a few species in the north-eastern United States. The only Old World representative is *Kissenia*, a monotypic genus, native of the dry country of south-west Africa, Somaliland and South Arabia.

Species of *Mentzelia*, *Loasa* and *Cajophora* are well-known in gardens.

The affinity of this family is also doubtful.

Order 5. *PEPONIFERAE*

(Cucurbitales)

Flowers generally unisexual, regular and generally pentamerous, with reduction to three in the pistil; stamens sometimes indefinite. Pistil inferior; the large placentas bear numerous ovules with a large persistent nucellus and two well-developed integuments. Endosperm absent or scanty.

Mostly tropical herbaceous plants, often climbing.

Allied to the previous order through Passifloraceae and neighbouring families.

Family I. CUCURBITACEAE

Flowers generally diclinous, regular and pentamerous, usually with reduction to three in the pistil. Calyx and corolla epigynous, generally inserted with the stamens at the edge of an epigynous outgrowth of the receptacle. Corolla generally sympetalous; stamens united in pairs, or into a single column, rarely free. Placentas thick, fleshy, bifid, bearing a number of ovules on each side. Fruit fleshy, rarely dehiscent. Seeds without endosperm; embryo straight with large oily cotyledons.

Annual or perennial herbs, with scattered long-stalked palminerved exstipulate leaves; generally climbing by means of tendrils.

Genera about 100; species about 800, in the warmer parts of the world.

The cotyledons appear above ground as the first green leaves of the plant. They are generally more or less oblong in shape, suborbicular in Bryony, and though entire in the seed often become emarginate from subsequent retardation of growth at the apex. In *Cucurbita* the escape of the cotyledons from the seed is aided by development of a peg upon the lower side of the hypocotyl by which the lower half of the testa is pressed to the ground, while the upper half is raised by growth cf the plumule.

The family consists chiefly of climbing herbaceous annuals, which contain a large amount of sap in their vegetative organs and have a remarkably rapid growth. *Dendrosicyos*, a genus of two species, confined to the island of Socotra, forms small

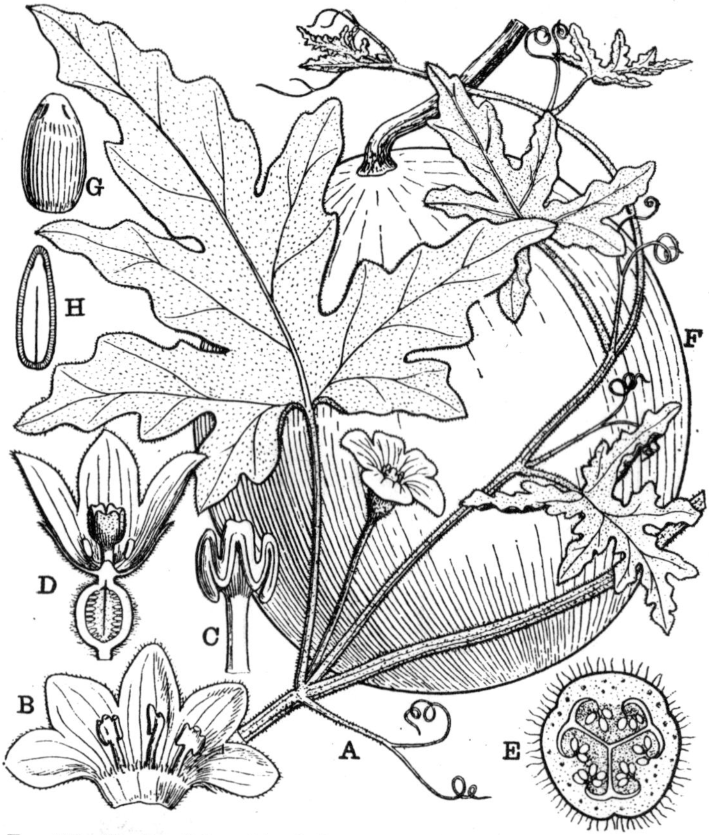

FIG. 106. *Citrullus Colocynthis*. A. Stem bearing a leaf with which are associated a male flower, a lateral branch and a branched tendril. B. Male flower cut open; of the five stamens four are joined in two pairs and one (the central) is distinct. C. One pair of stamens. D. Female flower in vertical section. E. Ovary in transverse section. F. Fruit. G. Seed. H. Same cut lengthwise. A, F, G, H somewhat reduced; B–E enlarged.

trees, but the stem is soft, like a turnip. Climbing is effected by tendrils, the morphological value of which has been much disputed. The leaves are arranged in a $\frac{2}{5}$ phyllotaxy on the often five-angled stem. If we trace the leaf-spiral upwards the position of the tendril is on the upper or anodic side of the leaf-base. In the leaf-axil are borne the flower, or inflorescence, and a branch; one or other may be absent in different genera or species. Thus in *Cucurbita Pepo* (Pumpkin, Marrow) we find associated with one leaf a flower, a leafy branch and a branched tendril (fig. 107, A; see also *Citrullus*, fig. 106); in *Cucumis sativus* (Cucumber) a median flower, an inflorescence, a leaf-branch and a tendril, or exceptionally a pair of tendrils (fig. 107, B; see also fig. 110.)

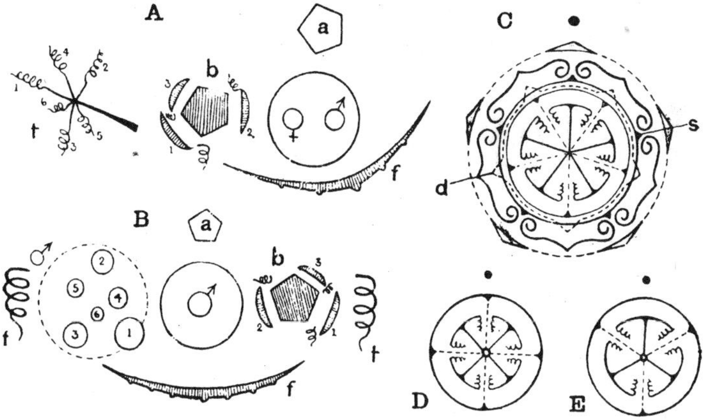

FIG. 107. A. Ground plan of arrangement of flower, shoot and tendril in *Cucurbita Pepo*; *a*, main axis; *f*, foliage-leaf subtending a male or female flower; *b*, leaf-shoot; *t*, tendril; the numbers indicate order of development. B. Same in *Cucumis sativus*; indicating letters as in A; to the left of the axillary male flower is a secondary male inflorescence, the circles indicating the order of development of the flowers. C. Diagram of female flower of *Cucurbita Pepo* with five carpels; *s*, whorl of staminodes; *d*, glandular ring. D, E. Arrangement of pistil of same with four and three carpels, in relation to the axis. (After Eichler.)

In *Ecballium*, the Squirting Cucumber of the Mediterranean region, a prostrate herb, there are no tendrils.

The explanation suggested by Braun and adopted by Eichler and others is the following. The flower is the axillary shoot of the foliage-leaf, the tendril is one of its bracteoles; the other bracteole is suppressed or exceptionally developed as a tendril as in *Cucumis sativus* (Cucumber, fig. 107, B). The leafy branch is a shoot borne in the axil of the tendril-like bracteole; the flowering branch is similarly a shoot borne in the axil of the second, usually aborted, bracteole. The tendril (bracteole) has become pushed out of its original position on the primary shoot (or flower) to its present position by the side of the foliage-leaf. When only one tendril is present the primary shoot (or flower) also becomes pushed towards the anodic side of the leaf-base; when two tendrils are present it preserves its median position. In sterile leaf-axils, the primary shoot (flower) remains undeveloped, its bracteole (tendril) and secondary branch (leaf-branch) alone being developed, or the tendril may also be absent. The leaf-spiral on the branches runs in the opposite direction (*antidromous*) to that on the main axis. When the tendril is branched as in *Cucurbita* or *Sechium* (fig. 110) the order of development of the branches is homodromous with the leaf-spiral of the main shoot.

According to this explanation, therefore, the tendril is a bracteole which has become pushed out of its place. Another theory, advanced by E. G. O. Müller in the *Pflanzenfamilien*, is based on the series of transitional structures between tendril and leaf found in *Cucurbita Pepo*. Sometimes the petiole twists round a support, sometimes the leaf-apex or its lateral ribs are tendril-like, or the tendril is a simple thread or a branched structure. Müller assumes that the twining part of the tendril is a leaf-structure, while the lower stiff portion is a stem-structure; the stem portion may be reduced to an inconspicuous or invisible rudiment. Other explanations have been advanced.

Engler[1] has shewn that in *Kedrostis spinosa* (Tanganyika territory) the explanation is a simple one; the leaves have thorn-like stipules one or other of which may grow out into a tendril.

The leaves shew considerable variety in form, but are very often palmately lobed or divided. A very constant character is found in the veining. The two strongest lateral nerves spring right and left from the base of the midrib, at about half a right angle. The other strong lateral nerves spring not from the midrib but from its first pair of branches, and are themselves the source of the next largest ribs (fig. 109, A).

Acanthosicyos is a thorny leafless dioecious shrub, three to five feet high, found in the sand-dunes round about Walfisch Bay in S.W. Africa. The thorns are modified stipules. This remarkable plant has a thick root, which may reach a length of 15 metres.

Stem and leaf shew a well-marked anatomical character. The vascular bundle is bicollateral, the xylem being covered both inside and outside by a broad phloem-band containing large sieve-tubes; it is separated from the outer band alone

[1] Engler's *Botan. Jahrb.* XXXIV, 362 (1904).

by cambium. The typical arrangement in the stem is two alternating circles of five bundles. Other arrangements may be derived from this by splitting, union, or disappearance of one or more bundles.

The diclinous, rarely bisexual, flowers are solitary or borne in inflorescences of various kinds, of which the male are more richly branched than the female. The androecium is often represented by staminodes in the female flower, and the gynoecium by an aborted pistil in the male. The flowers are regular. The perianth and stamens are attached to an often cup- or bell-shaped production of the receptacle, which is above the ovary in the female flower (fig. 110, D) (compare *Ribes*); the sepals are narrow and pointed; the corolla is generally gamopetalous, sometimes polypetalous. Another view of the structure of the flower is that what we have here called a production of the receptacle is a common base of the calyx and corolla, the stamens being then epipetalous. The first mentioned seems the more natural view, assuming the affinity of Cucurbitaceae with the perigynous Polypetalae. This affinity is illustrated by the tropical American genus *Fevillea* (fig. 109), which has five free petals and five free stamens alternating with them. The anthers are two-celled and dehisce by a longitudinal fissure. There is no indication, even in young anthers, of more than two cells. In *Thladiantha* (India to North China) two pairs of the stamens are closely approximated in the lower part of their filaments, the fifth standing apart. In *Sicydium*, the same pairs have their filaments united below, and the union between the filament and anther of each pair becomes more pronounced in various genera till, as in *Bryonia*, *Momordica* (fig. 111, A) and *Citrullus* (fig. 106, B), it is complete and the androecium apparently consists of three stamens, two with four cells and one with two cells. Further complication is introduced by the curving of the cells; in *Cucurbita* (fig. 108, E) they are much curved and by the cohesion of the connective the anthers are united into a central column. *Lagenaria* resembles *Cucurbita*, but the anthers are irregularly curved (fig. 108, D); in *Sicyos* (fig. 108, C) and in *Sechium* the filaments are united into a column bearing the anthers which are remarkably curved in the former. The most aberrant form occurs in *Cyclanthera* (warmer parts of America) (fig. 108, A, B), where the stamens are completely united into a central column with two ring-like pollen-containing chambers running round the top. The inferior ovary is of very various external form, from long and narrow to spherical or flattened; it may be cylindrical, or sharply angled or winged; smooth, hairy or prickly. There are generally three carpels, more rarely four or five; the generally numerous ovules are borne on as many thick fleshy bifurcating placentas which fill up the unoccupied space. In a few genera, as in *Sechium* (fig. 110, D), the ovary is unilocular with one seed.

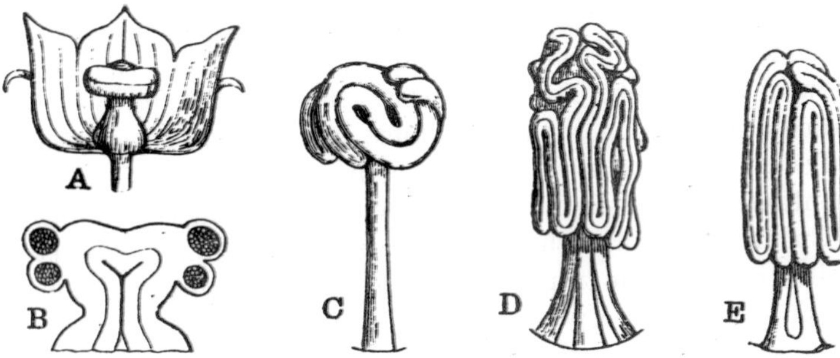

FIG. 108. Types of androecium. A. Male flower of *Cyclanthera* cut open to shew the androecium. B. Androecium in vertical section. C. *Sicyos*, filaments and anthers united. D. *Lagenaria* and E. *Cucurbita*, with curved anthers united into a column. All enlarged. (C–E after Müller.)

A transverse section of the fruit shews three (four or five) radiating lines with twice as many groups of seeds near the circumference (fig. 106, E). The simplest explanation of this appearance is that the edges of the carpels meet in the centre of the ovary and then curve outwards across the middle of each chamber, bifurcating near the circumference and bearing the ovules on the two incurving edges. The original carpel walls have become indistinguishable and form the pulpy mass of the fruit. The style is generally columnar bearing one forked stigma for each carpel. The stigmas are commissural, that is, above the dividing lines between the carpels, not, as we should expect, above their dorsal sutures. This position is explained by assuming that each is a joint structure composed of a branch of the stigmas of two adjacent

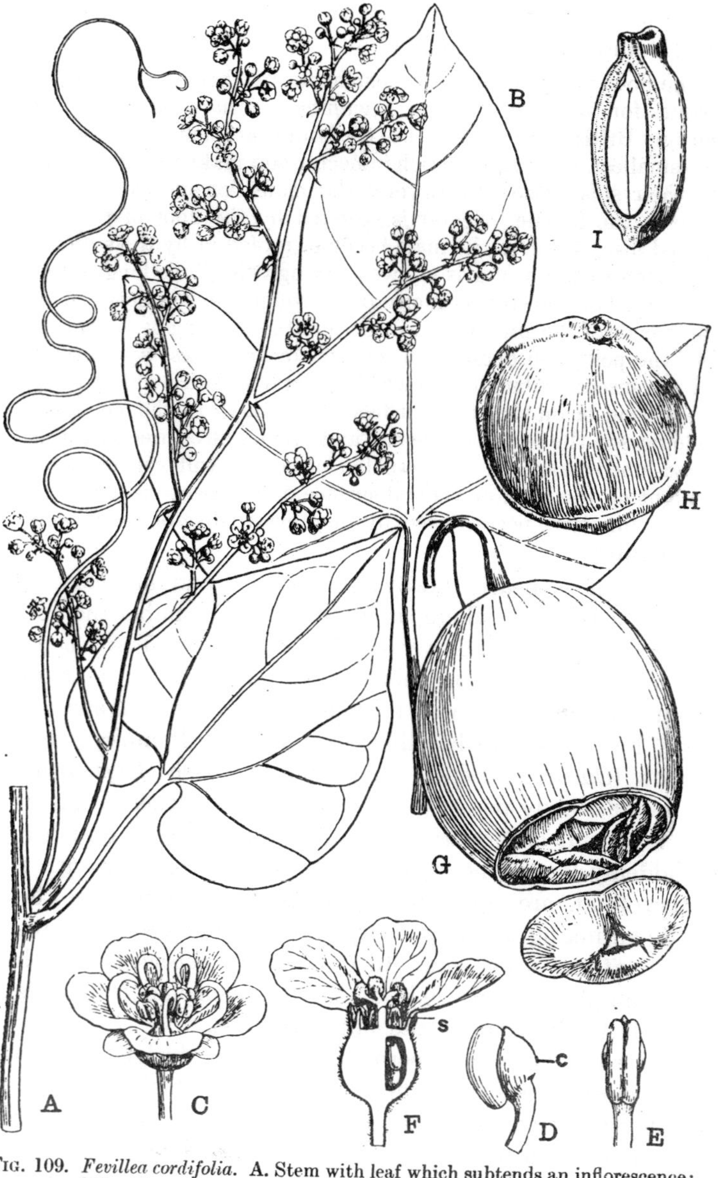

FIG. 109. *Fevillea cordifolia.* A. Stem with leaf which subtends an inflorescence; a tendril springs from one side of the leaf-axil. B. A lobed leaf. C. Male flower, the petals bear a median appendage, × 3. D. Side view, E, front view of a stamen, enlarged; *c*, connective. F. Female flower in vertical section; *s*, staminode. G. Fruit cut open. H. Seed, × $\frac{3}{5}$. I. Seed in vertical section.

carpels. The ovules are anatropous, with a short thick funicle and two integuments.

The flowers are sometimes conspicuous by their size and yellow colour (for instance, Marrow, Cucumber); others, as in Bryony, the only British representative of the family, are small and greenish. The smallest, however, are rich in nectar and make up by numbers for their small size. In Bryony the male flowers are twice as large as the female. Nectar is secreted by the floor of the cup-shaped disc, which in the male becomes roofed in by the stamens. Approach to the nectar is allowed through the three longitudinal apertures between the stamens and through the top between the upper ends of the anthers. As a result of the winding of the anther-cells the pollen on dehiscence collects towards these apertures so that an insect probing the flower for nectar will also collect pollen on its head or on the under surface of its body. In the female the style rises in the centre of the cup and divides into three spreading broad-lobed branches the papillae of which will collect pollen brought on the head or ventral surface of an insect-visitor. Thus cross-pollination, necessitated by the unisexuality of the flowers, is facilitated.

The fruit is nearly always soft, fleshy and indehiscent, forming a berry or where, as in Melon or Cucumber the epicarp forms a hard rind, a variety of berry known as a *pepo*. It sometimes reaches an enormous size as in Melon, Marrow or Squash, and varies much in shape, even in the same species. Note, for instance, the different forms of Pumpkin, and Marrow (*Cucurbita Pepo*), and of the Calabash (*Lagenaria*), the outer woody pericarp of which makes excellent flasks. The ripe fruit of *Ecballium* is highly turgid; when touched it suddenly leaves the stalk, and the seeds, together with a watery fluid, are squirted with considerable force, by the elastic shrinking of the pericarp, through the aperture thus formed at the lower end. In *Cyclanthera explodens* (South American Andes) one half of the zygomorphic fruit rolls back elastically and shoots out the seeds.

The seeds have no endosperm and contain a straight embryo with a short radicle and large flat cotyledons which are rich in oil; they are often flattened. The seed-coat generally consists of several (often ten) layers; the outermost is derived from the carpellary wall, and is, like the next layer (formed from the integument), capable of swelling in water. Then follow several hard and thick-walled layers, while the innermost are thin. *Sechium edule* (a monotypic genus native in the warmer parts of America, where, and in the Old World, it is

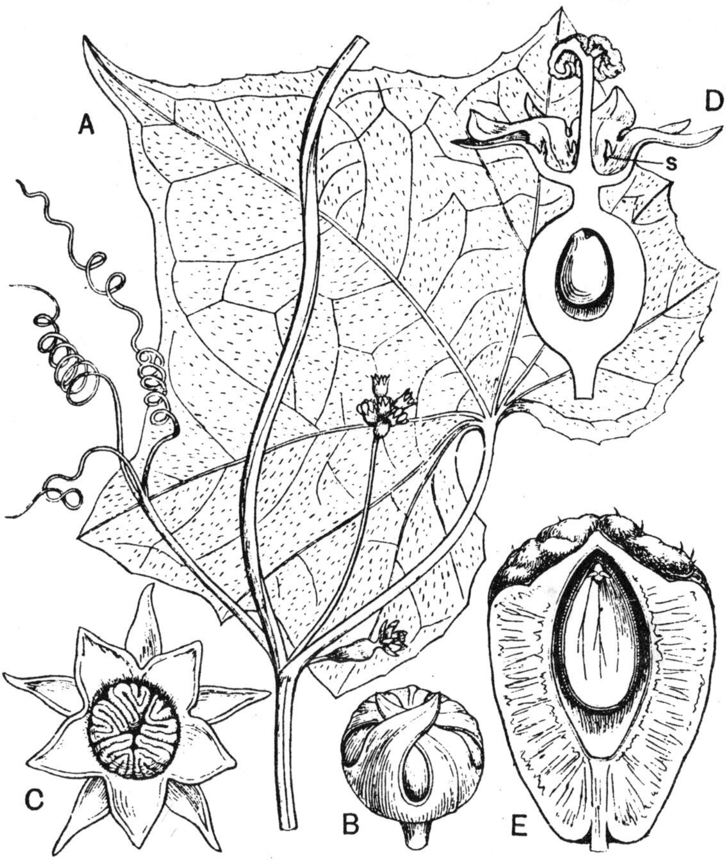

FIG. 110. *Sechium edule*. A. Stem bearing leaf which subtends a median female flower, a male inflorescence and a branched tendril, × $\frac{3}{4}$. B, C. Male flower in bud, and open exposing the five curved anthers, × 5. D. Female flower in vertical section, × 4; *s*, teeth of central disc. E. Fruit with seed cut vertically, × $\frac{2}{3}$. (After Baillon and *Flor. Brasil.*)

also cultivated) has a large fleshy one-celled one-seeded fruit (fig. 110, E). The seed, which is also large, conforming to the interior of the indehiscent fruit, germinates before the latter drops from the plant.

In *Momordica Charantia* (fig. 111) the orange-coloured fruit opens by three valves which bear the bright crimson seeds, the outer coat of which is pulpy. The fruit of *Fevillea* is a berry, often large, which finally becomes dry and somewhat hard. The large flat seeds have a thick integument enclosing a large embryo with thick cotyledons which are rich in oil (fig. 109).

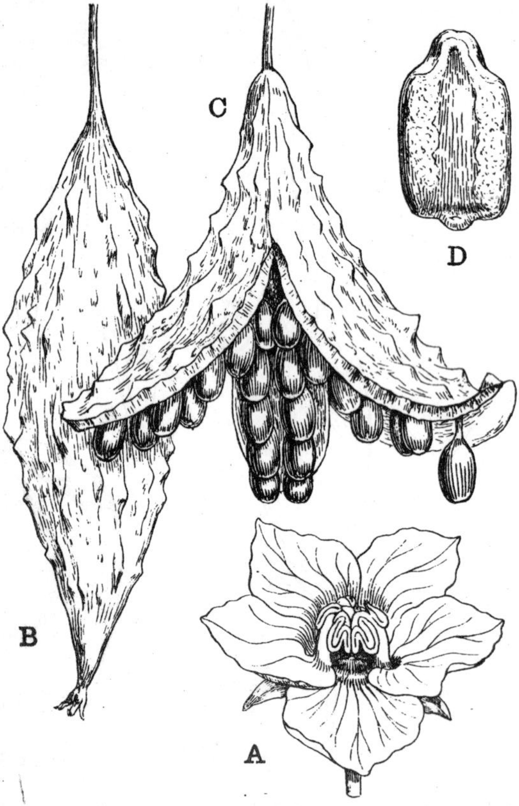

FIG. 111. *Momordica*. A. Male flower of *M. Balsamina*. B. Fruit of *M. Charantia*. C. Same opening, × $\frac{3}{4}$. D. Seed, × 3. (A after Baillon.)

Cucurbitaceae is a widely distributed family, most abundant in the tropics, and wanting in the colder regions. Three genera occur in Europe: *Ecballium* is monotypic and confined to the Mediterranean region, where occurs also the colocynth, *Citrullus Colocynthis*. Our sole British representative, *Bryonia dioica* (Bryony), spreads from the Mediterranean region into south and central Europe. A second species, *B. alba*, distinguished from Bryony by its black fruit and monoecious flowers, is found in central Europe; and a third, *B. cretica*, in the eastern Mediterranean. The genus contains five species besides, one in the Canary Islands, and four in further Asia. Three genera occur in temperate Eastern Asia. In America, *Sicyos angulatus* in Canada goes farthest north, while the two tropical American genera, *Cucurbita* and *Echinocystis*, reach as far north as California and Oregon. In the southern hemisphere, the family occurs outside the tropics at the Cape, in Australia, New Zealand and Argentina.

The Old World has the greater number of genera, but America contains the greater number of species. Excluding *Cucurbita*, which is probably not endemic in the eastern hemisphere, seven genera only are common to the East and West. Of these *Melothria* is one of the larger, with 60 species in the warmer parts of both Worlds, while *Cucumis* has 30 species, chiefly tropical African, but also generally distributed in warm regions. *C. Melo* (melon), a native of south Asia and tropical Africa, is widely cultivated; *C. sativus* (Cucumber) probably originated in India, and has been cultivated from earliest times. The fact that this, like *Cucurbita Pepo* and other tropical members of the family, runs through its life-history from seed to seed in a few months, allows of its successful cultivation in the summer in climates like our own. *Citrullus*, a genus nearly allied to *Cucumis*, has four species; two South African; a third, *C. vulgaris*, the Water-melon, native in tropical and south Africa, which was cultivated in Egypt and the East in earliest times, and had spread into southern Europe before the Christian era; and *C. Colocynthis*, which differs from the last in its bitter fruit and extends from N.W. India to tropical Africa and the Mediterranean region.

Many of the fruits are of use to man, while a few species are cultivated as ornamental plants, or as curiosities (*Ecballium*). The dried fibrous tissue of the rind of the fruit of *Luffa cylindrica* (Old World tropics) is the well-known loofah sponge.

There has been considerable difference of opinion as to the affinity of this family. The older botanists, such as Robert Brown,

De Candolle and Naudin, placed it with Passifloraceae among the perigynous polypetalous families, and this view was adopted by Bentham and Hooker. Eichler, on the other hand, placed it near Campanulaceae as an appendage to his series Campanulinae, basing his decision on the typically epigynous pentamerous flowers, the frequently gamopetalous corolla, the tendency to union of the stamens, and the form of the calyx with narrow yet plainly leaf-like points. This view has been followed by Engler who places the family by itself in an order Cucurbitales, next to the order Campanulales. There are, however, strong reasons for adopting the older view. The important characters of the ovule-structure which has a large persistent nucellus, frequently an extensive tapetal tissue, and two distinct integuments, are at variance with a position among the typical sympetalous families but find a parallel in Passifloraceae and allied families, in which we have also noted a tendency to union of the petals and epigyny. We have therefore adopted the position suggested by Hallier and regard Cucurbitaceae with Begoniaceae (and the small family Datiscaceae) as forming a distinct order next to the Passifloraceae group of families[1]. Recently P. Vuillemin[2] has maintained that Cucurbitaceae and Begoniaceae must be relegated to the apetalous group of orders, among which he classes them with Balanophoraceae, Rafflesiaceae, Datiscaceae, Nepenthaceae and Aristolochiaceae. He regards them as apetalous and the so-called corolla as an inner calyx.

Family II. BEGONIACEAE

A small widely distributed tropical family included almost entirely in the genus *Begonia* with more than 600 species in both Old and New Worlds but absent from Polynesia and Australia. There are four other genera, *Hillebrandia* and *Symbegonia*, each with a single species in the Sandwich Islands and New Guinea respectively, *Begoniella* with three species in Columbia and *Semibegoniella* with three in Ecuador.

They are mainly succulent herbs, generally erect, but often creeping, or acaulescent with an underground rhizome or tuber, the latter formed from the swollen hypocotyl or by local swelling of the rhizome. Many species of *Begonia* are root-climbers.

The leaves are alternate in two rows on the elongated

[1] See also J. E. Kirkwood in *Bull. New York Botan. Garden*, III, 313–402 (1904), and J. M. Coulter in *Botan. Gazette*, XXXIX, 73 (1905).

[2] *Annales d. Sciences Naturelles, Botan.* sér. X, V, 5 (1923).

stems; in creeping stems and root-climbers they are arranged dorsiventrally. They are generally asymmetrical, stalked and succulent, often brilliantly variegated, generally palmately nerved, with an entire, toothed or lobed margin or palmately divided. The stipules are generally large and often persist after the fall of the leaf. Hair-structures on stem and leaf shew a remarkable variety in size and form, and there are also transitional structures from true hairs to emergences in which the parenchyma beneath the epidermis takes part. Groups of small tubers are often found in the leaf-axils which correspond to the lateral branches of a suppressed axillary shoot. Adventitious buds are very readily produced, especially on isolated portions of the leaf, as in the method of propagation of the Rex Begonias (with large ornamental leaves) by buds which readily form when the cut surface of the leaf is kept moist in the soil. A callus is formed over the wound, and buds arise on the upper leaf-surface, at the point of section of the nerves, from a meristematic tissue which is developed from the callus, and also on the nerves beyond the cut, by division of a group of epidermal cells; they are specially associated with cells which bear trichomes.

In addition to the normal open or closed ring of vascular bundles in the stem, cauline bundles occur, especially in tuberous and thick-stemmed erect species, which run separately through the internodes and unite at the nodes with the vascular ring; in some species also cortical bundles occur. Crystals of calcium oxalate belonging to the quadratic system are present, either solitary or grouped in glands; and cystoliths are frequent, especially in the leaves, taking the form of double cystoliths, more or less spherical structures, on the common wall of two neighbouring cells. The inflorescence is generally axillary and forms dichasia with a tendency to pass into monochasia of the helicoid type. The primary axes end in a male flower; the female flowers are borne on the last or last axis but one.

The flowers are monoecious, generally somewhat zygomorphic, with a simple perianth; in *Hillebrandia* and *Begoniella* an inner whorl of small petals is present, alternating with the sepals. Except in *Begoniella* and the female flower of *Symbegonia*, where they unite to form a tube, the perianth-leaves are free. In the male they are generally two to four in number,

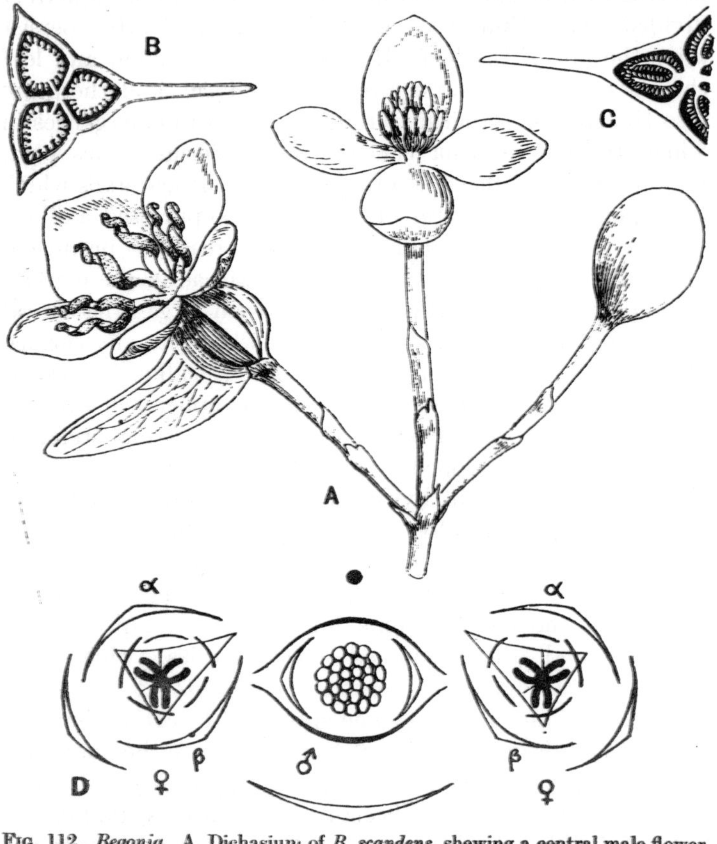

FIG. 112. *Begonia.* A. Dichasium of *B. scandens,* shewing a central male flower, a female flower and a male flower-bud, × 2½. B. Transverse section of ovary of same, × 5. C. Portion of transverse section of ovary of *B. acuminata* shewing a forked placenta, × 5. D. Diagram of dichasium of *B. acuminata* shewing a central male flower with four sepals in two pairs and two lateral female flowers with bracts and bracteoles (*a* and *β*) and five sepals. (D after Eichler.) (From *Flor. Jam.*)

in the latter case in opposite pairs, the large outer pair alternating with and covering the smaller inner pair. The numerous stamens (four in *Begoniella*) are in many whorls, and are free or with filaments more or less united to form a tube. The two-chambered anthers are basally attached and the connective is often produced; the anthers dehisce by means of longitudinal slits, rarely by pores. In the female flower the perianth is superior, with generally two to five (rarely six to eight) parts; the two outer members more or less cover the inner. Except in *Hillebrandia*, where the upper part is free, the ovary is quite inferior; it generally bears one to three, rarely six, wings, and is usually completely two- to three-, rarely four- to six-chambered, with generally axile placentas which are simple (fig. 112, B) or forked (fig. 112, C), or even twice forked. In *Hillebrandia* the ovary is unilocular and there are five forked recurving parietal placentas. The very numerous anatropous ovules have two integuments. The two to three, rarely four to six, styles are sometimes united at the base; they are generally deeply forked; the often twisted branches bear the stigmatic papillae. In spite of absence of floral nectaries and, except in a few cases, a distinctive scent, the flowers are probably insect-pollinated. The spreading, bright white pink or scarlet perianth-leaves and sometimes the additional attraction of coloured bracts suggest entomophily, as does also the later flowering of the female flowers which terminate the younger branches of the inflorescence, so that in the earliest stage the inflorescence is male.

The fruit is generally a horny capsule splitting lengthwise along the wings. The seeds are minute and very numerous; when ripe they contain no endosperm but are filled by a small straight thick poorly differentiated embryo, the cells of which are rich in oil.

The great majority of the species of *Begonia* inhabit damp districts, especially shady woods. The greatest development occurs in Brazil and in the Andine region and extends as far as Mexico; the next largest distribution centre is in the rain-forest district of the Eastern Himalaya, the mountains of further India, and the Malay Archipelago; eastern tropical Africa has few species, while the damp forests of the Cameroons and Gaboon are rich in species. The perennial tubers which characterise certain sections of *Begonia* are adaptations to the drier and cooler districts of the Andes, and a similar development occurs in South Africa (on the mountains of Natal), and also in the dry climate of Socotra where one species is found.

The affinity of the family is obscure. A parietal placentation is present in *Hillebrandia*, and the inferior ovary finds a parallel in Loasaceae among the Parietales. Its nearest ally is the small family DATISCACEAE (a few species in West Asia to India and the Malay Archipelago and one in Mexico and California) which agrees in having unisexual flowers, an inferior ovary with parietal placentation (as in *Hillebrandia*) and an oil-containing embryo, while the very scanty endosperm in the seed forms a transitional stage to the typically exendospermic seeds of Loasaceae and the Parietales generally. Begoniaceae has, however, much in common with Cucurbitaceae, namely the unisexual flowers, inferior ovary, the tendency to union of the stamens, the exendospermic seeds with an oily embryo and the palmate nervation of the leaf.

Order 6. *GUTTIFERALES*

Flowers regular, more rarely zygomorphic, generally bisexual, hypogynous. Calyx and corolla sometimes pentamerous but shewing much variety in number and arrangement of parts, which are usually free. Stamens often numerous, free or sometimes variously united. Pistil of two or more united (rarely free) carpels, superior; ovary generally multilocular with numerous ovules in the inner angle or when the ovary is unilocular on the ventral suture. Ovules with two integuments, generally anatropous. Endosperm when present containing oil and proteid granules.

Generally woody plants, often with intercellular secretory passages.

This order is closely related to Parietales, with which it is united by Engler, but may be distinguished by the generally axile placentation. Like the Parietales it may be regarded as derived from the Ranales, affinity with which is especially suggested by the Dilleniaceae with a frequently spirocyclic perianth, indefinite hypogynous stamens and sometimes free carpels.

Family I. DILLENIACEAE

Trees or shrubs, sometimes lianes, rarely herbaceous, with generally alternate simple entire evergreen leaves with or without stipules. The yellow or white often showy flowers are generally bisexual and regular, or sometimes zygomorphic. The arrangement of the perianth is often spirocyclic. There are usually five sepals (more rarely 3–∞) which often enlarge after flowering; generally five petals which fall early. The hypogynous stamens are indefinite (rarely 10 or fewer), free or variously united at the base; frequently some are staminodial; the form, position and mode of dehiscence of the anthers vary widely. There are one to many carpels which are free or more or less completely united, but the styles are generally free; ovules 1–∞ in each ovary,

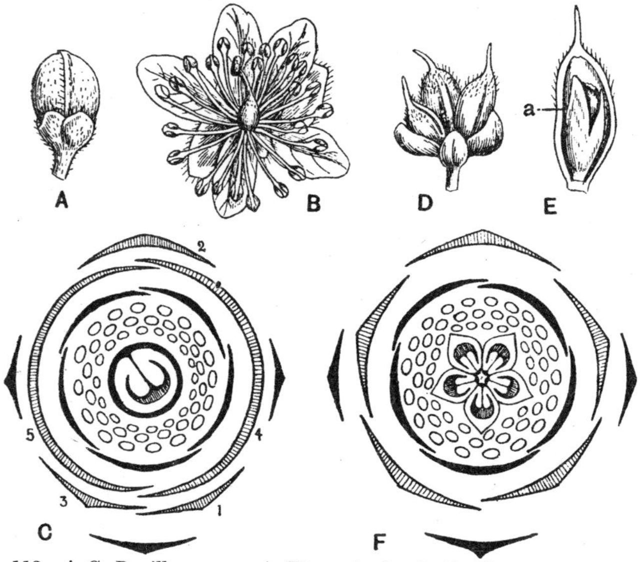

FIG. 113. A–C. *Davilla rugosa*. A. Flower-bud, × 2. B. Flower, × 2. C. Floral diagram, the numbers shew the order of development of the sepals. D–F. *Tetracera volubilis*. D. Fruit, each separate carpel splits along the ventral suture, × 1½. E. A single fruit cut open shewing the seed enveloped by an aril (*a*), × 2. F. Floral diagram. (A, B, D, E after A. Richard; C, F after Eichler.) (From *Flor. Jam.*)

anatropous, either erect from the base of the ovary, or ascending or horizontal when seated at the inner angle or on the ventral suture; the raphe is always ventral. The fruit is dry and dehiscent (capsular), more rarely indehiscent or a berry. The seeds are generally few or solitary, and have a conspicuous funicular aril closely united with the testa. They contain a copious fleshy or mealy endosperm and a generally very small embryo. (Fig. 113.)

Many of the lianes have anomalous secondary growth in the stems in the form of several concentric rings of secondary wood.

In species of *Saurauja* the flowers are borne on the old wood—

such plants are termed cauliflorous—a frequent phenomenon in woody plants of tropical forests.

There are about 300 species in 12 genera, widely distributed in the tropics, but especially Australian. The largest genus, *Hibbertia*, with about 100 species, is almost exclusively Australian; the plants are small shrubs with generally showy yellow flowers.

Dillenia indica, a large tree (India and Malaya), has very large showy flowers, and large edible fruits surrounded by the sepals which have become fleshy.

Family II. OCHNACEAE

Generally trees or shrubs, more rarely undershrubs and very rarely herbs; never climbing. The leaves are alternate, generally stiff, leathery, smooth and shining, with a large number of closely arranged secondary nerves running more or less parallel from the midrib to the margin; with few exceptions simple, with a toothed margin; stipulate with membranous subulate stipules which often fall early. A constant anatomical character is the presence of leaf-trace bundles in the cortex. The flowers are axillary, or in racemes or panicles; in *Sauvagesia* and allied genera often in axillary dichasia. They are bisexual, and usually regular, but may become more or less zygomorphic by the stamens becoming pushed to one side. There are generally five sepals (sometimes up to 10) which are free (rarely slightly united at the base), imbricate in the bud, and generally somewhat leathery. The five (rarely more, to 10) free petals are twisted in the bud, spreading in the open flower, and fall soon. The stamens are hypogynous, sometimes on an elongated floral axis, and vary widely in number and arrangement. In *Ochna* and other genera they are numerous, in three to five whorls, but various types of reduction occur, so that frequently there are only one or two whorls of fertile stamens, or one or two whorls may be represented by staminodes which are sometimes petaloid; the stamens often become pushed to one side of the flower; the filaments are generally very short and the two-chambered anthers long, opening by pores, or less often by longitudinal slits. The pistil also shews great differences. In *Ochna* and *Ouratea* (fig. 113*a*) there are 5–15 free carpels uniting above in a central common style and placed on a more or less elongated conical receptacle which becomes enlarged, thick and fleshy in the fruit, forming a cushion on which the drupes are borne. In the remaining genera there are only three to five carpels which unite to form an ovary, which is unilocular or becomes more or less completely three- to five-locular by the ingrowth of the placentas; when the placentas meet in the centre the placentation is axile, otherwise it is parietal. In *Ochna* and *Ouratea* there is a single ascending ovule in each chamber, but generally the ovules are numerous; the raphe is always ventral. The fruit consists of a cluster of drupes in *Ochna* and *Ouratea*, but is generally a few- to many-seeded dry fruit, indehiscent or splitting septicidally. In *Lophira*, a monotypic genus from tropical Africa, the two outer sepals become much elongated in the fruit forming a wing which ensures distribution (compare Dipterocarpaceae). The seeds are large and without endosperm (as in *Ochna*, *Ouratea*, *Lophira*), but more often small and usually winged, with a straight or bent embryo surrounded by endosperm.

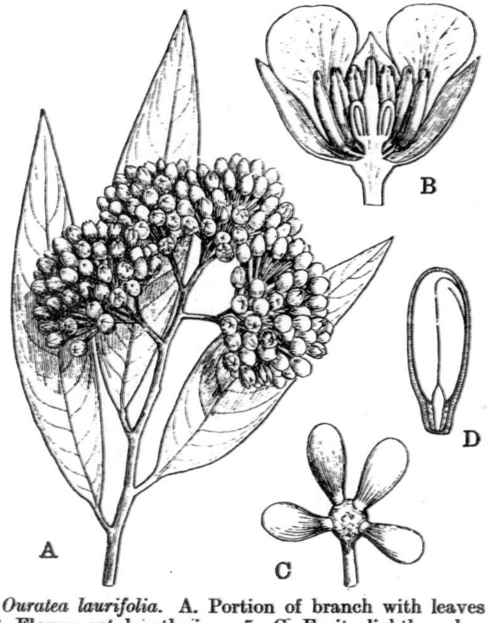

FIG. 113*a*. *Ouratea laurifolia*. A. Portion of branch with leaves and flowers, $\times \frac{4}{5}$. B. Flower cut lengthwise, $\times 5$. C. Fruit slightly enlarged. D. Seed cut lengthwise, $\times 2\frac{1}{2}$. (From *Flor. Jam.*)

There are 20 genera with about 400 species, widely distributed in the tropics, especially in Brazil. The largest genera are *Ochna*, tropical Asia and Africa, and *Ouratea*, tropics of both worlds. *Sauvagesia* and allied genera were placed by Bentham and Hooker in Violaceae, but are regarded by Engler as more suitably placed in Ochnaceae; they have the characteristic cortical bundles. The spirocyclic floral structure of *Ochna* and *Ouratea* suggests an affinity with the Ranales; but, as in Dilleniaceae, reduction has occurred in the androecium and gynoecium and most of the genera have a reduced type of flower.

Family III. MARCGRAVIACEAE

A small tropical American and West Indian family containing about 100 species in five genera, which are mostly climbing or epiphytic shrubs with pendulous terminal inflorescences. The leaves are simple, alternate and leathery, and in the largest genus *Marcgravia* are of two forms; the pendulous shoots which bear the terminal inflorescence have large stalked spirally arranged leaves, the sterile shoots bear smaller leaves in two rows which are appressed to the substratum, tree-trunks or rocks, to which the shoot clings by means of short adventitious roots. The flowers are bisexual. Except in *Marcgravia*, there are five small free persistent imbricate sepals, followed by five alternating petals which are free or more or less united below; in *Souroubea* the flower is pentamerous throughout, in *Norantea* there is an increase in the number of stamens, while in *Ruyschia* the pentamerous androecium is followed by a bicarpellary pistil. In *Marcgravia* the calyx consists of four small sepals arranged crosswise in alternating pairs; the four petals are united and fall like a cap when the flower opens; the numerous stamens are united below. The superior ovary is somewhat spherical with a short style or a sessile, more or less radiating, stigma; before pollination it is one-chambered but later becomes several-celled by the ingrowth of the parietal placentas which bear numerous anatropous ovules, each with two integuments. The fruit is a more or less spherical leathery capsule which opens from below in an irregular loculicidal manner, or remains closed; it contains numerous seeds; the straight or bent embryo is enveloped by a starchy endosperm.

The family is of special interest from the development of an upper bract above the two normal small bracteoles to form a coloured nectar-secreting generally hood- or pitcher-like structure. This reaches its highest development in *Marcgravia* where the central flowers of the umbellate inflorescence are abortive and the brightly-coloured bract which has become completely adnate to the flower-stalk is converted into a stalked nectar-containing pitcher, at the base of which an indication of the small sterile flower is seen. The pitchers stand beneath the fertile flowers which open successively and are said by Delpino to be strongly proterandrous, the anthers falling before the stigma is fully developed. They were described by Belt[1] as pollinated by humming birds which come in contact with the flowers while drinking the nectar and convey the pollen from one plant to another. I. W. Bailey[2], who has recently studied

[1] Belt, T. *The Naturalist in Nicaragua*, 1874.

[2] Bailey, I. W. "The pollination of *Marcgravia*." *Amer. Journ. Bot.* IX, 370 (1922).

two previously unknown species in British Guiana, throws doubt on this. He points out that the nectaries are so arranged that birds tend to approach the inflorescences from above and do not become coated with pollen. Furthermore, in the two species described and figured by him, the stamens dehisce before the corolla falls, at which time the stigma is coated with pollen; he suggests that the flowers are self-fertile.

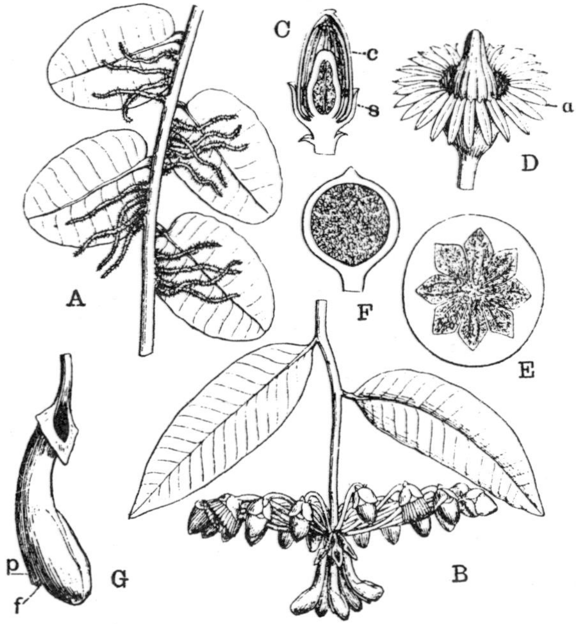

FIG. 114. *Marcgravia Brownei*. A. Portion of climbing shoot, shewing underside of leaves next the supporting trunk with rootlets, × ⅔. B. Portion of flowering shoot, × ⅓. C. Flower-bud cut lengthwise, nat. size; *s*, sepal, *c*, corolla. D. Flower after the fall of the corolla, nat. size; *a*, ring of spreading anthers. E. Ovary cut across, × 4. F. Fruit cut lengthwise, nat. size. G. Hollow bract containing nectar adherent to pedicel, *p*, which bears at its apex an abortive flower, *f*, × ⅔. (From *Flor. Jam.*)

The family was included by Bentham and Hooker as a tribe of Ternstroemiaceae, but its peculiar characters entitle it to be regarded as a separate family.

Family IV. TERNSTROEMIACEAE

(THEACEAE)

A family of 18 genera with about 250 species distributed through the warmer parts of the world. It comprises trees and shrubs, with alternate simple exstipulate, generally leathery, evergreen leaves, and generally solitary axillary regular, often showy, bisexual flowers. The perianth is often spirally arranged, and consists of five to seven sepals which are persistent, and generally five petals which are free or united at the base; both have a marked imbricate aestivation. In the Camellia (*Thea japonica*) a number of accessory bracteoles precede and gradually pass into the sepals, which number five to six, and themselves shew a transition to the petals

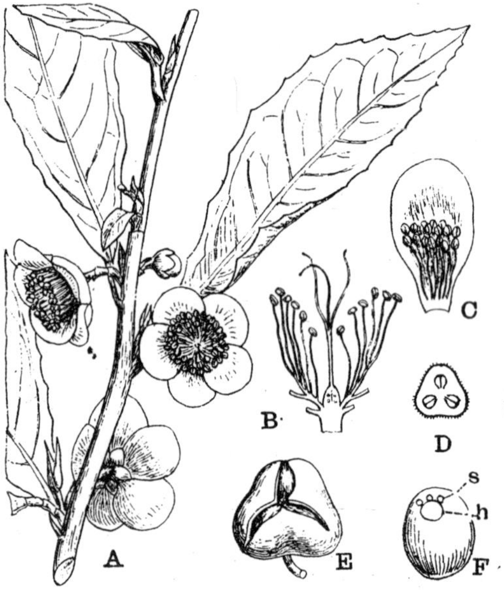

FIG. 115. *Thea sinensis*. A. Portion of plant, slightly reduced. B. Vertical section of a flower shewing some of the stamens and the pistil, × 2½. C. Single petal with bundle of stamens, × 2½. D. Ovary cut across, enlarged. E. Fruit shewing dehiscence, slightly reduced. F. Seed; *h*, hilum; *s*, scars of three ovules which have not formed seeds; slightly enlarged. (A, C, D, E after Bentley and Trimen; B, F after Szyszylowicz.)

(five to seven). The stamens are generally numerous, and are free or united into a tube at the base, or in five bundles which are opposite, and often united with, the five petals; they are sometimes slightly perigynous. The 2–∞ carpels unite to form a superior ovary of as many chambers and bearing as many free or more or less coherent styles. The ovules are generally numerous, borne on axile placentas, and anatropous. The fruit is a capsule, generally splitting loculicidally, or is indehiscent. The seeds have little or no endosperm and a large embryo; there is no aril.

A characteristic of the family is the presence of large thick-walled often much branched solitary cells (scleroids) in the tissue of stem and leaves. Secretory passages are not present.

The Tea-plant (*T. sinensis*) is an erect bushy, sometimes arborescent plant with smooth leathery more or less oval leaves, toothed at the margin and with well-marked veins. The white, scented flowers stand singly or two to three together in the leaf-axils. It has been cultivated from very early times in India and China (its use as a beverage is mentioned in the sixth century A.D.). It is perhaps a cultivated form of *Thea assamica*, which is also cultivated, but occurs wild in upper Assam. The various sorts of tea are the results of different methods in the process of manufacture.

The Camellia, *T. japonica*, a native of China and Japan, is a well-known greenhouse shrub.

The family as conceived by Bentham and Hooker in the *Genera Plantarum* contained a larger number of genera distributed among various tribes. Several of these groups are now generally regarded as representing distinct families, such as Marcgraviaceae, and some have been referred to other families. The family as described above is therefore restricted.

Family V. GUTTIFERAE

Flowers bisexual or unisexual, regular, hypogynous, with considerable variety in number and arrangement of the parts. There are generally two to six free sepals and an equal number of free petals. Stamens generally numerous, often more or less united. Ovary 2- to many-, generally 3- to 5-celled; ovules 1 to indefinite in each cell, amphitropous or anatropous; stigmas as many as the cells, generally sessile and radiating. Fruit various, fleshy and indehiscent or capsular. Seeds generally large, completely filled by the straight embryo.

Trees or shrubs, rarely herbs, containing resin or oil in schizogenous spaces or canals. Leaves simple, entire, opposite or sometimes whorled, generally exstipulate. Flowers in cymose inflorescences or solitary; often showy.

Genera 45; species about 900.

The plants are shrubs, sometimes lianes, or especially trees, evergreen, and inhabiting chiefly tropical areas with high rainfall. The genus *Hypericum* alone is more strongly developed outside the tropics and includes all stages from perennial herbs with a persistent rhizome to undershrubs and shrubs. In the primary forests of tropical America species of *Clusia* are epiphytic, with the habit of many epiphytic species of *Ficus*, developing numerous adventitious roots which form a close network round the supporting stem, which ultimately perishes.

The flower shews a remarkable diversity in the number and arrangement of parts. Bracteoles are often developed close beneath the calyx so that it is impossible to determine where the calyx begins and in many cases there is no sharp distinc-

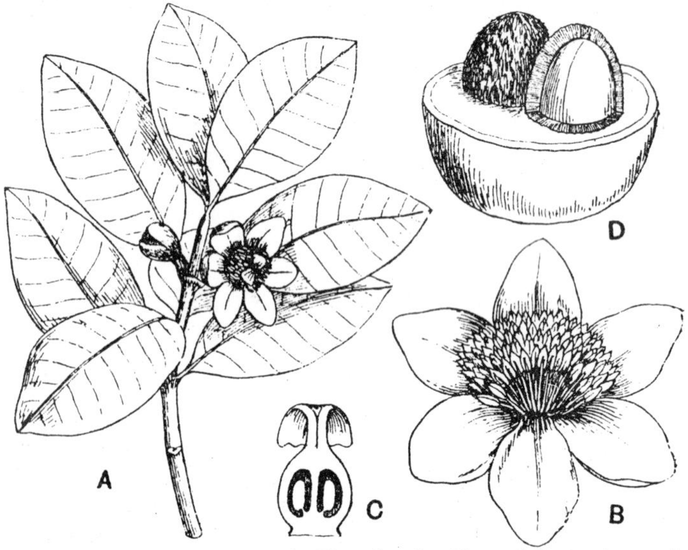

FIG. 116. *Mammea americana*. A. Shoot bearing bisexual flower, $\frac{1}{3}$ nat. size. B. Male flower, ×2. C. Pistil cut vertically. D. Drupe cut across shewing the two stones, in one of which the endocarp has been cut away exposing the seed with its two large fleshy cotyledons, $\frac{1}{2}$ nat. size. (A, C after Engler; B after *Flor. Brasil.*) (From *Flor. Jam.*)

tion between the sepals and petals. A cruciform arrangement occurs in *Havetiopsis* (Amazon district), where two pairs of sepals are followed by two pairs of petals and two pairs of stamens, the pistil forming a whorl of four carpels. In other genera a dimerous calyx and corolla are succeeded by two or three whorls of stamens and a whorl of carpels, or by numerous spirally arranged stamens. In other cases (e.g. species of *Clusia*) the cruciform arrangement characterises bracteoles and calyx, while petals and stamens are arranged spirally, or sepals, petals and stamens are all spirally arranged. In some cases, as in *Mammea* (fig. 116) the flowers are polygamous.

There is also great diversity in the androecium; the stamens are free or shew various degrees of union of the filaments, rarely forming a cup, more frequently a lobed synandrium, or arranged, as in *Hypericum*, in four or five bundles; staminodes also occur variously united or converted into secretory organs. The carpels are equal in number with the petals, or fewer, or twice or thrice as many; they are united into a single whorl. The styles are free or partly united, some times very short; the stigmas are generally broad. The ovules have two thick integuments, are either amphitropous or anatropous, and are attached by a short funicle to the ventral wall of the carpel or in multilocular ovaries to the inner angle; in *Calophyllum*, *Mammea* (fig. 116) and allies they are basal.

Entomophily is suggested by the showy and usually numerous flowers, but nectar-secreting organs are generally absent.

The fruit is generally a capsule opening septicidally or septifragally; after dehiscence the central column persists bearing the septa, with the seeds in the central angles. In *Garcinia* and allied genera it is a fleshy berry, in *Calophyllum* and *Mammea* a drupe containing one or several seeds.

The seed is often more or less enveloped by an aril which grows either from the funicle or micropyle. Considerable diversity occurs in the development of the embryo. In some cases (e.g. *Hypericum* and allied genera, and *Calophyllum*, *Mammea* and others) the cotyledons are well-developed but more frequently they are small or scarcely differentiated, while the hypocotyl is thick and large, as in *Clusia*, *Garcinia* and many others.

The family is represented in Britain by the genus *Hypericum* (St John's Wort) (fig. 117), perennial herbaceous or shrubby plants, the leaves of which are often dotted with translucent oil-glands. The yellow regular hypogynous flowers are bisexual and arranged in cymose, often dichasial, inflorescences. The five sepals are followed by five petals. The numerous stamens are arranged in three (as in *H. Androsaemum*, Tutsan) or five (as in most British species) bundles, in the latter case opposite the petals; a whorl of stamens is sometimes present opposite the sepals or its place is taken by five hypogynous glands, thus restoring the regular symmetry of the flower. There are three or five carpels, in the latter case alternating with the staminal bundles. The ovary is one-celled with three or five projecting parietal placentas, becoming three- or five-celled with axile placentas when the septa are completely united in the centre. The ovules are numerous. The fruit is a septicidal capsule or sometimes a berry as in Tutsan.

The widely open bright yellow flowers contain no nectar and are homogamous. They are visited by nectar-seeking insects for the sake of their pollen. The long styles radiate outwards between the staminal bundles, and if an insect bearing pollen from another

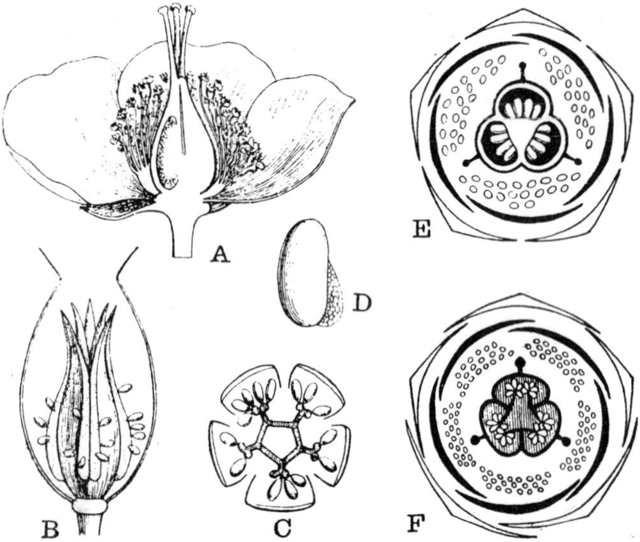

FIG. 117. *Hypericum.* A. Flower of *H. calycinum*, cut lengthwise. B. Fruit of *H. triflorum*, shewing the three placentas with a few seeds still attached to the revolute edges. C. Transverse sectional plan of same indicating method of dehiscence. D. Seed of *H. Androsaemum*. E, F. Floral diagrams of *H. quadrangulum* and *H. Androsaemum*. B, C, D enlarged. (A after Baillon; rest after Keller.)

flower comes first in contact with the stigmas cross-pollination is favoured; if, on the other hand, with the dehiscing anthers, pollen will be rubbed on to the stigma of the same flower. As the flower fades petals and stamens close up towards the centre and anthers and stigmas become closely interlocked so that, cross-pollination failing, self-pollination is ensured.

Hypericum contains about 300 species, distributed through temperate regions and on the mountains in the warmer parts of the earth. Some of the species have a wide range, such for instance as *H. humifusum* from Europe to further India and South Africa, *H. hirsutum* from Europe to Siberia and *H. japonicum* from Japan to New Zealand and Australia. Other species have a wide distribution in North or South America. It is suggested that a factor in this wide distribution may be the carriage of the small seeds, which are produced in large numbers, on the feet of birds.

The other genera are mainly tropical and restricted either to the Old or New World, though a few occur in both hemispheres. Such are *Vismia*, with a few species in west tropical Africa and a larger number in tropical America, and *Calophyllum*, with numerous species in tropical Asia and Australia and fewer in tropical America. *C. Inophyllum* is a widely distributed coast tree from Africa through the East Indies to Polynesia. Of the tropics Africa is the poorest in representatives of the family. The large genus *Garcinia* has 200 species in the tropics of the Old World, chiefly in tropical Asia and Polynesia; *G. Mangostana* is the Mangosteen, a widely cultivated fruit in the tropics; the edible portion is the fleshy aril enveloping the seeds. *Clusia*, with 100 species, is tropical American.

The family is of considerable economic importance. The wood of many species is hard and durable, and many yield valuable resins or gum-resins, especially in the genera *Calophyllum*, *Clusia*, and *Garcinia*; e.g. gamboge from *Garcinia Morella*. Others yield edible fruits such as the Mangosteen, and the West Indian Mammee-Apple (*Mammea americana*). A fatty oil is obtained from the seeds of *Calophyllum Inophyllum*, *Garcinia indica* and others; the thick sap of *Pentadesma butyraceum*, the West African "Tallow-tree," is used as butter.

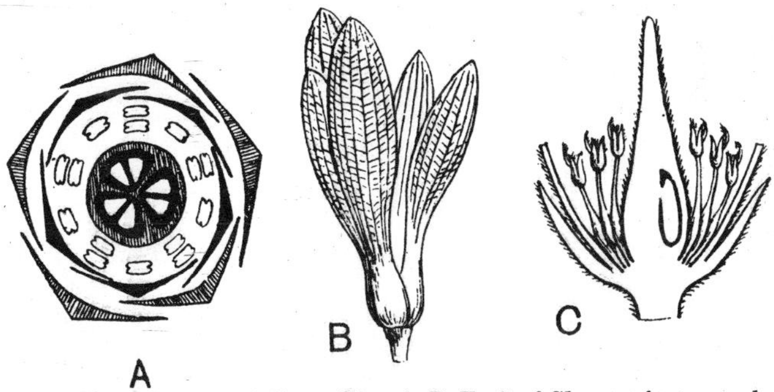

FIG. 118. Floral diagram of *Hopea Pierrei*. B. Fruit of *Shorea robusta* enveloped in the persistent calyx, about ½ nat. size. C. Longitudinal section of flower, the petals cut off above, × 6. (A after Pierre; C after Brandis.)

Family VI. DIPTEROCARPACEAE

Containing 16 genera with over 300 species of trees, rarely shrubs, inhabiting the hot damp forests of India and Malaya.

They have alternate entire leathery stipulate leaves and axillary panicles of regular bisexual pentamerous flowers, with often an increase (5, 10, 15 or ∞) in the number of the generally free stamens; the three carpels form a trilocular ovary with a pair of pendulous anatropous ovules in the central angle in each chamber. The fruit is a one-seeded nut enclosed in the persistent calyx, two or more of the segments of which may grow out to form wings which aid in distribution. The embryo fills the seed. (Fig. 118.)

They resemble Guttiferae in containing resin; species of *Shorea* yield Dammar-resin; *Dryobalanops aromatica* is the source of Borneo Camphor which forms yellow prisms between the elements of the wood. Many species are valuable timber-trees, such as *Shorea robusta*, the Sal tree of India.

Order 7. *MALVALES*

(COLUMNIFERAE of Eichler)

Flowers bisexual, regular, hypogynous, cyclic, pentamerous with a frequent increase in the number of members of the androecium and gynoecium. Sepals valvate in the bud; free or united. Petals free, aestivation various, often twisted. Stamens in two whorls, the members of the outer antesepalous whorl often absent or reduced to staminodes, of the inner generally multiplied and monadelphous, the united filaments forming a long column (hence Columniferae) round the pistil; or polyadelphous. Number of carpels very various, uniting to form a multilocular ovary, each with 1–∞ anatropous ovules with two integuments. Brunnthaler has demonstrated an endotropic course of the pollen-tube in a number of genera.

Herbs, shrubs or trees, with alternate simple stipulate leaves. Stellate hairs are very frequent on young parts and mucilage-sacs or -canals in the tissue of the cortex and pith.

The several families are closely allied, and it is very difficult or impossible to separate them by any single character. Different authors take different views as to their number. Thus Schumann in the *Pflanzenfamilien* recognised five, as also does Wettstein, and more recently Warming, Bentham and Hooker in the *Genera Plantarum*, three, and Baillon only two.

The Malvales are allied to the Guttiferales by their regular hypogynous flowers with pentamerous calyx and corolla, and especially by the androecium, which in both orders shews a marked tendency to union of parts. On the other hand, a marked affinity with Euphorbiaceae is suggested by the frequent occurrence of branched multicellular hairs, and especially by the structure of the ovary which often separates in the fruit into one-seeded parts; an endotropic course of the pollen-tube has also been demonstrated in both groups.

Family I. TILIACEAE

Tiliaceae contains about 400 species in 41 genera, the greater number of which are confined to the tropics. The most important extratropical genus is *Tilia* (Lime-tree or Linden) with about ten species in the northern portions of both hemispheres. It is represented in our own flora by two species, *T. cordata* (*parvifolia*) and *T. platyphyllos*. The plants are generally trees or shrubs with alternate, entire, dentate or lobed leaves and stipules, which often, as in *Tilia*, serve merely to protect the leaves in the bud and fall as the latter unfolds. Mucilage-cells occur in the pith and cortex. The regular bisexual flowers have a pentamerous, sometimes gamosepalous, calyx with valvate aestivation, followed by five free petals which are imbricate in the bud. The family is very closely allied to Malvaceae. The best distinction is found in the anthers which are two-celled in Tiliaceae, one-celled in Malvaceae. Moreover in Tiliaceae the stamens are, as a rule, connate at the base only or quite free. They are generally numerous and inserted at the base of the petals, or, as in *Grewia* (a large genus with 120 species widely distributed in the warmer parts of the Old World) and its allies, raised above the corolla by the development of an internode (androgynophore). The number of carpels shews considerable variation, the ovary being from two- to ten- or many-celled, with one or more ovules in the inner angle. The position of the ovules is also varied, and may be both ascending and pendulous in the same chamber. The fruit is two- to many-celled and dehiscent or indehiscent. *Tilia* has a five-celled ovary with two ovules in each cell. As it ripens to fruit one seed only develops, and four cells remain small and are crushed to one side; the ripe fruit is a globose indehiscent nutlet with one, rarely two, seeds. The embryo is surrounded by a fleshy endosperm, and has generally large leaf-like cotyledons, often, as in *Tilia*, more or less deeply lobed.

The flowers of *Tilia* are proterandrous, and pollination is effected by the visits of insects, especially bees and flies,

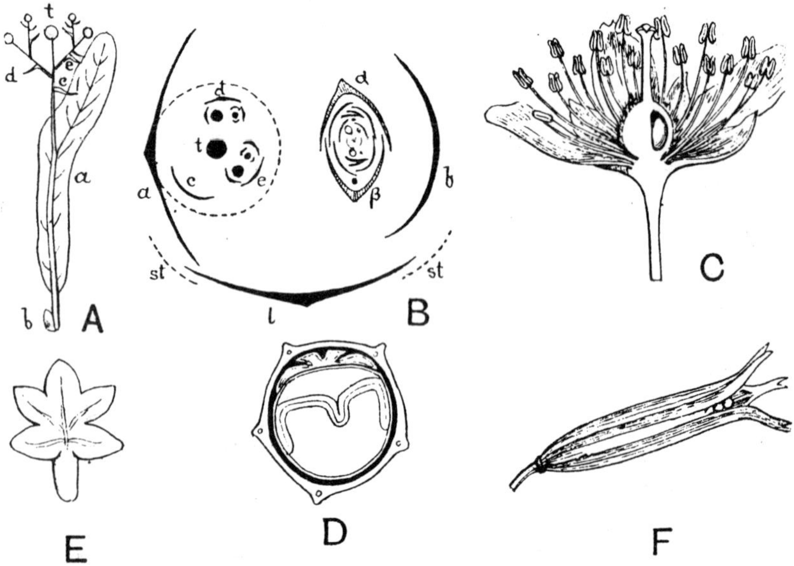

FIG. 119. A–E. *Tilia*. A. Scheme of flower-bearing shoot. B. Diagram of same, with subtending leaf, *l*; *st*, position of fallen stipules; *a*, *b*, first and second leaves of axillary shoot, subtending respectively inflorescence and leaf-bud; *t*, terminal flower; *c*, *d*, *e*, bracteoles, *d* and *e* each subtending a branch to the lower part of which they adhere. C. Flower in vertical section, × 3. D. Fruit cut across shewing the four suppressed carpels and the seed containing the embryo embedded in endosperm, enlarged. E. Embryo flattened out, enlarged. F. Fruit of *Corchorus capsularis*, nat. size. (A, B after Eichler; C after Warming; D after Wettstein; F after K. Schumann.)

which are attracted by the strong scent. Nectar is secreted in spoon-like glands at the base of the sepals. Distribution of the seed in this genus is facilitated by the membranous wing-like bract, which, when the fruit is ripe, becomes detached, bearing with it the inflorescence. It will be noticed that between the latter and the main stem is a leaf-bud (fig. 119, A).

The foliage-leaf apparently subtends both a flowering shoot and a vegetative bud. The explanation is as follows (see fig. 119, A, B). The inflorescence is a shoot borne in the axil of the subtending foliage-leaf (*l*). Its first leaf (*a*) is the membranous bract, to the midrib of which its axis adheres for some distance. Its second leaf (*b*) subtends a leaf-bud, is scale-like, and remains situated with its bud at the base of the inflorescence. The latter is dichasial, terminating in a flower (*t*) and bearing a younger lateral flower on each side; these secondary axes may again branch, so that the shoot becomes three- to seven-flowered.

In the other genera the fruit is generally dry, forming a loculicidal capsule (as in *Corchorus* (fig. 119, F) and in *Sparmannia*, familiar in the South African species *S. africana*, a favourite greenhouse plant), or sometimes separating into winged cocci. In *Grewia* it is indehiscent and drupaceous. Development of spines on the capsule or cocci, as in *Triumfetta*, *Sparmannia* and others, favours distribution by animals. In *Apeiba* (tropical America) the septa of the many-carpelled fruit form a pulp in which the seeds are embedded; the fruit opens above by teeth or by a pore.

Corchorus capsularis and *C. olitorius*, natives of India, are widely cultivated in the tropics for their tenacious bast-fibres which yield jute. The bast of the Lime is the popular "bast" of the gardener.

Family II. MALVACEAE

Stamens generally indefinite and monadelphous; anthers one-celled; pollen-grains large and spiny (*echinulate*). Leaves simple, palminerved.

Genera about 50; species about 1000.

The cotyledons are epigeal, spreading to the light and forming the first pair of green leaves. They are often cordate or reniform with an entire margin and when, as often happens, the leaves of the adult plant are palmately divided, those succeeding the cotyledons shew a gradual passage from the simple to the lobed form. Our native species are annual, biennial or perennial herbs often, as in Tree-mallow (*Lavatera*), growing very tall. In the warmer parts of the earth they are often shrubby or form large trees. The leaves are alternate, and often more or less palmately divided; it frequently happens that the lower leaves of a shoot are more or less rounded and entire, while the upper are palmately divided. Stipules are present but as a rule fall early.

The leaves and young shoots often bear stellate hairs; mucilage-sacs occur in the tissues, either singly or in rows.

The form of the inflorescence varies. The generally showy flowers are often solitary in the leaf-axils, the leaves being large and foliaceous. Or the latter may be small and bractei-

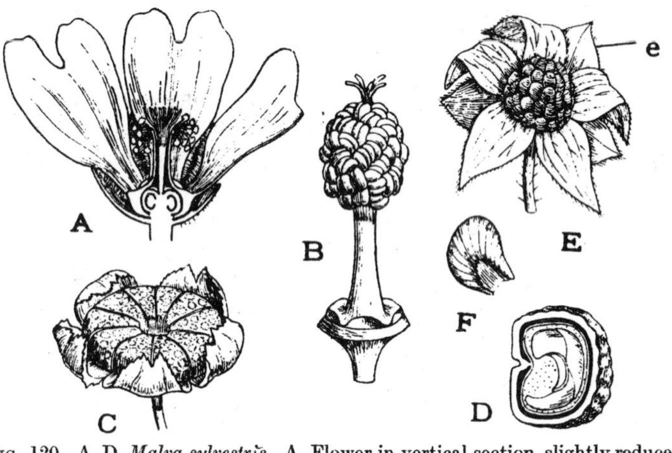

FIG. 120. A–D. *Malva sylvestris*. A. Flower in vertical section, slightly reduced. B. Flower after removal of calyx and corolla, × 2½. C. Fruit, × 2. D. A schizocarp with seed in vertical section, × 4; note the folded cotyledons. E. Fruit of *Malope malacoides* surrounded by persistent calyx and epicalyx (*e*). F. Single schizocarp of same, × 3. (A–D after Baillon; E, F after Gürke.)

form, and the inflorescence a raceme; or more complicated arrangements occur. In most of the genera an epicalyx is present, formed by a whorl of bracteoles just beneath the calyx and serving, like the latter, which it resembles in colour and texture, to protect the more delicate floral organs in the bud. The character of the epicalyx is useful in distinguishing genera. Occasionally, as in *Abutilon* (fig. 121), it is absent; when present the number, cohesion and position of the individual bracteoles afford distinctive characters, as in the case of the three British genera, *Althaea*, *Malva*, and *Lavatera*. The position of the median sepal varies according to the number of bracteoles. Its usual place in pentamerous dicotyledons is posterior; this holds good in those Malvaceae which have no epicalyx, and where, as in *Hibiscus*, the epicalyx consists of numerous bracteoles. Where, however, the latter are few, the odd sepal is anterior.

The petals, which are often large, forming a showy corolla, are twisted in the bud, and are more or less asymmetrical, a character which is generally correlated with twisted aestivation. They are free to the base, where they are attached to the staminal tube, and fall with the tube when the flower withers. The fertile stamens are generally very numerous and are considered to have arisen by the multiplication of five epipetalous members. This view is supported by the study of the development of the flower and also by the fact that in the two species in which only five stamens have been observed these are opposite the petals. The staminodes, which are represented by teeth on the summit of the staminal tube, represent an episepalous whorl. The reniform, one-celled anthers open by a slit which runs across the top and divides the anther into two valves; they are turned outwards at the time of dehiscence. The pollen-grains are large, spherical and covered with spines, and, together with the method of dehiscence of the anthers, afford the best diagnostic character of the family.

The number of the carpels varies from one to many. When there are five they are epipetalous as in *Abutilon* or episepalous as in *Hibiscus* and *Sida*. In *Pavonia* both arrangements occur in different species. In the three British genera, and in many others, the carpels are numerous and form a whorl round the summit of the axis in the centre of the flower, the styles rising from the centre as a single united axis dividing above into a corresponding number of stigma-bearing branches. In the small tribe *Malopeae* (fig. 120, E) (*Malope trifida*, a Mediterranean species, is a well-known garden plant) the numerous carpels are arranged one above another in more or less vertical rows.

One or more anatropous ovules are attached to the inner angle of the carpel. Their position varies; they are generally ascending, but may be pendulous or horizontal. In *Abutilon* and other genera the position varies in one and the same cell.

The flowers are proterandrous. When the flower opens the anthers occupy the centre and the unripe stigmas are hidden inside the staminal tube. After dehiscence of the anthers the filaments bend backwards and finally the matured stigmas spread out and occupy the same position in the centre of the flower.

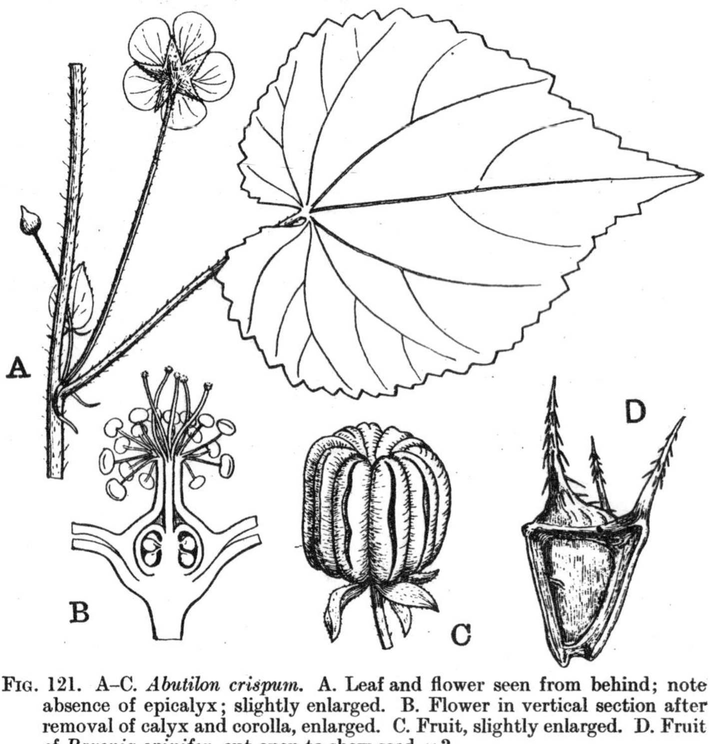

FIG. 121. A–C. *Abutilon crispum*. A. Leaf and flower seen from behind; note absence of epicalyx; slightly enlarged. B. Flower in vertical section after removal of calyx and corolla, enlarged. C. Fruit, slightly enlarged. D. Fruit of *Pavonia spinifex*, cut open to shew seed, × 3.

Pollination is effected by insects which come for nectar secreted in five pits, one of which occurs by the side of the base of each petal. The nectar is protected from rain by hairs on the lower margins of the petals. Self-pollination has been observed in small-flowered forms, such as *Malva rotundifolia*, the style-arms twisting so as to bring the stigmatic surfaces into contact with the anthers. Writing of *Malva sylvestris* and *M. rotundifolia* (both British species), H. Müller says: "These species often occur together and flower side by side for months at a time. In the struggle for existence *M. rotundifolia* has the advantage in being content with poorer soil, in the appearance of its flowers from one to several weeks earlier, and in the possibility of regular self-fertilisation; *M. sylvestris*, on the other hand, in its more vigorous growth, and much greater attractions for insects. These advantages seem to balance one another, for, about Lippstadt at least, both species grow together in equal abundance." In the larger brighter-flowered plant, which receives very numerous visits, the ends of the filaments curl outwards before the stigmas are mature so that self-pollination is impossible (fig. 120, A). *M. rotundifolia*, the smaller paler flowers of which attract few insects, regularly pollinates itself, its anthers remaining extended in such a position as to be touched by the papillate sides of the curling stigmas. Species of *Abutilon* and *Hibiscus* are visited by humming birds. Cleistogamic flowers occur in *Pavonia* and *Malvastrum*.

The fruits are mostly dry (*Malvaviscus* has berries), and their method of dehiscence also affords good diagnostic characters. In the Mallow tribe—*Malveae* (including *Malva, Althaea, Abutilon* and *Sida*)—they are generally one-seeded indehiscent schizocarps (fig. 120, C, D) separating from the persistent central column and from one another; in some genera the carpel and fruit are two- to many-seeded, as in *Abutilon* (fig. 121, C), and in some cases two-celled by the formation of a horizontal septum; the schizocarp then splits lengthwise into a pair of valves. In the *Hibisceae*, which includes *Hibiscus* (fig. 123, B) and *Gossypium* (Cotton-plant) (fig. 122, D), the fruit is a capsule dehiscing loculicidally.

The schizocarps sometimes bear barbed hook-like processes on their outer wall as in *Pavonia* (fig. 121, D); in other cases distribution is favoured by a hairy covering to the seed, an extreme case of which is seen in the Cotton-plant, the seeds of which are buried in a mass of long tangled hairs. The seed generally contains an embryo with large much folded cotyledons and a scanty supply of endosperm.

The species, which are very unequally distributed among the genera, occur in all regions except very cold ones; the number of species increases as we approach the tropics. *Malva rotundifolia* is the most northern, occurring at a latitude of 65° in Russia and Sweden, while species of *Hoheria* and

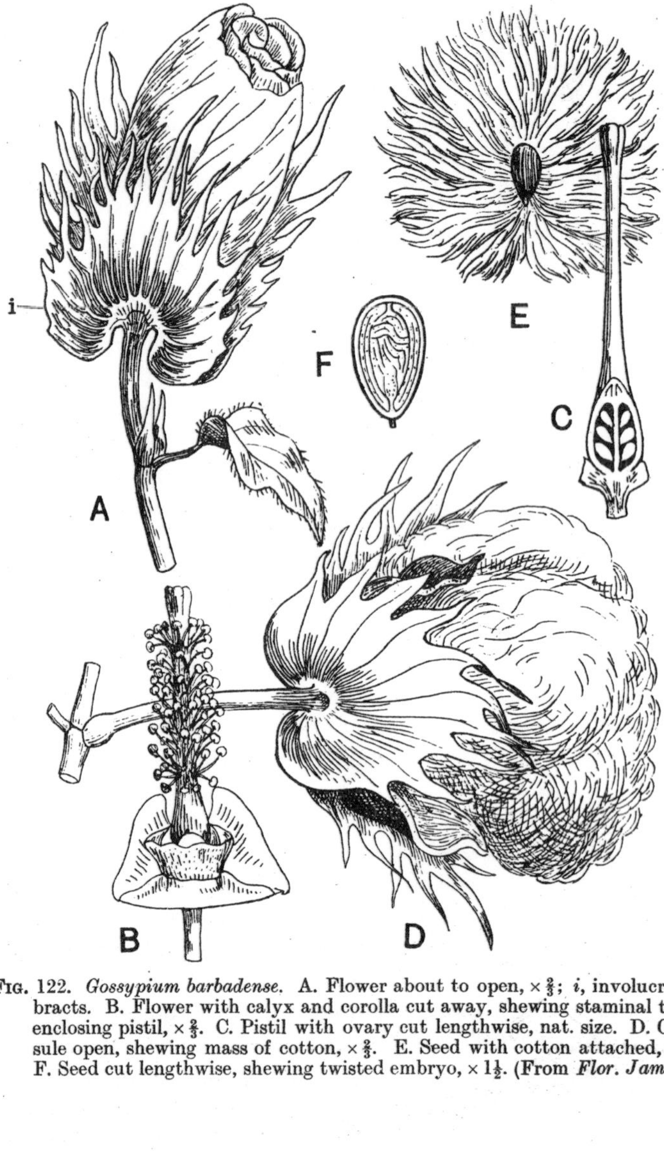

FIG. 122. *Gossypium barbadense.* A. Flower about to open, × $\frac{2}{3}$; *i*, involucre of bracts. B. Flower with calyx and corolla cut away, shewing staminal tube enclosing pistil, × $\frac{2}{3}$. C. Pistil with ovary cut lengthwise, nat. size. D. Capsule open, shewing mass of cotton, × $\frac{2}{3}$. E. Seed with cotton attached, × $\frac{2}{3}$. F. Seed cut lengthwise, shewing twisted embryo, × $1\frac{1}{2}$. (From *Flor. Jam.*)

Plagianthus reach as far as 45° south latitude in New Zealand. On the South American Andes, Malvaceae reach a considerable altitude, with a dwarf alpine habit—a thick root, and a crowded tuft of radical leaves from amongst which spring large showy short-stalked flowers[1].

Hibiscus, the largest genus, with 200 species, is widely distributed, chiefly in the tropics; two species, *H. Trionum* and *H. roseus*, occur in Europe. *H. Rosa-sinensis* is the scarlet Hibiscus generally cultivated as a garden flower throughout the tropics and often

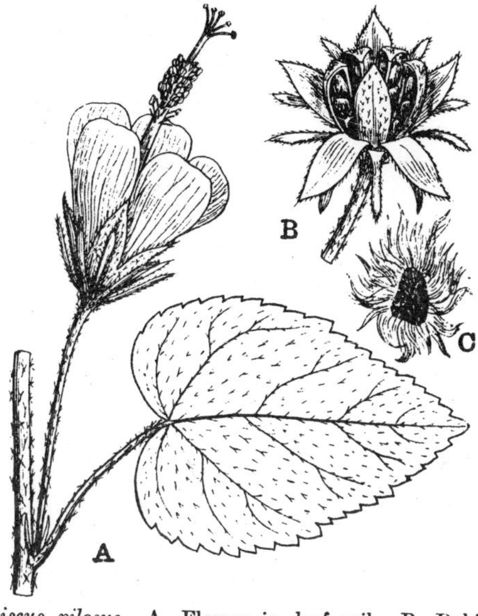

FIG. 123. *Hibiscus pilosus*. A. Flower in leaf-axil. B. Dehiscing capsule. C. Seed, × 2½. A and B slightly enlarged. (From *Flor. Jam.*)

found in greenhouses in temperate regions. *Abutilon* has about 100 species chiefly in the tropics of both hemispheres. *Lavatera* (20 species) is principally Mediterranean; *L. arborea* (Tree-mallow), a maritime plant, occurs on the coasts of Europe including Great Britain south of the Clyde; two species are found in the Canaries and one in Australia. *Althaea* has about 15 species in the temperate and warm regions of the whole world (two are British); *A. rosea*, the Hollyhock, is an eastern Mediterranean species. *Malva* has about 30 species in temperate Europe (three in Britain), Asia, North Africa and North America.

[1] See Hill, A. W. "The Acaulescent Species of *Malvastrum*." *Journ. Linn. Soc.* (*Botany*), XXXIX, 216 (1909).

Many genera are largely or exclusively American, *Howittia* (one species) and *Plagianthus* (10 species) are confined to Australasia, *Hoheria* (four species) is endemic in New Zealand, *Lagunaria* (one species) in Norfolk and Howe Islands and East Australia, while *Julostylis* is a monotypic genus in Ceylon. Many forms of *Gossypium* are cultivated as sources of cotton throughout the tropics; *G. barbadense* (Sea-island Cotton), *G. peruvianum* (South American Cotton), and *G. hirsutum* (Short Staple Cotton) are natives of tropical America; *G. herbaceum* is a native of the East Indies and *G. arboreum* (Tree Cotton) of tropical Africa.

Family III. BOMBACACEAE

An exclusively tropical family, especially American, containing 140 species in 22 genera, closely allied to Malvaceae but differing in the following points. The anthers contain one, two or sometimes more cells, and the pollen-grains are generally smooth, never spiny; staminodes are often present. The ovary is formed from two to five carpels (when five are present they stand opposite the petals), with two to numerous erect, anatropous ovules. The smooth seeds are sometimes embedded in a pith-like tissue or in "wool" which is formed by an outgrowth from the pericarp-wall, sometimes they have an aril; endosperm is absent or forms a very thin layer.

The plants are trees, generally tall and with often very thick trunks as in the Cotton-tree, *Ceiba* (*Eriodendron*) *pentandra* (tropics of Old and New Worlds), or the Baobab, *Adansonia digitata*, of tropical Africa, the stem of which reaches an enormous thickness, but the wood is very light and soft. The leaves are entire or digitate (Baobab), with stipules which soon fall; hairs when present are stellate or form stalked scales.

The principal genera are *Adansonia* (10 species, tropical Africa, Australia), *Bombax* (50 species, mainly American), *Ceiba* (*Eriodendron*) (ten species mainly American) and *Durio* (15 species in the Indo-Malayan region); *D. zibethinus* is the Durian, the fruit of which has a delicate flavour associated with an unpleasant smell.

Family IV. STERCULIACEAE

An almost exclusively tropical family in both hemispheres containing about 700 species in 58 genera. The plants are trees (sometimes cauliflorous, fig. 124 A), shrubs or herbs, sometimes lianes, with simple, entire, more rarely lobed or digitate, alternate leaves with generally short-lived stipules. The flowers, which are borne in complicated inflorescences, are generally bisexual but sometimes unisexual by abortion, as in *Sterculia* and *Cola*. They are regular, or sometimes zygomorphic, and pentamerous. The petals are often reduced in size and sometimes wanting as

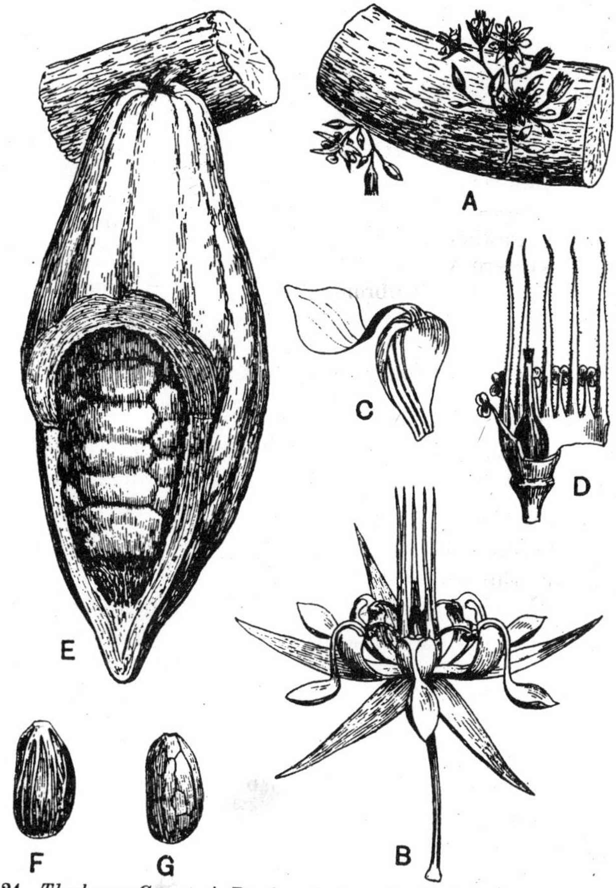

FIG. 124. *Theobroma Cacao*. A. Portion of a branch with flowers, × $\frac{2}{3}$. B. Flower, × 4. C. Petal, × 8. D. Staminal column partly cut to shew pistil, × 5. E. Pod cut open, × $\frac{1}{2}$. F. Seed, × $\frac{2}{3}$. G. Seed with testa removed, × $\frac{2}{3}$. (C–G after Bentley and Trimen.)

in *Sterculia* and *Cola*. The stamens are typically in two whorls; those of the outer whorl (antesepalous) are staminodial or wanting, those of the inner (antepetalous) are fertile; the filaments are more or less united to form a tube; the anthers are two-celled. The generally five-chambered ovary is sometimes carried up along with the stamens above the petals by development of an internode (androgynophore); the carpels are generally antepetalous and contain two to numerous anatropous ovules with an outwardly directed micropyle in each chamber. The fruit is generally dry, often separating into cocci. The seed contains a fleshy endosperm and an embryo with flat, folded or rolled leaf-like cotyledons.

Theobroma Cacao, the Cocoa tree, a native of tropical America, is now widely cultivated in the tropics. It bears large reddish-yellow fleshy fruits; in each of the five chambers is a double row of almond-like seeds containing an embryo with a pair of very large folded oily cotyledons. The last are the source of chocolate and cocoa. Another useful plant is *Cola acuminata* (Cola-nut), a native of western tropical Africa, the seeds of which yield the alkaloids Theïn and Theobromin.

Order 8. *TRICOCCAE*

This order comprises families with unisexual, rarely bisexual, hypogynous flowers, which are generally regular with a single whorl of free perianth-leaves, sometimes naked. The stamens are generally equal in number to and opposite the perianth-leaves, and pollination is generally anemophilous. Growth of the pollen-tube is often endotropic. The most constant floral character is the pistil of three carpels forming a trilocular ovary, each chamber containing in the inner angle one or two pendulous anatropous ovules with a ventral (rarely dorsal) raphe. The great majority of plants belonging to this order are included in the large family Euphorbiaceae.

Family I. EUPHORBIACEAE

Flowers monoecious or dioecious, sometimes much reduced by abortion, regular or slightly irregular. Calyx and corolla both present, or the latter or both absent. Male flower: stamens as many as the perianth-leaves, or double as many, or very numerous or few or one (in *Euphorbia* the whole flower consists of a single naked stamen); filaments free or united, anthers two-celled; rudiment of ovary present or absent. Female flower: staminodes present or absent; ovary generally three-celled, sometimes two- or four-celled;

ovules one or two in each chamber, collateral, pendulous, anatropous, with a ventral raphe; the micropyle carunculate. Fruit generally a capsule splitting into three cocci which separate from a persistent central column. Seeds with abundant endosperm and a large central straight or bent embryo.

The habit shews great variety including small annual herbs and trees. The inflorescence is usually compound, and in its ultimate ramifications cymose.

A large family, containing more than 220 genera with about 4000 species, chiefly tropical.

Our British representatives, species of *Euphorbia* (Spurge) and *Mercurialis*, are herbs, but the greater number are woody plants and often trees. Even in the same genus, however, the utmost diversity occurs. For example, several of our native Spurges are small annual herbs, other species of *Euphorbia* form erect sturdy bushy plants several yards high, while in the desert regions of tropical Africa and elsewhere species occur with a Cactus-like habit, the stem thick and fleshy and the leaves reduced to spines. In Australia and at the Cape a suffruticose heath-like habit is common. Many species of the widely distributed tropical genus *Tragia* are climbers. *Phyllanthus* contains, besides annual herbs, also trees, while many members forming the section *Xylophylla* have their ultimate branches leaf-like (phylloclades), the leaves being reduced to scales.

The form and position of the leaves are equally varied. The leaves are generally alternate but may be opposite or whorled; in many cases upper branches bear opposite leaves, while in the rest of the plant a scattered arrangement occurs. They are simple and entire or deeply cut (as in *Ricinus*, Castor-oil plant) or even palmately-compound. The venation is pinnate or palmate.

The inflorescence shews considerable variation. Generally the first branchings are racemose, the later cymose. The partial inflorescences are closely crowded cymes arranged in spikes, or standing in the axils of the upper leaves. They are generally unisexual, the male frequently containing numerous flowers, while the female flowers are solitary. The partial inflórescence or cyathium of the *Euphorbieae* has a superficial resemblance to a bisexual flower (fig. 125). It is a cyme the terminal flower of which is female and naked; below the latter are four or five bracts which are connate and form a calyx-like involucre. In the axil of each bract is a scorpioid cyme

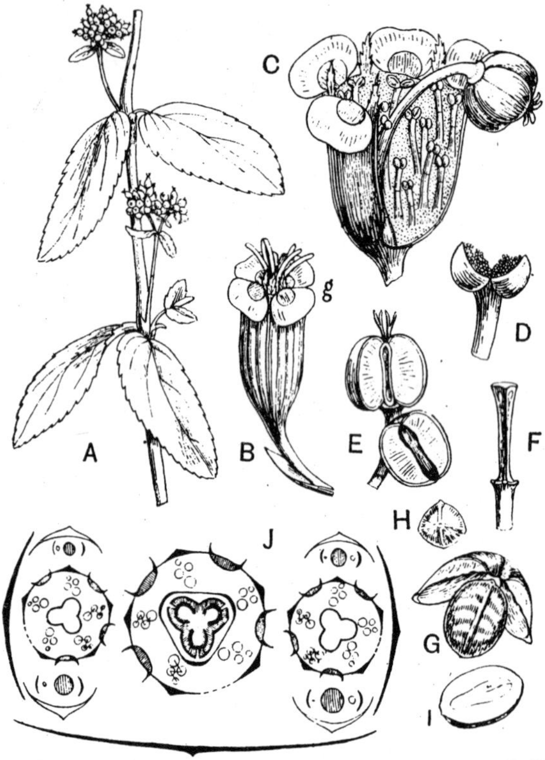

FIG. 125. *Euphorbia hypericifolia*. A. Part of flowering branch, $\times \frac{2}{3}$. B. Cyathium, $\times 16$; *g*, glands of involucre. C. Cyathium cut open, $\times 24$. D. Stamen, $\times 100$. E. Capsule, one coccus separated, $\times 10$. F. Columella of capsule, $\times 15$. G. Coccus with seed, $\times 12$. H. Seed viewed from the end. I. Seed cut lengthwise shewing embryo. J. Diagram of partial inflorescence of *E. Peplis* shewing details of three cyathia. (J after Eichler.) (From *Flor. Jam.*)

of male flowers each consisting of a single stamen. Between the segments of the involucre are large, frequently oval or crescent-shaped, glands. In *Euphorbia* the glands are always free, in the African genus *Synadenium* they are united into a ring. The glands are often brightly coloured, and in some

Euphorbias bear large bright-coloured petaloid appendages. In *Anthostema* both male and female flowers have a tubular perianth (fig. 126, A, B); in the remaining genera of this tribe the male are completely naked and the female have only occasionally a rudimentary perianth. There is, however, a joint in the axis which bears the anther, indicating a distinction of an upper portion, the stamen, from a lower, the pedicel; the external appearance of the two portions is also frequently different, for instance, owing to presence or absence of hairs.

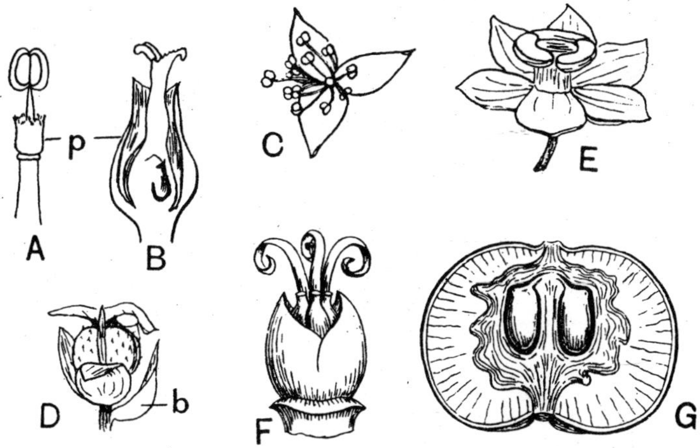

FIG. 126. A. Male flower of *Anthostema*. B. Female flower of same; *p*, perianth. C. Male, D. Female flower of *Mercurialis*; *b*, bract. E. Male flower of *Phyllanthus cyclanthera*. F. Female flower of *Hippomane* with three styles removed. G. Fruit of same in section. All enlarged except G which is slightly reduced. (A, B after Warming; C, D after Le Maout and Decaisne; E after Baillon.)

The flowers in the family generally are regular and hypogynous; perigyny is rare (e.g. in *Bridelia*, a wide-spread tropical Old World genus).

The form of the flower varies widely. Thus *Wielandia*, a monotypic genus from the Seychelle Islands, has pentamerous flowers; the two outer whorls, forming calyx and corolla respectively, are followed in the male by a regularly alternating whorl of five stamens, and in the female by five antepetalous carpels. In the male flower the carpels are represented by five rudimentary structures.

In the large tropical genus *Croton* (nearly 700 species) a pentamerous calyx and corolla are generally present, but the

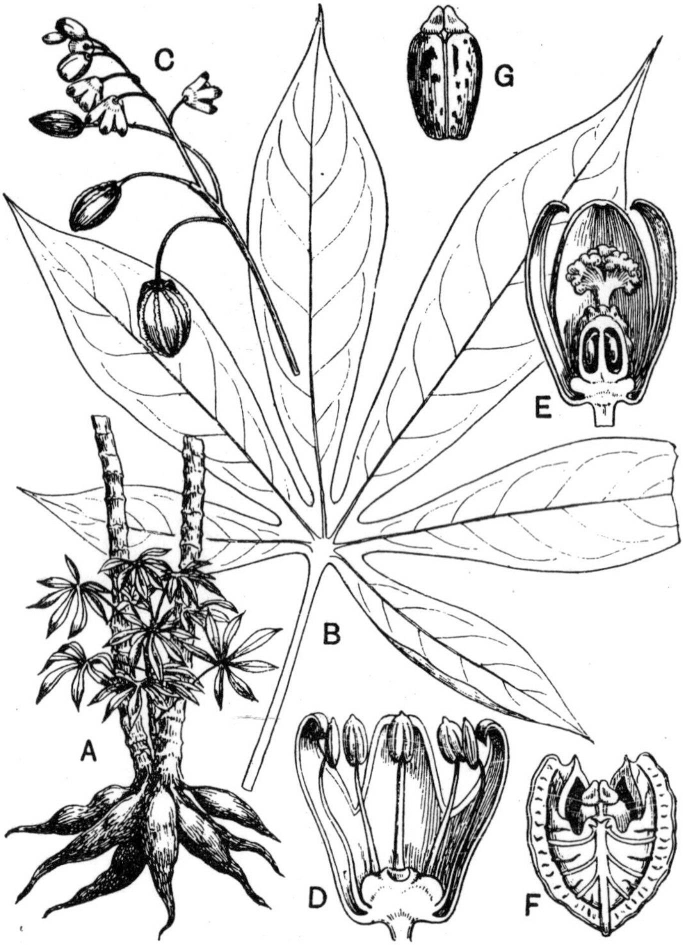

FIG. 127. *Manihot utilissima.* A. Lower part of stem shewing young shoot and tubers, much reduced. B. Leaf, nat. size. C. Portion of inflorescence, × $\frac{1}{3}$. D. Male flower cut lengthwise, × 5. E. Female flower cut lengthwise, × 2. F. One carpel of the ripe fruit shewing the seed, nat. size. G. Seed, nat. size. (C—G after Tussac.) (From *Flor. Jam.*)

latter may be inconspicuous or absent in the female flower. The stamens are often very numerous, some species having 80–100. With very few exceptions there are three carpels.

The two perianth-whorls are isomerous, and though they are very frequently pentamerous, other numbers, such as three, or more rarely four, occur.

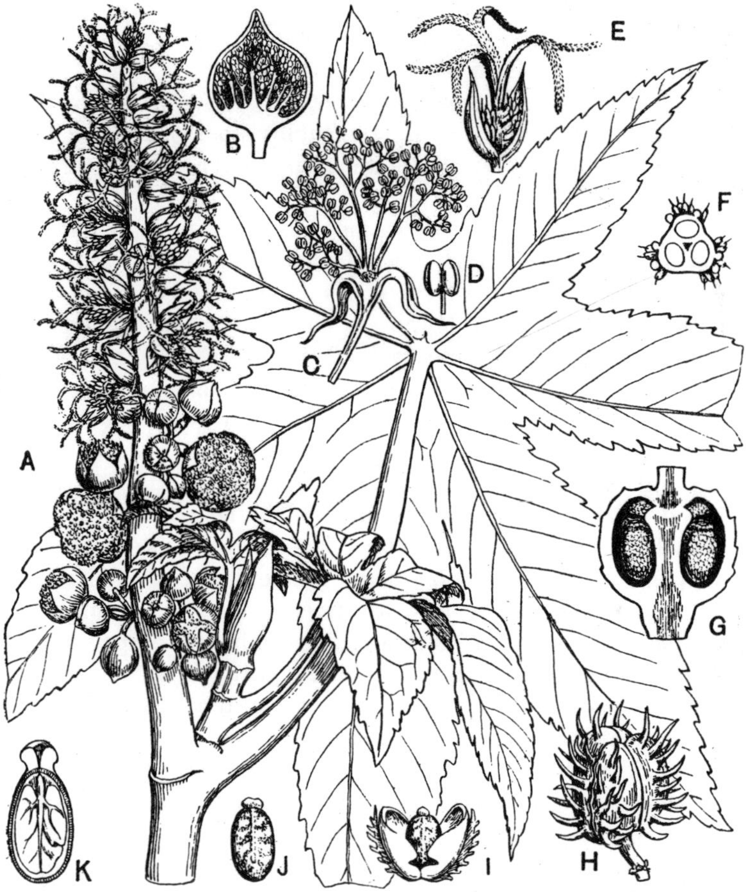

FIG. 128. *Ricinus communis*. A. Upper part of flowering branch, × ⅔. B. Unopened male flower, × 2. C. Ditto open, × 2. D. Anther, × 10. E. Female flower, × 2. F. Ovary cut across, × 4. G. Ovary cut lengthwise shewing ovules with obturator, × 10. H. Capsule, nat. size. I. Coccus split open, nat. size. J. Seed, nat. size. K. Seed cut to shew embryo, × 1½. (Partly after Baillon and Bentley and Trimen.) (From *Flor. Jam.*)

A great many genera are apetalous. In *Manihot* (fig. 127), a large tropical American genus (140 species) containing *M. utilissima* (Manioc or Cassava), the calyx is often large and petaloid. Our native Dog's Mercury (*Mercurialis*) has minute flowers with three sepals, followed in the male by 8–20 stamens, in the female by a bicarpellary pistil (fig. 126, C, D). The greatest amount of reduction occurs in *Euphorbia*, where, as already mentioned, the male consists of a single stamen, the female of a tricarpellary ovary.

The aestivation of the calyx, whether imbricate or valvate, supplies useful characters for the subdivisions of the family. When present the corolla is generally polypetalous, rarely gamopetalous.

The stamens are free or more or less monadelphous, especially below. In *Ricinus* (Castor-oil plant) and allied genera the androecium forms a branching tree-like structure, with the anther-cells borne on short ultimate branchlets (fig. 128). In *Phyllanthus cyclanthera* (fig. 126, E) not only are the filaments united but also the anthers, the latter forming a closed ring, as in *Cyclanthera*, a genus of Cucurbitaceae. The divisions between the two chambers of each half-anther are generally absorbed so that the anther is two-celled. The ovary, in conformity with the number of the carpels, is generally trilocular. The three styles may be simple, sometimes even large and petaloid, but are generally more or less bipartite; they are free or more or less united. *Mercurialis*, as already indicated, has only two carpels. The ovules are generally provided with a caruncle or obturator which persists in the seed. It is an outgrowth from the placenta at the base of the funicle, which covers the micropyle like a lid or hood and plays an important part in the conduction of the pollen-tube[1]. Where there are two ovules in a chamber, each may have a caruncle or one may cover both.

Outgrowths of the floral axis, forming a disc, are common; in the female generally in the form of a ring or cup round the ovary, in the male generally as free or united extrastaminal, rarely intrastaminal, glands.

Owing to the complete separation of the sexes, cross-pollination is necessary. *Mercurialis* and others with long thread-like styles are anemophilous; but in many cases insects

[1] See Schweiger, J. "Beiträge zur Kenntnis der Samenentwicklung der Euphorbiaceen." *Flora*, XCIV, 339 (1905).

are the pollinating agents. The brightly-coloured glands or bracts such as occur in many Euphorbias (e.g. *E. pulcherrima*, the Poinsettia of greenhouses), *Dalechampia* and others, the petaloid calyx of *Manihot*, or even the petaloid disc-segments as in the Mascarene *Petalodiscus*, serve as a means of attraction, while nectar-secretions are frequent, as in the involucral glands of *Euphorbia*.

The fruit is generally a capsule, splitting into three cocci (figs. 125, 127, 128) which separate from a central column and split lengthwise into two valves. Berries and drupes also occur. The Mancinil (*Hippomane mancinella*) of Central America and the West Indies, a tree with a very poisonous latex, has a several-seeded drupe (fig. 126, G); and *Bischofia*, another monotypic genus, widely distributed through tropical Asia and the Pacific islands, has a fleshy berry. *Hura crepitans*, the Sand-box of tropical America, has numerous carpels which, in the ripe capsule, separate with great violence into as many woody cocci. The caruncle on the seed has been shewn in species of *Mercurialis* and *Euphorbia* to function as an elaiosome (oil-body) attracting ants which aid in dissemination.

The cotyledons lie flat in the endosperm, as in the Castor-oil seed, or are bent or folded. The relative breadth of the cotyledon affords a character for distinguishing the two primary subdivisions of the family, viz. *Platylobeae* with cotyledons much broader than the radicle (by far the larger division) and *Stenolobeae* with narrow cotyledons about as broad as the radicle.

Alchornea (*Coelebogyne*) *ilicifolia*, a native of Australia, is of interest as one of the few instances of the adventitious production of embryos from cells of the nucellus. Polyembryony, or the presence of more than one embryo in a seed, is a frequent result.

With the exception of the arctic and cold alpine zones, Euphorbiaceae are distributed over the whole earth. They occupy widely different localities. On the one hand are the very characteristic desert types, on the other marsh plants, such as *Caperonia*, which spreads from Brazil to Central America and the West Indies, and is also represented in tropical Africa. Not a few are tropical forest-trees, while many, such as the Spurges and Dog's Mercury, are widely distributed weeds of cultivation. Others, such as Manioc (*Manihot utilissima*) and the Soap-tree (*Sapium sebiferum*), are now spread by cultivation far beyond their original area. *Euphorbia* (British) is cosmopolitan and contains about 1000 species, chiefly in subtropical regions, occurring more sparsely in tropical and temperate zones. *Mercurialis*, also British, has only seven species, chiefly Mediterranean; one is found in Eastern Asia.

The chief centre of distribution of the family in the Old World is the Indo-Malayan region; in the New World, Brazil. The *Stenolobeae*, characterised by narrow semi-cylindrical cotyledons and generally heath-like habit, are, with the exception of one genus, *Dysopsis*, exclusively Australian. *Dysopsis* occurs on the South American Andes and in Juan Fernandez.

Many of the species are poisonous; some, as the South African *Toxicodendron*, are among the most poisonous known plants. Others have been or are still widely used as medicines, for instance the Spurges. Castor-oil is obtained from the oily endosperm of *Ricinus communis*, a plant probably native in tropical Africa, but now, as a result of cultivation, spread through all the warmer parts of the world. *Aleurites moluccana* and *Sapium sebiferum* are oil- and fat-yielding plants; species of *Hevea*, *Mabea*, *Manihot* and *Sapium* yield caoutchouc, and species of *Croton* and *Euphorbia*, resin. *Manihot utilissima* is one of the most important food-plants of the tropics, having a thick tuberous root very rich in starch; it is the source of Brazilian arrowroot. Species of *Codiaeum* (the Crotons of hot-house cultivation are leaf-varieties of *C. variegatum*), *Phyllanthus*, *Jatropha* and *Euphorbia* are widely grown as ornamental plants. In the tropics the thorny Euphorbias make excellent living fences.

The following is an outline of the subdivision of the family adopted by Pax in the *Pflanzenfamilien*.

A. *PLATYLOBEAE*. Cotyledons broad.

Subfamily I. PHYLLANTHOIDEAE. Two ovules in each ovary-chamber. Laticiferous tissue and internal phloem absent.

(*a*) Embryo large.

α. Calyx of male imbricate. Tribe 1. *Phyllantheae*.

β. Calyx of male valvate. ,, 2. *Bridelieae*.

(*b*) Embryo small. ,, 3. *Daphniphylleae*.

Subfamily II. CROTONOIDEAE. One ovule in each chamber. Laticiferous tissue present or absent; typical internal phloem occasionally present.

(*a*) Partial inflorescence not a cyathium.

α. Stamens bent sharply inwards in the bud. Calyx of male imbricate or valvate; corolla generally present. Tribe 1. *Crotoneae.*

β. Stamens erect in bud.

1. Calyx of male valvate. Flowers generally apetalous. Inflorescence racemose, spicate or paniculate; axillary or terminal. Tribe 2. *Acalypheae* (includes *Mercurialis*).
2. Calyx of male valvate or almost imbricate, male flowers with or without a corolla. Flowers in a dichasium. Tribe 3. *Jatropheae.*
3. Calyx of male valvate, more rarely imbricate. Flowers always apetalous, in simple, terminal spikes or racemes. Tribe 4. *Manihoteae.*
4. Calyx of male imbricate; male flowers with a corolla. Tribe 5. *Cluytieae.*
5. Calyx of male imbricate. Flowers apetalous. Laticiferous tubes segmented. Tribe 6. *Gelonieae.*
6. Calyx of male imbricate. Flowers apetalous. Laticiferous tubes unsegmented. Tribe 7. *Hippomaneae.*

(*b*) Partial inflorescence a cyathium. Tribe 8. *Euphorbieae.*

B. *STENOLOBEAE.* Cotyledons narrow.

Subfamily III. PORANTHEROIDEAE. Two ovules in each chamber.

Subfamily IV. RICINOCARPOIDEAE. One ovule in each chamber.

Family II. BUXACEAE

Buxaceae is a small family containing six genera and about 60 species, about two-thirds of the latter being included in the genus *Buxus*. It was formerly included in Euphorbiaceae but differs in the dorsal, not ventral, raphe of the anatropous ovule, and also in the loculicidal dehiscence of the fruit. They are all perennial evergreen plants, generally shrubs or trees. *Buxus sempervirens* (Box) (fig. 129, A–D) grows wild on the chalk hills of Kent, Surrey (Box-hill), Bucks and Gloucester. It is a native of Central and Southern Europe and North Africa and extends through Asia to the temperate Himalayas, China and Japan. Its leaves have been found fossil in Tertiary deposits in France.

The species of *Buxus* fall into two sections, *Eubuxus* confined to Europe, Asia and Africa and *Tricera* in the West Indies.

Of the other genera, *Sarcococca* (four species) is tropical Asiatic; *Pachysandra* has two species, one in the Alleghany mountains, the other in Japan; *Styloceras* has three species in the tropical regions of the South American Andes; *Notobuxus*, one species in Central Africa, and one in Natal; and *Simmondsia* is a monotypic Californian genus.

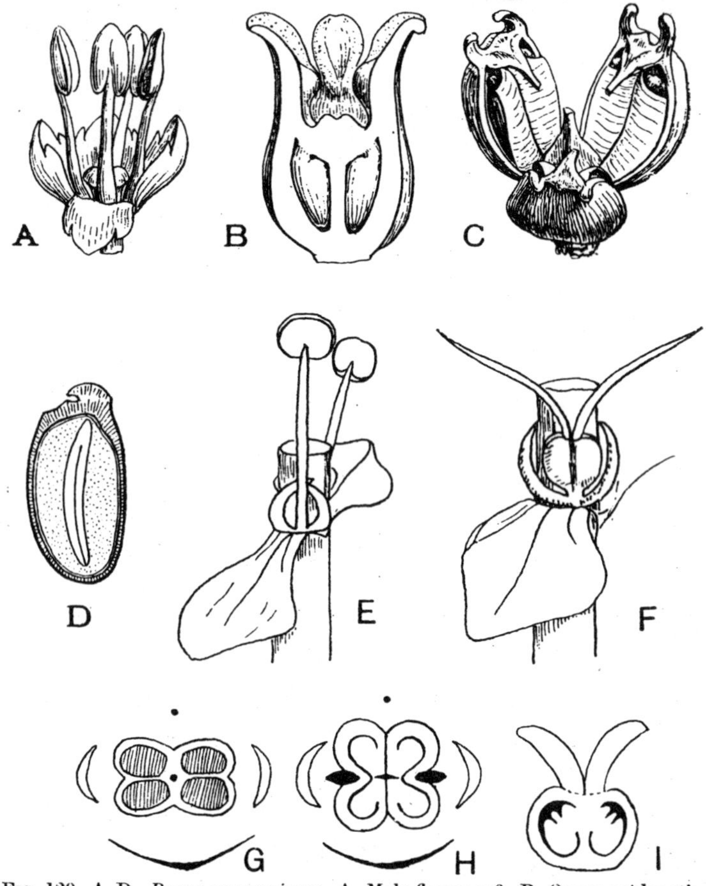

FIG. 129. A–D. *Buxus sempervirens*. A. Male flower, × 6. B. Ovary cut lengthwise, enlarged. C. Capsule dehiscing, × 2½. D. Seed cut lengthwise, × 4. E–I. *Callitriche verna*. E. Portion of shoot bearing two axillary male flowers. F. The same with female flower. G. Diagram of male flower. H. Diagram of female flower. I. Vertical section through a young female flower. E, F, I much enlarged. (E, F after Le Maout and Decaisne; G, H, I after Eichler.)

Family III. CALLITRICHACEAE

The family contains a single genus, *Callitriche*, with 26 species of slender terrestrial or aquatic plants, extremely reduced in both vegetative and floral structure. The stem is thin and delicate, and the vascular bundle is greatly reduced. The narrow leaves are opposite and entire, and the upper often form a rosette on the surface of the water, as in *C. verna*. In the terrestrial forms they bear characteristic stellate hairs. The plants are monoecious and the flowers are solitary in the leaf-axils, consisting of a single terminal stamen or of a sessile or shortly-stalked bicarpellary ovary, between a pair of delicate bracteoles (fig. 129, E–I), which are absent in *C. autumnalis*. The ovary is two-lobed and each cell becomes divided by a false partition into two, in each of which is a pendulous anatropous ovule with a ventral raphe and only one integument. There are a pair of long styles which, like the two carpels, are placed transversely. An accessory shoot is sometimes borne in the leaf-axil beneath the principal shoot, and may be vegetative or floral. When both the principal and accessory shoots are floral the upper is usually a male, the lower a female flower. As the lower flower has generally no bracteoles the whole has the appearance of a single bisexual flower.

Pollination may occur above the water, or beneath the surface when, as in *C. autumnalis*, the plant is quite submerged. *C. verna* is proterogynous. The lobes of the fruit, which are keeled or winged at the back, eventually separate into four one-seeded indehiscent drupels. The genus is distributed over the greater part of the earth.

The affinity of *Callitriche* is very doubtful. According to Robert Brown, De Candolle, Hegelmaier, Bentham and Hooker and others it is with Haloragaceae. Others suggest a relation with the Sympetalae, which is doubtfully supported by the single integument of the ovule. The position indicated by Lindley, Eichler, and Baillon is the one adopted here. Affinity with Euphorbiaceae is found in the number and position of the ovules, while for the reduction in the flowers we may compare the genus *Euphorbia*.

Baillon includes *Callitriche* in Euphorbiaceae but Pax (in the *Pflanzenfamilien*) separates it as a distinct family on the ground of the septum in each carpel and the single integument of the ovule.

Order 9. *GERANIALES*

Flowers bisexual, regular, more rarely zygomorphic, hypogynous, cyclic and pentamerous, with often obdiplostemonous androecium. Sepals imbricate in the bud, free or sometimes united at the base, often persistent. Petals in the bud imbricate, sometimes twisted, free. Stamens typically in two obdiplostemonous series, sometimes five by abortion of one whorl, sometimes more numerous, free, or filaments united at the base into a ring or tube. Carpels, often reduced to three, united to form an ovary with as many chambers. Ovules one, two or more in the inner angle of each chamber, pendulous and anatropous with the micropyle directed outwards, with two integuments. Endosperm present or absent.

Generally herbaceous plants, sometimes woody, or lianes, with generally simple alternate, sometimes opposite leaves; stipules present or absent.

A well-marked order, allied to the Malvales, from which it is distinguished by the typically obdiplostemonous androecium and the prevalence of herbaceous forms. The flowers are typically regular, but zygomorphy, generally median, occurs in derived forms, sometimes characterising the family, as in Tropaeolaceae and Balsaminaceae, and attaining its highest development in Malpighiaceae where the flower is obliquely zygomorphic.

Family I. GERANIACEAE

Flowers regular or slightly zygomorphic, androecium typically obdiplostemonous, the antepetalous whorl sometimes staminodial, rarely 15 stamens. Ovary generally five-celled with one or two ovules in each chamber, the micropyle directed outwards and upwards. Fruit a capsule, or separating into five one-seeded beaked portions. Embryo straight or curved, completely filling the seed, or with a thin layer of endosperm.

Herbs, or woody below, with alternate or opposite, usually palminerved, stipulate leaves.

About 650 species in 11 genera widely distributed throughout the temperate zones, with comparatively few in the tropics and generally at high levels.

The plants are mostly herbs, or become woody below, and stems and leaves are generally thickly covered with simple or glandular hairs. The stems of many of the South African

Pelargoniums are thick and fleshy and the plant often persists through the long dry season by means of a tuberous thickening (fig. 131, A). The most strikingly xerophytic type is the small South African genus *Sarcocaulon*, the fleshy stem of which bears the persistent spine-like leaf-stalks (fig. 131, B, C).

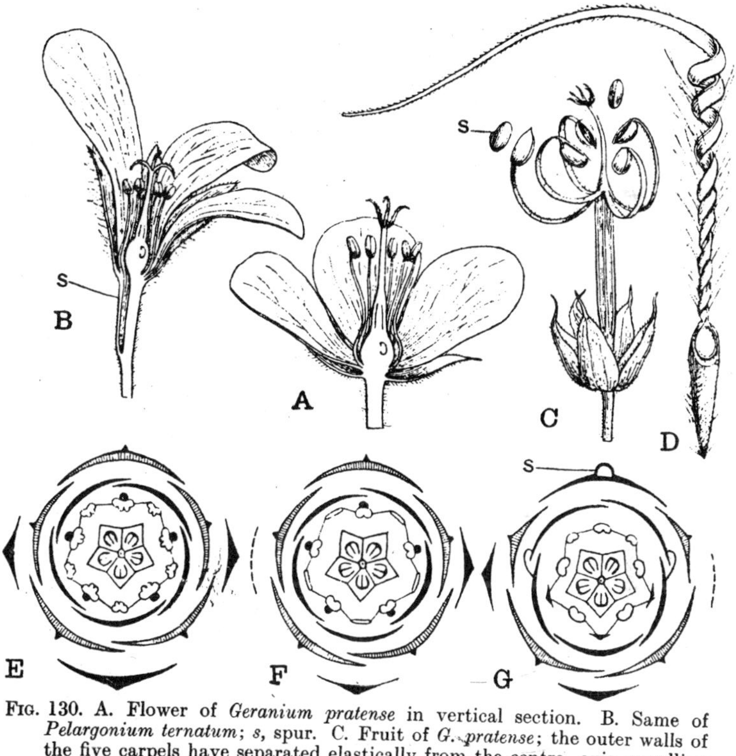

FIG. 130. A. Flower of *Geranium pratense* in vertical section. B. Same of *Pelargonium ternatum*; *s*, spur. C. Fruit of *G. pratense*; the outer walls of the five carpels have separated elastically from the centra axis, expelling the seeds, *s*. D. Coccus of *Erodium moschatum*. E, F, G. Floral diagrams of *Geranium pratense*, *Erodium cicutarium* and *Pelargonium zonale*; *s*, spur on upper sepal. A, B slightly reduced; C, D × 5. (A, B after Reiche; E, F, G after Eichler.)

The leaves are opposite or alternate, with generally a pair of small stipules at the base of the long stalk and a palmi-nerved blade. The inflorescence is generally cymose, and each flower has a pair of bracteoles. The flowers are bisexual, hypogynous, and, except in *Pelargonium* and to a less extent in *Erodium*, regular. The five sepals are imbricate in the bud and free, except in the small South American genus *Viviania*, and in the monotypic genus *Dirachma* from Socotra, where they are united into a tube with valvate limbs. Alternating with the sepals are five free petals, which also are imbricate in the bud. *Pelargonium* has a zygomorphic flower with spur-like tubular hollowing of the floral axis below the posterior sepal, and petals differing in size or shape. Indica-

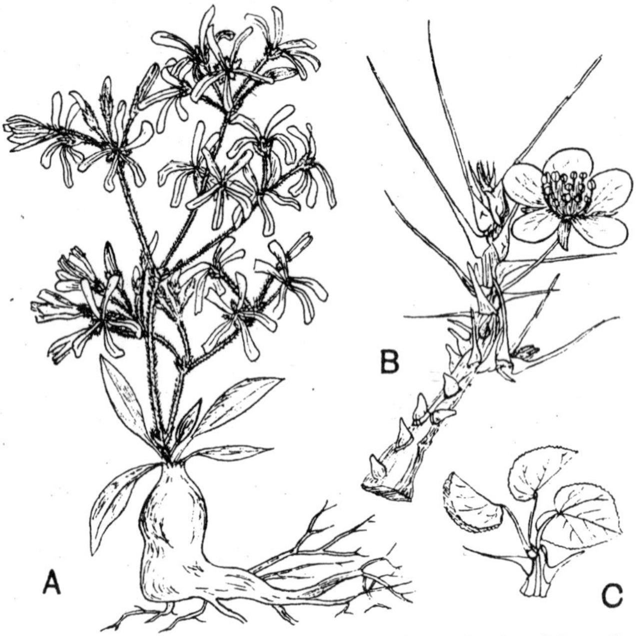

FIG. 131. A. *Pelargonium undulatum*, plant in flower developed from the persistent tuberous stem. B, C. *Sarcocaulon Marlothii*. B. Stem in flower shewing persistent spine-like leaf-stalks. C. Young leaf-bearing shoot. All reduced. (A after Andrews; B, C after Engler.)

tions of zygomorphy occur also in *Erodium*. In *Geranium* the androecium is obdiplostemonous (fig. 130, E); the stamens of the inner whorl are longer than the outer ones, and at the base of each of the former is a nectary. In *Erodium* the stamens of the outer whorl are reduced to scale-like staminodes; while in *Pelargonium* only from two to four are fertile. No satisfactory explanation of the obdiplostemony in Geraniaceae has been given. We can only state the fact that above the whorl of petals there is a break in the regular alternation of the members of successive whorls, the stamens of the outer whorl being opposite and not alternate with the petals. The inner stamens and the carpels follow in regular order. There is no indication of the suppression of a whorl between the corolla and androecium, and the outer whorl of stamens originates before the inner so that there is no question of subsequent displacement. The five (sometimes fewer) carpels unite to form an ovary with as many chambers and pass above into a long well-developed style or "beak," which divides at the top into a corresponding number of slender stigmas. In each chamber, attached to the central column, are one or two, rarely more, pendulous anatropous ovules so arranged that the micropyle is towards the outside and the raphe turned towards the placenta. When ripe the carpels separate along the septa into five one-seeded portions (cocci), which also break away from the central column, either rolling elastically outwards and upwards (fig. 130, C) or becoming spirally twisted above the seed-case. In most Geraniums the cocci are open on the inside so that the seeds escape, often being shot to a considerable distance by the suddenness of the elastic up-rolling. In *Erodium* and *Pelargonium* each coccus is a closed schizocarp (fig. 130, D), and the dispersal and planting of the contained seed are promoted by the long twisted upper portion which becomes separated from the central column and forms an awn. Numerous bristles or hairs on the upper part of the awn favour distribution by currents of air, while the very hygroscopic coiled portion bearing stiff upward pointing hairs, by becoming more or less closely wound with the varying amount of dampness in the air, works the pointed base containing the seed into the ground.

In *Geranium*, *Erodium*, *Pelargonium* and two allied genera the embryo fills the seed and is more or less bent, while the cotyledons are rolled or folded upon each other. In other genera a thin layer of endosperm surrounds the straight or curved embryo. The cotyledons are green while still enclosed in the seed.

The larger-flowered species of *Geranium* are markedly pro

terandrous, the outer stamens, the inner stamens and the stigmas maturing successively. Thus in *G. pratense* each flower passes through three well-marked stages. Each whorl of stamens in turn ripens, standing up prominently in the centre of the flower and shedding the pollen. As the anthers wither the filaments bend outwards, and when all the anthers have diverged the stigmas, hitherto incapable of pollination, become ripe and expanded. By this arrangement self-pollination is prevented. The visits of numerous bees, which come for the nectar secreted by the glands at the base of the inner stamens, ensure cross-pollination. H. Müller refers to it as generally the most conspicuous and abundantly visited plant in the meadows where it grows. In the species with smaller, less conspicuous and less visited flowers, means are taken for rendering possible self-pollination. Thus in *G. molle*, the flowers of which are only $\frac{1}{3}$ to $\frac{1}{2}$ inch across, the divisions of the stigma begin to separate even before the first five stamens have all dehisced and owing to the proximity of dehiscing anthers and stigmas the chances of self-pollination are great.

Erodium cicutarium occurs in two forms, the one having regular homogamous flowers, the other larger flowers which, like *Pelargonium*, are zygomorphic from the difference in size between the upper two and the lower three petals; they are also markedly proterandrous and adapted for insect-pollination.

Of the eleven genera among which the species are distributed, *Geranium* covers the largest area. Its species (250 in number) spread over the temperate regions and a few occur in the mountains in the tropics; they are mostly herbs. Three of our British species, *G. sylvaticum*, *G. pratense* and *G. Robertianum*, also extend into the Arctic zone; while, in the southern hemisphere, *G. patagonicum* and *G. magellanicum* reach the Antarctic. The majority of our eleven native species are confined to Europe, West or North Asia and North Africa, *G. pyrenaicum*, *rotundifolium*, *Robertianum* and *lucidum* extend to India, while *G. dissectum* occurs in North America.

Erodium has 60 species (three in Britain), most of which are confined to the Mediterranean region and West Asia, but some are found in Central and temperate America, at the Cape and in West Australia. *Pelargonium* (250 species), distinguished from *Geranium* by its more or less shrubby habit and zygomorphic flower, has its centre in South Africa, with a few scattered species in other parts of the Old World.

Of the remaining genera, including about 80 species, two, *Monsonia* (mainly tropical and South African) and *Sarcocaulon*—a small xerophytic genus with thick fleshy stems and small leaf-development (fig. 131, B), confined to South and South-West Africa—constitute, with *Geranium*, *Erodium* and *Pelargonium*, the tribe *Geranieae*. The remaining six genera (including 44 species) differ from the *Geranieae* in the absence of a characteristic beak to the ovary and form four tribes distinguished by characters of the calyx, which is sometimes tubular or bell-shaped, and the fruit; *Biebersteinia* occurs in Central and West Asia, four genera are South American, mainly Andine, and *Dirachma* is a monotypic genus from Socotra.

The so-called Geraniums of gardens are artificially produced hybrids belonging to the genus *Pelargonium*, the species of which are readily crossed. Many are products of crossing the large-flowered *P. grandiflorum* with other Cape species. The "zonal Pelargoniums" are produced from crosses of *P. zonale* with *P. inquinans*.

Family II. OXALIDACEAE

The family is represented in Britain by the Wood-sorrel, (*Oxalis Acetosella*). It differs from Geraniaceae in having the ten stamens united at the base, in its five free styles, and in the mode of dehiscence of its fruit, which is a capsule splitting along the dorsal suture of each of the five carpels, or a berry. In *Oxalis* an outer fleshy layer (aril) of the seed-coat separates elastically from an inner hard layer (fig. 132, F), and shoots the seed out of the chamber. The embryo is straight and embedded in a fleshy endosperm.

The plants are generally herbs, which are perennial by means of fleshy tubers or bulbs. The leaves are alternate, stalked, often compound, and generally exstipulate. As may be seen in Wood-sorrel, the leaflets are folded and bent back in the bud and in the sleep-position assumed at night. Phyllodes occur in some species of *Oxalis*.

Considerable variety exists in the relative length of the stamens and styles in the genus *Oxalis*. Hildebrand, who investigated a large number of species, found that twenty were trimorphic, having short-, median- and long-styled forms respectively, with five stamens at each of the two remaining levels (fig. 132, D), fifty-one were dimorphic, while thirty had only the one form of flower. In the last class are *O. Acetosella* and the two species *O. corniculata* and *O. stricta* which, though found in England, are not generally considered to be indigenous.

Experiments with several trimorphic species shewed that productiveness was greatest after a legitimate cross, that is, by the

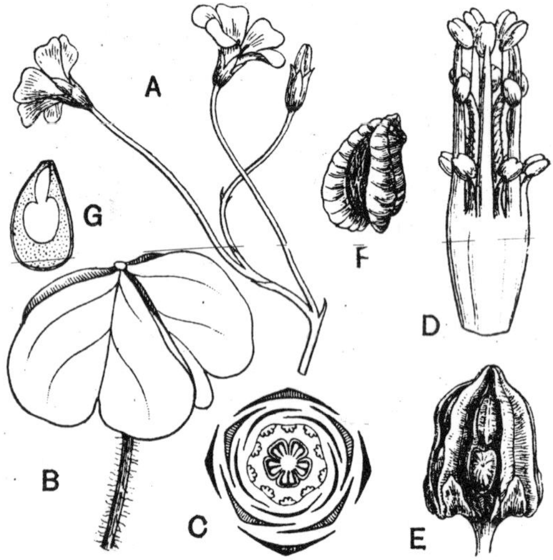

FIG. 132. *Oxalis.* A, B, D. *O. corymbosa.* A. Portion of inflorescence, × $\frac{2}{3}$. B. Leaf, × $\frac{2}{3}$. D. Stamens, long and short, and styles in median position, × 6. C, E—G. *O. Acetosella.* C. Diagram of flower. E. Ripe fruit, much enlarged. F. Seed with aril, much enlarged. G. Seed cut lengthwise, much enlarged. (C, E–G after Baillon.) (From *Flor. Jam.*)

pollination of a stigma from anthers in a similar position; in some cases union between flowers of the same form proved useless. Pollination is effected by bees where the corolla is funnel-shaped, by butterflies where the lower part of the petals forms a narrow tube. In Wood-sorrel, besides the ordinary open flowers, cleistogamic flowers occur, and also transition forms between the two, in which the anthers of the outer stamens become gradually reduced and those of the inner more and more closely applied to the shortened styles.

The family contains about 900 species, the greater number of which are distributed through the warmer parts of the earth. Of the seven genera, *Eichleria*, with two species, is a native of Brazil, *Dapania* (2 species) and *Sarcotheca* (6 species) are Malayan, and *Hypeocharis* has seven species on the high Andes of Bolivia and Peru; *Averrhoa* (2 species) is a tree cultivated in the tropics for its fruit (a berry) which tastes like a gooseberry, and *Biophytum*, which has leaves sensitive to the touch, has about 50 species in the tropics of the three great continents. The remaining 800 species are contained in *Oxalis*, which has one centre of distribution in South Africa and another in South America.

Family III. BALSAMINACEAE

This family is represented in our flora by one species only, *Impatiens Noli-tangere* (Touch-me-not).

The plants are generally annual herbs with succulent stems full of watery juice, and simple leaves which are alternate, opposite, or in whorls of threes. Stipules are absent or represented by a pair of glands at the base of the petiole. The flowers are borne singly or several or many together on axillary peduncles, and are strongly zygomorphic in the median plane. They are pentamerous and represented by the formula S5, C5, A5, G(5). The sepals are petaloid, the posterior is very large and spurred, the two lateral are small and pushed forward to the anterior aspect of the flower, while the two anterior are minute or suppressed. The slender pedicel is often twisted so that the spur becomes anterior, when the flower is termed *resupinate*. The anterior petal is large and external in the bud, the lateral and posterior are generally more or less united in pairs on each side of the flower so that there are apparently only three petals. The five stamens have short broad filaments and anthers cohering and covering the pistil like a cap. Ultimately they rupture at the base and are lifted up by the lengthening pistil. The ovary is oblong, five-celled, and terminates in a sessile five-toothed stigma. Each cell contains a single axile row of pendulous anatropous ovules.

The fruit is a capsule with loculicidal dehiscence. It springs open at a touch when ripe, the five valves separating from the placentas and becoming elastically coiled while the seeds are shot out on all sides. The slightly compressed seeds are completely filled with the embryo, which lies straight along the axis and consists of two large cotyledons face to face and at the upper end a small radicle. Small cleistogamic flowers occur in *I. Noli-tangere* and other species, in addition to the coloured ones; the latter are proterandrous and pollinated by bees as they suck nectar from the spur.

Impatiens contains about 400 species, most of which inhabit the mountains of tropical Asia and Africa. A few occur in Central and Northern Asia, South Africa and North America. *Impatiens Noli-tangere* is the only European species. A second genus, *Hydrocera*, is represented by a single species, a marsh-plant found from India to Java. *I. parviflora*, an annual with very small yellow flowers, a native of Siberia, has become naturalised in Europe (including Britain) as a garden escape. *I. fulva*, a North

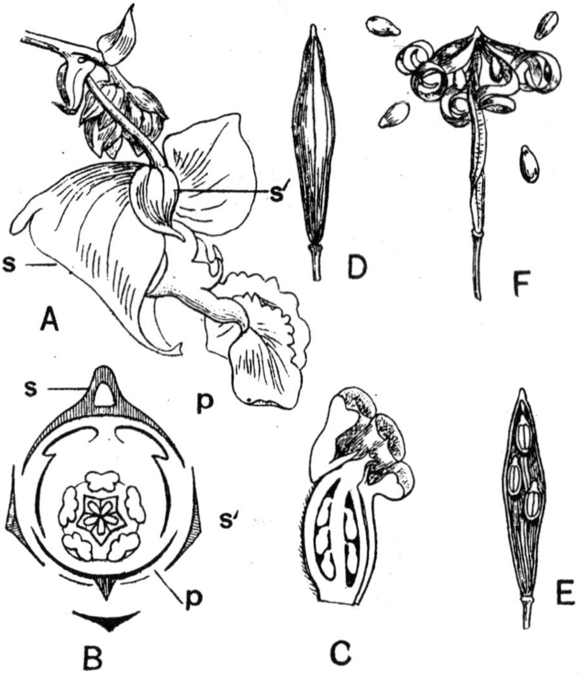

FIG. 133. *Impatiens.* A. Flower of *I. Balsamina*, nat. size. B. Floral diagram of *I. Roylei*; *s*, spurred sepal; *s′*, lateral sepal; *p*, large petal. C. Vertical section of androecium and gynoecium of *I. Balsamina*, × 5. D. Fruit of *I. Noli-tangere*, nat. size. E. Same in vertical section. F. Same dehiscing. (A after *Botan. Mag.*; B after Eichler; C, D, E after Warburg; F after Baillon.)

American species resembling *I. Noli-tangere* but with orange flowers, has become naturalised on river-banks in Britain. *I. Balsamina* (Balsam), an East Indian species, is well known in gardens. The tall rapidly growing *I. Roylei*, a Himalayan species, is also grown, and has become naturalised in South Wales.

Family IV. TROPAEOLACEAE

This family contains one genus, *Tropaeolum*, with about 60 species on the mountains of Mexico and the Andes as far south as Chile; three occur also in South Brazil. *T. majus* and *T. minus*, natives of Peru, are the "Nasturtiums" of our gardens.

The plants are juicy herbs, sometimes perennial by possession of tubers. The scattered peltate or lobed leaves are generally exstipulate and their long stalks are often sensitive to contact, twisting round objects presented to them and thus enabling the plant to climb. The long-stalked conspicuous flowers are solitary in the axils

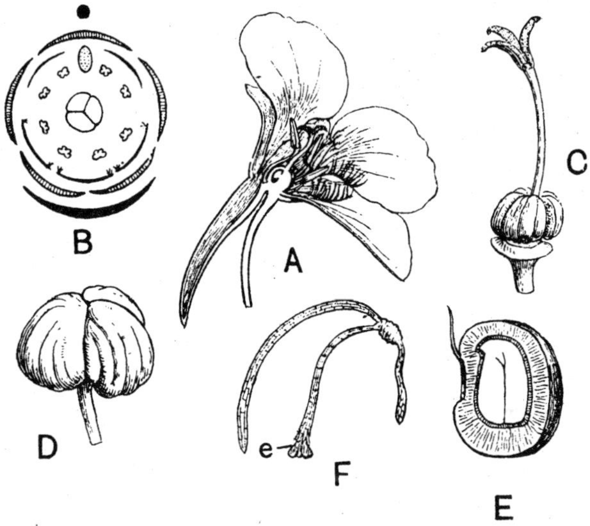

FIG. 134. *Tropaeolum majus*. A. Flower in vertical section, slightly reduced. B. Diagram of flower. C. Gynoecium, enlarged. D. Fruit. E. One coccus of fruit cut lengthwise shewing seed in section, enlarged. F. Developing embryo, *e*, at the end of the suspensor, and the two lateral haustoria, much enlarged. (F after Schacht.)

of the foliage-leaves, and are zygomorphic, with the formula S5, C5, A8, G(3). The sepals are petaloid; the posterior and two lateral, together with a unilateral growth of the receptacle, form a large dorsal spur. The three anterior petals also differ from the two posterior in having a claw and a ciliated margin above it. The stamens are arranged in two rows of four, one row on each side of the ovary; they may be compared with the two pentamerous whorls of Geraniaceae, in each of which the median stamen has been suppressed. Zygomorphy in the median plane is disturbed by the oblique position of the carpels, of which the odd one is not exactly posterior but a little to the right or left of this position. There is a single pendulous anatropous ovule in the inner angle of each of the three ovary-chambers. The central filiform style divides at the top into three stigma-bearing branches. The fruit divides into three one-seeded fleshy schizocarps. The embryo with its large cotyledons fills the seed-coat.

The flowers are proterandrous; the anthers one by one in the order of their development are brought opposite the entrance to the nectar-containing spur, shed their pollen and then resume their original position. Finally the style bends over and brings the stigmas into the same position. The stiff cilia on the lower margins of the anterior petals form a fence against the entrance of those insects which would be too small to effect pollination and the visits of which would result only in a loss of useful pollen.

The embryology is peculiar. A three-legged structure results from the development of the oospore. The suspensor becomes much elongated, grows out of the micropyle and bears two long appendages (haustoria) one of which grows into the tissue of the placenta and absorbs nourishment for the developing embryo while the other grows out into the ovary-cavity and may serve as a respiratory organ; at the other end of the suspensor the embryo is developed.

Family V. LINACEAE

A small family of about 200 species, more than half of which belong to the genus *Linum*, the remainder being very unequally distributed among eight other genera. The two British genera, *Linum* and *Radiola*, with two others, are herbs or small shrubs with the fertile stamens equal in number to the petals, and form a section, *Eulineae*, closely allied to the Geraniaceae and Oxalidaceae. The remaining five genera form a second section—*Hugonieae* (after the largest genus *Hugonia*)—consisting exclusively of tropical trees and shrubs, often climbing, which in their many-membered androecium (stamens two to four times as numerous as petals) recall the Ternstroemiaceae.

The flowers are regular and generally pentamerous, with a quincuncial calyx followed by a whorl of regularly alternating petals, which are imbricate or twisted in the bud. The filaments of the stamens are united at the base, forming a ring on the outside of which are nectar-secreting glands. The carpels are generally five in number and opposite the petals, forming a five-chambered ovary, with two pendulous anatropous ovules in the inner angle of each chamber; the latter often become halved by the ingrowth of a "false" septum from the dorsal suture. The styles are free, and the stigmas terminal. The fruit is a capsule (except in some of the *Hugonieae* where it is drupaceous) surrounded at the base by the persistent calyx, and splitting septicidally into as many valves as there are carpels, or into double the number when false septa divide each carpel into two chambers. The seeds have a fleshy endosperm in which the usually straight embryo is embedded.

The leaves are alternate, or rarely opposite, simple, entire, with small stipules or exstipulate. The form of the inflorescence varies, and is cymose (*Linum* and *Radiola*) or racemose.

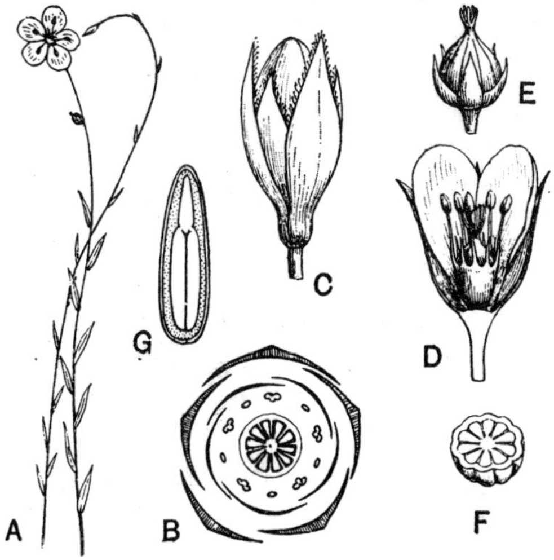

FIG. 135. *Linum jamaicense*. A. Upper part of stem, × ⅔. B. Diagram of flower. C. Flower bud, × 7. D. Bud opened with part of calyx and corolla removed, × 5. E. Capsule, enlarged. F. Capsule cut across. G. Seed of *L. usitatissimum*, cut lengthwise. (From *Flor. Jam.*)

Linum is generally distributed in temperate and subtropical regions; three species are native in the British Isles, the commonest of which, *L. catharticum*, a small annual, with the lower leaves opposite, extends into Arctic Europe. *L. usitatissimum* (Flax) occurs as an escape wherever the plant is cultivated for the sake of its long tough bast-fibres or its oily seeds (linseed oil). It is a native of the country lying between the Persian Gulf, and the Caspian and Black Seas[1].

The five fertile stamens are opposite the sepals, while opposite the fugacious petals the ring formed by the cuneate bases of the filaments is prolonged into five teeth which may be regarded as staminodes. The American, South African and some of the

[1] De Candolle, A. *Origin of Cultivated Plants.*

European species (as *L. catharticum*) and also *L. usitatissimum* are homomorphic; in these the visits of insects may result in cross- or self-pollination according to the method of entrance. If insect-visits do not occur, self-pollination is effected by the bending inwards of the stamens which previously stood out at some distance from, though on the same level with, the stigmas. Dimorphic species, with long-styled and short-styled flowers, are confined to Europe, North Africa and Asia. Experiments have shewn that pollination of a flower with pollen from the same form leads in many cases to little or no production of seed, while full fertility follows a legitimate crossing. On germination the seeds become covered with mucilage, produced by the degeneration of cell-walls in the outer layer of the seed-coat.

Radiola, a monotypic genus, is a very small annual herb with slender repeatedly forking stem, opposite exstipulate leaves and a tetramerous flower. It grows in damp sandy places in Europe (including the British Isles) and North Africa, in the mountains of tropical Africa and in temperate Asia.

Family VI. ZYGOPHYLLACEAE

There are about 200 species in 25 genera in dry districts in the warmer parts of the world, especially in salt-deserts, where they form a characteristic feature of the vegetation. They are rarely annual herbs, generally undershrubs or shrubs, with opposite, rarely alternate, generally paripinnate leaves; stipules are present. The regular bisexual flowers are terminal and solitary, or cymose. They are pentamerous, or sometimes tetramerous, with obdiplostemonous androecium (rarely 15 stamens). The stamens generally bear outgrowths at the base which unite to form an appendage standing inside the staminal ring. The ovary is generally four- or five-chambered, with one to several ovules in each chamber, pendulous from an axile placenta. The single angular or furrowed style bears a terminal stigma. Pollination is presumably effected in the majority by insect-agency. The fruit is a loculicidal or septicidal capsule or divides into indehiscent one-seeded portions; more rarely a berry or drupe. Endosperm is present or absent.

Guaiacum officinale, a small tree of the dry coast-area from Florida, West Indies and equatorial America, yields guaiacum-wood—used medicinally for the bitter resin contained in the heavy greenish-brown heart wood. (Fig. 136.)

Tribulus terrestris, a herb, native in dry and sandy districts in South Europe to Central Asia and in tropical and South Africa, has a thorny schizocarp which is readily carried by animals.

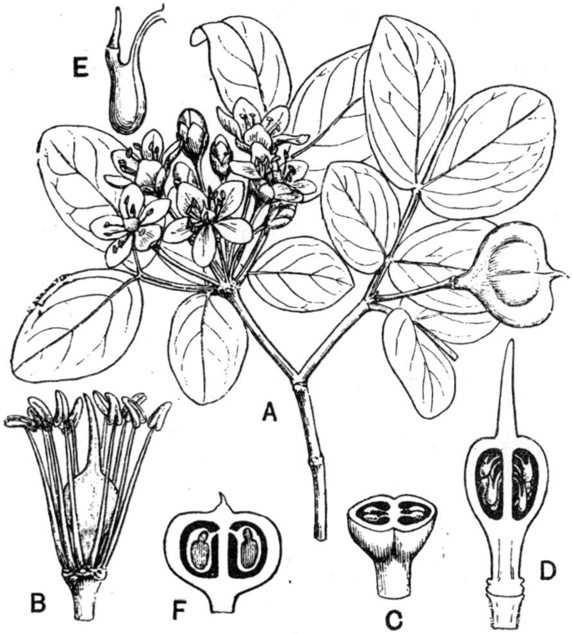

FIG. 136. *Guaiacum officinale.* A. Small shoot bearing inflorescence and one fruit, × ⅔. B. Flower with sepals and petals removed, × 3. C. Ovary cut across, × 4. D. Pistil cut lengthwise, × 4. E. Ovule much enlarged. F. Fruit cut lengthwise, nat. size. (After Berg and Schmidt.) (From *Flor. Jam.*)

Family VII. MALPIGHIACEAE

A tropical family of woody plants, mostly lianes, but including also small trees and shrubs, and in some cases, as in the genus *Camarea* (dry parts of Brazil), low-growing small-leaved xerophytic shrubs with a large swollen root. Some of the finest lianes of the tropical forests belong to this family, and many of these, especially in the American species, shew abnormal secondary thickening due to localised growth of xylem and phloem, forming, similarly to what occurs in many Bignoniaceae, star-shaped figures of wood with corresponding ingrowths of phloem and parenchyma, sometimes associated with phloem-islands in the xylem. The plants are generally more or less densely covered with hairs; in some species with stinging hairs. The hairs are one-celled and variously branched. The leaves are generally opposite and entire, often with glands at the base or on the petiole; stipules are present varying in form and position. The inflorescence is racemose and usually compound; the flower-stalks are jointed and bear a pair of bracteoles below the joint. The often large showy flowers are generally bisexual; they are pentamerous, obdiplostemonous and, at any rate in the gynoecium, obliquely zygomorphic. The receptacle is convex or flat, rarely slightly hollowed, when the flower becomes perigynous. The sepals are generally united at the base and often provided with nectaries of varying form; they persist in the fruit. The petals have generally a claw, and a fringed or toothed margin, and form a regular corolla; the corolla is, however, sometimes obliquely zygomorphic. The androecium is typically obdiplostemonous and regular, but frequently becomes

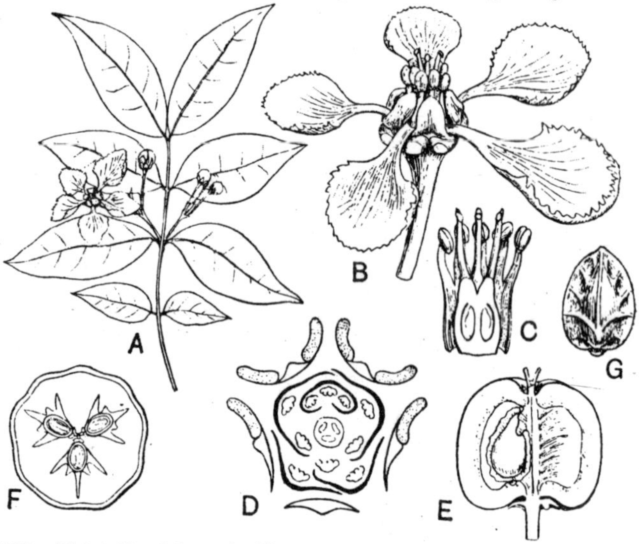

FIG. 137. *Malpighia glabra.* A. Flowering shoot, ½ nat. size. B. Flower, × 3. C. Flower with sepals and petals removed, cut lengthwise, × 5. D. Diagram of flower; the shaded portions on the sepals indicate the glands. E. Drupe cut lengthwise shewing seed and embryo, × 2. F. Drupe cut across shewing the three pyrenes, × 2. G. One pyrene, × 2. (After *Flor. Jam.*)

zygomorphic by one or more members becoming staminodial or completely aborted. The filaments are generally united below, forming a longer or shorter tube. The anthers are introrse with often an enlarged connective. The pistil consists generally of three (rarely two, four or five) united carpels, placed obliquely, the plane of symmetry passing through the third sepal in order of development. The inner angle of each chamber bears a single pendulous ovule with upwardly directed micropyle and a ventral funicle. The structure of the flower implies pollination by aid of insects; cleistogamic flowers also occur. The fruit divides into one-seeded portions which are sometimes winged or nutlike or split open on the dorsal surface; more rarely it is a nut or drupe (*Malpighia*). The embryo completely fills the seed.

The family is widely distributed in the tropics, especially of the New World, and contains about 800 species in 56 genera.

Order 10. *RUTALES*

Closely allied to Geraniales but differing in the occurrence of disc-formations in the flower below the ovary, and the very general presence of oil-glands. The plants are also more generally woody.

Family I. RUTACEAE

Flowers bisexual, rarely unisexual, regular or sometimes slightly zygomorphic, with an annular or cushion-like disc-formation generally between or above the stamens. Sepals and petals four or five, androecium generally obdiplostemonous, stamens sometimes equal in number to the petals, rarely indefinite. Carpels generally four or five, rarely fewer or more, often free below and united above, with usually one or two ovules in each chamber of the ovary. Fruit very various. Seed containing a large straight or curved embryo, with or without endosperm.

Herbs, shrubs or trees with alternate or opposite, simple or compound, generally smooth gland-dotted exstipulate leaves. A characteristic feature is the presence of oil-glands in the ground-tissue.

Genera 120; species about 1200. Widely distributed in temperate and warm regions.

The typical floral diagram is the same as that of Geraniaceae and is expressed by the formula S5, P5, A5 + 5, G(5). As in Geraniaceae modifications also occur, such as reduction or suppression of the antepetalous stamens, or a tendency to zygomorphy of the flower. An important difference in the floral structure is the characteristic development of the receptacle between the stamens and ovary into a ring-, cushion- or cup-like disc (fig. 139, A), or even into a more or less elongated gynophore (fig. 139, C). An important vegetative character is the presence of glands, generally lysigenous, containing a volatile, often scented oil, which, in the leaves, appear as pellucid dots over the whole surface or at the margin only.

The plants are mostly shrubs or trees inhabiting warm countries. The leaves are generally alternate, more rarely opposite, and simple, more or less divided, or compound. They often conform strikingly to the foliage of other plants characteristic of the vegetation of given areas. Thus the *Diosmeae*, with their small narrow leaves and heath-like habit, resemble many other of the components of the shrubby vegetation at the Cape, while the narrow linear or lanceolate leaves of many of the *Boronieae* (an exclusively Australasian tribe) have this form in common with many other Australian shrubs, and the pinnately-cut leaves of others of the tribe

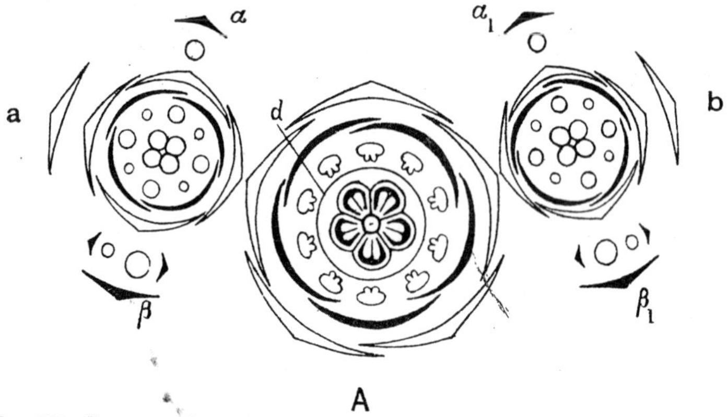

FIG. 138. *Ruta graveolens*. A. Pentamerous terminal flower; *a*, *b*, the two uppermost foliage leaves, each subtending a monochasial cyme with tetramerous flowers; *d*, disc. (After Eichler.)

recall the equally characteristic Australian Proteaceae. The pinnately compound leaves of many of the north-temperate and tropical trees and shrubs strikingly resemble those of the Simarubaceae, Burseraceae, Meliaceae, Anacardiaceae or Sapindaceae found associated with them. This is merely an expression of the fact that the family has a wide distribution and that its members are adapted to very various conditions, climatic and otherwise. An interesting form of leaf is that found in *Citrus* (Orange, Lemon, etc.), where a simple blade is separated by a joint from a winged petiole (fig. 140); since this form occurs in genera which have also compound leaves it is regarded as a compound leaf reduced to one leaflet. In the tribe *Aurantieae*, which includes the Orange, Lemon,

Citron and others, reduction of the leaf frequently goes still further, the first one or two leaves of a bud in the young plant being reduced to thorns.

The flowers are sometimes solitary, but generally arranged in an inflorescence which is rarely a simple raceme, more often involving some cymose arrangement (fig. 138).

The great majority have obdiplostemonous flowers with an isomerous pistil. Where only one whorl of stamens is present, it is antesepalous; many of the *Diosmeae* and others shew an intermediate condition in the replacement of the antepetalous stamens by staminodes. In *Citrus* pleiomery occurs, the numerous stamens being associated in bundles (polyadelphous) derived by splitting of simple primordia. In this genus the number of carpels is also generally increased; the Orange (*C. Aurantium*) has 6–20. The number of carpels is sometimes reduced to one.

As already stated the flowers are typically pentamerous; the odd sepal is posterior and the odd petal anterior. Tetramerous and trimerous flowers also occur. In *Ruta* (Rue) only the terminal flower is pentamerous, the lateral being tetramerous, with the first sepal next to the bracteole and the petals placed diagonally in relation to it (fig. 138). In trimerous flowers the odd sepal is posterior and external. The flower of *Dictamnus* (Dittany) is slightly zygomorphic in the median plane (fig. 139, B). As a rule, however, where zygomorphy occurs it is oblique. The calyx is regular with imbricate aestivation, and the sepals are united at the base. In the American tribe *Cusparieae*, which is characterised by zygomorphic flowers and a gamopetalous corolla, the sepals are often united into a tube. The petals are generally free, conspicuously larger than the sepals and white, red or yellow in colour. The Australian genus *Correa*, a species of which (*C. speciosa*) is well known in cultivation, has a short cup-like calyx, and a long narrowly bell-shaped gamopetalous corolla (fig. 141).

In most genera the carpels are only slightly united at the base or sides, the ovary forming a deeply lobed structure with the united styles rising from the centre. In the *Aurantieae* and the closely allied *Toddalieae*, the union of the carpels is complete, the ovary being multilocular and entire or only slightly grooved with a terminal undivided style. There are sometimes several ovules arranged in two rows (figs. 139, 140) in each carpel, but usually there are only two, which are collateral or superposed (fig. 141, D). The ovule has a ventral raphe and an upwardly pointing micropyle, but where two are superposed the micropyle of the upper is often directed

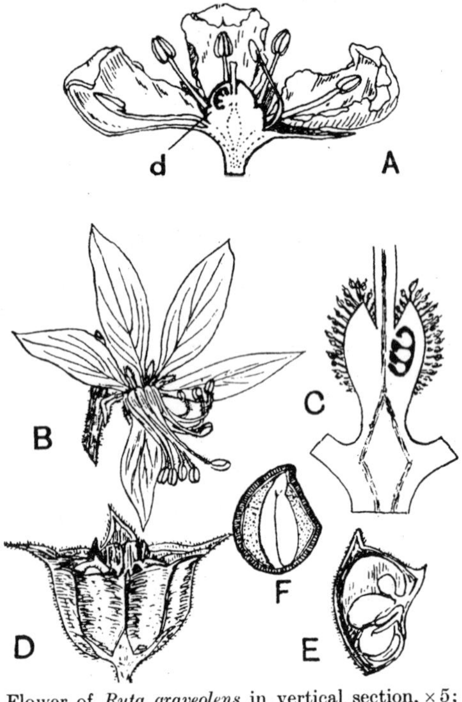

FIG. 139. A. Flower of *Ruta graveolens* in vertical section, × 5; *d*, disc. B–F. *Dictamnus albus*. B. Flower. C. Vertical section shewing ovary and gynophore, × 5. D. Fruit dehiscing. E. Chamber of fruit in section. F. Seed, × 4. (After Engler.)

downwards. Sometimes only one ovule is present. The genus *Feronia* affords a striking exception in having a syncarpous ovary with many parietal placentas on which numerous ovules are arranged in many rows.

The flowers of most Rutaceae are well-arranged for pollination by insect-agency; the coloured corolla rendering them conspicuous and the disc secreting a supply of easily accessible nectar. Cross-pollination is often favoured by the marked proterandry of the flowers. Thus in *Ruta graveolens* (Rue) the stamens rise successively to the centre of the flower, shed their pollen, wither and fall back again. After all the stamens have bent back the stigma ripens, but H. Müller observed that before it withers the stamens again rise up, and if, through lack of insect-visitors, they still retain a

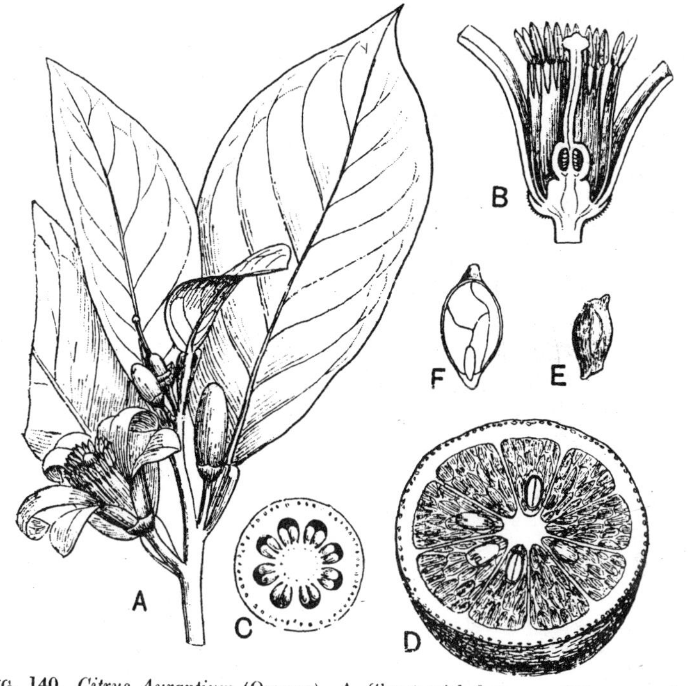

FIG. 140. *Citrus Aurantium* (Orange). A. Shoot with leaves and flowers, $\times \frac{2}{3}$. B. Flower cut lengthwise, with petals cut, $\times \frac{2}{3}$. C. Ovary cut across, much enlarged. D. Fruit cut across, $\times \frac{1}{2}$. E. Seed, $\times \frac{2}{3}$. F. Seed cut lengthwise, shewing folded cotyledons. (From *Flor. Jam.*)

stock of pollen, some of it is shed upon the stigma, thus effecting self-pollination. The flowers are visited by many flies and bees.

The fruit varies considerably, but the form is more or less related to the degree of union of the carpels. Where the latter are only partially united in the flower, as in *Dictamnus*, the fruits generally become isolated and dehisce along the ventral suture, the dry papery endocarp becoming at the same time elastically separated from the exocarp and exposing the seeds, which have a dry crustaceous smooth or warty coat (fig. 139). On the other hand, where the carpels are completely united in the flower, the fruit is generally a drupe or berry. The berry of the Orange-group has a smooth endocarp or encloses a fleshy pulp which consists of large-celled juicy emergences growing into the chambers from their walls and gradually filling up the cavity (fig. 140, D). The well-known slippery feeling of the seeds of *Citrus* is due to the mucilaginous nature of the outer membrane of the epidermis, which protects the seed from being crushed when the pulp is eaten by animals.

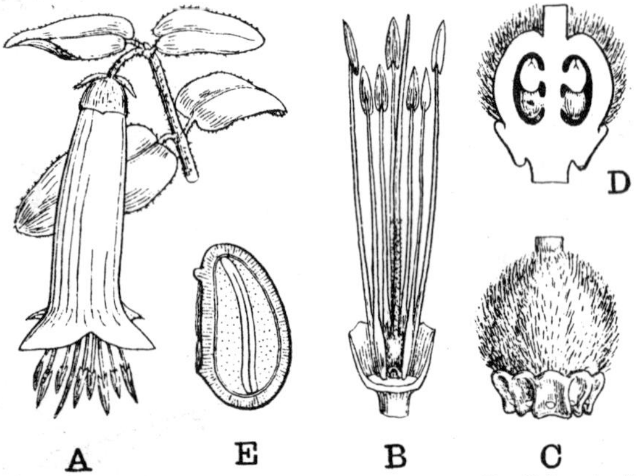

FIG. 141. *Correa speciosa*. A. Flower-bearing twig. B. Calyx (cut), stamens (one removed) and pistil, × 2 (nearly). C. Disc and ovary, × 6. D. The same in vertical section. E. Seed cut lengthwise, × 4 (nearly).

The seeds contain a large embryo, which is straight or bent with flat plano-convex convolute or folded cotyledons. Endosperm is present or absent. Polyembryony is frequent in *Citrus*, and results, as Strasburger shewed, from an adventitious production of embryos from cells of the nucellus above the embryo-sac. As many as thirteen may occur in one seed but seldom more than three germinate.

Engler distinguishes the following subfamilies which differ in the character of the fruit:

I. RUTOIDEAE. Carpels generally four or five, more or less distinct in the fruit and opening on the inside with usually separating endocarp. It includes the following tribes:

Xanthoxyleae (25 genera). A widely distributed group of woody plants with usually small greenish or greenish-white, always regular, sometimes unisexual flowers. Carpels rarely with more than two ovules. *Xanthoxylum* (temperate East Asia and North America); *Fagara* (200 species in the tropics) includes useful timber-trees. *Choisya ternata*, a favourite greenhouse or garden shrub, is a native of Mexico.

Ruteae (7 genera). Herbs or undershrubs, more rarely shrubs, with median-sized always bisexual flowers. Carpels generally with more than two ovules. Natives of the North temperate zone. *Dictamnus* is a monotypic genus spreading from Central and South Europe to the Caucasus, Northern China and Amurland; the flower is zygomorphic. *Ruta* has about 60 species spreading through the Mediterranean region to Eastern Siberia; *Ruta graveolens* (Rue) is a strongly smelling herb with bipinnate leaves and dull yellow flowers often cultivated in Britain.

Boronieae, with 18 genera (*Boronia*, *Correa*, etc.), is exclusively Australian. Undershrubs or shrubs with regular bisexual flowers and seeds with a fleshy endosperm.

Diosmeae (11 genera) is South African. Shrubby plants (rarely trees) with simple leaves and seeds without endosperm.

Cusparieae, 16 genera. Shrubs and trees with regular or zygomorphic flowers. About 100 species in tropical America.

II. FLINDERSIOIDEAE. Fruit a loculicidal or septicidal capsule. A small group of woody plants, mainly Australian.

III. SPATHELIOIDEAE. Fruit a winged drupe. One genus only, in the West Indies.

IV. TODDALIOIDEAE. Fruit a drupe or dry and winged. Trees or shrubs widely distributed in temperate and tropical countries. *Ptelea trifoliata*, a North American tree, and *Skimmia japonica*, a shrub (East Asia), are often grown in gardens.

V. AURANTIOIDEAE. Fruit a berry. 14 genera in the eastern hemisphere, mainly tropical. Includes the genus *Citrus*, several species of which have so long been cultivated as to render difficult the systematic arrangement of the various races. *Citrus Aurantium* includes the various Oranges—the more important races are var. *amara*, the Bitter or Seville Orange, var. *Bergama*, the Bergamotte, var. *sinensis*, the sweet or China Orange; *C. medica* includes the Citron and Lemon (var. *Limonum*), and Lime (var. *acida*); *C. nobilis* is the true Mandarin, *C. grandis* the Shaddock, and *C. paradisi* the Grape Fruit; they are mostly natives of China and Cochin China.

Family II. SIMARUBACEAE

A small family of about 200 species and 30 genera occurring in the warmer parts of the world. The plants are shrubs or trees containing bitter principles in the bark and wood; with generally scattered pinnate leaves. The flowers are small, regular and bisexual or generally unisexual by abortion, with 3- to 7-merous calyx and corolla, and as many or double as many stamens (obdiplostemonous) often provided with a scale-like appendage at the base, and four to five or fewer carpels completely united to form a one-celled ovary or united only by the style or stigmas; ovules usually one in each chamber. Between stamens and carpels is a disc which assumes various forms. The fruit is a schizocarp with dry, sometimes winged, or drupaceous mericarps, or drupaceous with two to five one-seeded chambers. Endosperm is very thin or absent; the embryo has thick cotyledons.

Ailanthus glandulosa (Tree of Heaven), a native of China, is an ornamental tree with pinnate leaves and winged fruit. *Picrasma excelsa* (West Indies) yields Jamaica Quassia-wood, and *Quassia amara* (tropical America) Surinam Quassia-wood.

The family is closely allied to Rutaceae, differing mainly in the absence of oil-glands and in the marked tendency to unisexuality in the flowers.

Family III. BURSERACEAE

A tropical family, chiefly American, with about 600 species in 13 genera. The plants are shrubs or trees with scattered compound gland-dotted leaves and small flowers; balsams and resins are contained in passages formed by separation of cell-layers or destruction of cells. The small flowers are generally unisexual, with 5- to 4-merous calyx and corolla and haplostemonous or obdiplostemonous androecium. A disc is present, and the five to three carpels form a compound multilocular ovary with generally two ovules standing side by side in each cell. The fruit is a two- to five-seeded drupe or a capsule. Endosperm is absent, and the straight or rolled embryo fills the seed. Many species are valuable as sources of aromatic resins and balsams. Such are species of *Boswellia* from Somaliland and Arabia which yield frankincense or olibanum; and species of *Commiphora* from the same countries and from Northern Abyssinia which yield myrrh.

Family IV. MELIACEAE

An almost exclusively tropical family containing about 45 genera and 750 species, mostly trees and shrubs with generally alternate pinnately compound exstipulate leaves, and axillary panicles of regular bisexual or polygamous flowers. The calyx and corolla are 4- to 5-merous, the 8–10 stamens are generally united to form a longer or shorter staminal tube; a disc is present between stamens and pistil or absent. The small ovary is generally two- to five-chambered, with one or two, seldom more, ovules in each chamber. The fruit is a capsule, berry or drupe; the seeds are often winged. Endosperm may be present and the cotyledons are then often leaf-like, or absent when the cotyledons are fleshy and often united.

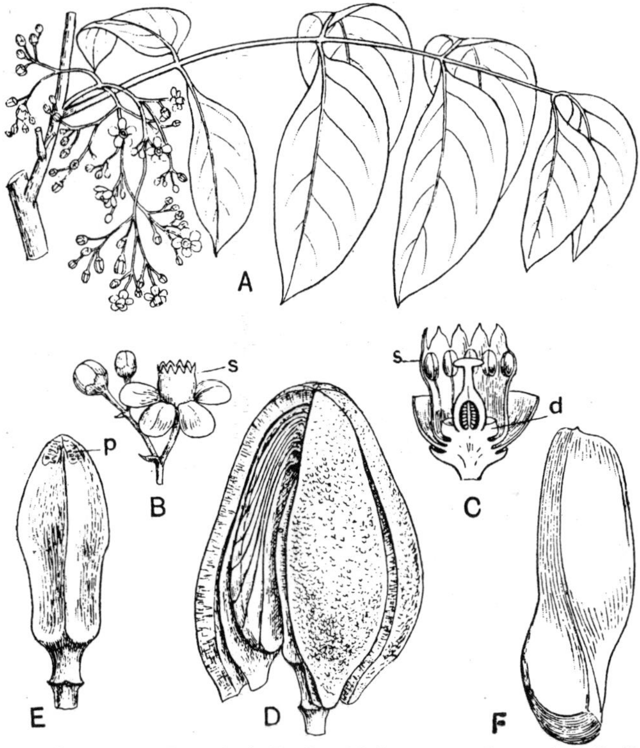

FIG. 142. *Swietenia mahagoni*. A. Leaf and inflorescence, ½ nat. size. B. Buds and open flower; *s*, staminal tube; × 1½. C. Flower cut lengthwise; *s*, staminal tube; *d*, disc; × 5. D. Fruit opening, one valve has been removed exposing the seeds, × ½. E Central axis of fruit; *p*, points of attachment of seeds; × ½. F. Seed, × ⅔. (After *Flor. Jam.*)

Many are valuable timber trees. *Swietenia mahagoni* (West Indies) is the true Mahogany; African Mahogany is the product of *Khaya senegalensis* and other species. *Cedrela odorata* (West Indies) yields the cedar-wood of commerce. *Melia Azedarach*, an ornamental tree, also yields valuable timber.

Order 11. *SAPINDALES*

Flowers bisexual or unisexual by reduction, regular or more often zygomorphic, hypogynous, often with a disc-formation below the ovary; generally pentamerous, sometimes tetramerous, with a diplostemonous androecium, and reduction in the ovary generally to three or two carpels. Reduction is also frequent in the androecium. Ovules with two integuments, one or two in each chamber of the ovary, position various, pendulous with a dorsal raphe or ascending with a ventral raphe (except in Polygalaceae where it is pendulous with a ventral raphe as in Geraniales). Fruit various, often one-seeded. Embryo large; endosperm generally absent.

Generally trees or shrubs (herbs in Polygalaceae), with simple or compound, alternate or opposite leaves and numerous small flowers.

Closely allied to the two previous orders.

Family I. ANACARDIACEAE

A family of 66 genera with about 500 species, chiefly tropical but extending into southern Europe and temperate Asia and America. They are trees or shrubs containing resin-passages with alternate exstipulate simple or compound (unequally pinnate) leaves, and generally numerous small flowers in terminal or axillary panicles. In habit they closely resemble other families of the order, especially Sapindaceae (and also Rutaceae in the previous order) but are distinguished by the presence of schizogenous resin-passages and the solitary pendulous or ascending anatropous ovule attached to the ventral face of the carpel with the raphe dorsal.

The flowers are typically bisexual but often unisexual by reduction of the androecium or gynoecium; they are generally regular and pentamerous, with reduction to three in the gynoecium. The form of the floral axis shews great variety. The more usual form is a short convex receptacle with hypogynous corolla and stamens, often with small nectar-secreting outgrowths between the stamens. In other cases a disc, often cup-like in form, is present between stamens and pistil, as in *Rhus* (fig. 143), or the axis is lengthened or thickened;

thus in *Mangifera indica* (Mango), a thick cushion is developed between corolla and stamens (fig. 144, A), while in *Melanorrhoea* an elongation of the floral axis between the corolla and pistil is correlated with an increase in the number of staminal whorls. The ten stamens are all fertile (*Buchanania*) or reduction and loss of function may take place; for instance *Rhus* has sometimes only five, *Anacardium* has 10–7, of which only one is fertile and much exceeds the rest in size; *Mangifera* has at most five (antesepalous) stamens; here also

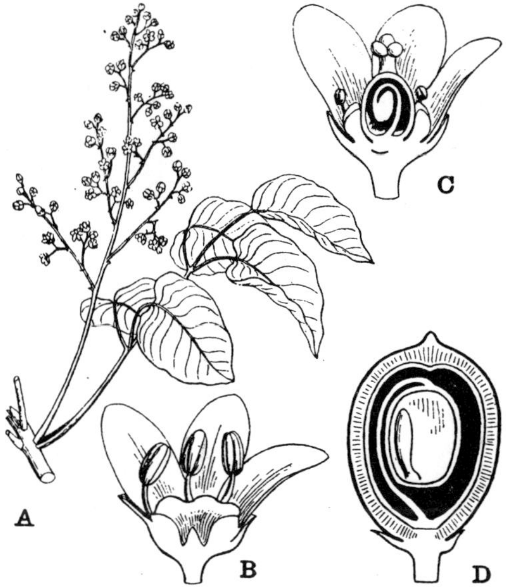

FIG. 143. *Rhus Metopium.* A. Panicle of male flowers with leaf, × $\frac{1}{3}$. B. Male flower cut lengthwise, × 4. C. Female flower cut lengthwise, × 4. D. Drupe cut lengthwise, × 2. (After Sargent.) (From *Flor. Jam.*)

reduction to a single fertile one may occur. Often only one of the three carpels contains an ovule, or where each is ovuliferous, only one ripens; *Buchanania* has five distinct carpels but only one ovule and seed. In *Mangifera*, *Anacardium* and allied genera only one carpel is present. The fruit is dry or drupaceous; the seed contains a more or less bent embryo with flat or thick plano-convex cotyledons and little or no endosperm. *Cotinus Coggygria* (formerly known as *Rhus Cotinus*), the Wig-tree, native from southern Europe through temperate Asia to China, has polygamous flowers, the stalks of which in the fruiting season lengthen and become hairy rendering the infructescence, which becomes detached as a whole, light and easily carried by the wind.

The largest genus, *Rhus*, contains 120 species, which are widely distributed in the warmer parts of the world. *Rhus Coriaria* is the Sumach, a native of southern Europe; *R. Toxicodendron* is the North American Poison-oak or Poison-ivy; the Japanese *R. vernicifera* yields lacquer. In *Anacardium occidentale* (Cashew-nut), a native of tropical America, the solitary carpel forms a nut with a

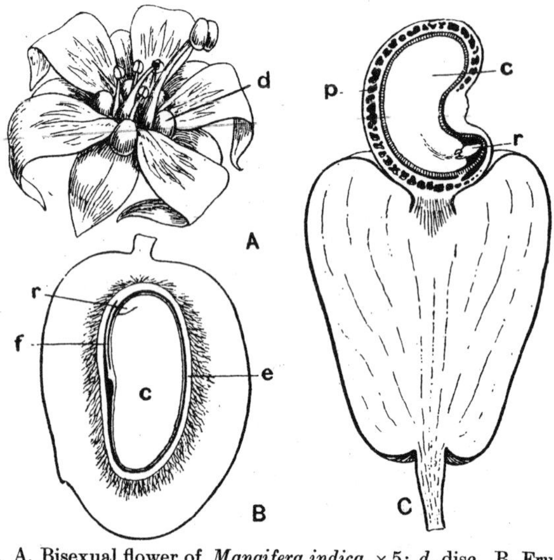

FIG. 144. A. Bisexual flower of *Mangifera indica*, × 5; *d*, disc. B. Fruit of same cut vertically, × ½; *e*, endocarp; *f*, funicle; *c*, cotyledon; *r*, radicle. C. *Anacardium occidentale*. Section of fruit and seed on the swollen fruit-stalk, × ⅔; *p*, pericarp with numerous oil-containing cavities; *c*, cotyledon; *r*, radicle. (C after Engler.) (From *Flor. Jam.*)

hard bitter-resinous pericarp containing an oily edible seed; the fruit-stalk swells to form a large fleshy pear-like edible structure (fig. 144, C). The solitary carpel of *Mangifera indica* (Mango) (fig. 144, B) produces a large drupe with a luscious mesocarp. The seeds of *Pistacia* yield oil; those of *P. vera* (Mediterranean region) are eaten as Pistachio nuts. *P. Terebinthus* (the Terebinth) and *P. Lentiscus* are widely distributed in the Mediterranean area; the latter, which yields mastic resin, is a conspicuous feature of the Maqui vegetation; the former yields a turpentine which exudes from cuts in the stem; the pod-like galls, rich in tannin, borne on the vegetative organs and flower-stalks are an article of commerce in the East. *Schinus molle* (Pepper tree), a native of the Andes and subtropical South America, is a graceful tree, widely cultivated in the Mediterranean region and subtropical climates.

Family II. SAPINDACEAE

An important tropical and subtropical family with about 130 genera and 1000 species. They are trees or shrubs, sometimes lianes climbing by tendrils and with remarkable stem-structure, the result of anomalous secondary growth in thickness. In addition to the central vascular bundle-system, cortical systems are developed, often shewing considerable complication, as in the stems of *Serjania*, *Paullinia*, *Thinouia* and others. The tendrils are axillary and represent modified inflorescence-axes; they are forked at the apex, and the branches are often flat and rolled like a watch-spring. The leaves are alternate, generally compound and pinnate, with or without a terminal leaflet; only in climbing species are small stipules present. Latex or resin is often contained in special sacs or cells which appear in the dried leaves as transparent points or streaks. The numerous inconspicuous flowers are borne in unilateral cymes arranged in racemes or panicles. They are unisexual, though apparently polygamous, with obvious male and female, and apparent bisexual flowers; in the last the stamens though present bear permanently closed and functionless anthers. They are regular or obliquely zygomorphic, generally pentamerous, sometimes tetramerous. The sepals are generally free, and five in number with imbricate aestivation, in symmetrical flowers apparently four by union of the third and fifth; the pentamerous corolla in symmetrical flowers becomes similarly tetramerous by suppression of a petal; the petals are free and often provided on the inside with scales and hair-tufts which cover the nectaries. Between petals and stamens the floral axis is developed to form a disc, which is generally ring-like and bears glandular swellings opposite the petal-insertions.

The stamens are usually in two pentamerous whorls, often reduced to eight or even fewer, by suppression (see fig. 145); they are inserted inside the disc around the pistil or pistil-rudiment. The pistil is usually trimerous, with a terminal style and a three-celled ovary, and usually one ascending ovule with a ventral raphe in each chamber. The large fruit

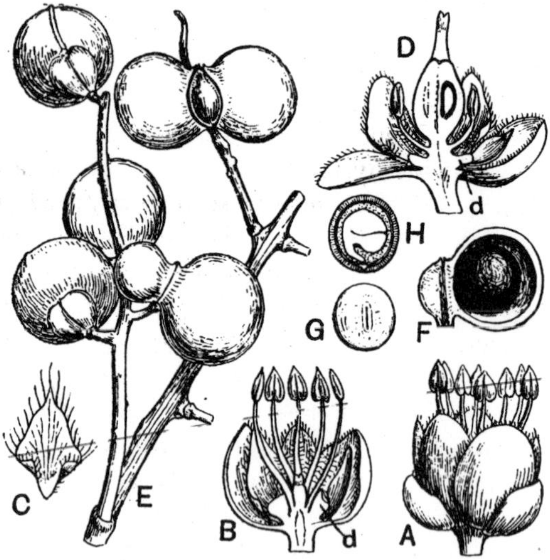

FIG. 145. *Sapindus Saponaria.* A. Male flower. B. Ditto cut lengthwise. C. Petal. D. Female flower cut lengthwise. E. Fruiting branch. F. Coccus cut open, shewing the seed. G. Seed shewing the hilum. H. Seed cut lengthwise. *d*, disc. A–D, × 10; E–H, nearly nat. size. (After *Flor. Bras.*) (From *Flor. Jam.*)

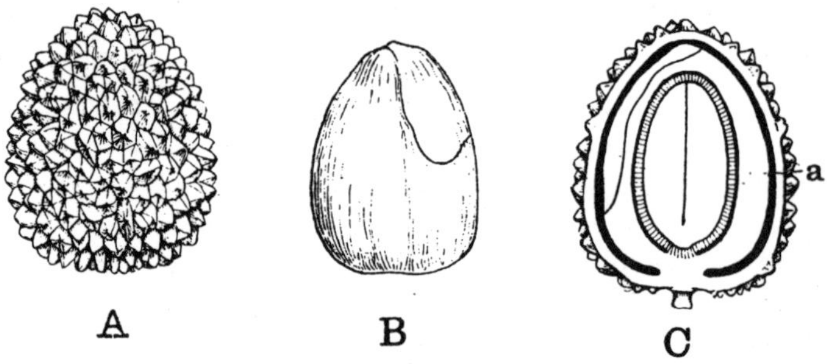

FIG. 146. *Litchi chinensis.* A. Nut. B. Aril enclosing seed. C. Nut in section: the aril (*a*) completely surrounds the seed.

is dry, forming a capsule or nut, or fleshy and a berry or drupe; it is sometimes a schizocarp and frequently winged. The seed, which often bears an aril (e.g. the fleshy aril of the Litchi), is exendospermic, containing only the curved embryo.

Serjania and *Paullinia* are large genera (containing about 350 species) of tropical American lianes with watch-spring-like tendrils and generally winged fruits. The berries of species of *Sapindus*, a genus with 11 species, occurring in the warmer parts of Asia and America, contain saponin and yield a lather with water. *S. Saponaria* (fig. 145), Soapberry tree (tropical and subtropical America), has only one or two of the segments (cocci) of the fruit developed. *Litchi chinensis* (Litchi), a native of China, is widely cultivated for the fleshy sweet aril which surrounds the seed and is contained in a brittle nut-like fruit. A similar structure occurs in species of the allied genus *Nephelium* (tropical Asia); *N. lappaceum* is the Rambutan. Other genera yield valuable timber.

Family III. ACERACEAE

Flowers regular, unisexual, or often polygamous, generally conforming to the formula S5, P5, A4 + 4, G(2). Petals free. Stamens hypogynous or perigynous, inserted on or inside an annular, often lobed, disc. Ovary two-lobed, with two collateral or superposed ovules in each chamber, ovules orthotropous to almost anatropous, attached by a broad base with a dorsal raphe. Fruit a schizocarp of two one-seeded samaras. Seed exendospermic, cotyledons flat, folded or rolled.

Trees with opposite exstipulate leaves, generally palmately lobed and veined. Inflorescence racemose.

Genera two, with about 120 species in the northern hemisphere, chiefly north-temperate, or on mountains.

Germination is epigeal; the green strap-shaped cotyledons of *Acer Pseudo-platanus* (Sycamore) are familiar objects. A North American species, *A. dasycarpum*, is exceptional in having hypogeal cotyledons. Owing to the regular development of the buds in the axils of each pair of leaves young Sycamores shew for some time a very regular branching.

The leaves are generally simple and palminerved; different species of *Acer* shew an interesting series from a simple to a compound form. In the Himalayan *A. oblongum* they are simple and entire; in certain east Asiatic species the margin is serrate; *A. monspessulanum* and others have a trilobed blade; while in most of the species, as the Common Maple (*A. campestre*) and Sycamore, the two lateral lobes branch again, forming a five-lobed blade. In the widely cultivated *A. palmatum* (Japan) and its allies the branching is carried further and the leaves are generally eleven-lobed. A few species from China and Japan are trifoliolate, while the section *Negundo* (American), often separated as a distinct genus, has imparipinnate compound leaves (with three to five leaflets), as also has the second genus *Dipteronia*.

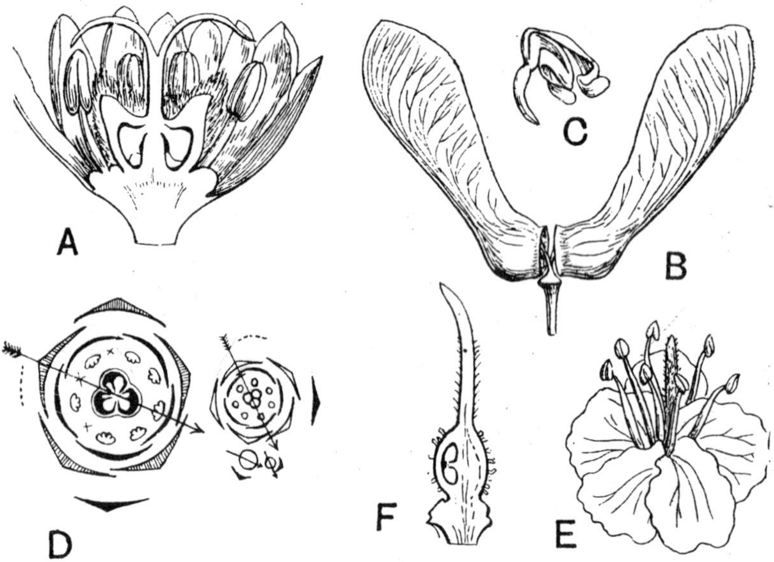

FIG. 147. *Acer Pseudo-platanus*. A. Vertical section of flower, enlarged. B. Fruit. C. Embryo, enlarged. (After Le Maout and Decaisne.) *Aesculus Hippocastanum*. D. Diagram of a three-flowered cyme. The arrow indicates the plane of symmetry. E. Flower (inverted in comparison with D). F. Pistil cut vertically, enlarged. (D after Eichler; E, F after Le Maout and Decaisne.)

The leaves are plicate in the bud and protected by scale-leaves which consist of the broadened leaf-base, the upper portion being undeveloped. This relation can often be traced by observing the transitional forms between scales and foliage-leaves as the bud opens in spring.

The inflorescence is generally terminal on a short leafy shoot; the main axis ends in a flower. It is a panicle, as in *A. Pseudo-platanus*, which is sometimes corymbose in form.

or by shortening of the main axis becomes umbellate, or a simple raceme (female plants of *A. Negundo*), a spike (the Chinese *A. Henryi*) or an umbel (*A. japonicum*).

In andro-monoecious species the lower flowers of an inflorescence are male, the upper (younger) bisexual. The different forms of flowers may be derived from the formula S5, P5, A5 + 5, G(2).

Rarely, as in the North American *A. grandidentatum* and the Japanese *A. carpinifolium*, the petals are suppressed; in the commonly cultivated *A. dasycarpum*, apetaly is sometimes incomplete. *A. rubrum* and others are haplostemonous from suppression of the inner whorl of stamens. Usually the two median stamens are suppressed, leaving eight in all. The filaments are free and inserted within or upon the fleshy disc (fig. 147, A). Pleiomery occurs occasionally in all the whorls, but is most common in the gynoecium; three carpels are often found, especially in the terminal flower, while four- to eight-carpellary fruits sometimes occur.

The tendency to a separation of the sexes is general. The most common forms are andro-monoecism and -dioecism through a reduction of the ovary in the earlier flowers. *A. Negundo* and its allies are dioecious; the inflorescence is umbellate in male, a simple raceme in female plants. Pollination is effected by insects, with either long or short probosces, which suck the nectar copiously secreted on the disc. Undoubted hybrids occur; one between *A. monspessulanum* and *A. campestre* growing wild in Herzegovina. Each carpel forms in the fruit a dry one-seeded winged pericarp (samara). In *Dipteronia* the wing surrounds the seed, in *Acer* it is one-sided. The embryo is more or less bent, folded or rolled, the cotyledons often being much coiled. The manner of folding and position of the embryo vary in different species; sometimes the backs, sometimes the edges of the cotyledons face the placenta, while the tips may be in the centre of the coil or on the outside.

The single species of *Dipteronia* occurs in Central China. The species of *Acer* are mountain- and hill-loving plants of the northern hemisphere. In the Himalayas they ascend to 8000 to 10,000 feet. The greatest number of species inhabit an area stretching from the Eastern Himalayas to Central China. Japan is also rich in species. *A. niveum* is found in Assam, Java and Sumatra. The Western Himalayas have fewer representatives than the Eastern and these are related to those of the Mediterranean region, where the centre of distribution lies in the east, the Balkan peninsula and the forest region of the Western Caucasus being richest. Six species pass into Central Europe, and three of these, *A. campestre*, *A. platanoides* and *A. Pseudo-platanus*, reach high latitudes, the second occurring at 61° to 62° North in Scandinavia. *A. campestre*, native in England in thickets and hedgerows, is a small tree. *A. Pseudo-platanus*, extensively planted in Great Britain, is a native of Central Europe and West Asia. In the New World Maples are found from South Canada and Oregon to Mexico, chiefly in the mountain ranges; the species on the Atlantic are distinct from those on the Pacific side. Leaves shewing the characteristic form and nervation, as well as flowers and fruits closely allied to those of recent species, are very frequent in Tertiary strata. Fossil Maples are found throughout the whole Arctic region and from their circumpolar distribution in the Oligocene, and the relation of the species to those inhabiting Europe and North America at successive periods, Pax concludes that the genus is of Arctic origin. Travelling southward, it was much more generally distributed during Tertiary times than at the present day.

Maples are useful timber-trees, the wood being hard and durable and taking a high polish. Sugar occurs in the cortex of many species; in the North American Sugar Maples (*A. saccharinum* and others) in sufficient quantity to make it worth extracting. *A. dasycarpum*, *Negundo*, *platanoides* and the Sycamore are widely planted as ornamental trees.

Family IV. HIPPOCASTANACEAE

A small family nearly allied to the Aceraceae but distinguished by the irregular obliquely zygomorphic flowers and tricarpellary ovary. It contains two genera with 18 species distributed through the north temperate zone, but chiefly American. They are trees with opposite exstipulate digitately compound stalked leaves. The large winter-buds of the Horse-chestnut, covered with resinous scale-leaves, contain the young shoot with its terminal inflorescence in an advanced condition and open very rapidly in the spring. The bud-scales, as in the Maples, are the equivalent of the leaf-base. The large pyramidal inflorescence is a mixed one consisting of a number of scorpioid cymes arranged in a panicle; it is known as a *thyrsus*. The plan of a single cyme and its flowers is shewn in the diagram (fig. 147, D).

Only one of each pair of bracteoles is developed, and subtends the next flower. The floral formula is S5, P5, A8 – 5, G (3). The sepals are united (free in two species forming the second genus *Billia*) and form a regular or irregular calyx with imbricate aestivation. The two upper petals differ in shape and colour from the three lower, the middle one of which is often absent. The stamens are hypogynous and inserted inside a disc which is often one-sided. The variation in the number of stamens results from the loss of two or more of the members of the episepalous whorl. One of the three carpels lies anteriorly in the plane of symmetry. Each of the three ovary-chambers contains two ovules, both with two integuments; the lower ovule is descending with a dorsal raphe, the upper ascending with a ventral raphe (fig. 147, F), or horizontal. The flowers are andro-monoecious, the male flowers, in which a rudimentary ovary is present, often open first. Generally some of the bisexual flowers are biologically female from the premature dropping of the anthers. The bisexual flowers are proterogynous; in the first stage the stamens are bent sharply downwards while the style projects in a long ascending curve, in the second, or male stage, the stamens rise almost to a horizontal position. Nectar is secreted on the disc and the petals are large and conspicuous. In the Horse-chestnut the upper petals bear yellow spots at the base which become red after dehiscence of the anthers. Humble-bees play the chief part in the transport of pollen. Generally only one ovule in the ovary develops into a seed, two out of the three ovary-chambers becoming crushed by the considerable growth of the one containing the very large seed. The ripe fruit is a leathery capsule, spiny or smooth according to the species, and opening loculicidally by three valves. The seed is large, roundish, with a smooth shiny leathery testa, and a large pale-coloured hilum. A curved embryo occupies the whole interior; the large thick cotyledons are inseparable (conferruminate), and the radicle lies in a fold of the testa.

The species are scattered over the north-temperate zone; North

America is the richest region. *Aesculus Hippocastanum* (Horse-chestnut) is the only European representative at the present day, growing wild in the mountains of Northern Greece and Albania. In earlier times it was more widely distributed; fossil seeds are recorded from the upper Pliocene at Frankfort-on-the-Main, and seeds and leaves of other species have been described from various parts of Central Europe. The tree was introduced into cultivation in Vienna by Clusius in 1576 and was first planted in England early in the next century.

Of the other species, seven inhabit the United States of North America, different ones occurring on the Atlantic and Pacific sides, one only reaches as far north as Canada; two species occur in the Himalayas, two in Japan and one in Northern China. Two species, comprising the genus *Billia*, grow, the one in the mountains of Mexico, the other in Guatemala, New Granada and Venezuela.

Many hybrids occur in cultivation; the red Horse-chestnut, *A. carnea*, is a hybrid between *A. Hippocastanum* and *A. Pavia*. The Horse-chestnut and several North American species (*A. glabra* with yellow, and *A. Pavia* with red flowers) are well known ornamental trees. The seeds are used as fodder, and the oil contained in them was formerly a specific for gout and rheumatism. The roots contain saponin and those of several American species are crushed and used in washing woollen stuffs.

Family POLYGALACEAE

(Of doubtful position)

A small family represented in Britain by a few species of *Polygala* (Milkwort), and variously placed by different systematists. In the arrangement of Bentham and Hooker it is placed between the cohorts Parietales (Papaveraceae, Cruciferae, Resedaceae, etc.) and Caryophyllinae; Warming places it in the series Sapindales after Aceraceae, while in Engler's system it appears among the Geraniales. An explanation of these various views on the systematic position is supplied by Chodat[1], the most recent exponent of the family, who describes it as "a very natural family, not closely allied with any others."

The plants are herbs, shrubs or small trees with simple entire alternate, opposite or whorled leaves, which are

[1] *Pflanzenfamilien*, III, 4, 323.

generally exstipulate. The flowers are borne in racemes, spikes or panicles, and both bracts and bracteoles are present.

The floral structure supplies well-marked characters. The flower is medianly zygomorphic with five generally free sepals of which the inner are often (as in Milkwort) large and petaloid like the wings of the flower of *Papilionateae*. There are typically five petals, but only three, the two upper and the

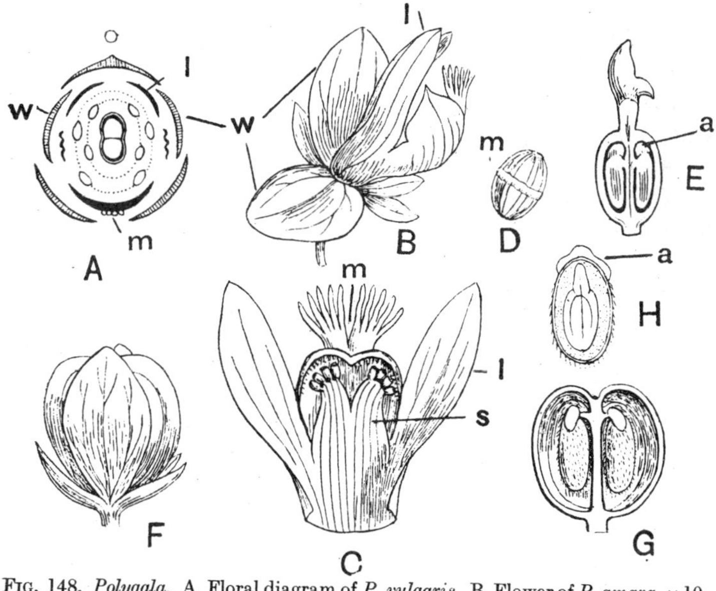

FIG. 148. *Polygala*. A. Floral diagram of *P. vulgaris*. B. Flower of *P. amara*, × 10. C. Corolla and androecium of same spread open, × 10. D. *P. Chamaebuxus*. Pollen-grain, highly magnified. E–H. *P. amara*. E. Pistil shewing ovary in vertical section, × 10. F. Fruit with persistent calyx, × 10. G. Fruit in vertical section, × 10. H. Seed, × 15. *a*, aril; *l*, lateral, *m*, median petal; *s*, staminal tube; *w*, wing-sepal. (A, D after Chodat; others after Berg and Schmidt.)

lower median, are usually present; the latter forms a keel and often bears a dorsal fimbriated appendage. The petals are generally more or less adnate with the staminal tube, which is open at the back and formed by the union of the filaments of eight stamens. The androecium is derived from two pentamerous whorls by suppression of the median member in each. Reduction is sometimes carried further and seven, five, four or three stamens only occur. The pistil consists of two median carpels forming a bilocular ovary, each chamber containing a pendulous anatropous ovule. The style and stigma shew very great variety in form, and afford good diagnostic characters. According to Chodat, the stigmas appear to be adapted for self-pollination, forming a pocket into which the anthers directly open; where the pocket is absent the anthers attach themselves directly to the stigmatic papillae. But this does not imply the exclusion of cross-pollination which often results from the visits of bees and other insects.

The form of the pollen-grain is very characteristic and affords "the surest mark of distinction of the family" (Chodat). The grains are ellipsoidal with a coarse pitting at the poles, and longitudinal bands, broken in the middle by an equatorial ring. The fruit is generally a capsule dehiscing loculicidally into two one-seeded portions. Endosperm is present or absent; differences occurring even in the same genus. The seed-coat often bears a micropylar aril (caruncle) which in species of *Polygala* functions as an elaiosome and the seeds are myrmecochorous.

The family contains about 10 genera with about 680 species, of which 450 belong to *Polygala*. It has a very wide distribution, occurring in all parts of the world except New Zealand, Polynesia, and the Arctic provinces of North America and Asia. It is represented in Britain by a few species of *Polygala*: the common *P. vulgaris* is a small wiry perennial found in heaths and meadows, with white, pink, blue or purplish flowers; *P. calcarea* is closely allied to the former and grows on dry calcareous slopes in the South-east of England. *Polygala senega* (North America) yields the officinal Radix Senegae. The members of the widely distributed tropical genus *Securidaca* are mainly lianes. The genus *Epirrhizanthes* (two species in Malaya) consists of small chlorophyll-free saprophytic plants with scale-like leaves and dense terminal spikes of flowers.

The genus *Xanthophyllum*, with about 40 species extending from India to North Australia, comprises trees sometimes reaching 50 feet in height.

Order 12. *CELASTRALES*

Flowers bisexual or unisexual by abortion, regular, hypogynous, cyclic, usually 4- or 5-merous, with one whorl of stamens, and generally reduction in the gynoecium. The petals are free or sometimes united at the base. The stamens alternate with the petals, and are generally associated with a disc. The ovary is superior and multilocular, and contains in each chamber one or more ovules with upwardly directed micropyle and dorsal raphe, or with micropyle directed downwards and ventral raphe. The embryo is generally surrounded by fleshy endosperm.

Generally woody plants with simple leaves and small green or whitish flowers. The order is closely related to the Sapindales in which it is included by Engler as a subdivision characterised by the regular flower, with stamens generally equal in number to the petals. The frequent occurrence of tetramery and a tendency to union of the petals is to be noted.

Family I. CELASTRACEAE

Contains about 45 genera and 450 species generally distributed except in the colder regions of the earth. The members are chiefly trees or shrubs with simple, membranous or leathery, opposite or alternate leaves, and generally a cymose inflorescence of small greenish or white flowers. The flowers are regular and bisexual or become unisexual by abortion; they conform to the formula S4–5, P4–5, A4–5, G(2–5). The sepals are small, free or united at the base, inferior and generally persistent. The petals are generally free and inserted below a well-marked disc which assumes very different forms. The stamens, which alternate with the petals, are free and situated upon or on the edge of, or beneath, the disc. The carpels form a two- to five-celled ovary situated upon the disc or surrounded by, or more or less sunk within it; each chamber contains generally two erect anatropous ovules, more rarely are they pendulous. The short style ends in a capitate, often lobed stigma. The fruit is various, forming a loculicidally dehiscing capsule, an indehiscent dry fruit, a drupe or a berry. The seeds are erect, more or less completely enveloped in a large brightly-coloured aril, and contain an embryo with large green cotyledons generally embedded in a fleshy endosperm, or, rarely, filling the seed.

Nectar is secreted on the disc and pollination is effected in the European species of *Euonymus* chiefly by the visits of flies, or sometimes by small Hymenoptera or ants.

Euonymus europaeus (Spindle-tree) is the only British representative; its crimson loculicidal capsule, exposing orange-coloured arils, renders it a conspicuous and beautiful object when in fruit. *E. japonicus* is a commonly grown hardy evergreen.

The genus contains about 70 species occurring chiefly in India, the Himalayas and Eastern Asia, a few in the Sunda Islands, and in the Philippines, one in Australia, about four in Central America and the same number in North America and in Europe. A few species of *Euonymus* are lianes; the genus *Celastrus*, which is chiefly developed in the mountains of India and China, extending to Japan, Malaya and Australia, with one species, *C. scandens*, in North America, consists mainly of woody twiners. The leaves of *Catha edulis* (Arabia and East Africa) are used as a sedative when dried.

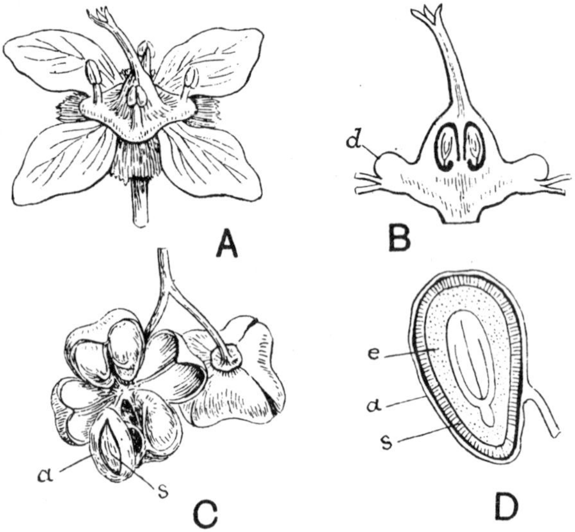

FIG. 149. *Euonymus europaeus*. A. Flower, × 4. B. Vertical section through the ovary and floral axis, × 6; *d*, disc. C. Fruit shewing dehiscence, nat. size; *a*, aril; *s*, seed. D. Seed cut lengthwise, × 4; *a*, aril; *s*, testa; *e*, endosperm. (A, B after Loesener.)

The small family (II) STAPHYLEACEAE is distinguished by its compound generally imparipinnate leaves. *Staphylea* contains eleven species distributed through the north temperate zone; several are known in cultivation.

Family III. AQUIFOLIACEAE

A small family of three genera and 300 species, almost wholly included in the genus *Ilex*. The plants are shrubs or trees having alternate simple leathery leaves with minute stipules. The inflorescence is a few-flowered axillary cyme. The small, regular flowers are bisexual or unisexual through abortion (the plants being dioecious) and 3- to 6-merous. The Holly (*Ilex Aquifolium*) and many other species conform to the formula S4, P4, A4, G(4). The sepals are small, hypogynous and often persistent; the inconspicuous petals are free or connate at the base, hypogynous and deciduous. There is no disc; the stamens are equal in number and alternate with the petals, to the extreme base of which they are attached. A rudimentary pistil occurs in the male flowers and staminodes with the form of stamens but with barren anthers in the female flowers. The ovary is superior, globular or ovoid in

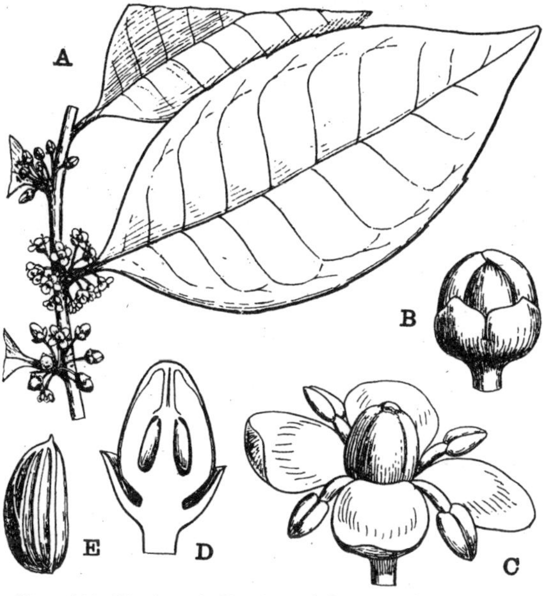

FIG. 150. *Ilex nitida* Maxim. A. Portion of flowering branch, × ⅔. B. Flower-bud, × 5. C. Flower, × 5. D. Ovary in vertical section, × 5. E. Pyrene, × 4. (From *Flor. Jam.*)

form, often four-celled, with one or two pendulous anatropous ovules springing from the inner angle of each chamber. The style is absent or short; the stigma lobed or capitate. The fruit is a drupe with often four stones (pyrenes). The embryo is very small at the top of a copious fleshy endosperm, and has the radicle pointing upwards.

Our only British species is the Holly which grows wild from southern Norway to central Turkey, and from England through Central Europe to the Caucasus and northern Persia. The chief centre of distribution of the genus is in Central and South America; Asia has about half as many species; Africa, Australia and Europe a few only. The leaves and small twigs of the South American *Ilex paraguariensis* are the source of Paraguay tea or Maté.

The family contains also two other genera closely allied to *Ilex* —*Nemopanthes,* with one species in the mountains of eastern North America, distinguished from *Ilex* by having linear petals free from each other and from the stamens, and *Phelline,* distinguished by the valvate aestivation of the corolla and the lobed fruit, with several species in New Caledonia.

Family IV. EMPETRACEAE

A small family of doubtful affinity containing a few species in frigid and temperate zones. The plants are small heath-like shrubs with linear strongly revolute exstipulate leaves and small unisexual regular flowers in few-flowered heads. The flowers have generally rudiments of the aborted stamens or pistil. They are trimerous or dimerous, sepals, petals and free stamens alternating regularly. There is no disc. The pistil consists of a two-, three- or six- to nine-chambered superior ovary, bearing a short style with as many branches; each chamber contains a single ovule borne erect on the central placenta with a ventral raphe and a single integument. The fruit is a drupe with the same number of stones as carpels. The embryo lies in the axis of a fleshy endosperm.

Empetrum nigrum (Crowberry) has a circumpolar distribution, in Arctic and subarctic regions, and occurs on the high moors of Central Europe and Siberia and in the mountains of the north temperate zone. Closely allied species occur in the Andes of Chile, and in Antarctic America and Tristan d'Acunha.

The structure of the leaves, which have a well-developed cuticle on the upper outer face while the stomata are contained in the deep groove lined with hairs formed by the rolled-back margins, is well adapted to xerophytic conditions.

Order 13. *RHAMNALES*

Flowers as in Celastrales but with a marked tendency to unisexuality by abortion, and to perigyny or epigyny. The single whorl of stamens is antepetalous. Petals small sometimes united at the base or above, sometimes absent. Intrastaminal disc well-developed. Carpels five to two (rarely more), united, ovary with as many chambers, each containing one to two ascending ovules with a dorsal, lateral or ventral raphe and two integuments. Embryo filling the seed or surrounded by endosperm.

Shrubs or trees, often lianes, with simple or palmately compound leaves, and small inconspicuous flowers.

Probably a parallel series with the Celastrales. Each may be regarded as derived from some diplostemonous type resembling the Rutales, the antepetalous or antesepalous whorl having disappeared.

Family I. RHAMNACEAE

Contains about 45 genera with 500 species, and occurs in all parts of the world in which climatic conditions allow of the growth of woody plants. *Rhamnus*, two species of which are the representatives of the family in the British flora, is the most widely distributed genus. It includes 100 species and has its chief centre of development in Europe and extratropical Asia, is sparsely represented in Africa beyond the Mediterranean region and is wanting in Australia and Polynesia. *Zizyphus* (40 species) is chiefly Indo-Malayan, but spreads into the Mediterranean region, Africa, Australia and tropical America; the stipules are often thorny as in *Z. Spina-Christi*; the fruits are often edible—*Z. Lotus*, a Mediterranean species. has the credit of being the Lotus plant of the ancients—while those of *Z. vulgaris* occur in commerce as Spanish or French jujubes. *Gouania* has 30–40 species distributed throughout the tropics of both hemispheres. Most of the other genera are confined to limited areas.

The plants are shrubs or trees, often spiny and sometimes climbing. The leaves are simple, with small stipules. The flowers are small and inconspicuous, generally in cymes. They are bisexual (or sometimes unisexual by abortion as in Buckthorn), regular with a pentamerous (more rarely tetramerous) calyx, corolla, and androecium, the stamens being opposite to, and usually enclosed by, the small concave petals, which are, however, sometimes absent. The flowers are perigynous or epigynous, and there is generally a well-developed intrastaminal disc. The carpels form a three- (sometimes two- or one-) celled ovary which is superior or inferior according to its freedom from, or union with, the receptacular cup. Each chamber contains one erect basal anatropous ovule with two integuments. Entomophilous pollination is indicated by the disc on which nectar is secreted. Hymenoptera and Diptera have been noted as visitors to *Rhamnus Frangula* and *Paliurus aculeatus*, which are proterandrous.

The fruit assumes various forms, the pericarp being adapted in different ways to facilitating distribution of the seed. Nearly all the genera fall under one of the following types. In *Rhamnus* and many other genera the fruit is a drupe with two to four stones. In others it is dry and separates elastically into mericarps, thereby flinging out the seeds. Or wind may

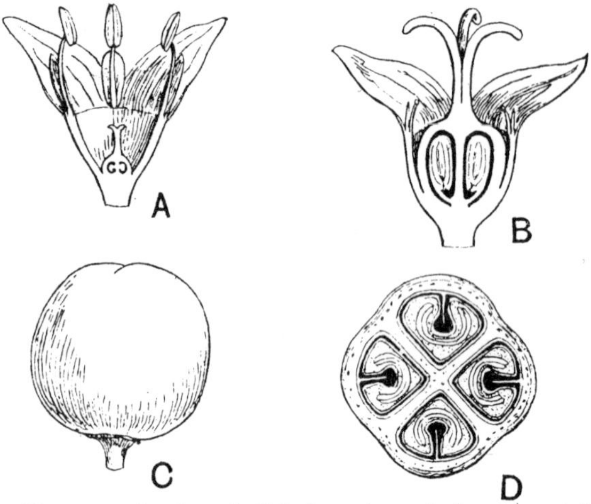

FIG. 151. *Rhamnus catharticus.* A. Male flower in vertical section, × 5. B. Female flower in vertical section, × 6. C. Fruit, × 2½. D. Fruit cut across, shewing the one-seeded stones. (A, B, C after Weberbauer; D after Wettstein.)

be the agent of dispersal, the indehiscent pericarp forming a large vertical or horizontal (*Paliurus*) wing, or splitting into several winged mericarps. The large straight embryo consists chiefly of the two flat or concave greenish cotyledons, the radicle and plumule being small. It may occupy the whole interior of the seed but is usually surrounded by a thin layer of endosperm. An aril is occasionally present but only in the monotypic Australian and Polynesian genus *Alphitonia* is it large and brightly coloured.

Of our two British species, *Rhamnus catharticus* (Buckthorn) grows from Westmorland southward, chiefly on chalk; it occurs through the north temperate zone of the Old World and in North Africa. *R. Frangula* has a similar distribution but grows wild in

Scotland; the wood yields a fine charcoal and is known by gun-powder-makers as Dogwood.

The tribe *Colletieae*, a group mainly of extratropical South American genera, are shrubs with branches reduced to stiff, often flattened spines, and small leaves which often fall early. Species of the North American genus *Ceanothus* are grown as ornamental plants; the small long-stalked flowers are crowded in dense compound inflorescences, the colour generally extending to the calyx and stalk.

Family II. VITACEAE

A generally distributed mainly tropical and subtropical family containing 11 genera and about 600 species, more than 300 of which belong to the widely distributed tropical and subtropical genus *Cissus*. We have no British representative, but *Vitis vinifera* (Vine), *Parthenocissus quinquefolia* (Virginia creeper) and *P. Veitchii* (Ampelopsis) are well known in cultivation. The plants are tendril-climbing (rarely erect) shrubs closely allied to Rhamnaceae, having a similar floral structure, namely four to five sepals, petals and stamens, the stamens being antepetalous, and a generally bicarpellary ovary seated on, or more or less sunken in, a well-marked disc. It is, however, distinguished by the climbing habit, which is rare in Rhamnaceae, by the berried fruit, and seeds with copious endosperm and small embryo. The flowers are hypogynous or slightly perigynous, but never epigynous.

The leaves are scattered and stipulate, sometimes palminerved, palmilobed or palmately compound, but shewing great variety. The tendrils appear opposite the leaves, and are apparently extra-axillary. Recent comparative study by Max Brandt[1] in several genera confirms the view expressed by Eichler that the tendril represents the main axis which has been pushed aside by the stronger growth of the branch borne in the axil of the opposed leaf. The stem of the plant is therefore a sympodium. The inflorescence occupies the same position as the tendril (fig. 152, A) and structures which are partly flower-bearing and partly tendril occur; they may develop into leafy branches. The tendrils become attached by

[1] Brandt, Max. "Untersuchungen über den Sprossaufbau der Vitaceen." Engler's *Bot. Jahrb.* XLV, 509–563 (1911).

coiling or by the development of cushion-like adhesive discs as in the so-called "*Ampelopsis Veitchii*," a favourite wall-climber.

The small greenish flowers are borne in compound dichasial inflorescences, or in panicles passing in the ultimate branches into dichasia. In *Pterisanthes* (south tropical Asia) the

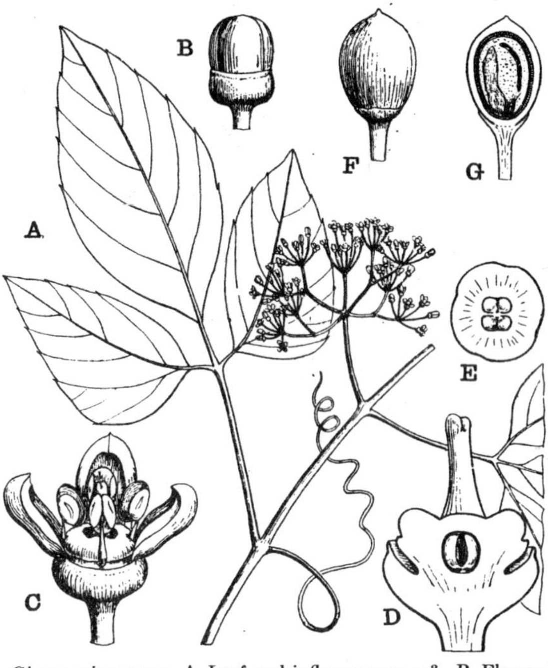

FIG. 152. *Cissus microcarpa.* A. Leaf and inflorescence, $\times \frac{2}{3}$. B. Flower bud, $\times 4$. C. Flower, $\times 5$. D. Flower with petals removed in vertical section, $\times 11$. E. Ovary cut across, $\times 11$. F. Fruit, $\times 2$. G. Same cut lengthwise shewing seed with ruminated endosperm and small basal embryo. (From *Flor. Jam.*)

inflorescence axis or its branches are flat and ribbon-like, bearing the generally stalked male flowers on the edges and the bisexual, or apparently bisexual, flowers sunk on both faces. The calyx is usually only very slightly developed, often forming merely a ring round the base of the corolla. The petals often fall off like a cap when the flower opens. There is no trace of an antesepalous whorl of stamens. The stamens are free, except in *Leea*, where they are joined at the base to form a tube which is united to the base of the corolla. The anthers are introrse. Inside the androecium is a well-developed glandular disc, varying in form in different groups of species, by which the pistil is more or less surrounded. The two carpels (three to eight in *Leea*) are joined and contain as many chambers, in each of which are generally two collateral anatropous ovules (rarely only one) ascending from the base of the chamber, with a ventral raphe and a downward pointing micropyle. The short or long style bears generally a small stigma.

The wild species of *Vitis* are polygamo-dioecious, that is, one plant bears flowers which are functionally male only, another flowers which are functionally female only. In many of the cultivated forms the male flowers are also fertile. According to Rathay, the glandular disc in *Vitis vinifera* secretes little or no nectar and is merely a scent-organ, and Loew considers that this species is derived from entomophilous ancestors but has gradually, with progressing gyno- and andro-dioecious differentiation, become more and more anemophilous. In many other members of the family entomophily is certainly prevalent; for instance, in many species of *Cissus* a large amount of nectar is secreted on the disc.

The fruit is very uniform, forming a more or less juicy berry containing one to four (in *Leea* to six) seeds with a strong or crustaceous coat. They are well adapted for distribution by birds which eat the fruit and drop the seeds undamaged. The endosperm, which contains oil and is generally ruminate, surrounds the small axile embryo.

The species are mainly found in damp, hot situations especially as lianes in primary forest. They occur, however, at considerable altitudes on tropical mountains, as in the Himalayas. In Africa and also in the South American pampas numerous species of *Cissus* form typical steppe- and desert-plants, the vegetative portions becoming more or less fleshy and serving as water-reservoirs. In many the root is tuberous, in others the internodes of the stem are fleshy and swollen, in others again the leaves. In many cases the fleshy stem is erect and bears no tendrils. A remarkable development occurs in the cactus-like lianes of the African steppes which are almost or quite leafless. The 30 species of *Vitis* are temperate to subtropical in the northern hemisphere, chiefly American and East Asiatic. *V. vinifera*, the Grape-vine, is wild in the Mediterranean region, spreading eastward to the Caucasus and northward to the Rhine valley, but Engler has recently shewn that it was formerly, in prehistoric times, wild throughout the south of Europe and in part of Central Europe. The European cultivated Vine consists of numerous forms of *V. vinifera*. In the Eastern United States native species, chiefly *V. Labrusca* (Fox-grape), have been cultivated and recently some of the American varieties have been introduced into Europe to serve as stocks for the better European varieties as they are more resistant to the attacks of the Phylloxera.

Dried currants are the dried fruit of a seedless variety of the grape-vine which is cultivated in various parts of Greece. They were originally brought from Corinth, whence their name. Raisins are the dried fruits of certain varieties, comparatively rich in sugar, which grow principally in the warm climate of the Mediterranean coasts.

Order 14. *ROSALES*

Flowers bisexual, rarely unisexual by abortion, generally cyclic, regular to zygomorphic, hypogynous to epigynous, usually pentamerous with frequent increase in the number of stamens and decrease in the number of carpels. Sepals and petals generally free, stamens free or sometimes united. Carpels free or united, but styles generally free; ovules numerous, more rarely few, generally anatropous and inserted on the ventral suture or on thick marginal or axile placentas; integuments two, sometimes single. Seeds large or small with or without endosperm.

The floral axis is often more or less hollowed so that the insertion of the sepals, petals and stamens is perigynous; epigyny may result from the union of the floral axis with the sunken ovaries.

Plants of very different habit. Leaves simple or compound, with or without stipules.

A very natural group the families in which are connected by transitional forms. Two tendencies are noticeable, the passage of the regular into the zygomorphic flower and of the hypogynous into the epigynous flower. The family Podostemaceae is remarkable for the striking adaptation of its vegetative organs to life in rapidly flowing water.

Family I. CRASSULACEAE

Flowers bisexual, actinomorphic, generally 5-merous, but varying from 3- to 30-merous. Pistil isomerous with the calyx and corolla; carpels free or united below, generally bearing a scale-like glandular appendage at the base. Ovules generally numerous in two rows on the ventral suture, more rarely few or solitary. Fruit generally of follicles. Seeds minute; endosperm much reduced or absent.

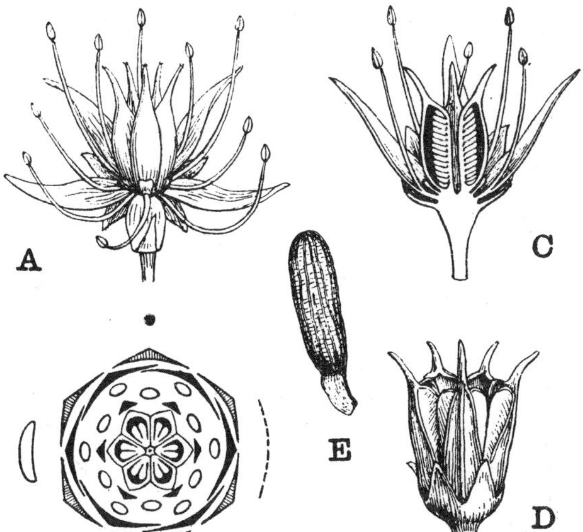

FIG. 153. *Sedum.* A. Flower of *S. spectabile.* B. Floral diagram (after Eichler). C. Flower of *S. spectabile* in vertical section. D. Fruit of *S. altissimum.* E. Seed of same. A, C, D barely three times nat. size; E × 6.

Herbs or small shrubs, generally with thick fleshy stems and leaves, adapted for life in dry, especially rocky places. Leaves exstipulate, alternate or whorled, generally simple and entire. Inflorescence cymose. A cosmopolitan family which finds its chief development in South Africa.

Genera about 30; species about 1300.

Crassulaceae are very closely allied to Saxifragaceae, the most distinctive characters being the fleshy habit and the nectar-secreting appendages of the carpels, though even these may fail. Many species of *Crassula* and others have flat and only slightly fleshy leaves, while *Penthorum* (three species from N.E. America, China and Japan) with membranous leaves and no carpellary appendages may be put in either family; Van Tieghem has regarded it as the type of a distinct family, Penthoraceae. It has the characteristic isomerous pistil of Crassulaceae, but this occurs exceptionally in Saxifragaceae.

So deeply impressed has the fleshy habit become that it extends to the cotyledons, which appear above ground as a pair of succulent green leaves. The plants are generally perennial and shew marked adaptation for life in dry and exposed positions. The fleshy leaves are often reduced to a more or less cylindrical or subulate form as in species of *Sedum* (Stone-crop). In other cases they form closely crowded radical rosettes as in House-leek (*Sempervivum*). The bulk of the tissue is parenchymatous with a copious sap rich in calcium oxalate and free organic acids. Mechanical and conducting elements are poorly developed. Transpiration is also checked by the secretion of a layer of wax on the epidermis giving a glaucous appearance to the plant, and by sinking of the stomata below the general level. Most members of the family are rich in tannin, which occurs generally in the parenchymatous tissues, or more or less localised in spots characteristic of definite species. Means of vegetative propagation are general. Many species spread by a creeping branched rhizome, the branches giving rise to new individuals by death of the older parts. Others, such as House-leek, have a more specialised means, sending out runners (offsets) which perish after producing a new terminal rosette of leaves; or bulbils may be formed. In *Sedum dasyphyllum* the formation of separating leaf-rosettes in place of flowers has been observed. In some cases small portions of the stem or leaves give rise to new plants by adventitious budding; as in *Bryophyllum*, where adventitious buds spring from the edge of the leaf and develop into new plantlets.

The flowers are generally arranged in cymose inflorescences at the end of the leaf-shoot, or in lateral cymes. They form dichasia with a tendency to pass into monochasia, or are purely monochasial. Dichasia and monochasia may be arranged in racemes, corymbs, umbels or panicles. In *Cotyledon* the flowers form terminal spikes or racemes, well seen in *C. Umbilicus* (Penny-cakes, Navelwort), with round peltate leaves, which grows on dry rocks and walls, especially on the west coasts of Great Britain. Both bracteoles of the flower are present, or the sterile one is less developed or suppressed, or both are absent.

The flowers are markedly regular and isomerous, though the number of parts is very varied and not constant even in the same species, sometimes varying even on the same plant, while in one and the same genus wide differences occur. Thus in *Sempervivum* the flowers are from 6- to 30-merous. In *Sedum* the number of parts varies from three to seven; of our British species, *S. Rhodiola* has 4-merous flowers, which are dioecious with petals smaller or wanting in the female; in *S. Telephium* they are 5-merous, in *S. reflexum* often 6-merous. In *Cotyledon* the number varies from five to six, in *Crassula* from three to nine, in *Rochea* pentamery is constant, and in *Bryophyllum* (fig. 154) and *Kalanchöe* tetramery. The sepals are generally free or nearly free, but are united almost to the tip in *Bryophyllum*. The petals are hypogynous or shortly perigynous, and are free, or united only at the base, except in a few genera such as *Cotyledon*, *Kalanchöe* and *Bryophyllum* where they form a tube.

In *Crassula* and some species of *Cotyledon*, there are five stamens alternating with the petals. Often, however, an inner antepetalous whorl is present as in most species of *Sedum* and *Cotyledon* (e.g. *C. Umbilicus*). In some species of *Sempervivum* the inner whorl is sterile.

When the corolla is polypetalous the stamens are hypogynous or slightly perigynous, in gamopetalous flowers they are situated on the corolla-tube (fig. 154, B), the antepetalous slightly above the antesepalous. Except in a Mexican species of *Cotyledon* they are free. The carpels are free or sometimes united at the base; the styles are free. The hypogynous scales, which correspond in number with the carpels and are regarded as appendages of these, generally function as nectaries, but in *Monanthes*, a small genus in the Canary Islands and Morocco, they are large and petaloid.

Proterandry prevails among the majority of the Crassulaceae. In flat open flowers, such as *Sedum acre*, a great variety of insects can reach the nectar; the number becomes more restricted where the flower is less open or where, as in *Cotyledon* and *Bryophyllum*, the corolla is tubular. According to Delpino, the richly nectar-bearing hanging flowers of *B. pinnatum* are pollinated by the help of humming-birds. In the golden-yellow flowers of *Sedum acre* the five antesepalous stamens stand erect in the fully opened flower and shed their pollen, the antepetalous are bent outwards and closed while the stigmas are not mature. As the stamens of the outer whorl wither the antepetalous take their place in the

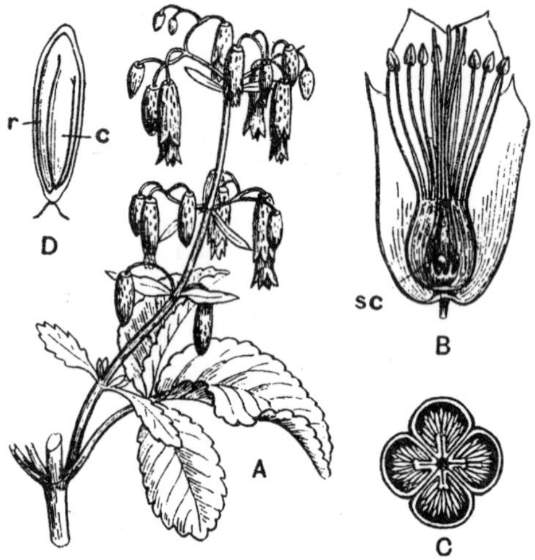

FIG. 154. *Bryophyllum pinnatum*. A. Inflorescence and leaf, × $\frac{1}{6}$. B. Flower cut lengthwise, nat. size; *sc*, scale. C. Ovary cut across, × 3. D. Seed cut open shewing embryo, × 40; *c*, cotyledon; *r*, radicle. (After *Flor. Jam.*)

centre and dehisce, and the full development of the stigmas soon follows. In sunny weather, when plenty of insects are about, the pollen is removed before the stigmas are ready to be pollinated; but in dull weather, pollen remains on the anthers of the inner whorl till the stigmas are ripe, so that self-pollination is possible. In *Sedum album* chances of self-pollination are almost excluded by a still more marked proterandry, while other species are more or less proterogynous.

In the fruiting stage the follicles are often united at the base, thus forming a transition to the capsular fruit which occurs in a few genera. The minute seeds are suited for wind-distribution.

Crassulaceae are almost entirely absent from Australia and Polynesia, and but few occur in South America; otherwise they are very generally distributed. *Sedum*, the largest genus, contains about 140 species in temperate and cold parts of the northern hemisphere, chiefly in the Old World; one species occurs in the Peruvian Andes. *Crassula* has 300 species chiefly at the Cape; the section *Tillaeaoideae*, which is often separated as a distinct genus *Tillaea*, is, however, ubiquitous; to it belongs *T. muscosa*, of our British flora, a rare annual on sandy heaths. *Cotyledon* has 90 species in Africa (chiefly south), West and South Europe, temperate Asia as far as Japan, Mexico, and South America (a few species). *Sempervivum* has about 75 species in the mountains of Central and South Europe, the Caucasus, the Himalayas, Abyssinia and the Canaries and Madeira; in the two latter it is especially developed. The common House-leek, *S. tectorum*, is a native of Central and southern Europe. *Kalanchöe* (100 species) is especially a tropical African and Malagasy genus, but spreads through the tropics of the Old World and into South Africa. One species, *K. brasiliensis*, occurs in Brazil, but is also found in the East Indies and tropical Africa. *Bryophyllum* has more than 20 species in Madagascar, one of which, *B. pinnatum*, occurs in the warmer parts of both the east and west hemispheres. *Rochea* is a small South African genus known in cultivation. Four genera contain one or a few species only in Turkestan, South Africa, Carolina and Sikkim respectively, and *Penthorum* three species in N.E. America, China and Japan. The family is represented in Britain by nine species of *Sedum*, *Tillaea muscosa* and *aquatica* and *Cotyledon Umbilicus*; the House-leek, *Sempervivum tectorum*, is not indigenous.

Family CEPHALOTACEAE. See Appendix

Family II. SAXIFRAGACEAE

Flowers usually bisexual, typically dichlamydeous, and actinomorphic. Floral receptacle polymorphic, shewing every degree between hypogyny and epigyny of the flower. General formula, S5, P5, A5 + 5 or 5 + 0, G (2). The swollen placentas bear several rows of anatropous ovules which have sometimes only one integument. Fruit a capsule or berry. Seeds small with a copious endosperm surrounding a small embryo.

Mostly herbs with alternate, more rarely opposite leaves, and generally small or moderate-sized flowers in various inflorescences.

Genera 80; species about 1100. Cosmopolitan but having its greatest development in temperate regions.

The vegetative organs shew great variety, affording nothing that we can regard as strikingly characteristic of the family. The greater number of the species are perennial herbs; such, for instance, as nearly all our native Saxifrages, species of *Chrysosplenium*, and all the species of *Parnassia*; annuals are less frequent, as species of *Saxifraga*, *Chrysosplenium* and others. Several genera are shrubby, such as *Ribes* (Currant and Gooseberry), *Philadelphus* (Mock Orange), *Hydrangea* and *Escallonia*; trees also occur.

The leaves are of very various forms, and stipules may be present or absent in closely allied genera and even in the same genus. Similarly there are no widely characteristic anatomical features, though investigation shews minor points of interest. Thus all the species of *Saxifraga* are said to have a well-marked endodermis in the stem; this is also present in *Chrysosplenium* but absent in *Astilbe*, *Heuchera* and others. It has been found also that in many cases sections of *Saxifraga* and allied genera which have been founded on differences in the structure of the flower and the form of the leaf also shew anatomical differences such as presence or absence of a ring of sclerenchyma. The form of the hairs is a help in distinguishing sections, genera and even subfamilies. The Escallonioideae, for instance, generally have one-celled hairs, the Saxifragoideae several-celled. Water-pores, often exuding a chalk-secretion, occur on the edges of the leaves, especially in species of *Saxifraga*. Vegetative propagation is well marked in certain Saxifrages; for instance, by production of bulbils in the leaf-axils as in *S. bulbifera* and others, or by long slender stolons as in *S. flagellaris* and *S. sarmentosa* (the commonly cultivated pot-plant known as Mother-of-thousands) bearing small terminal leaf-rosettes. In *S. stellaris* var. *comosa* adventitious buds are developed in the inflorescence.

The inflorescence is very varied. In many genera, as *Ribes* and *Escallonia*, it is a raceme, the two bracteoles are generally present and compound inflorescences arise by production of branches in the axil of one or both bracteoles. The lateral branches often develop as dichasial cymes, as in *Chrysosplenium*, where the same arrangement holds throughout; generally in the ultimate branchings one bracteole only of each pair is fertile and monochasia are formed. In *Parnassia* the branching is monochasial from the first. The large spreading inflorescence in *Hydrangea* (fig. 155) is made up of dichasia with monochasial endings. In the small tropical American genus *Phyllonoma* the cymose inflorescence springs from the upper surface of the leaf. In the great majority of the genera the flower is pentamerous with generally a reduction to two in the gynoecium. The two carpels are median in *Ribes* (fig. 157), *Escallonia* and others, but more often oblique,

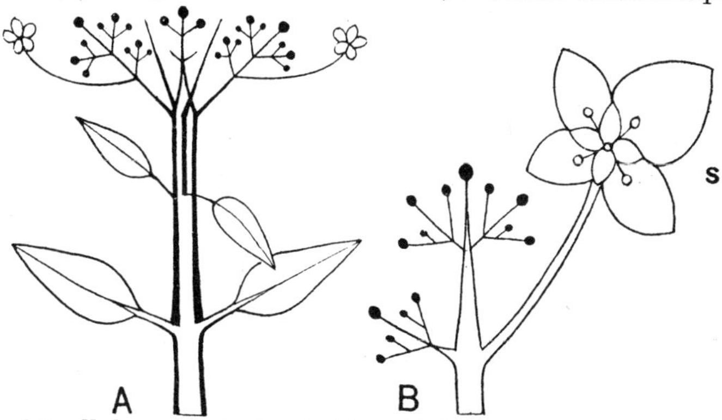

FIG. 155. *Hydrangea hortensia.* A. Plan of inflorescence: two of the lateral branches are elaborated and the right-hand one is shewn in detail in B. The large sterile flower, *s*, corresponds with a dichasium (on the opposite side of the axis) the lateral branches of which are monochasial. The branch systems indicated by a black line in A arise in the axils of the upper pair of foliage-leaves and are united with the main axis above their insertion. (After Eichler.)

lying in the plane of the first-developed sepal (fig. 156); rarely are they lateral as in *Ribes alpinum*, a species indigenous in the north of England. The androecium is obdiplostemonous (*Saxifraga*), or there is a single whorl of stamens which alternate with the petals (*Ribes*); in *Parnassia* the outer whorl is represented by five large, generally fringed, scales. Tetramery also occurs, but is rarer; *Francoa* has the flower tetramerous throughout, but the arrangement is not constant, pentamery occasionally occurring in one and the same species. *Philadelphus* (Mock Orange) has a tetramerous calyx, corolla and gynoecium, but indefinite stamens; the pleiomery of the androecium arises, according to Payer, from splitting of four rudiments alternating with the petals.

Instances of suppression are found in apetalous and unisexual flowers. The latter occur in *Astilbe* and are frequent at the periphery of the large inflorescence of *Hydrangea* and its allies. *Ribes alpinum* and allied species are dioecious. In *Saxifraga sarmentosa* the flowers become zygomorphic by increase in size of two adjacent petals. The position of the carpels relative to the other whorls varies with the shape of the thalamus and is not constant in the same genus. The carpels are sometimes free (species of *Astilbe*) though this is rarely the case, and generally when superior they are united below and free above. It is interesting to note the occurrence in a single genus, *Saxifraga*, of hypogyny, perigyny and epigyny. Its 320 species are divided into 15 sections[1]. Two of these, *Hirculus* and *Robertsonia*, are characterised by a flat thalamus; the former is represented in our flora by the

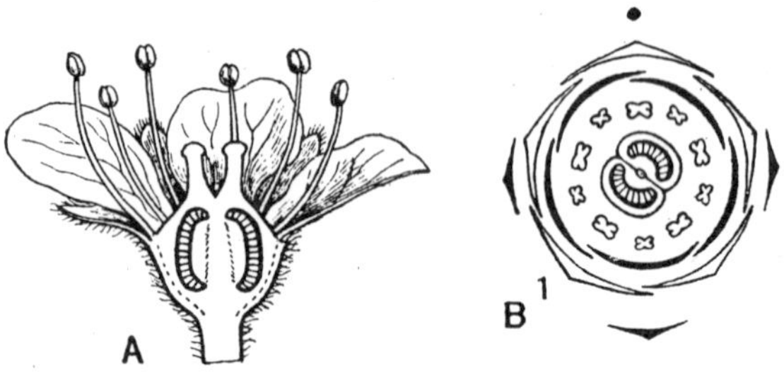

FIG. 156. *Saxifraga granulata*. A. Flower in vertical section, enlarged (after Warming). B. Floral diagram (after Eichler); 1, the first developed sepal.

rare *S. Hirculus*, the latter by *S. umbrosa* (London Pride), *S. Geum* (south and west Ireland) and other closely allied species. Another section (*Trachyphyllum*), to which belongs *S. aizoides*, an alpine species (British), has the torus flat, cup-shaped or bowl-shaped. In the rest, comprising by far the greater number of species, the receptacle is more or less concave and coherent with the ovary. In *Francoa* there are large outgrowths of the thalamus, functioning as nectaries, between the stamens, while in some sections of *Saxifraga*, and in *Escallonia*, the inferior ovary bears epigynous discs which have a similar function. In *Ribes* the small sepals and petals and the stamens are borne on the edge of a frequently

[1] Engler and Irmischer, *Pflanzenreich*, 1916–19 (*Saxifraga*).

petaloid prolongation of the receptacle above the top of the inferior ovary (fig. 157).

Contrasting with this great variety in position of the ovary and adhesion of the carpels a very constant character is afforded by the generally swollen placentas bearing several rows of anatropous ovules. The position of the placenta is parietal or axile, differing often in nearly-allied genera, according as the edges of the carpellary leaves project less or more towards the centre of the ovary. Occasionally they are developed only at the base of the ovary-chamber, and in *Escallonia* are remarkable in hanging from the apex.

Pollination is effected by the help of insects which are induced to visit the flowers for the nectar secreted on the thalamus or, in epigynous flowers, on the top of the ovary. In *Parnassia* the fimbriated staminodes bear a pair of flat

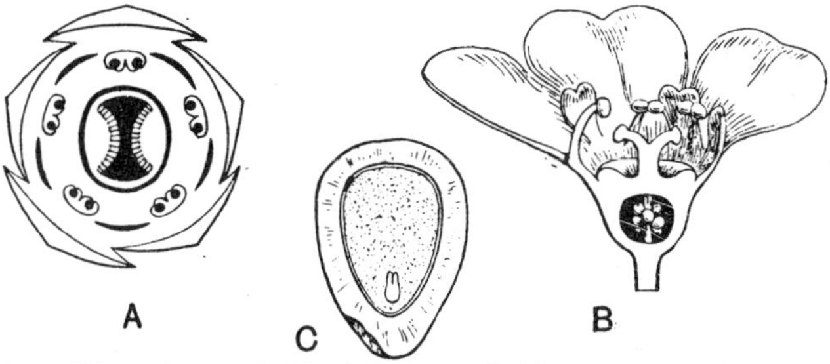

FIG. 157. *Ribes rubrum*. A. Floral diagram. B. Flower in vertical section, × 4. C. Seed in vertical section, much enlarged. (After Baillon.)

nectaries on the surface facing the ovary. The petals are mostly white, sometimes yellow or reddish, and are generally rendered more conspicuous by association of several or many flowers in the inflorescence. In *Hydrangea* and its allies the peripheral flowers of the head are large and sterile, while the fertile flowers are much less conspicuous; where, as in *Chrysosplenium* (Golden Saxifrage), the petals are suppressed, the calyx is light-coloured. According to Engler, homogamy or proterogyny are most common. Of the Saxifrages, however, three species are strikingly proterogynous, *S. oppositifolia* and *S. tridactylites* (both British) are sometimes proterogynous, sometimes proterandrous, while the rest, so far as the phenomenon has been studied, are proterandrous. *Parnassia* is strikingly proterandrous. In *Ribes*, the Red and Black Currants (*R. rubrum* and *R. nigrum*) are homogamous, while Gooseberry (*R. Grossularia*) is proterandrous.

Most of the ovules develop into seeds, and the fruit is a capsule. Some genera, however, produce berries, such as the Currant and Gooseberry, where the seeds have a succulent outer coat. The seeds are generally small with a copious endosperm surrounding a small embryo.

Engler recognises seven subfamilies, the characters, geographical distribution and principal genera of which are as follows:

I. SAXIFRAGOIDEAE. Herbs; leaves generally alternate without stipules or with stipule-like outgrowths of the sheath. Calyx and corolla generally pentamerous, more rarely tetramerous. Pistil of two, more rarely three to four, carpels; ovary one- to four-celled, superior to inferior. About 33 genera with about 600 species distributed through the Arctic and north temperate zone, often alpine. In America they stretch down the western mountain chain to the Andes, Southern Chile and Cape Horn. The small genus *Vahlia* occurs in subtropical and tropical Africa and Asia. *Saxifraga* is by far the largest genus; its 320 species are distributed over the mountains of the Arctic and north temperate zone, and found also in the Andes; fifteen species are British. *Chrysosplenium* has 85 species; its distribution is very similar to that of *Saxifraga*. *Heuchera* has about 50 species in North America and the mountains of Mexico. *Parnassia* has 50 species in the extratropical regions of the northern hemisphere, especially in mountain meadows. *P. palustris* (Grass of Parnassus) occurs in wet moors and bogs. *Astilbe japonica* is a common ornamental plant. The genus contains about 24 species in the Himalayas, Eastern Asia, and south-eastern North America. The remaining genera are small and generally confined to limited areas.

The genus *Parnassia*, characterised by its conspicuous whorl of staminodes and ovary with three to four parietal placentas, is somewhat anomalous in the subfamily. It has been included in Droseraceae by some botanists and Lula Pace (*Bot. Gaz.* liv, p. 306) finds it more closely related with this family in the characters of ovule and embryo-sac development. Diels, the monographer of Droseraceae in the *Pflanzenreich*, refuses to admit it in that family, and the position adopted here is that favoured by Bentham and Hooker (*Genera Plantarum*) and Engler (*Pflanzenfamilien*).

II. Francöoideae. Perennial herbs with radical leaves and a scape with a racemose or spicate inflorescence. Flowers normally tetramerous. Two small genera, *Francoa* and *Tetilla*, from the mountains of Chile. *Francoa* is cultivated.

III. Hydrangeoideae. Shrubs or trees with simple, generally opposite, exstipulate leaves. Perianth generally pentamerous. Stamens generally epigynous. Ovary half-inferior or inferior, generally three- to five-celled. About 19 genera with about 250 species chiefly in temperate North America and Eastern Asia. *Philadelphus*, *Deutzia* and *Hydrangea* are the largest genera, and are all well known in cultivation. *Philadelphus coronarius* is the sweet-scented shrub known popularly as Syringa or Mock Orange.

IV. Pterostemonoideae. Shrubs with alternate simple leaves with small stipules. Ovary inferior, five-celled, with a few ovules on the axile placenta. One genus (*Pterostemon*) with two species from the highlands of Mexico.

V. Escallonioideae. Generally shrubs, or trees, with simple, alternate, often leathery and glandular-toothed, exstipulate leaves. Stamens isomerous with the corolla. Ovary superior to inferior; ovules with a single integument. About 21 genera with about 100 species, half of which belong to *Escallonia*, a South American genus found chiefly on the Andes and mountains of Southern Brazil; *E. rubra* and *E. macrantha* are cultivated as ornamental plants. The rest are small (often monotypic) genera; several have a restricted distribution in the southern hemisphere. *Phyllonoma* occurs in New Granada and Mexico, *Polyosma* spreads from India to tropical Australia, and *Itea* occurs in the warmer parts of Eastern Asia and on the Atlantic side of North America.

VI. Ribesioideae. Shrubs with simple, alternate, exstipulate leaves, and racemed haplostemonous flowers. Ovary inferior, one-celled, with two parietal placentas. Fruit a berry. One genus, *Ribes*, with about 130 species in the north temperate zone, the mountains of Central America and along the Andes to the Strait of Magellan. Four are British—*R. Grossularia* (Gooseberry), with spinous branches and one- to three-flowered peduncles, native in the north of England, and *R. alpinum*, *R. rubrum*, and *R. nigrum*, with branches not spinous and many-flowered racemes, natives in parts of England and in Scotland.

R. aureum and *R. sanguineum*, natives of North-west America, are spring-flowering ornamental bushes.

VII. BAUEROIDEAE. Shrubs with opposite trifoliolate exstipulate leaves, and solitary axillary flowers. Ovary half-inferior with two parietal placentas and numerous ovules. Fruit a loculicidal capsule. One genus, *Bauera*, with three species in New South Wales and Tasmania.

These subfamilies are sometimes separated as distinct families. It is, however, more instructive to group them together under Saxifragaceae, which thus affords a good example of variation among groups obviously very closely allied.

Allied to the Saxifragaceae is the small family (III) CUNONIACEAE, which consists of woody plants with opposite or whorled stipulate leaves, and small flowers crowded in heads or in racemose or paniculate inflorescences. The flowers resemble those of Saxifragaceae but the ovules are arranged in two series in each chamber of the bilocular ovary. The 25 genera contain about 240 species, more than half of which belong to *Weinmannia*, and are almost confined to the southern hemisphere.

Family IV. PITTOSPORACEAE

A small family[1] occurring in the warmer parts of the Old World and specially developed in Australia. It is closely allied to the subfamily Escallonioideae of Saxifragaceae, from which it is distinguished by an anatomical character, the presence of schizogenous resin-passages in the cortex. They are woody plants sometimes climbing, with alternate generally leathery exstipulate leaves. The showy flowers are bisexual, generally regular, and conform to the formula S5, P5, A5, G(2). The petals are sometimes connate below, forming a more or less tubular corolla. The stamens are inserted below the ovary; three to five carpels are sometimes present and the ovary is one- to several-chambered. The ovules are generally numerous and inserted in two rows on the parietal or axile placentas; they are anatropous and have a single integument. The fruit is a capsule or a berry. The small embryo is excentrically placed in the copious hard endosperm. Species of *Pittosporum* and other genera are grown as greenhouse plants.

Family V. PODOSTEMACEAE

A small family (43 genera, about 130 species) of submerged herbs highly modified in association with their aquatic habit. They are widely distributed in the tropics, growing attached by special organs (haptera) to stones or rocks in rapidly flowing water.

[1] About 200 species in 9 genera, *Pittosporum* includes 160.

The characteristic feature of the plants is the green dorsiventral thallus-like structure, shewing wonderful diversity in form, upon which leaf- and flower-bearing secondary shoots are developed. The minute seeds give rise on germination at the beginning of the rainy season to primary axes which are rarely tall and floriferous, generally more or less considerably reduced; there is no primary root; from the primary axis buds out endogenously the creeping thallus which, with a few exceptions (e.g. *Lawia*), is of root-nature. In the simpler cases the root-nature is obvious, as in *Tristicha*, or *Podostemum Ceratophyllum* (fig. 158, A), where it is a thin thread-like organ, endogenously branched, with its tip covered by an ordinary root-cap, but bearing acropetally developed endogenous leafy shoots which reach a high degree of complexity; the structure of the vascular cylinder is slightly dorsiventral but otherwise not markedly different from that of an ordinary root. A similar thread-like thallus is characteristic of most of the American forms. In the Indian *Podostemum subulatum* the thread-like thallus is exogenously branched with a remarkable collenchymatous root-cap; the dorsiventrality is more marked and the secondary shoots are smaller. Other Indian genera shew a remarkable series of forms (Willis (2)). In *Dicraea* (fig. 159, A) and *Griffithella* they are flat and expanded, often widely drifting and highly polymorphic, recalling in form the thallus of *Fucus* and other brown Algae which live under similar conditions of varying submergence and exposure. They are endogenously developed from the base of the hypocotyl of the seedling, but are exogenously branched, markedly dorsiventral in external and internal structure, and have a collenchymatous root-cap. In *Hydrobryum* (fig. 159, E) the thallus is branched and ribbon-like or deeply lobed like a liverwort, with scarcely any trace of root-character; it is developed from the base of the hypocotyl, sometimes endogenously but often exogenously, and increases by marginal, not apical growth. This flattening and increase in importance of the thallus as an assimilatory organ is accompanied by great reduction in the importance of the secondary shoots. In *Tristicha* these are large and complex and ultimately bear many flowers; in *Podostemum* and *Hydrobryum* they are short, more or less prostrate though branched and bear a few flowers (fig. 159, E), while in *Dicraea* and the other Indian genera mentioned, the secondary shoots are at first mere tufts of leaves endogenously formed in acropetal succession on the thallus (fig. 159, A), which towards the flowering season elongate and bear a few bracts, a spathe, and a terminal flower.

The leaves are similar in structure on the primary axis and the secondary shoots. In *Tristicha* and allies they are small, entire and very delicate, recalling the leaves of a moss; in the Indian species of *Podostemum* they are simple and usually subulate or linear, generally

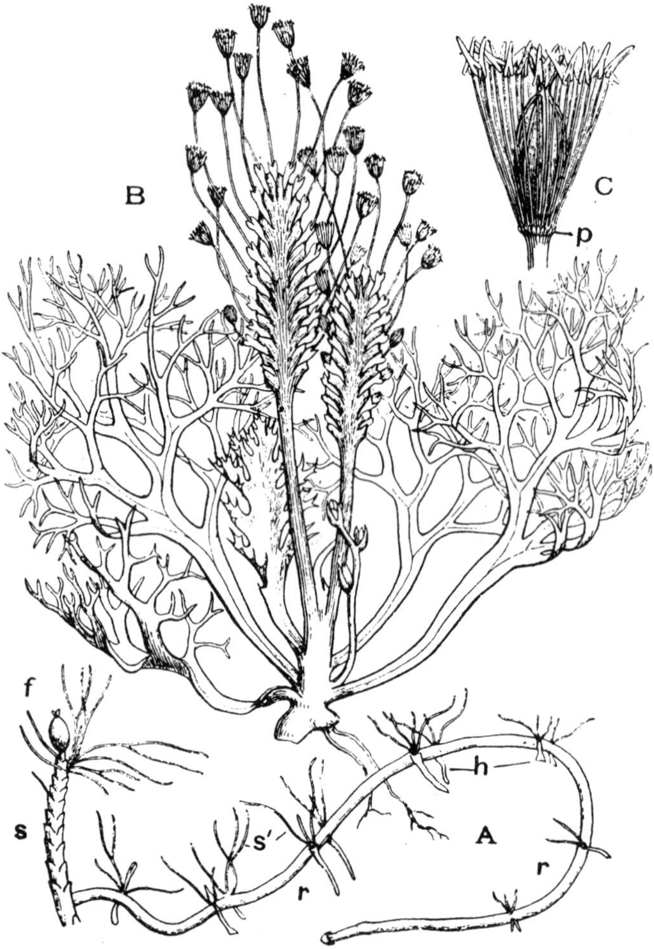

FIG. 158. A. *Podostemum Ceratophyllum.* A shoot *s*, from which springs a long creeping root, *r*, bearing new shoots above, *s′*, and haptera below, *h*: *f*, flower; $\times 1\frac{2}{3}$. B. *Mourera Weddelliana.* A creeping rhizome-like axis bearing large much-cut leaves and erect flowering shoots, the flowers being carried up above the spathes on long pedicels; $\times \frac{1}{2}$. C. Flower of same, $\times 2\frac{1}{2}$; *p*, perianth. (A after Warming; B after Baillon.)

very small, but in *Mourera* and other South American genera they are often very large and exhibit forms like those of many marine Algae (fig. 158, B). The leaf-base is slightly sheathing in the vegetative stage, but during development of the flower becomes much enlarged, forming a sheathing scaly bract while the tip falls away. The leaves are generally arranged in two rows on the secondary shoots; the arrangement on the primary axis is more complex.

The fixing organs (haptera) are generally formed exogenously from the surface tissues of thallus or shoot and have a growing point rather like that of the roots. On reaching the substratum they flatten out upon it.

The internal structure consists of fairly uniform parenchyma generally without intercellular spaces, often collenchymatous, especially in the neighbourhood of the scattered simple conducting bundles. In the latter the xylem is much reduced, the bast, on the other hand, contains broad sieve-tubes and companion-cells. The absence of intercellular spaces is explained by Goebel as correlated with the habitat in rapidly moving well-aerated water; large intercellular spaces to carry oxygen to lower portions in mud or stagnant water being unnecessary. In many species silica is secreted, often in considerable quantities, in the peripheral portions, and acts as a protection against drying up when the plants are exposed by fall in level of the water.

The very small bisexual flowers are borne in cymose inflorescences (fig. 158, B) or are solitary and terminal. In the most perfect form the flower is regular with a perianth of three to five free or more or less united leaves (fig. 159, B). In others the protecting perianth is replaced by a spathe and the perianth is represented by a ring of small scales (fig. 158, C). In these cases the flowers are, as in the former, entomophilous and carried on a long stalk above the water-level. The hypogynous stamens are in one or several whorls—with normal four-chambered introrse anthers—and there is a symmetrical pistil of three or two carpels united to form a three-, two- or one-celled ovary, with as many free styles. From these forms can be traced a progressive series of reduction in the size and conspicuousness of the flowers which then become dependent on the wind or on self-pollination. This is accompanied by dorsiventral development owing to suppression of the perianth and the upper stamens. Finally in the *Eupodostemeae* (*Dicraea*, *Hydrobryum*, *Mniopsis*, *Podostemum* and allied genera) the flowers are small, markedly dorsiventral and commonly inverted within the spathe, the perianth is absent or represented only by a pair of filiform structures at the sides of the androecium, which is reduced to a single or forked stamen on the lower side of the flower, and the two lobes of the bicarpellary fruit are unequal (fig. 159, C, D, E).

The numerous anatropous ovules are borne on a stout axile placenta (fig. 159, B). The fruit is a capsule opening septifragally containing generally very small seeds; the outer layer of the seed-coat is mucilaginous. The seed contains an embryo with two thick cotyledons (fig. 159, F).

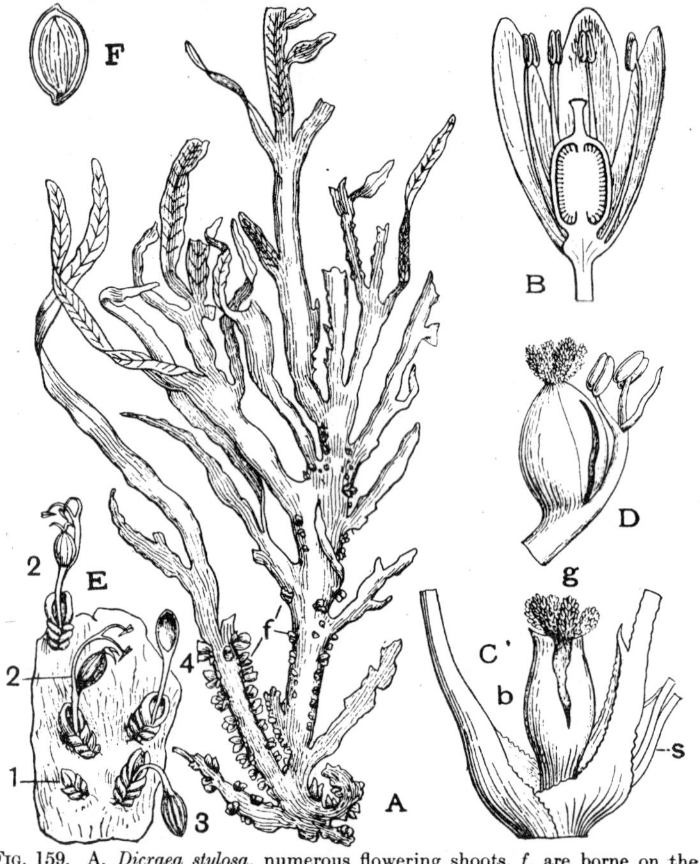

FIG. 159. A. *Dicraea stylosa*, numerous flowering shoots, *f*, are borne on the lower part of the branched ribbon-like thallus, × ½. B. Flower of *Weddelina squamulosa* in vertical section, × 5. C. *Mniopsis Saldanhana*, apex of shoot bearing a flower, subtended by two leaves, from the sheath of the upper leaf springs a lateral shoot, *s*; *b*, spathe; *g*, the projecting styles; × 8. D. Flower of same after escaping from the spathe, × 8. E. *Hydrobryum olivaceum*. The flat thallus bears short shoots with a few closely appressed leaves and a terminal flower; the spathe opens laterally to release the stalked flower: 1, flower still enclosed in spathe; 2, flowers after escaping from the spathe; 3, fruit; 4, the smaller valve of the fruit has fallen and the seeds have been scattered; × 2½. F. Seed cut open shewing embryo, × 33. (A, C, D after Warming; B after Baillon.)

The opening of the flowers and shedding of the seeds coincide with the fall of the water and the consequent exposure of the plant, the vegetative portions of which may perish. The flower and fruit are typically those of a land plant, and the seed has no special adaptation for dissemination by water. The mucilaginous character of the outer seed-coat favours distribution by the feet of wading birds to which the seed may cling.

W. Magnus (3) has studied the embryology in five species, representing as many genera, and finds that, notwithstanding certain striking differences, it has much in common with that of other Angiosperms. Shortly after fertilisation the embryo breaks through the embryo-sac and by means of a long suspensor passes into the nucellar cavity, and then sends out a large haustorial cell into the tissues of the funicle and outer integument. This direct feeding of the embryo probably accounts for the rapid ripening of the seed.

Widely differing views have been held as to the systematic position of the Podostemaceae. Warming(1) regards the family as most nearly allied to the Saxifragaceae, with which it has in common the hypogynous, often bicarpellary, ovary, numerous anatropous ovules and free styles, and this view is now generally accepted.

REFERENCES

(1) Warming, E. "Podostemaceae" in *Pflanzenfamilien*, III, 2*a* (1891).
See also "Familien Podostemaceae." A large and well illustrated series of studies on the family (in Danish) in *Kgl. Danske Vidensk. Selsk. Skrifter* (Copenhagen), 1881–1901.

(2) Willis, J. C. "Studies in the Morphology and Ecology of the Podostemaceae of Ceylon and India." *Ann. Roy. Bot. Gard. Peradenyia*, I, 267 (1902). (Includes good illustrations of habitat, etc.). Also p. 181 in same volume, "A revision of the Podostemaceae of India and Ceylon."

(3) Magnus, W. "Embryology of the Podostemaceae." *Flora*, n.s. v, 275–336 (4 pls. and 41 figs.) (1913).

Family VI. HYDROSTACHYACEAE

The family Hydrostachyaceae, containing one genus, *Hydrostachys*, with a few species in Madagascar, tropical and South Africa, formerly included as a tribe of Podostemaceae, has been separated by Warming as a distinct family. They are submerged water plants with thick tuber-like stems bearing long simple or pinnately divided leaves, often covered with numerous small scale-like emergences, and long-stalked spikes of *naked dioecious* flowers. The male consist of a single stamen protected by a bract, the female of two similarly protected carpels united to form a unilocular ovary with median placentas bearing numerous ovules each with a single integument; there are two free styles. The fruit is a capsule opening at the ventral sutures of the carpels.

Family VII. HAMAMELIDACEAE

A small family of trees and shrubs occurring in subtropical and temperate climates, especially in Asia, Atlantic North America and a few in Madagascar and South Africa. The leaves are generally alternate, simple and stipulate, and the flowers inconspicuous and arranged in spikes or heads which are sometimes surrounded by coloured bracts. The flowers are of great variety of structure in the different genera. In *Liquidambar* (four species in Asia Minor, Eastern Asia, Central and Atlantic North America) and *Altingia* (China to Java) they are monoecious; the male have no perianth but consist merely of spikes of stamens with no distinction into individual flowers, the female consist of numerous flowers united by the ovaries into spherical heads and bearing an inconspicuous epigynous calyx and a few barren stamens. In *Distylium* (Himalayas, mountains of Java and East Asia) the flowers are hypogynous and apetalous; sepals are absent or present in varying number up to five, and the stamens vary from two to eight. *Hamamelis* and other genera have complete bisexual regular hypogynous flowers with four or five sepals, petals and stamens; *H. virginiana* (Witch Hazel), a bush, native of eastern North America and often seen in cultivation, has long strap-shaped yellow petals, the flowers appearing before the leaves (fig. 160). A constant character is found in the two united carpels, the ovary containing one to many pendulous ovules and bearing two slender styles. The fruit is a two-chambered capsule splitting loculicidally and sometimes also septicidally; the pericarp consists of a woody or leathery exocarp and a more or less horny endocarp. The more or less oval, sometimes winged seeds contain a large straight embryo with flat cotyledons, surrounded by endosperm.

The hairs shew a characteristic structure recalling that found in the subfamily Hydrangeoideae of the family Saxifragaceae. They are one-celled and awl-shaped with brownish contents, and are generally associated in clusters and usually united at the base forming a two- to many-limbed bundle-hair. They are generally distributed, or more or less closely crowded, especially on the ovary and young parts of the stem.

There are 23 genera containing about 100 species. *Bucklandia populnea* (Himalayas and mountains of Java and Sumatra) and *Altingia excelsa* (south-west China and Java) are fine timber-trees; species of *Liquidambar* yield a gum-resin, storax.

This and the next family Platanaceae are grouped together by Wettstein as the order Hamamelidales, which he places in the Monochlamydeae near Urticales on account of the "many primitive characters" represented by the genera. The great variety of floral structure suggests a very old type, and the resemblance to the Rosales in the structure of the gynoecium Wettstein regards as an indication that the origin of this group may be sought in forms resembling the Hamamelidaceae. The more general view adopted by Warming, Baillon, Niedenzu (who has elaborated the family in the *Pflanzenfamilien*) and Engler is that the Hamamelidaceae are nearly allied to the Saxifragaceae and especially to the small family Cunoniaceae with which it agrees in the woody habit, the two-valved capsule and the winged seed; the anatomy of the wood is also similar.

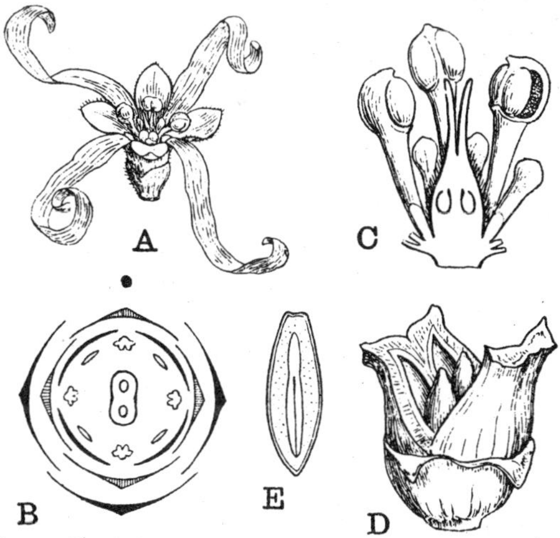

FIG. 160. *Hamamelis virginiana.* A. Flower, × 3. B. Floral diagram. C. Vertical section shewing stamens and pistil, × 10. D. Fruit dehiscing, × 2. E. Seed cut lengthwise, × 3. (C after Köhler; B, D, E after Sargent.)

Family VIII. PLATANACEAE

The family consists of the single genus, *Platanus*, comprising a few species of large deciduous trees, with alternate long-stalked palmate leaves generally toothed on the margin; the upper edges of the leaf-stalk join at the base forming a cap which protects the developing bud; the stipules are large and embrace the stem above the leaf-insertion. The cork-cambium arises in the subepidermal layer; in *P. occidentalis* the bark peels in large thin sheets. The young shoots are densely clothed with woolly hairs consisting of a long jointed hair ending in a sharp point and bearing at each node a three- to five-rayed whorl of pointed cells; the walls are very thick. Similar hairs, but with much reduced branches, occur on the calyx and the ovary; in the latter case they develop to form the long slender jointed hairs which cover the fruit.

The unisexual flowers are closely crowded in spherical heads, several of which are borne on a long pendulous stalk which ultimately splits into fibres. The inflorescences and heads are monoecious, but the frequent presence in the flowers of rudiments of the other sex indicates that they are derived from a bisexual flower.

FIG. 161. *Platanus acerifolia*. A. Shoot bearing male (*a*) and female (*b*) inflorescences, reduced; *s*, stipules. B. Female flower, enlarged. C. Fruit in vertical section, × 3. (After Niedenzu.)

The flowers are slightly perigynous, cyclic and 3- to 8-merous, with regularly alternating sepals, petals, stamens or carpels. The sepals are small, free and hairy, the petals larger, thin and spathulate. The stamens have a very short filament and a long anther, the halves of which split lengthwise; the connective is produced to form a shield-like cover. The carpels are free; the elongated ovary passes into a long style, which is hooked above and bears the stigma along the inner side. There is one ovule, rarely two, orthotropous and pendulous from the ventral suture, with two integuments.

The flowers are wind-pollinated. The fruit is a close aggregation of caryopses, which through lateral pressure become four-angled and inverted-pyramidal; they are densely covered with hairs and bear the remains of the style on the top. A scanty endosperm surrounds the long thin straight embryo, which has linear, often unequal, cotyledons.

Platanus orientalis is native from the eastern Mediterranean to the Himalayas, *P. occidentalis* from Mexico to Canada; there are three other species in Mexico and North America; *P. acerifolia*, the Plane of the London streets and open spaces, is said to be a hybrid between *orientalis* and *occidentalis*. In later Tertiary times the genus was distributed throughout Europe, North Asia and North America.

In the structure of the style, stigma and anther, *Platanus* resembles the Hamamelidáceae, but it is more nearly allied to the Rosaceae as indicated by the perigynous flower with completely free carpels, and especially to the subfamily Spiraeoideae. It is, however, distinguished from the other families of the order by the almost or quite orthotropous ovule. Niedenzu, who has elaborated the family in the *Pflanzenfamilien*, regards it as representing, in the position of the ovule and the floral structure generally, a type which has remained in the lowest stage of development as compared with the other families of the order.

The account of the floral structure given here is based on that of S. Schonland[1] who describes and figures the development and adult structure of the flower. More recently this account has been called in question by R. F. Griggs[2] who found no traces of a perianth beyond an insignificant and transitory circular scale surrounding each group of stamens. In the female flower the carpel-group was protected by three or four hairy staminodes. In the opinion of this author *Platanus* is apetalous and should be removed to the neighbourhood of the Urticales where it was placed by Bentham and Hooker and is retained by Wettstein.

Family IX. ROSACEAE

Flowers generally bisexual and actinomorphic with a pentamerous calyx and corolla, two, three, or four times as many or indefinite stamens, and one to indefinite carpels; carpels one-celled, with mostly two pendulous or ascending anatropous

[1] Schonland, S. "Ueber die Entwicklung d. Blüten u. Frucht bei den Platanen." Engler's *Jahrbuch*, IV, 323 (1883).

[2] Griggs, R. F. "On the characters and relationships of the Platanaceae." *Bull. Torrey Botan. Club*, XXXVI, 389 (1909).

ovules. Calyx, corolla and stamens more or less perigynous. Fruit very various, often associated with the thalamus in a pseudocarp. Seed generally without endosperm; embryo with planoconvex often fleshy cotyledons.

Herbs, shrubs or trees with alternate simple or compound stipulate leaves.

A large cosmopolitan family containing about 90 genera with 2000 species.

The cotyledons are epigeal in germination and in the seedling stage are nearly all of the same general outline, varying between oval and oblong; sometimes they are emarginate. The greatest divergence from the general type occurs in the tribe *Sanguisorbeae*, which form a natural group with reduced flowers, and having the cotyledons deeply emarginate with a cordate base. The leaves succeeding the cotyledons shew transitional stages to the adult form. This is well seen in species of *Potentilla*, as *P. Anserina* (Silverweed).

The plants are of very different habit. Low-growing herbs such as Strawberry, the closely allied genus *Potentilla*, and *Alchemilla*; scrambling bushes, as the Brambles or Rose; or trees, as Apple, Pear, Cherry, Plum, and other fruit trees. Both root and stem may take part in vegetative propagation. Leaf-buds are commonly produced on the roots of Cherry and other fruit trees; the Strawberry spreads by slender runners which take root and produce a new acaulescent plant at their apex, while Raspberry is propagated by suckers or branches from the base of the erect leafy shoots which, after running for some distance horizontally beneath the soil, grow upward and form aerial leafy stems. In shrubby forms a change of branches into thorns is frequent, as in *Prunus spinosa* (Sloe or Blackthorn) and *Crataegus Oxyacantha* (Hawthorn).

The prickles of the Roses and Brambles are emergences; by their means the plant is able to scramble over surrounding vegetation. Stipules are rarely absent, as in some species of *Spiraea*; they are small and caducous as in *Pyrus* (Apple, Pear, etc.) and *Prunus* (Plum, Cherry, etc.), or persistent and adnate to the leaf-stalk as in *Rosa*, *Rubus* and nearly all our British genera. The prominent part played by the leaf-base in bud-protection is well illustrated by young shoots of *Rosa canina* (Dog-rose) (fig. 162); the lowest bud-scales are simply leaf-bases, the tridenticulate apex indicating the presence of a pair of lateral stipules (*s*) adnate to the leaf-stalk, the upper bear a small pinnately compound leaf (*l*) affording a transition to the perfect form. A similar reverse transition to leaf-base can be traced in the bracts (*b*). The leaves are simple or pinnately or more rarely palmately compound. The subfamily Prunoideae has simple leaves; in other cases

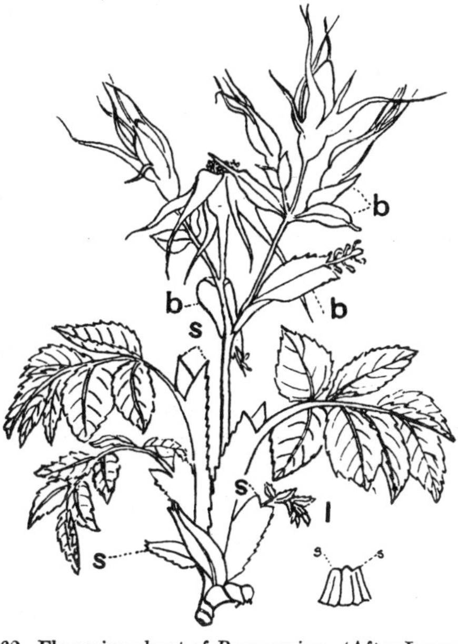

FIG. 162. Flowering shoot of *Rosa canina*. (After Luerssen.)

simple and compound occur in the same genus, as in *Pyrus*, where *P. Malus* (Apple) and *P. communis* (Pear) have simple leaves, *P. Aucuparia* (Mountain Ash) pinnate. In the species of the section *Aria* the leaves are more or less pinnately cut, and hybrids of these with *P. Aucuparia* shew leaves pinnately compound below and pinnately cut above. In *Potentilla* they are generally palmately, but sometimes (*P. Anserina*) pinnately, compound.

In warmer climates the leaves are often leathery and

evergreen. Woody members of the family shew certain points of agreement in anatomical characters. In the Pomoideae, periderm originates in the epidermis, in the Prunoideae in the hypodermal layer. The primary cortex generally has a collenchymatous hypoderma and numerous crystal-sacs in the parenchyma. Stone-cells are generally absent both from the primary and secondary cortex, but are characteristic of the small tropical subfamily Chrysobalanoideae. Medullary rays are broad in Rosoideae and Prunoideae, narrow in Pomoideae. In Prunoideae gum is formed by disorganisation of the wood (Cherry-gum, etc.)

The flowers are sometimes solitary as in *Rubus Chamaemorus* (Cloudberry); often, however, branching takes place in the axil of the bracteoles, giving rise to an indefinite or more rarely a definite inflorescence. In *Agrimonia* it is a simple raceme; in *Poterium* the small flowers form dense heads or spikes; in *Rosa* the flowers are solitary or a few together and corymbose; the many-flowered inflorescence of *Ulmaria* (Meadow-sweet) consists of a number of cymes arranged in a corymb.

The sepals are almost invariably greenish and sepaloid; in some species of *Rosa* the outer members are more or less foliaceous. In *Potentilla*, *Fragaria*, *Geum* and others there is an outer calyx (epicalyx) of generally smaller alternating members representing the stipules of the sepals proper, adjacent ones having united in pairs (compare the frequent union of adjacent foliaceous stipules in our British Rubiaceae). The petals are generally large and petaloid; white or red are the prevailing colours, while among the Rosoideae (see below) yellow is common. Blue flowers occur only in a few of the Chrysobalanoideae. Sometimes, as in *Alchemilla*, *Poterium* and allied genera, petals are absent.

Between the stamens and carpels lies a cushion-shaped or ring-like nectar-secreting disc easily accessible to visiting insects, which are also attracted by the large amount of pollen.

In the tribe *Sanguisorbeae* many genera have flowers eminently suited for wind-pollination, with a greenish inconspicuous calyx, no corolla or nectary, but large projecting often brush-like stigmas and long exserted stamens. There is also a tendency towards separation of the sexes, especially noticeable in *Poterium Sanguisorba*, where the upper flowers are female, the lower often exclusively male. An endotropic course of the pollen-tube has been observed in *Alchemilla* and *Sibbaldia*, and parthenogenesis is frequent in the former. Parthenocarpy (formation of fruit without previous fertilisation) occurs in the genera with fleshy fruits.

The stamens are bent inwards in the bud. They are very variable in number. Occasionally, as in *Alchemilla*, there is a simple whorl of four alternating with the four sepals (fig. 169, C); *A. arvensis* (fig. 169, D) has only one. Generally

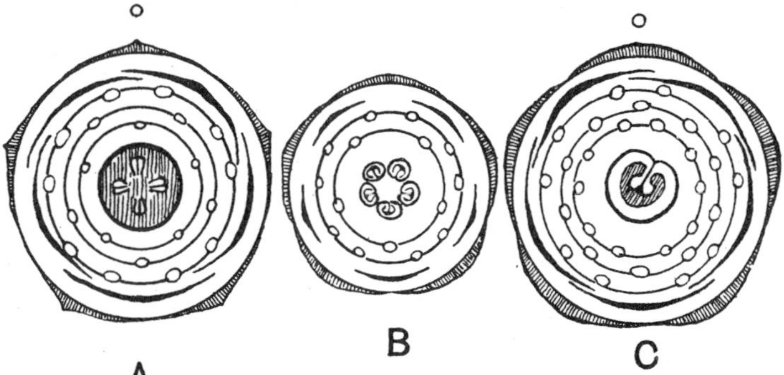

FIG. 163. Floral diagrams of A, *Pyrus communis*; B, *Nuttallia*; C, *Prunus Padus*. (After Eichler.)

they are two, three, or four times as numerous as the members of the perianth-whorls, or indefinite. A common arrangement is one with an outer whorl of ten in five antesepalous pairs. In Pomoideae (fig. 163, A) this is generally followed by a whorl of five antepetalous, and a third whorl of five antesepalous. In Prunoideae the second whorl contains ten members alternating with those of the outer (fig. 163, C); there is sometimes a third alternating whorl. In many species of *Potentilla*, *Geum*, and *Rubus* the outermost whorl of ten, which may be regarded as consisting of either antesepalous or antepetalous pairs, is followed by a second whorl of ten alternating with those of the outermost whorl.

In *Potentilla fruticosa* an inner whorl of five pairs of antesepalous alternates with an outer whorl of five antepetalous pairs, while a third whorl alternates with the individual members of each pair of the second (fig. 164). In *Rosa* five antepetalous pairs are generally followed by a whorl of ten alternating with the individual members of the outer ring, and these again by successive 10-merous whorls.

The shape of the torus and the relative position and number of the stamens and carpels, as well as the structure of the fruit, vary widely and will best be considered by a brief account of the different subfamilies of which Focke in the *Pflanzenfamilien* recognises six.

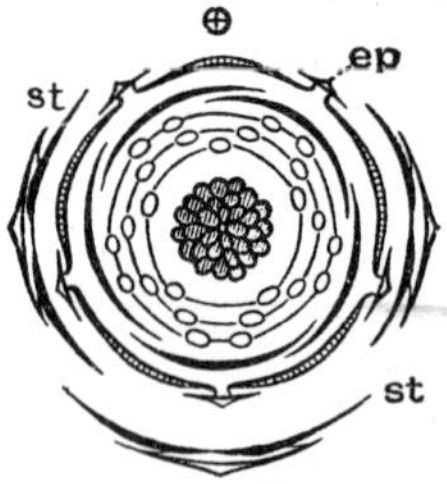

FIG. 164. Diagram of *Potentilla fruticosa*; *ep*, epicalyx; *st*, stipules of bract and bracteoles. (After Eichler.)

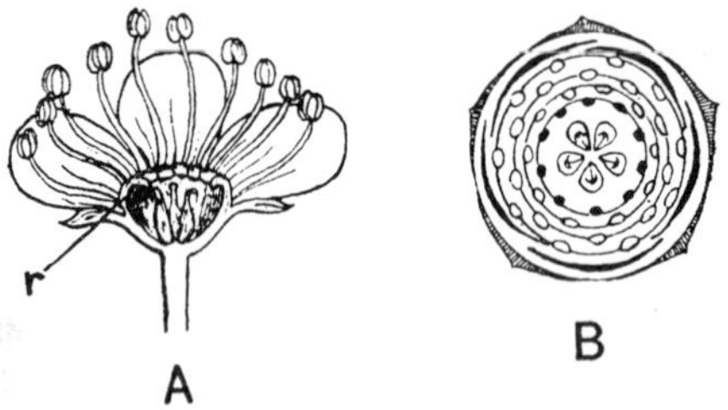

FIG. 165. A. Flower of *Spiraea decumbens* in vertical section; *r*, the well developed nectar-secreting ring partially roofs over the shallowly concave receptacle; enlarged (after Focke). B. Floral diagram of *S. hypericifolia* (after Eichler).

Subfamily I. SPIRAEOIDEAE. Is nearly allied to the Saxifragaceae. The torus is flat or slightly concave, never forming a convex carpophore or a deep cup; the carpels form a central whorl, frequently of five, and are free as in *Spiraea*, or united below. The ovary contains several ovules (more rarely only two) and the fruit consists typically of two- to several-seeded follicles. The flowers have a 5-merous calyx and corolla, and the stamens vary from 10–∞. The plants are generally unarmed shrubs with simple or compound often exstipulate leaves, and numerous small white, rose or purple flowers arranged in racemes or panicled inflorescences. The 17 genera are chiefly north temperate. Numerous Spiraeas are grown in gardens, while *S. salicifolia* occurs in Britain apparently wild in plantations, though not indigenous. Our native Meadowsweet and Dropwort are now placed in the genus *Ulmaria*, and included in the Rosoideae on account of their one-seeded fruit. *Quillaja saponaria*, the bark of which contains saponin, is the Chilian Soap-tree; in an allied genus *Lindleya* (Mexico) the ovary is syncarpous and the fruit a capsule of several cells. This indicates an affinity with the next subfamily Pomoideae, of which the small east Asiatic genus *Stranvaesia* has the carpels separate when ripe and dehiscent.

Subfamily II. POMOIDEAE. The floral axis forms a deep cup with the inner wall of which the five or fewer carpels are more or less completely united as well as with each other (fig. 166, A). Each carpel contains generally two ovules. The fruit (fig. 166, B) is a pseudocarp consisting of the large fleshy torus surrounding the ripe ovaries the endocarp of which is coriaceous or stony, and surrounds the generally one-seeded chamber. *Stranvaesia* above-mentioned is exceptional.

Pomoideae are shrubs or trees with simple or pinnate stipulate leaves. The flowers are white or rose-coloured, often showy. The calyx and corolla are pentamerous; the stamens generally 20 or more.

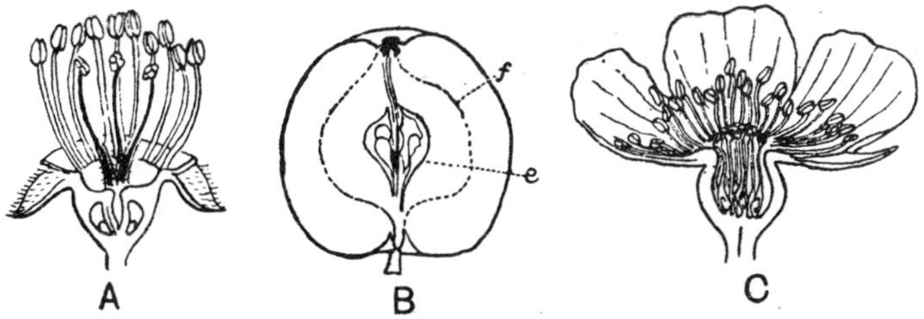

FIG. 166. A. Flower of Apple (*Pyrus Malus*); petals removed. B. Fruit of same; *e*, endocarp; *f*, vascular bundle. C. Flower of *Rosa spinosissima*. All in vertical section. (A, B after Focke; C after Baillon.)

The 14 genera are distributed over the north temperate zone extending in western America through Mexico to the Andes of Peru and Chile. *Pyrus*, the largest genus, has about 50 species in the north temperate zone. *P. communis* (Pear) and *P. Malus* (Apple) have been cultivated since prehistoric times; *P. torminalis* (Service) has a hard, almost stony, endocarp; in *P. Aucuparia* (Rowan, Mountain Ash) and *P. Aria* (White-beam) the endocarp is leathery. *P. Aucuparia* reaches the limit of tree-vegetation in Europe and North Asia. *Mespilus germanica* (Medlar) is a native of Greece and Western Asia. *Crataegus* has numerous species, especially in North America; *C. Oxyacantha* (Hawthorn) is British; many species are grown in gardens. *Cotoneaster*, with 20 or more species in the north temperate zone of shrubs or small trees, is represented in Great Britain by *C. integerrima*, which is found on the limestone cliffs of the Great Orme, in North Wales.

Subfamily III. ROSOIDEAE is characterised by numerous carpels which are either situated upon a swollen receptacle as in

Fragaria or *Rubus*, or at the bottom of the receptacular cup as in *Rosa*. Each carpel contains one or two ovules and the fruit is one-seeded and indehiscent. It is by far the largest subfamily, including 39 genera, which fall into smaller natural tribes distinguished by the form of the torus and of the fruit. The chief are: (1) *Ulmarieae*, containing one genus, *Ulmaria*, with eight to nine species in the north temperate zone, of which two, *U. palustris* (Meadow-sweet) and *U. Filipendula* (Dropwort),

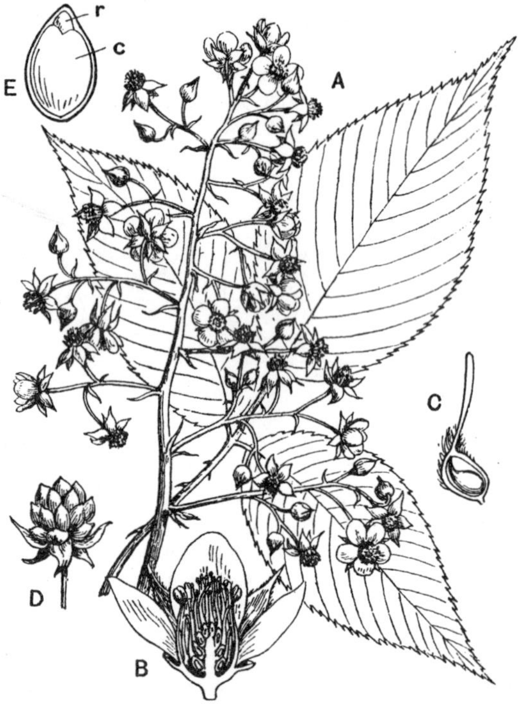

FIG. 167. *Rubus jamaicensis*. A. Leaf and inflorescence, × ⅔. B. Flower cut lengthwise, × 3. C. Unripe carpel, × 10. D. Ripe fruit, nat. size. E. Seed cut lengthwise, × 10; *c*, cotyledon; *r*, radicle. (From *Flor. Jam.*)

are British. The torus is flat or shallowly concave and the usually ten carpels ripen into one-seeded indehiscent dry fruits.

(2) *Potentilleae*. The carpels are generally numerous, on a large rounded or convex outgrowth of the centre of the torus. In *Rubus* (Bramble) there is no epicalyx and the fruits are drupels situated upon the dry torus. *Rubus* is a large genus found almost everywhere except in the driest and hottest parts of the earth. The greater number of species occur in the forest-

region of the north temperate zone and in the high mountains of tropical America. Besides about 180–200 undoubted species, a very large number of subspecies and varieties have been described. *R. fruticosus* (Blackberry) is a very polymorphic plant; *R. Idaeus* is the Raspberry; *R. Chamaemorus* (Cloudberry) is found from Derbyshire northwards. In *Fragaria*, *Potentilla*, and a few allied genera, the sepals are stipulate, and an epicalyx is formed by the union of adjacent stipules. As in *Rubus*, stamens and carpels are indefinite and the fruits are achenes. In *Fragaria* the torus becomes much enlarged and juicy in the fruiting stage, in the other genera it remains dry. *Dryas* and *Geum* represent a third section, in which the style is terminal

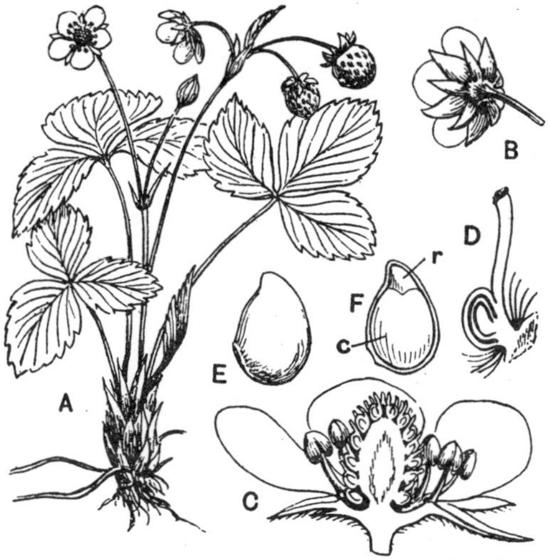

FIG. 168. *Fragaria vesca* (Strawberry). A. Plant, $\times \frac{2}{3}$. B. Flower seen from below, nat. size. C. Flower cut lengthwise, $\times 4$. D. Carpel cut lengthwise, $\times 4$. E. Achene, $\times 10$. F. Ditto cut lengthwise, $\times 10$; *c*, cotyledon; *r*, radicle. (From *Flor. Jam.*)

and persistent in the fruit, forming in *Dryas* a slender feathery appendage to the achene, recalling the fruit of *Clematis*, and in *Geum* often a barbed awn. *Potentilleae* are chiefly north temperate, Arctic and alpine plants. *Geum* also occurs in the south temperate zone.

(3) *Roseae* contains only *Rosa*, and is characterised by an urn-shaped or tubular torus, enclosing the numerous carpels and becoming fleshy and bright-coloured in the fruiting stage. The general habit recalls that of *Rubus*; in both cases the plants are prickly shrubs while many Roses have the scrambling habit characteristic of Brambles. The species of *Rosa* also are

extremely variable and so many species, subspecies, varieties and forms have been described that, as with the Brambles, the identification of Roses is extremely difficult[1].

(4) *Sanguisorbeae* are a reduced group of Rosoideae. They have the persistent urn-shaped torus, completely enclosing the fruit, as in *Rosa*, but with a great reduction in the number of the carpels (often only one is present) is associated the absence of the fleshy bright-coloured torus, which in *Rosa* ensures the distribution of the large number of achenes. The torus remains dry, often becoming hard. An epicalyx is present in *Alchemilla* (fig. 169), and a few other genera. The flowers are generally apetalous (*Alchemilla*, *Poterium*), often unisexual, and frequently anemophilous, as in *Poterium Sanguisorba*, where the small flowers are crowded on long-stalked heads, the upper being female with protruding feathery stigmas, and the lower male (or bisexual) with numerous exserted stamens; the pendulous anthers are attached to long

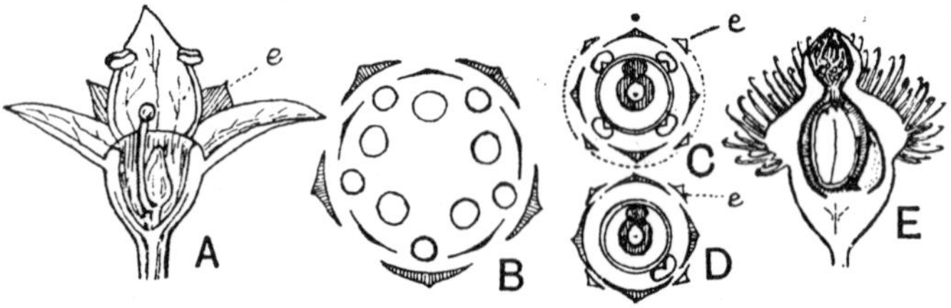

FIG. 169. A. Flower of *Alchemilla alpina* in vertical section, enlarged. B. Floral diagram of *Agrimonia pilosa*, shewing arrangement of sepals, petals and stamens. C, D. Floral diagrams of *Alchemilla vulgaris* and *A. arvensis*. E. Fruit of *Agrimonia Eupatorium* in vertical section, enlarged. *e*, epicalyx. (B after Goebel; C, D after Eichler.)

slender filaments. *Agrimonia* has small yellow-petalled nectarless flowers borne in a long terminal spike; in the fruiting stage the hardened torus is crowned by numerous hooked bristles (fig. 169, E), which are of service in the distribution of the one or two enclosed achenes. Such development of barbs or prickles on the torus is a frequent device in the tribe for ensuring distribution. The number of stamens is very variable, even in the same genus or species; sometimes numerous, but more often few, sometimes only one (fig. 169, B, C, D).

Its success is demonstrated by the wide area over which many members of the tribe are found. The species of *Acaena* occur chiefly in extratropical South America but spread along the Andes to Mexico and California and are found also in the Sandwich Islands, New Zealand, Tasmania, South Australia,

[1] See Matthews, J. R. "Hybridism and classification in the genus *Rosa*." *New Phytologist*, XIX, 153 (1920).

South Africa, Tristan d'Acunha and other islands of the southern hemisphere. The barbed thorns are eminently adapted for carriage in the feathers of sea-birds such as the Albatross. *A. adscendens* is found in New Zealand, Tierra del Fuego, and the Falkland Islands.

Subfamily IV. NEURADOIDEAE contains only two small genera. They are desert-loving herbs with yellow flowers. The 5–10 carpels are united with each other and with the base of the cup-shaped torus which enlarges and forms a dry covering round the one-seeded fruits. *Neurada procumbens* spreads through North Africa to the Indian desert. *Grielum* occurs in the sandy and salt deserts of South Africa.

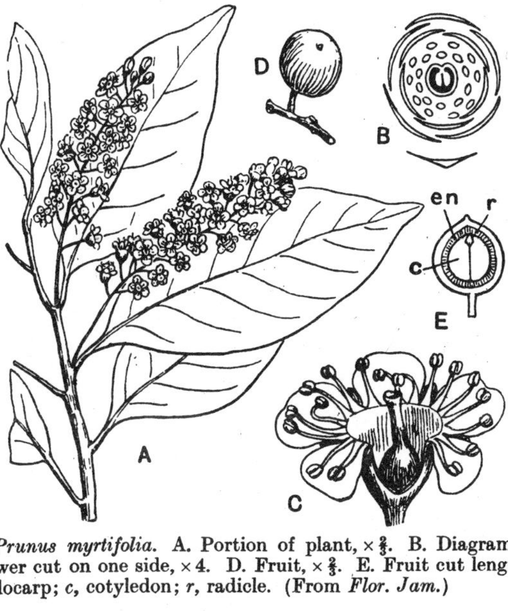

FIG. 170. *Prunus myrtifolia.* A. Portion of plant, × $\frac{2}{3}$. B. Diagram of flower. C. Flower cut on one side, × 4. D. Fruit, × $\frac{2}{3}$. E. Fruit cut lengthwise, × $\frac{2}{3}$; *en*, endocarp; *c*, cotyledon; *r*, radicle. (From *Flor. Jam.*)

Subfamily V. PRUNOIDEAE contains five genera, the species of which are distributed through the north temperate zone, passing into the tropics; they are characterised by a free solitary carpel with a terminal style and a pair of pendulous ovules, and the fruit a one-seeded drupe. With few exceptions the torus forms a more or less deep cup at the edge of which spring the five sepals, five alternating petals and 10, 20 or more stamens. The stony endocarp bears a longitudinal seam along which it splits in germination. The seed (kernel) consists chiefly of two large fleshy planoconvex cotyledons; there is no endosperm. The plants are deciduous or evergreen (*Prunus Laurocerasus*, Cherry-laurel) trees or shrubs with simple undivided leaves, and often small and caducous stipules. Flowers are often showy, and white or pink, generally arranged in racemes (fig. 170). The monotypic genus *Nuttallia* from north-west America is exceptional in having a whorl of five carpels (fig. 163, B); through it the subfamily is related to Spiraeoideae. The large genus *Prunus* is divided into subgenera, the distinctions being based on the arrangement of the leaves in the bud, the character of the mesocarp and endocarp and the shape of the torus.

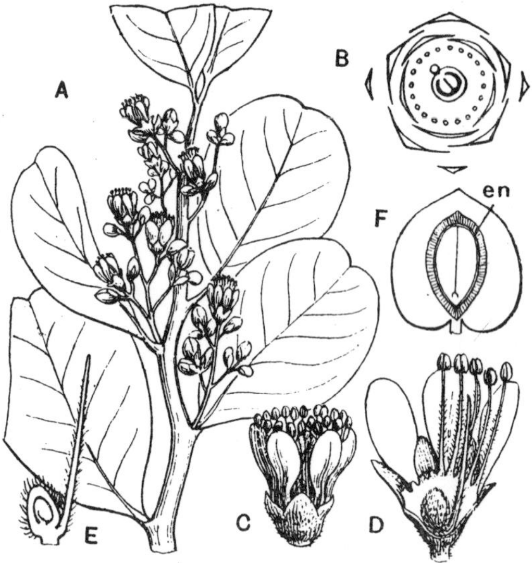

FIG. 171. *Chrysobalanus icaco.* A. Portion of plant, × ⅔. B. Diagram of flower. C. Flower, × 2. D. Flower cut lengthwise and four stamens removed, × 3. E. Pistil, a vertical section of ovary removed, × 4. F. Fruit cut lengthwise. × ⅔; *en*, endocarp. (From *Flor. Jam.*)

In *P. communis* (Plum) and allied species the leaves are rolled in the bud. *P. communis* includes several well-marked subspecies, *P. spinosa* (Sloe or Blackthorn), *P. insititia* (Bullace) and *P. domestica* (Wild Plum). The last mentioned has been cultivated since classic times but is not certainly known to be indigenous anywhere at the present day. The other subgenera have conduplicate leaves. The subgenus *Amygdalus* contains the Almond (*P. Amygdalus*), which grows wild in Central Asia, and the closely related Peach (*P. persica*), the native country of which is unknown. *Amygdalus* has the mesocarp generally not juicy, but covered with hairs, and the endocarp often furrowed; the torus is short and wide-mouthed. The sub-genus *Cerasus* differs from *Amygdalus* in having a fleshy mesocarp and a smooth or wrinkled (not deeply furrowed) stone. *P. Cerasus* (Wild Cherry), probably a native of the Balkans and Asia Minor, is the origin of many of our cultivated forms (Morello, Duke and Kentish cherries); others are derived from *P. Avium* (Gean). The sub-genus *Laurocerasus*, to which belongs *P. Padus* (Bird-cherry), is distinguished from *Cerasus* by its entire stigma, unfurrowed style and flowers in racemes. *P. Laurocerasus* (Cherry-laurel) is a well-known evergreen with nectaries on the back of the leaves.

Fossil remains of the genus *Prunus* are frequent in Tertiary strata.

Subfamily VI. CHRYSOBALANOIDEAE (genera 13, species 200) resembles Prunoideae in having a solitary free carpel and the fruit a drupe, but differs in the basal style, the ascending (not pendulous) ovules and frequently zygomorphic flowers (figs. 171, 172). The plants are tropical evergreen trees or shrubs; largely South American. In several genera (*Hirtella*, *Parinarium* and others) the flower is markedly zygomorphic, the stamens being pushed to the anterior face, while the torus is hollowed on the posterior face into a sac or spur below the ovary. Such flowers are adapted for pollination by insects with long probosces. This subfamily is of great interest from its affinity on the one side with Prunoideae, as already indicated, and, on the other, in the zygomorphic flowers just described, with the family Leguminosae. This second affinity is emphasised in *Acioa* where the filaments of the 10 to 15 or more stamens are united into a long flat band becoming free only near the top (fig. 172).

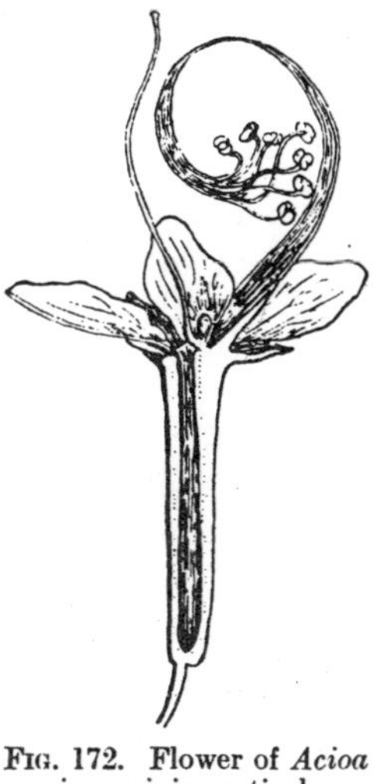

FIG. 172. Flower of *Acioa guianensis* in vertical section, shewing the solitary carpel with a long style, and the stamens united into a band. (After Focke.)

H. O. Juel[1] finds that in the early stages of development of a species of *Parinarium* there are three rudimentary carpels corresponding to the tripartite stigma. In normal flowers both the posterior chambers are undeveloped, but not infrequently abnormal flowers arise in which there are three distinct chambers.

[1] *Arkiv Bot.* XIV, No. 7, 1–12 (1915).

Juel considers that the *Parinarium* type of gynoecium is based upon a trimerous form, and suggests that this may apply to the whole subfamily, the relationship of which to the Prunoideae is doubtful. He considers that the evidence justifies the removal of the Chrysobalanoideae from the Rosaceae but not from the order.

Family X. CONNARACEAE

A small tropical family containing about 250 species in 10 genera. They are mostly woody climbers, sometimes shrubs or trees, with alternate imparipinnate exstipulate leaves and small regular generally bisexual flowers arranged in racemes or panicles. The flowers are pentamerous; the sepals are generally persistent and become hardened round the base of the fruit; the petals are imbricate and free; the stamens are generally in two whorls, the five antepetalous being shorter than the five antesepalous; the filaments are generally united at the base. The five (rarely fewer) carpels are free and one-celled, containing two collateral ovules, with two integuments, ascending from the base; only one carpel develops fruit and forms a one-seeded follicle, opening generally on the ventral suture. The solitary erect seed is usually surrounded at the base with an aril. Endosperm is present or absent.

This family is most nearly allied to Leguminosae but is distinguished by the typically pentamerous pistil and absence of stipules.

Family XI. LEGUMINOSAE

Flowers bisexual, generally zygomorphic, sometimes (subfamily Mimosoideae) actinomorphic. Calyx inferior, generally of five sepals, with the odd one anterior, generally more or less united, with imbricate or valvate aestivation, the latter especially in the regular flowers. Petals free, or in regular flowers more or less united, equal in number and alternating with the sepals; corolla generally markedly zygomorphic; aestivation valvate or variously imbricated. Stamens typically ten, inferior or slightly perigynous, free or more or less united into a tube, in which case they are monadelphous, or diadelphous by separation of the posterior stamen. Carpel one, with the ventral suture posterior, ending in a simple style and stigma which surmounts a unilocular ovary containing generally numerous ovules arranged in two alternating rows along the ventral suture. Ovules amphitropous, or anatropous, or sometimes campylotropous, obliquely ascending or pendulous. Fruit generally a legume containing many seeds and dehiscing by both dorsal and ventral sutures. Seeds generally with a leathery testa, on a short or elongated funicle, which often forms a more or less fleshy aril. Endosperm generally absent or sparsely developed, the embryo consisting of two large flat cotyledons with a generally superior or ventrally placed radicle.

Trees, shrubs or herbs of very various habit. Leaves generally alternate, compound and stipulate.

The second largest family of flowering plants, containing about 550 genera with more than 12,000 species, forming a very natural cosmopolitan group in spite of the considerable variation in vegetative and floral characters.

In most members of the family the cotyledons come above the ground and function as the first green foliage-leaves. They are generally more or less oblong in shape; sometimes with an unequal, cordate or auricled base. Cordate cotyledons are frequent in the subfamily Caesalpinioideae where the seed generally contains endosperm and the embryo is straight with the radicle lying between the auricles of the cotyledons. In the genus *Acacia* (subfamily Mimosoideae) they shew great variety and may be oblong and entire, or slightly cordate, or have a deeply auricled or sagittate base. In the tribe *Vicieae* of the subfamily Papilionatae the fleshy planoconvex cotyledons fill the seed-coat and are hypogeal, while in the closely allied *Phaseoleae* both epigeal and hypogeal occur. *Scorpiurus sulcata*, a Mediterranean plant with a twisted indehiscent pod, is remarkable in having cylindrical, fleshy, furrowed cotyledons which are twisted about in the albuminous seed. During and after germination they become much larger but retain their original form.

Interesting examples of exceptional forms of germination are seen in species of *Medicago* and *Hedysarum*. In the former the numerous seeds germinate within the indehiscent spirally-coiled pod, causing severe competition among the seedlings. In *Hedysarum* the pod (lomentum) splits transversely into a number of indehiscent pieces which are forced open on germination by the swelling of the cotyledons and the elongation of the hypocotyl, the wall of the fruit becoming pinned to the ground while the seedling rises clear above it. A remarkable method of germination has been described[1] in the genus *Inga*, where the naked embryo

[1] Borzi, A. *Atti d. Reale Accadem. d. Lincei*, ser. v. Rendiconti, XII, I, 131 (1903).

falls from the pod and germinates in the earth. The seed-coat forms a shining soft wool-like mass, from which, when the pod has opened, the embryo escapes on the slightest pressure as a thick bean-like body. Its escape is effected by visits of birds which are attracted by the conspicuous seed-coat.

The transition from cotyledon to the adult form of leaf may be gradual, as in *Trifolium*, where the leaf succeeding the cotyledons is simple, and the subsequent ones are trifoliolate, or the compound form is at once assumed, as in *Lupinus*, where the leaves immediately succeeding the cotyledons are digitately compound, though with fewer leaflets than in the adult form. In Gorse (*Ulex europaeus*) there is much variation in the shape of the primary leaves before the adult spinous form is assumed. The leaves are all simple, or those immediately succeeding the cotyledons are trifoliolate passing above into bifoliolate and simple forms; or other variations occur. In *Acacia* the primary leaves are pinnately compound, while in those species where the ultimate form is phyllode-like this is gradually assumed by development of the main petiole and diminution in number and size of the leaflets.

The family contains plants of very different habit. Our British genera, all of which belong to the subfamily Papilionatae, include a few shrubs such as *Genista*, *Ulex* (Gorse) and *Cytisus* (Broom). The majority, however, and this applies to the subfamily as a whole, are herbaceous, as *Trifolium*, *Medicago*, *Melilotus*, *Lotus*, while the species of *Lathyrus* and *Vicia* (Vetch) climb or trail by help of tendrils which are modified leaf-structures. Others, such as *Phaseolus multiflorus*, have an herbaceous twining stem. Species of *Bauhinia* (Caesalpinioideae) with curiously flattened twisted woody stems are characteristic lianes of tropical forests, and *Entada gigas* (Mimosoideae) is also a common tropical climber in both hemispheres, its thick woody stem scrambling to great heights. The subfamilies Mimosoideae and Caesalpinioideae, confined to the warmer parts of the earth, consist chiefly of trees and shrubs. To the former belong the Acacias and Mimosas, to the latter the Tamarind, the Judas tree of South Europe (*Cercis Siliquastrum*) and others. The False Acacia (*Robinia Pseud-acacia*) and Laburnum (*Laburnum vulgare*) are examples of an arborescent habit in the Papilionatae. True water plants are rare, but occur in the tropical genera *Neptunia* and *Aeschynomene*. The roots, especially in the

Papilionatae, frequently bear numerous small tubercles, which contain nitrogen-giving bacteria. Hence the value of crops of Clover, Beans, etc., in a rotation or as green manure.

The leaves are alternate, rarely opposite or whorled, and generally compound and stipulate; sometimes the individual leaflets have stipels at the base as in many members of the tribe *Phaseoleae*. One of the commonest leaf-forms is the digitately trifoliolate, as in *Trifolium*; pinnate and imparipinnate (*Robinia*, *Laburnum*) are also frequent. The Mimosoideae have generally bipinnate leaves. The petiole falls with the leaflets, or later, or persists, becoming indurated, to form spines as in *Astragalus*, *Oxytropis*, *Onobrychis* and others. Various reductions and modifications for special purposes occur. In the Pea and Vetch, the end of the rachis or a greater or less number of leaflets are developed as tendrils. In Gorse the leaves of the adult are reduced to simple spines (sometimes bearing a single minute leaflet) or small scales subtending a spine-like branch. The leaves of many Australian Acacias consist of a vertically placed flattened phyllode-like structure representing the petiole; a similar reduction takes place in some species of *Mimosa*.

Stipules are generally free, more rarely, as in *Medicago*, *Lupinus* and *Trifolium*, adnate to the petiole. In *Pisum* (Pea) they are leaf-like, replacing as assimilating organs the leaflets, which have become tendrils. In *Cytisus* (Broom) they are minute, and absent in *Ulex*.

Frequently the stipules take the form of thorns, as in *Robinia*. In species of *Acacia* the thorns are hollow and often much swollen, affording shelters for ants, which protect the plant from attack by leaf-cutting ants and similar foes. *Acacia sphaerocephala*, a Central American species, is one of the best known examples. The ants live entirely on the plant, obtaining food from extrafloral nectaries on the leaf-stalks and from oblong yellowish food-bodies on the tips of the leaflets.

Leaf-movements are very general. Such are the sleep-movements in *Trifolium*, *Phaseolus*, *Robinia*, *Acacia* and *Desmodium*, in which the leaflets assume a vertical position at night-fall. Spontaneous movements independent of light-changes also occur, as in *Desmodium gyrans* (Telegraph-plant) of tropical Asia, where the two small lateral leaflets move up and down every few minutes, if the temperature is sufficiently high. In species of *Mimosa* (as *M. pudica*, Sensitive-plant) (fig. 173) the leaves assume a sleep-position in response to a touch. The seat of the movement in these cases is the pulvinus or swollen base of the leaf-stalk or petiolule. Extrafloral nectaries are found on the stipules or leaf-stalks of many members of the family; *Vicia sepium* is an example from our own flora.

The origin of the periderm shews great variety. It seldom arises in the epidermis (*Cytisus*), more frequently in the subepidermal

FIG. 173. *Mimosa pudica* shewing leaves expanded and closed, flower-head in bud and open, and pods in various stages of dehiscence; the thickened borders of the pod separate and the pod breaks up into one-seeded joints; reduced. (From *Flor. Jam.*)

layer (*Hymenaea*, *Bauhinia*) or from a deeper layer of the cortex (*Gleditschia*, *Robinia*) or finally from the pericycle (as in *Ulex*). According to Moeller, the three subfamilies may be distinguished by the structure of the sieve-tubes. In Papilionatae the individual members are short, only a little wider than the parenchyma-cells, and have simple transverse plates. In Caesalpinioideae they are much shorter, very much wider than the parenchyma-cells, and have several coarsely fitted plates on the moderately inclined ends; often, too, they are connected laterally with adjoining tubes.

In Mimosoideae they are much longer, but no broader than in Papilionatae, and have narrow scalariform plates.

According to Saupe, the subfamilies cannot be separated upon characters derived from the anatomy of the wood. On the other hand, the smaller groups and the genera have certain distinctive peculiarities, especially in the form of the medullary rays in tangential section, and in the distribution of wood-parenchyma and its relation to the vessels.

The form and structure of the stem of the lianes are of special interest. In Papilionatae anomalous secondary thickening is common and arises from the production of successive new cambium-zones outside the original ring; either in the pericycle (*Mucuna*) or in the phloem-zone (*Wistaria sinensis*, *Rhynchosia*), concentric vascular bundle-rings or broader or narrower strands are formed. Where the successive cambiums are active only at two opposite points a flat ribbon-like stem is the result as in *Rhynchosia*. In the Caesalpinioideae the climbing *Bauhinia*-species shew very complicated anomalous secondary thickening, as well as remarkable external conformation, the stem being flattened and at the same time shewing basin-like undulations. In some species growth in thickness is normal; in others successive cambium-zones arise concentric with the primary zone, as described for Papilionatae; while in others new and distinct centres of growth, each with its cambium-zone, arise outside the primary zone.

No anomalous growth in thickness has been found in climbing Mimosoideae; in some cases the stem becomes strongly winged. *Entada gigas* has a very soft parenchymatous wood; large gum-passages occur in both wood and parenchyma.

The tissue of most Mimosoideae and Caesalpinioideae and of many Papilionatae is rich in tannin-sacs, and gum-passages in the pith and medullary rays are frequent, especially in species of *Acacia* and *Astragalus*; gum-arabic is an exudation from the branches of *Acacia Senegal*; *Astragalus gummifer* and other species yield gum-tragacanth. The thick walls of the wood-fibres are deeply coloured in the heart-wood of *Haematoxylon campechianum* (Logwood), *Pterocarpus santalinus* (Red Sandal-wood) and others.

The inflorescence shews great variety, but is always of the indefinite type. The simple raceme is common, as in *Laburnum*, *Robinia*, *Indigofera* (fig. 180), also the spike or head (*Trifolium*, fig. 174). Compound inflorescences are frequent in Mimosoideae. In Lupins a whorled arrangement results from local secondary growth in length. Dorsiventral racemes also occur, especially in the tribe *Vicieae*.

In the great majority flowers are borne on lateral branches of the third order; more rarely on those of the second (as in *Cytisus*, *Genista*, *Lupinus*) or fourth (*Trifolium pratense*, *Phaseolus*). The most constant characters are the hypogynous

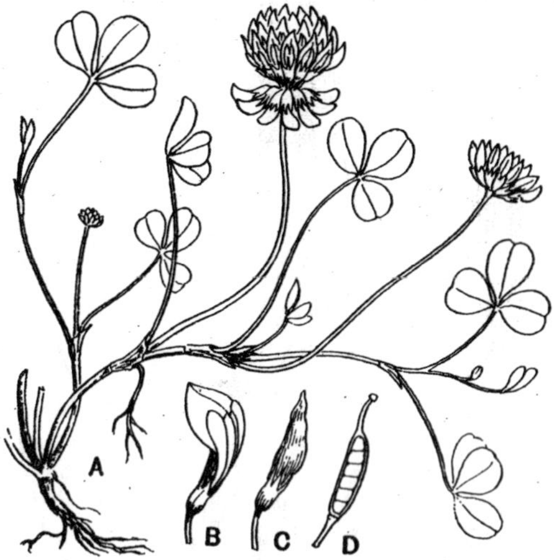

FIG. 174. *Trifolium repens.* A. Portion of plant, $\times \frac{2}{3}$. B. Flower, $\times 2$. C. Pod enveloped in the persistent calyx and corolla, $\times 2$. D. Pod with one valve removed, $\times 2$. (From *Flor. Jam.*)

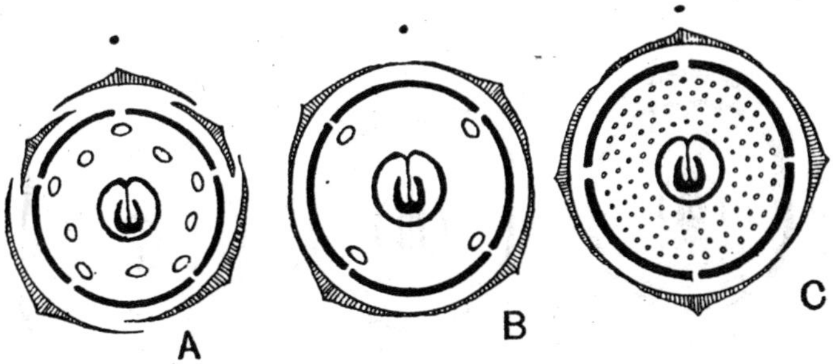

FIG. 175. Floral diagrams of A, *Parkia africana*; B, *Mimosa pudica*; C, *Acacia latifolia*. (After Eichler.)

or slightly perigynous arrangement of the parts, the anterior position (fig. 181) of the odd sepal, which is the first developed, the others following in ascending order (from front to back), the polypetalous corolla, and the gynoecium of one median carpel, with a terminal style ending in a simple stigma, and two alternating rows of ovules on the ventral suture of the ovary which faces the back of the flower.

The arrangement and form of the perianth-segments, especially of the corolla, and the number and cohesion of the stamens, shew considerable variation. Three well-marked types are recognised, each characteristic of a subfamily.

In Mimosoideae, the smallest subfamily with about 40 genera, the flower is regular and the aestivation of both calyx and corolla valvate; *Parkia* and an allied genus have imbricate sepals; (fig. 175). The perianth is generally pentamerous, but 3-, 4- or 6-merous flowers occur: *Mimosa pudica* is tetramerous, and all four forms are found in the genus *Mimosa*; while in *Acacia* many species are 3- or 4-merous. The sepals are more or less united below into a cup, and the petals may also cohere at the base. The stamens vary greatly in number; their number and cohesion afford distinctive tribal characters. Thus they are indefinite and free in *Acacieae* (containing the single large genus *Acacia*) (figs. 175, C, 176), indefinite and more or less monadelphous in *Ingeae.* In the four other tribes they are as many or twice as many as the petals. The long slender filaments are crowned by small bilocular anthers which dehisce longitudinally. The pollen-grains are often united into little packets. Occasionally the anther-chambers are divided by several transverse septa. In many cases the long exserted yellow stamens are, as in *Mimosa,* the most conspicuous part of the flower.

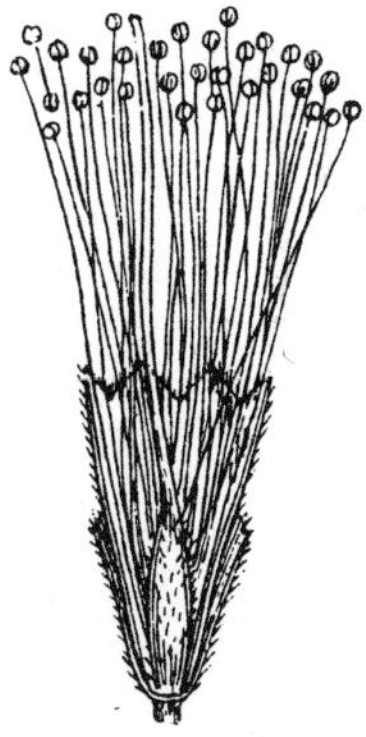

FIG. 176. Flower of *Acacia farnesiana* with calyx and corolla partly removed, × 7. (From *Flor. Jam.*)

In Caesalpinioideae (133 genera) the flowers are medianly zygomorphic, and generally 5-, more rarely 4-merous. The two upper sepals are sometimes united, as in *Tamarindus* (fig. 179, B), otherwise the members of the calyx are generally free; their aestivation is imbricate, rarely, as in *Cercis Siliquastrum* (Judas-tree) (fig. 179, A), valvate. The typically pentamerous corolla shews great variation. Its aestivation is ascending imbricate, the posterior petal being innermost. In

Cercis (fig. 177) it resembles a papilionaceous corolla, the two lower (anterior) petals forming a larger pair enclosing the essential organs, while the posterior pair are reflexed and wing-like, and the odd petal is erect. In *Cassia* (fig. 178) all five are subequal and spreading. In the handsome flower of *Amherstia nobilis* (a plant of further India, not uncommon in cultivation) the anterior pair are small or suppressed, while the three posterior are well-developed, the odd petal being the largest. In *Krameria* the anterior pair are represented by glandular scales, while in *Tamarindus* (fig. 179, B) they are absent. Finally apetalous genera (or species) occur, as in *Copaifera* and *Ceratonia* (fig. 179, C, D). The stamens are generally ten in number, rarely indefinite, or fewer by abortion

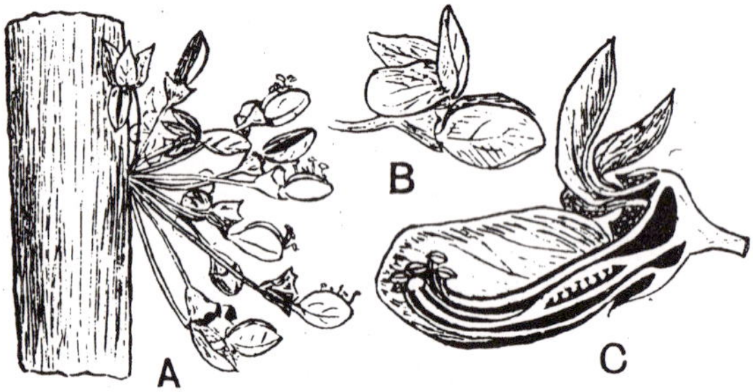

FIG. 177. *Cercis Siliquastrum.* A. Inflorescence growing from the old wood, reduced. B. A flower. C. Flower in longitudinal section, enlarged. (After Taubert.)

(figs. 178, 179). They are free, as in *Cercis*, or more or less united; thus in *Amherstia* they are diadelphous, the posterior one free, the rest united for more than half their length; and in *Tamarindus* the three fertile stamens are similarly united into an open sheath on the top of which are indicated the small staminodes.

The third and largest subfamily, Papilionatae, is characterised by the markedly zygomorphic true papilionaceous flower. The five sepals are as a rule coherent, with an ascending imbricate arrangement (fig. 181, A); often the two upper and three lower segments are respectively united, forming a two-lipped structure. The corolla consists typically of five unequal petals with a descending imbricate aestivation; the outermost (posterior) petal is the largest and forms the broad free

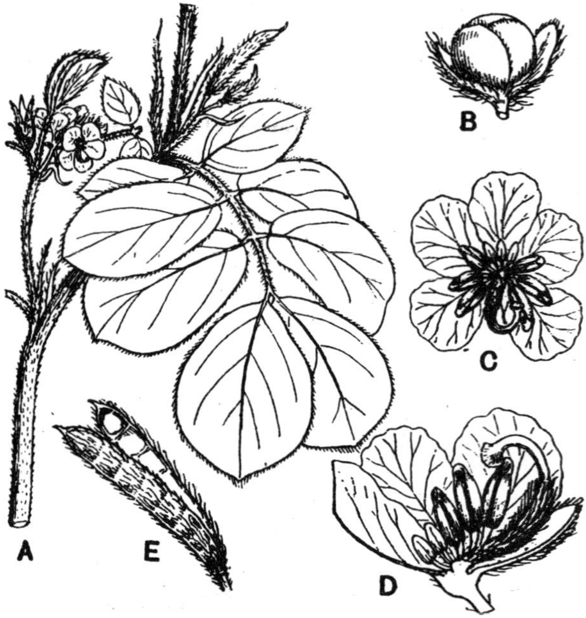

FIG. 178. *Cassia uniflora.* A. Branch with flower and young fruit, $\times \frac{1}{2}$. B. Flower-bud, with pair of bracteoles, $\times 2$. C. Flower from above, $\times 2$; the three upper stamens are imperfect. D. Flower in section, $\times 3$. E. Pod, nat. size. (After *Flor. Jam.*)

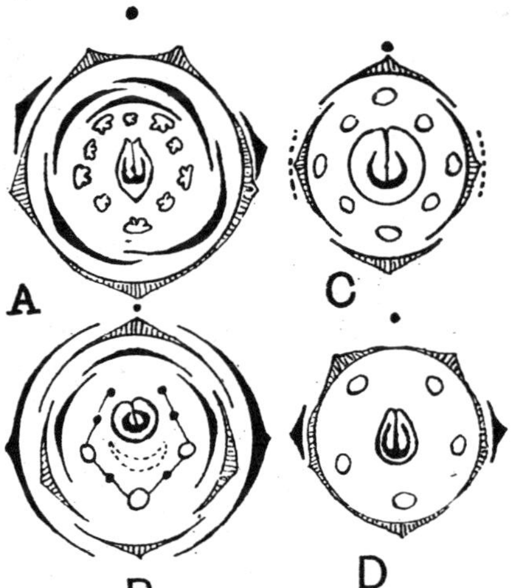

FIG. 179. Floral diagrams of A, *Cercis Siliquastrum*; B, *Tamarindus*; C, *Copaifera*; D, *Ceratonia*. (After Eichler.)
Note. In B the posterior petal is represented as outermost, an exceptional case.

standard (*vexillum*), the lateral pair, which are also free and generally long-clawed, form the wings (*alae*), while the anterior pair are closely appressed, and often more or less coherent, and form the keel (*carina*) in which are enclosed the stamens and pistil. In the North American *Amorpha* (fig. 181, C) wings and keel are absent. The ten stamens are inserted at the same height as the corolla. They are almost or quite free, or monadelphous (figs. 181, B, 182, F) or diadel-

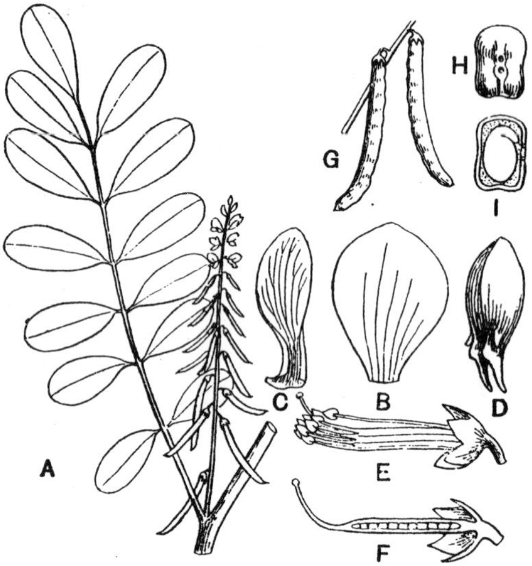

FIG. 180. *Indigofera tinctoria.* A. Raceme and leaf, × ⅔. B. Standard, × 6. C. Wing, × 6. D. Keel, × 6. E. Flower with corolla removed, × 7. F. Ovary and calyx cut lengthwise, × 7. G. Ripe pods, × ⅔. H. Ripe seed, × 3. I. Ditto cut lengthwise, × 3. (From *Flor. Jam.*)

phous, the posterior one being free (figs. 181, A, 180, E). Occasionally the posterior one is absent, as in the West Australian *Chorizema* (fig. 181, D), or as many as five may be absent.

The ten stamens of the Caesalpinioideae and Papilionatae, though arranged in a single whorl in the adult flower, are diplostemonous in origin, the five antesepalous arising before the antepetalous; the former are generally longer than the latter, and other differences, as in size of anther, or presence or absence of a terminal gland, often distinguish the two series (fig. 182, F).

The single median carpel is sometimes borne on a longer

or shorter stalk, and often girt at the base by a hypogynous nectar-secreting disc. In the second and third subfamilies the zygomorphy of the flower is often expressed in the terminal style by a difference between back and front, and a well-marked bending; more rarely is it straight, or rolled as in *Phaseolus*. Departures from a monocarpellary pistil are rare. *Tounatea dicarpa* (Caesalpinioideae) has two carpels, while in several genera of the tribe *Ingeae* (Mimosoideae) there are more than one. Occasionally the number of ovules is reduced; some species of *Trifolium* and *Medicago* have only one. The ovules are amphitropous or anatropous, sometimes campylotropus, and obliquely ascending or pendulous, with one or two integuments; the number varies even in the same genus.

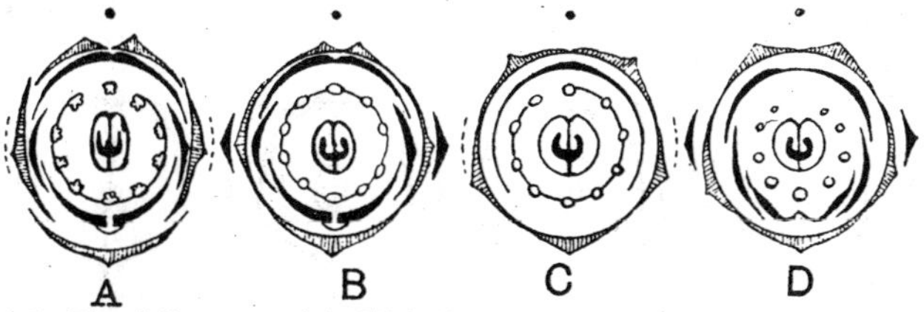

FIG. 181. Floral diagrams of A, *Vicia Faba*; B, *Laburnum vulgare*; C, *Amorpha fruticosa*; D, *Chorizema cordatum*. (After Eichler.)

In Mimosoideae and Caesalpinioideae the stamens and stigma are freely exposed and the stamens as well as the petals serve to attract insects. Frequently, as in *Mimosa*, the flowers are massed in heads.

The relation of insects to the flower has been carefully studied only in Papilionatae, and chiefly in European species. H. Müller has described the structure of the flower and its pollination in numerous species. In nectar-containing species nectar is secreted from the inner sides of the bases of the filaments and accumulates round the base of the ovary, lying at the bottom of a tube formed by the filaments of the stamens and rendered more rigid by the claws of the petals and the stiff, more or less tubular, calyx. It is accessible therefore only to an insect with a long proboscis, and observation shews that Papilionatae are essentially bee-flowers. In such species the posterior stamen is free and a passage to the nectar is made by an arching outwards of its base or of the bases of the adjoining filaments on either side. The parts of the corolla also play an important part in the process. The flowers stand more or less horizontally, and while the keel encloses the stamens and protects them from rain and attack by unbidden pollen-eating insects, the wings serve as a platform on which the bee alights and the large erect coloured standard renders the flower conspicuous. The wings and standard have other uses, as can be seen by watching a bee at work. The way to the nectar opens in the centre of the flower at the base of the standard, which serves as a fulcrum against which the bee pushes its head whilst standing on the wings. The latter are firmly morticed into the keel by interlocking of processes in the adjoining surfaces of both, or by projections of the wings fitting closely

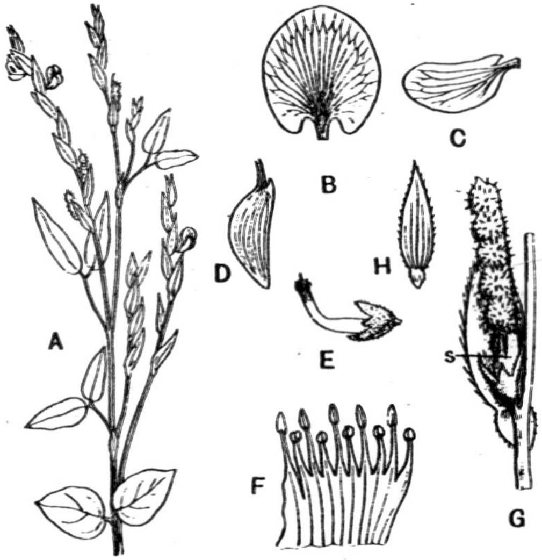

FIG. 182. *Zornia diphylla*. A. Upper portion of branch, with leaves and flower-spikes, × ⅔. B. Standard, × 3. C. Wing, × 3. D. Keel, × 3. E. Flower with corolla removed, × 3. F. Upper portion of staminal sheath more highly magnified. G. Ripe pod with one of the bracts removed, × 3: *s*, persistent staminal sheath. H. Bract slightly magnified. (From *Flor. Jam.*)

into hollows in the keel, the result being that the pressure exerted on the wings by the feet of the insect depresses the keel and exposes the stamens or stigma which come in contact with the under surface of the insect. The keel is kept in its place and, if repeated visits are necessary, is brought back into position by means of a lobe at the base of each wing which embraces the central rigid column of which the ovary forms the axis.

Müller distinguishes four types of structure according to the way in which the pollen is applied to the bee. First those in which stamens and stigma return within the keel after a visit. Such flowers, of which *Melilotus, Trifolium, Onobrychis* and *Laburnum* are examples, admit of repeated visits. Secondly those where the essential organs are confined under tension and the visit causes the flower to explode. Thus in *Genista tinctoria* when the flower is ready for pollination, the keel is much compressed and closely surrounds the curved style and the pollen which is heaped above it, the anthers having dehisced and withered before the expansion of the standard. The tightly closed keel is kept in a horizontal position by the action of two equal and opposite forces; the column of stamens with the enclosed style has an upward tension, so that if freed it springs erect and parallel to the standard; the claws of the wings and of the two petals forming the keel have a downward tension and if freed from the style spring vertically downwards. When a bee alights on the wings and thrusts its head beneath the standard, the lobes at the base of the wings, which kept the keel in place, are pushed aside; at the same time the union between the upper borders of the keel is severed from behind forwards by the pressure of the staminal column, and as soon as the splitting reaches the tip the opposing forces are given full play and the staminal column springs upwards scattering a cloud of pollen, while the wings and keel spring downwards. The presence of the insect prevents the column springing to its full height, but the ascending style forces the stigma and instantly afterwards a mass of pollen against the under surface of the insect. If the visitor has been dusted with pollen in another flower, cross-pollination results; as the bee creeps backwards out of the flower the stigma gets dusted with its own pollen. The flowers contain no nectar and spend all their pollen in a single explosion, after which the essential organs are enfolded by the standard. A similar mechanism occurs in *G. anglica* and *pilosa*, and also in *Medicago* and the Broom (*Cytisus*).

A third type involves a piston-mechanism and is seen in *Lotus corniculatus* (Bird's-foot Trefoil). In the mature flower the conical tip of the keel is tightly packed with pollen in which the stigma is embedded. The five inner stamens which have withered after dehiscence lie in the lower part of the keel, the filaments of the five outer, on the contrary, have continued to grow and now tightly close in the base of the pollen-containing cone with their swollen club-shaped ends, on which the withered anthers are still seen. At the tip of the cone is a narrow opening. A slight downward pressure to the keel forces the thick ends of the filaments (that is, the piston) further into the cone and squeezes a narrow ribbon of compressed pollen through the opening; a greater pressure causes the protrusion of the stigma. When the pressure ceases the thickened filaments tend to spring apart, and this, assisted by its own elasticity, raises the keel into its original position. The bee by its efforts to reach the nectar gets its under surface smeared with pollen, while the stigma also rubs against it and will probably acquire pollen from flowers previously visited as well as from its own stamens. *Anthyllis*, *Ononis* and *Lupinus* are other examples of a piston-mechanism.

Lathyrus, *Pisum*, *Vicia* and *Phaseolus* exemplify a fourth type, the style having a brush of hairs which sweeps the pollen in small portions out of the apex of the keel. As the young stigma becomes receptive only after being rubbed, autogamy does not necessarily occur where it is surrounded by pollen in the flower.

Enantiostyly, or the existence of two kinds of flowers, often found on the same plant, in which the style projects on the right or left-hand side respectively, occurs in *Cassia*. Cleistogamy is common among Papilionatae (as in species of *Ononis*, *Trifolium*, *Vicia*, etc.); this is often associated with geocarpy or production of the fruit underground, as is especially well seen in the Groundnut (*Arachis hypogaea*) and *Trifolium subterraneum*, where the flower-stalks grow downwards and bury the young fruits which ripen underground. In these species amphicarpy frequently occurs, that is, production of both aerial and subterranean fruit; the latter generally contain only one seed.

The typical fruit is a pod (legume), splitting in most cases along both dorsal and ventral sutures into a pair of membranous, or leathery, or more rarely fleshy valves. Dehiscence is often explosive, the valves separating elastically and twisting spirally, thereby shooting out the seeds, as in Gorse, Broom, Laburnum, etc. In *Entada* (fig. 183), some Mimosas, *Desmodium* and others the pod is constricted between the seeds, forming a lomentum which does not dehisce but breaks into one-seeded joints. In *Astragalus* it is divided by a secondary (or false) longitudinal septum.

The pods shew very great variation in size and form; some measure only a fraction of an inch, others, like *Entada gigas* (fig. 183) (Mimosoideae), are great woody structures more than a yard long and several inches wide. They are generally flattened but sometimes round and rod-like, as in *Cassia fistula*, or more or less bent, or spirally coiled as in *Medicago*. The tribe *Dalbergieae* (Papilionatae) is characterised by indehiscent pods which are often only one-seeded and winged; and winged indehiscent pods occur in other genera. In *Colutea* the pod is an inflated several-seeded bladder which bursts under pressure; it becomes detached and is blown some distance before bursting, or breaks and scatters the seeds while swaying on the plant (*C. arborescens*, Bladder Senna, is

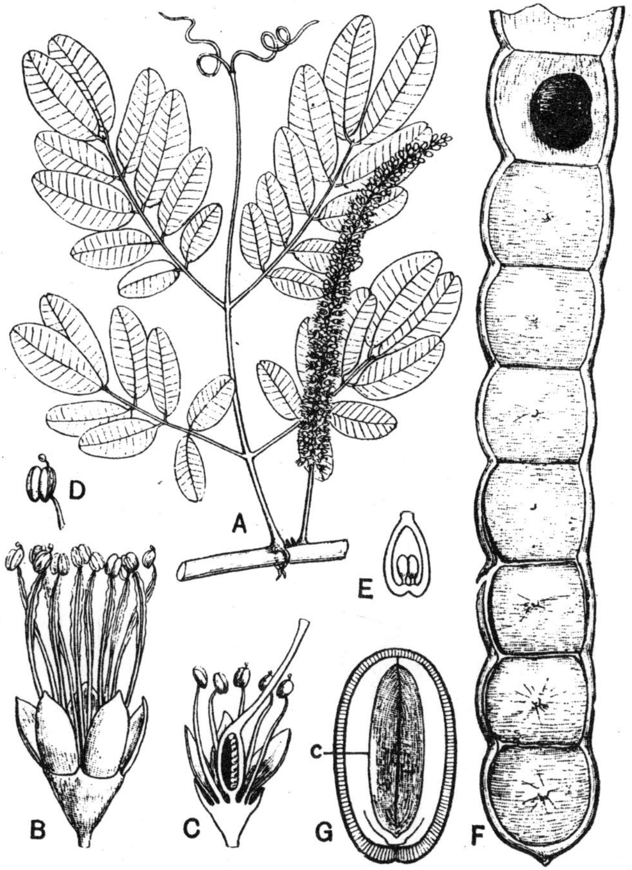

FIG. 183. *Entada gigas* (*E. scandens*). A. A leaf and inflorescence, × ½. B. Male flower, × 4. C. Fertile flower in section, × 4. D. Anther with gland, × 8. E. Ovary cut across, × 7. F. Portion of pod, with the valve of one joint removed to shew seed, × ⅙. G. Seed cut across shewing the cavity, *c*, between the cotyledons. (From *Flor. Jam.*)

a commonly-grown shrub). The funicle often develops an aril, which is sometimes brilliantly coloured and, by contrast with the dark testa, renders the seed very conspicuous; or the seed-coat is bright coloured, as in the scarlet seeds of

Abrus precatorius. The testa is generally smooth, sometimes leathery or bony, sometimes thin and membranous.

In Mimosoideae and Papilionatae endosperm is wanting or present in small quantity, generally filling up the space round the radicle. In Caesalpinioideae it may be absent or present in greater or less quantity, forming in the tribe *Bauhinieae* a thin layer round the embryo, while in the *Cassieae* it is copious and cartilaginous.

The embryo has generally flat leaf-like or fleshy cotyledons with a short radicle which is straight in the first two subfamilies, incurved and accumbent to the cotyledons in the third.

Seed-distribution is effected in very various ways. (1) By elastic dehiscence of the pod, to which we have already referred. In the well-known wall-climber, *Wistaria sinensis*, the seeds are said to be shot more than nine yards. (2) By the wind. Besides the winged or inflated pods, as in *Colutea*, the enlarged bladder-like persistent calyx may act as a carrier, as in *Trifolium fragiferum*; while some one- or few-seeded fruits (as species of *Medicago*) have a light spongy pericarp enabling them to be rolled considerable distances over the ground. (3) Animals are the agents in the case of fleshy pods containing seeds with a hard testa which will pass uninjured through the body. Such are the Tamarind and the fruits of the Carob-tree (*Ceratonia siliqua*). The various fleshy red, orange or yellow arils attract birds. In other cases pods are themselves falcate in shape, or crowned with a hooked persistent style, or bear hooks or spines, which enable them to cling more or less readily to fur or wool.

Leguminosae is the second largest family of seed plants, Compositae alone exceeding it. They occur wherever climate admits of growth of seed plants, and are often characteristic features of the vegetation. The tropical rain-forests are rich in genera and species of Mimosoideae and Caesalpinioideae; there are more than 200 species of *Cassia* in the Brazilian flora. Papilionatae are mostly herbaceous, and fewer in number, but in subtropical forests arborescent forms (especially of the tribe *Dalbergieae*) occur along with those of the other subfamilies. In the regions of deciduous forests where growth is interrupted by the cold of winter, leguminous trees are rare; Mimosoideae are absent from Europe; Caesalpinioideae are represented by *Cercis Siliquastrum* (Southern Europe to Persia), which occurs as far north as the Southern Tyrol, and *C. chinensis* and *C. canadensis*, while *Gymnocladus* and *Gleditschia* also occur in both North America and China; *Robinia* (N. America) represents Papilionatae. On the other hand, herbaceous Papilionatae become a more important feature of the vegetation, ascending into Arctic and high alpine regions to the limit of seed plants. Shrubs and undershrubs such as *Ulex*, *Cytisus* and *Genista* form a characteristic feature in north European and Mediterranean areas. The evergreen Australian bush-vegetation is characterised by numerous Acacias and genera of the tribe *Podalyrieae* (Papilionatae) such as *Chorizema* (West Australia), *Oxylobium* and others. Papilionatae play an important part in steppe-formations, doubtless partly due to their nitrogen-fixing properties. A great many species of *Astragalus*, as well as species of *Oxytropis*, *Hedysarum*, *Onobrychis* and others, occur in the steppe areas of Eastern Europe and Western Asia; while *Podalyrieae* and *Genisteae* are characteristic of those of South Africa and Australia.

But little is known of the history of the family. From numerous leaf-remains which are found it is probable that all three subfamilies existed in Tertiary periods.

Leguminosae is one of the most important families from an economic point of view. The seeds, rich in starch and proteids, are a widespread source of food, as in the various Beans, Pea, Vetch, Ground-nut (*Arachis*), Lentil (*Lens esculenta*) and others. *Phaseolus multiflorus* is the Scarlet Runner. Species of *Trifolium*, *Medicago*, *Melilotus*, *Vicia*, *Onobrychis* and many others are fodder-plants. Many of the tropical trees yield useful woods; species of *Crotalaria*, *Sesbania*, *Aeschynomene* and others yield fibre; numerous species of *Acacia* and *Astragalus* gums; *Copaifera*, *Hymenaea* and others balsams and resins; the seeds of the Ground-nut and a few others yield oil. Dyes are obtained from *Genista* (yellow), *Indigofera* (blue), *Mucuna* and others; *Haematoxylon campechianum* is Logwood. Species of *Trifolium* and *Lupinus*, and *Pisum sativum* form a valuable green manure, enriching the soil in the nitrogen which they have acquired from the air by means of the root-tubercles. Species of *Acacia*, *Cassia* (Senna leaves), *Astragalus*, *Tamarindus indica* (Tamarind), *Glycyrrhiza glabra* (Liquorice-root) are or have been used in medicine. *Physostigma venenosum*, the Ordeal Bean of Calabar, contains a strong poison.

Among ornamental trees and shrubs are *Cercis* (*C. Siliquastrum*, Judas-tree), *Gleditschia*, *Genista*, *Cytisus*, *Colutea*, *Robinia*, and *Acacia*; *Wistaria sinensis* is a well known wall-climber; *Lathyrus* (Sweet and Everlasting Pea), *Lupinus*, *Galega* (Goat's-rue) and others are garden plants.

As already indicated, the family is divided into three subfamilies: the subdivision into tribes adopted by Taubert in the *Pflanzenfamilien* follows closely that of Bentham and Hooker.

Subfamily I. MIMOSOIDEAE. Flowers regular; calyx generally gamosepalous, sepals valvate; petals valvate; stamens as many or twice as many as petals or ∞, free or cohering; pollen-grains often united in packets. Generally trees or shrubs with bipinnate leaves and small flowers which are sometimes sessile in spherical heads or cylindrical spikes, sometimes stalked in lax racemes or umbels.

Subdivided into six tribes which are distinguished by differences in the androecium, including 40 genera; two-thirds of the species are contained in the two largest genera, *Acacia* (500 species) and *Mimosa* (350 species); they are distributed throughout the warmer parts of the world; *Acacia* is more especially developed in Africa and Australia (the various Wattles); *Mimosa* is chiefly tropical and subtropical American.

Subfamily II. CAESALPINIOIDEAE. Flowers zygomorphic; sepals generally free (sometimes the two upper are united) and imbricate in bud; petals imbricate, the uppermost inside; stamens 10 or fewer by abortion, free or more or less coherent; ovary sometimes stalked; embryo with generally a straight radicle. Trees and shrubs, more rarely herbs, with pinnate or bipinnate, more rarely simple leaves; stipels absent. Flowers various, sometimes large and brightly coloured, sometimes small and inconspicuous, generally in racemes, more rarely spicate. It is very difficult to delimit tribes in this subfamily. In the latest revision nine are recognised, in three of which the calyx is markedly gamosepalous, while in the rest the sepals are almost or quite free. Other distinctive points are found in the leaves, in the number and character of stamens and the relation between ovary and receptacle.

Thus *Bauhinieae* have a gamosepalous calyx and simple leaves; *Eucaesalpinieae* have sepals free, leaves all or some bipinnate, while *Cassieae* differ in having only simply pinnate leaves, and are further distinguished by having more or less basifixed anthers, with a terminal porous dehiscence (fig. 178, D); *Kramerieae* have the anterior pair of petals modified into fleshy, scale-like glands; *Amherstieae* have leaves simply pinnate, and the ovary-stalk is adherent at the back with the receptacular cup.

Subfamily III. PAPILIONATAE. Flowers zygomorphic, papilionaceous; calyx generally gamosepalous, segments ascending imbricate, often bilabiate from the greater union of the two upper and three lower segments respectively. Petals generally very unequal with descending imbrication; stamens inserted on the same level as the petals, generally ten, monadelphous, or diadelphous, or almost or quite free; ovary sessile or stalked; radicle generally curved and accumbent. Herbs, or more or less shrubby, rarely trees, with simple, or palmately, or simply-pinnately compound leaves; stipels frequently present. Flowers various, mostly of moderate size and brightly coloured, sometimes very small and inconspicuous, solitary, or in spikes, racemes or heads, rarely cymose.

In this subfamily also it is difficult strictly to delimit tribes. Habit and especially the leaf-character are the best guides to their recognition. Ten may be distinguished, of which six are represented in our own flora.

Tribes (1) *Sophoreae* and (2) *Podalyrieae* have free stamens; in the former the leaves are pinnate, in the latter simple or palmate. *Sophoreae* has 33 genera, the two largest of which, *Sophora* and *Ormosia*, are distributed through the tropics of both hemispheres; the rest are small, often monotypic, mainly tropical genera, with a restricted distribution, half of them being confined to the warmer parts of America. *Podalyrieae* has 37 genera, 30 of which are confined to Australia, especially the west and north; *Podalyria* and one other are South African, the remainder are North American and temperate Asiatic, and one (*Anagyris*) is restricted to the Mediterranean region and Teneriffe.

The remaining tribes have monadelphous or diadelphous stamens.

Tribe (3). *Genisteae*. Shrubs or sometimes herbs, with simple or digitately compound leaves and entire leaflets; stamens generally monadelphous. The 43 genera contain 950 species. A small group of genera is confined to Australia and another to South Africa. The largest group contains *Crotalaria*, with 350 species generally distributed in warm regions, and 18 other genera. The section *Spartiinae* (9 genera) includes *Lupinus*, a large principally American genus; *Spartium* with one species, *S. junceum* (Spanish Broom), a characteristic Mediterranean species; *Genista* (100 species) in Europe (three British), N. Africa and W. Asia; and *Laburnum* with three species in S. Europe; the remainder occur principally in the Mediterranean region or South Africa. *Cytisinae* includes *Ulex* (Gorse) and *Cytisus* (Broom), which occur chiefly in Europe, the Mediterranean region and West Asia; the seeds of Gorse and Broom bear an elaiosome, a large, orange-coloured appendage rich in oil, in the region of the hilum, and are distributed by ants.

Tribe (4). *Trifolieae.* Herbs or rarely shrubs, with pinnately or rarely palmately trifoliolate leaves, the veins generally ending in teeth; stamens diadelphous or sometimes monadelphous. Of the six genera, five are British: *Ononis* (70 species), chiefly in the Mediterranean region extending to the Canary Islands with a few in North and Central Europe (two British). *Trigonella*, 70 species chiefly in the eastern part of the Mediterranean region, spreading north to Central Europe with one species (*T. ornithopodioides*) in Britain, several in North Africa and temperate Asia, one in S. Africa and one in Australia. *Medicago*, about 50 species in Europe, temperate Asia, North Africa, and at the Cape; five are British, and the Lucerne (*M. sativa*), an Eastern Mediterranean plant which has become widely naturalised, is also common. *Melilotus*, about 20 species in the temperate and subtropical zones of the Old World; two are British, while *M. officinalis* is a doubtful native. *Trifolium*, 300 species, chiefly north temperate, but with a few species on the mountains of tropical Africa, at the Cape, on the Andes and in subtropical South America; eighteen are native in Britain and others (as *T. incarnatum*) are cultivated.

Tribe (5). *Loteae.* Herbs or undershrubs with pinnately three- to many-foliolate leaves, and entire leaflets; stamens di- or monadelphous. Genera eight. *Lotus*, 90 species in temperate Europe and Asia, especially in the Mediterranean region, and a few in South Africa and Australia; *L. corniculatus* (Europe westwards to India) and four less widely distributed species are British. *Anthyllis*, 30 species in Europe, North Africa and Asia; *A. Vulneraria* is British. *Hosackia*, 30 species in Western North America. The remaining genera are small and occur chiefly round the Mediterranean area.

Tribe (6). *Galegeae.* Herbs, erect shrubs or more rarely trees or climbing shrubs, with pinnate leaves, and generally entire leaflets; stamens generally diadelphous. Genera 65. More than half the species (about 1600) belong to *Astragalus*, and the majority of these inhabit the temperate regions of the northern hemisphere of the Old World, especially the Asiatic steppes; a few ascend into the Arctic area; three occur in Britain; a few are found in the mountains of tropical Africa and one in Natal; more than 200 are American, chiefly North American, but also occur in the Alpine regions of the tropical Andes in Chile. There are none in Australia. They are annual or perennial herbs, sometimes densely branched undershrubs or shrubs; many species become thorny by the hardening of the persistent leaf-stalk. *Oxytropis*, a closely allied genus, has 150 species in the temperate and colder regions of the northern hemisphere; two occur in Scotland. *Robinia* (six species) is North American; *R. Pseud-acacia* is the "Acacia" commonly cultivated in southern England. *Colutea* (ten species), a South European genus, includes *C. arborescens* (Bladder-senna of gardens).

Of the remainder the majority inhabit the warmer parts of the world, especially of the eastern hemisphere. *Indigofera* (fig. 180), *Psoralea* and *Tephrosia* are large tropical and subtropical genera, especially African.

Tribe (7). *Hedysareae*, with fruit a lomentum, has 47 genera (over 700 species); about two-thirds are tropical or subtropical, the remainder occur chiefly in Europe, the Mediterranean region and temperate Asia, or the Andine region of temperate South America. The three British genera, *Ornithopus*, *Hippocrepis* and *Onobrychis*, belong to the Europe-Mediterranean-West Asiatic centre of distribution. *Arachis hypogaea* (Ground-nut), a native of tropical America, is widely cultivated in warm countries for the oily seed.

Tribe (8). *Vicieae*. Herbs, with paripinnate leaves, the rachis ending in a tendril, or point; stamens diadelphous. Six genera (two British). *Vicia* (Vetch, Tare), 120 species in the north temperate zone and southern and western South America; generally climbing by leaf-tendrils; 11 British; *V. sativa* (Vetch) and others are useful fodder-plants; *V. Faba* (Broad Bean). *Lathyrus*, 100 species, chiefly in the northern hemisphere with a few in South America and on the mountains of tropical Africa; nine British; *L. Nissolia* has grass-like phyllodes; *L. odoratus* (Sweet-pea). *Pisum*, six species in the Mediterranean area and Western Asia; *P. sativum* (Pea). *Lens*, six species with a similar distribution to *Pisum*; *L. esculenta* (Lentil) has been cultivated since the bronze age. *Cicer*, 14 species in Western Asia; *C. arietinum* (Chick-pea) is cultivated for food in Southern Europe and India. *Abrus*, six species in the tropics; the hard scarlet seeds of *A. precatorius* are well-known.

Tribe (9). *Phaseoleae*. Climbing herbs or more rarely erect or shrubby, very rarely trees. Leaves pinnate, generally trifoliolate; leaflets usually with stipels; stamens generally diadelphous. Genera 47, generally distributed through the tropics and warmer parts of the earth. *Phaseolus*, the largest genus, has 150 species in tropical and warm temperate regions; *P. multiflorus* (Scarlet Runner) and *P. vulgaris* (French or Kidney Bean), natives of South America, are widely cultivated.

Erythrina crista-galli (Brazil) is a handsome garden plant. *Physostigma venenosum* (West Africa), Calabar or Ordeal Bean.

Tribe (10). *Dalbergieae.* Trees or shrubs, sometimes climbing. Leaves pinnately five- to many-foliolate; stamens monadelphous or diadelphous; pod indehiscent, sometimes drupaceous. Genera 27, almost exclusively tropical; more than half are confined to the New World. *Dalbergia* (100 species) occurs in the tropics of both Old and New Worlds. Some are lianes climbing by means of short branches which are sensitive to contact. *Pterocarpus santalinus* yields red sandal-wood.

Order 15. *MYRTIFLORAE*

Flowers regular, typically bisexual and cyclic, polypetalous, perigynous to epigynous; generally 4- or 5-merous (6-merous in Lythraceae), with diplostemonous or obdiplostemonous androecium and frequent reduction in the gynoecium. Stamens sometimes ∞ (Myrtaceae). Ovary multilocular to unilocular, with a simple style (except Haloragaceae) and generally numerous ovules on axile placentas. Endosperm usually absent. Herbs, shrubs or trees with opposite, less often alternate, simple generally entire and exstipulate leaves; internal phloem is often present in the stem.

This order is closely allied to the last. It is characterised by an increased tendency to perigyny culminating in the complete union of the ovary with the floral axis in the typically epigynous families. Other characters are the simple style, the frequency of tetramery in the flower and of intraxylary phloem in the stem.

Family I. THYMELAEACEAE

The plants are generally shrubs with much branched stems. Some species, inhabitants of dry countries where the vegetation above ground is destroyed by periodical fires, have a stout ascending underground woody rhizome from which new growths arise in the form of thick clumps of erect switch-like branches, which terminate in showy densely-flowered heads (species of *Gnidia* and *Stellera*). High trees are rare, as in the Asiatic *Aquilaria* and allied genera, as also are herbs. The leaves are generally alternate; they are entire, shortly stalked or sessile, usually narrow or needle-like, and often closely crowded on the stem, giving the plant a heath-like habit. Sometimes they are broad and flat, and then often have numerous parallel lateral nerves. Stipules are absent. The plants are generally glabrous. The family is represented in Britain by two species of *Daphne*, *D. Laureola* (Spurge Laurel) and *D. Mezereum* (Mezereon), shrubs with few branches and flattened more or less obovate leaves and small tubular fragrant axillary flowers.

An anatomical character, absent only in the genus *Drapetes*, is the presence of internal phloem; in some genera also (as *Aquilaria* and allies) phloem-islands occur in the wood. The very strongly developed and layered fibrous bast gives a characteristic toughness to the stems. This is especially well seen in the West Indian lace-bark tree, *Lagetta*, in which the bark separates on maceration in thin lace-like layers.

The inflorescence is racemose and of very various forms; sometimes a true raceme or spike; frequently a few- or many-flowered head, which is enveloped by bracts, recalling the head of a Composite; umbels also occur and occasionally solitary terminal or axillary flowers. The flowers are bisexual or sometimes unisexual, regular or rarely somewhat zygomorphic.

The most remarkable feature of the flower (fig. 184) is the long tubular or urn-shaped coloured receptacle, the further development of which shews great variety. It may fall early or persist and envelop the ripe fruit. In some cases the upper portion falls off at a constriction leaving the lower portion which surrounds the fruit. The four to six sepals spring from the edge of the receptacle and are coloured like it; petals are absent or more or less developed in the form of scales alternating with the sepals round the mouth of the receptacle. The stamens are generally double the number of the sepals and arranged in an upper and lower series on the receptacular tube. Dimorphic or trimorphic flowers, differing in the relative length of stamens and style, are of frequent occurrence. The pistil consists generally of a single carpel, with a terminal or laterally placed style bearing a terminal stigma, and containing a single large anatropous ovule suspended from near the apex. In a few genera (*Aquilaria* and allies) there are two united carpels bearing a single style and containing a two-celled ovary, each chamber with one ovule.

Insect-pollination is indicated by the brightly coloured, generally many-flowered inflorescences, the sweet scent, and the frequent occurrence of nectaries at the base of the ovary; also by the numerous cases of dimorphy or trimorphy of the flower, and the occurrence of dicliny or dioecism.

The nature of the fruit varies with the manner of development of the pericarp, which is thin, forming a dry nutlet, or the exocarp is thick and leathery, or sometimes pulpy, when the fruit is a drupe (fig. 184). A drupe-like fruit is also formed by the fleshy development of the persistent base of the receptacle. In these latter cases the fruit is distributed by the agency of birds; the dry fruits which are very light and sometimes bear hairs, resembling a pappus, are carried by the wind. Endosperm is generally scanty or absent, and the embryo is large with thick convex cotyledons and upwardly pointing radicle (fig. 184, B).

The family contains 40 genera and nearly 500 species. It is cosmopolitan with the exception of the polar regions, but specially developed in South Africa and Australia, and in the northern hemisphere in the Mediterranean region and the Asiatic steppes, forming a characteristic element of the vegetation of the latter. Generally they are dry-country plants. The genera are, as a rule, of limited distribution, with exceptions, such as our British representative *Daphne*, which spreads throughout Europe and Central Asia to China and Japan and southwards to the Indo-Malayan Archipelago. The small and somewhat anomalous genus *Drapetes*, a low-growing plant with a moss-like habit, has a wide distribution in the southern hemisphere, being represented by species in Tierra del Fuego, New Zealand, Tasmania and S.W. Australia, Borneo and New Guinea.

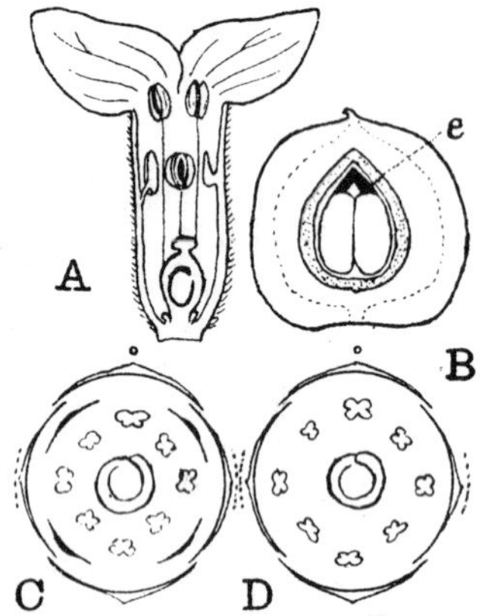

FIG. 184. A. Flower of *Daphne Mezereum*. B. Drupe of same (*e*, endosperm); both in vertical section; enlarged. C. Floral diagram of *Gnidia aurea*. D. Floral diagram of *Daphne*. (A, B after Wettstein; C, D after Eichler.)

Family II. ELAEAGNACEAE

A small family of generally much branched erect shrubs, sometimes trees, densely clothed with silvery, brownish or golden-coloured scale- or stellate hairs. The leaves are alternate or opposite, and entire, without stipules. The flowers are axillary, either solitary or in clusters of a few, or in short racemes. In *Elaeagnus* they are bisexual or male, in *Hippophaë* and *Shepherdia* (North America) dioecious or sometimes diclinous. The bisexual and female flowers have a more or less tubular receptacle, the lower part as in *Elaeagnus*, or the whole as in *Hippophaë*, enclosing the ovary; in the former case there is a constriction above the ovary and the upper portion falls after flowering while the lower remains enveloping the fruit; in the latter the whole persists round the fruit. In the male the receptacle is cup-shaped or nearly flat. The two (*Hippophaë*), four, or sometimes more sepals spring from the edge of the receptacle; they are valvate in the bud. Petals are absent. The stamens are borne at the mouth of the receptacle, and are either equal in number and alternating with the sepals (*Elaeagnus*), or twice as many (*Hippophaë*); the anthers are attached dorsally and open lengthwise. Rudiments of the suppressed pistil or stamens are generally absent. A disc in the form of glandular developments of the receptacle, which alternate with the stamens, is sometimes present. The ovary is one-celled, containing a basal anatropous ovule with two integuments, and bears an elongated thread-like style ending in a capitate stigma. The fruit is berry-like or drupaceous (*Elaeagnus*) from the development of the persistent receptacular tube; the true pericarp is thin. The fleshy covering is often edible and of pleasant taste and attractive to birds. The seed has a hard testa and contains a straight embryo with thick, fleshy cotyledons and a radicle pointing downwards: there is little or no endosperm. The small flowers of *Hippophaë* are wind-pollinated; those of the other two genera have nectar-secreting glands, while the fairly large yellow or white sweet-scented flowers of *Elaeagnus* are obviously adapted for insect-pollination. The fruits are admirably adapted for distribution by birds owing to their fleshy coat enveloping a well-protected seed.

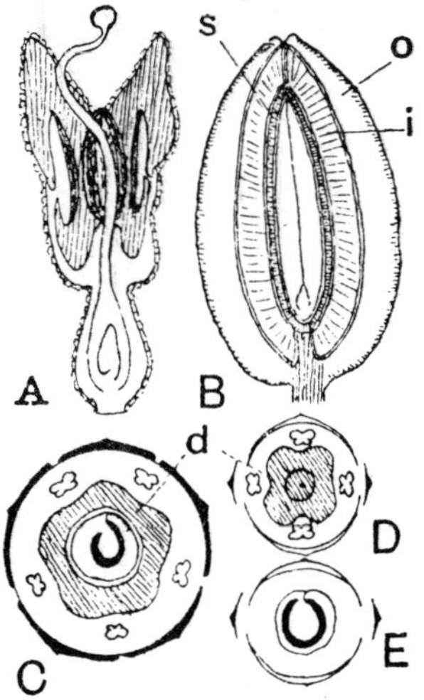

FIG. 185. A. Bisexual flower of *E. angustifolia*, in vertical section. B. Fruit of *Elaeagnus*, in vertical section; *o*, outer fleshy layer; *i*, inner hard layer of persistent receptacular tube; *s*, hard seed-coat. The true pericarp is a thin layer between *i* and *s*. Floral diagrams of C, *E. angustifolia*, bisexual flower; D, male, and E, female flower of *Hippophaë rhamnoides*: *d*, disc. (A, B enlarged, after Wettstein; C, D, E after Eichler.)

There are three genera and about 20 species (more than half of which belong to *Elaeagnus*) chiefly in the north temperate and subtropical zone of both Old and New Worlds; a few species of *Elaeagnus* occur in the tropics in the Indo-Malayan region. They are generally xerophytes inhabiting steppes or sea-coasts, or dried sandy river-beds. The family is represented in Britain by *Hippophaë rhamnoides* (Sea-buckthorn), which occurs in places on sandy sea-shores on the east and south-east coasts and has been planted as a sand-binder elsewhere. It is a shrub with slender branches, some of which may be short and spinescent; the leaves are silvery beneath. The flowers are borne on the old wood; the globose or oblong fruit is orange-yellow in colour.

The family is nearly related to Thymelaeaceae, with which it has a similar development of the receptacle.

Family III. LYTHRACEAE

A family of 23 genera and 450 species, chiefly tropical, the number rapidly diminishing as we approach temperate regions; absent from the cold zones. It is represented in Great Britain by *Lythrum Salicaria* (Purple Loose-strife), frequent in damp places from Argyll and Perth southwards, a second very local species (*L. hyssopifolia*) confined to a few southerly English counties, and *Peplis Portula* (Water-purslane), a small herb growing in moist places from Caithness southwards.

In temperate climates the plants are annual or perennial herbs, but in warmer are often shrubby, and sometimes form trees, as in the Old World genus *Lagerstroemia*, and the tropical American *Lafoënsia*. A constant anatomical character is the presence of internal phloem, the vascular bundles of the stem being bicollateral. The leaves are generally opposite, and always simple and entire, with very small stipules. The flowers are borne in racemes, each flower springing from the axil of a leaf- or scale-like bract; the two bracteoles may be barren, as in the subgenus *Hyssopifolia* of *Lythrum*, or, as in the subgenus *Salicaria*, branching occurs in their axils and a number of axillary dichasia are produced. In both cases further branching may occur and large compound often pyramidal inflorescences result.

As in our British species, the flowers are generally 6-merous, a smaller proportion are 4-merous, and the rest shew variations between 3- and 15-merous. In the tropical American *Lafoënsia* the flowers are 9–15-merous. They are bisexual and regular or, as in *Cuphea*, zygomorphic. The receptacle is hollow, generally tube-like, and the sepals spring from the margin; they are valvate in aestivation. An epicalyx is frequently present (as in *Lythrum*) formed by the union in pairs of sepal-stipules. The petals, which are also free, alternate with the sepals on the cup-margin; they have a crumpled aestivation, are generally fugacious and often red or purple in colour. They are sometimes absent, as in *Peplis* and some allied genera with small flowers, such as *Ammania* and *Rotala*, many species of which are cleistogamic. The stamens are inserted on the inside of the receptacle-cup, often low down, and are generally twice as many as the sepals

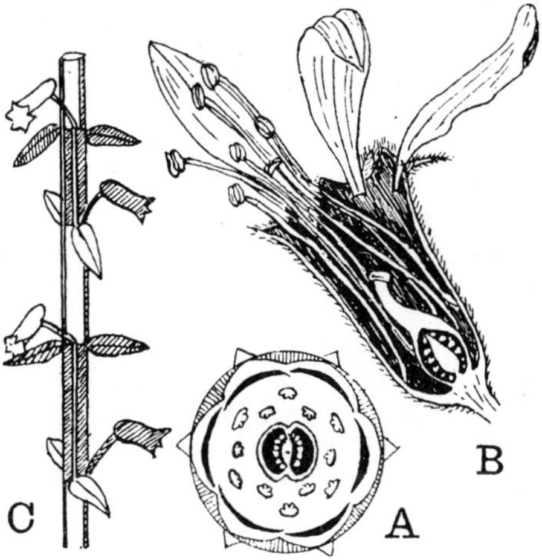

FIG. 186. A. Floral diagram of *Lythrum Salicaria*. B. Short-styled flower of same in vertical section, enlarged. C. Plan of inflorescence of *Cuphea*. (A after Eichler; B and C after Koehne.)

or petals; the inner whorl (antepetalous) may be absent, or further reduction occurs, as in *Rotala*, a small-flowered genus allied to *Peplis*, where there is sometimes only one.

There are two to six carpels united to form a superior two- to six-chambered ovary with a simple style and generally capitate stigma. The septa between the chambers sometimes disappear in the upper part of the ovary. The ovules are generally numerous with axile placentation; they are anatropous and ascending. The fruit is a capsule with various dehiscence. The embryo is straight with flat cotyledons (folded in *Lagerstroemia*) and completely fills the seed.

Cuphea is the largest genus, with 200 species in America, chiefly in Brazil and Mexico; a few species are known as greenhouse plants. The flowers are zygomorphic from the development of the receptacular cup into a spur on the posterior side of the flower, accompanied by reduction in size of sepals and petals towards the anterior side and abortion of the posterior stamen. The plants are often covered with sticky glandular hairs, and the flowers are borne singly between the leaves (which are opposite-decussate) at each node. This anomalous position of the flower is explained (fig. 186, C) by assuming its displacement from the axil of the leaf in which it originates upwards through one internode, where it springs from the stem vertically above its own leaf but between the leaves at the node in question. *Lythrum* has 25 species, which are found mostly in damp places, in Europe, Asia (chiefly bordering on the Mediterranean and in the steppe region), Africa (chiefly North Africa) and America; two occur in Australia and one in New Zealand. *L. Salicaria* is one of the few species found in both Old and New Worlds, spreading from Europe through temperate Asia to Japan, and occurring also in the north-eastern United States and southern Canada, as well as in south-east Australia. Its flowers, which are of a fine purple colour, are trimorphic, having long-, mid-, and short-styled forms, with the two whorls of stamens in the other two positions respectively. Cross-pollination is effected by bees which seek the nectar secreted in the base of the receptacle. *Peplis* has only three species, one North American, our native species, *P. Portula*, which is European, and a third in temperate Asia. Some genera yield colouring matter; the most important is the tropical Old World *Lawsonia inermis*, from which the red dye, Henna, is prepared, much used in the East for colouring finger-nails, etc. *Lafoënsia* (tropical America) and *Lagerstroemia* are useful timber-trees; the tropical South American *Physocalymma scaberrimum* yields a much prized rose-coloured wood.

Family IV. PUNICACEAE

The genus *Punica*, regarded by Bentham and Hooker as an anomalous genus of Lythraceae, is now generally placed in a distinct family, Punicaceae, which comprises only two species *P. Granatum* (Pomegranate) and *P. protopunica*. The Pomegranate is found wild from the Balkan Peninsula to the Himalayas, but has been cultivated from the earliest times, and is now widely distributed in warm regions. The second species is a native of Socotra. *Punica* is distinguished from Lythraceae by the union of the ovary with the receptacle. The numerous stamens spring from the inner face of a thickening of the receptacle above the ovary; *P. protopunica* has one series of carpels, but in *P. Granatum* there are two or three series which are originally concentric, but in the course of development of the ovary the outer become carried up so that the series become ultimately superposed. The placentation of the numerous anatropous ovules is originally axile but following the alteration of position of the ovary-cells the placentas of the upper series come to occupy the middle line of the outer wall of the cells (compare the similar change of position of the placenta from axile to parietal in *Mesembryanthemum*). The fruit is a berry with a thick leathery coat, divided into chambers by the thin walls of the carpels; the "flesh" of the fruit is the pulpy outer coat of the seeds which contains a refreshing acid juice. The inner seed-coat is horny and is filled by the embryo which has large cotyledons rolled one on the other.

The plants are small trees or sometimes bushes, the twigs often ending in a spine; the leaves are generally opposite and crowded on short shoots; stipules are absent. The showy, generally scarlet, flowers are borne solitary or a few in a cluster (a cyme) at the end of short leafy shoots.

Family V. SONNERATIACEAE

Sonneratiaceae is a small family (3 genera, 12 species), allied to Lythraceae, of tropical (Asiatic and African) trees, including *Sonneratia*, species of which occur in the Mangrove vegetation of tropical Asia and East Africa; these bear the characteristic negatively geotropic respiratory roots which grow a yard or more in height from the mud in which the roots creep. It differs from Lythraceae in the partial union of the ovary with the receptacular cup and in the parietal or subbasal placentation. Internal phloem is present in the stem.

Family VI. RHIZOPHORACEAE

A small family (17 genera, 60 species) of woody plants, generally shrubs or small trees, with simple evergreen often leathery generally opposite leaves with caducous stipules. Many of them form characteristic components of the Mangrove vegetation of muddy coasts and estuaries in the tropics. At high tide their lower portions are flooded by the sea, at low tide they rise from a soft evil-smelling mud. The plants shew remarkable adaptations to their habitat. In *Rhizophora* the short stem, the base of which perishes early, is supported by large downwardly curving roots, the exposed portions of which bear numerous large lenticels which allow ready communication through intercellular spaces between the atmosphere and the submerged portions of the roots. In species of *Ceriops* aeration is similarly effected by slender negatively geotropic lateral branches which spring from the horizontally creeping submerged roots. The species of *Bruguiera*, which grow on less exposed portions of the Mangrove, have a normally developed stem, sometimes reaching a considerable height, with correspondingly fewer prop-roots. Aeration of the submerged roots is effected by knee-like upgrowths of the horizontally creeping submerged roots.

The leaves possess in a marked degree the xerophytic structure of plants associated with salt-water, having a well-developed water-storing tissue below the strongly cuticularised upper epidermis.

The flowers are borne singly or in few-flowered cymose clusters in the leaf-axils. They are bisexual or sometimes unisexual by abortion, regular, and perigynous or more or less completely epigynous. The floral envelope is generally 4- or 5-merous, comprising an outer whorl of fleshy or leathery persistent sepals and an inner whorl of petals, which are generally shorter than the sepals, free, and often have a claw and a more or less lobed or cut blade. The stamens are 8 to ∞ in number, and situated on the outer edge of a lobed perigynous or epigynous disc, generally in one series. The anthers are four-celled (except in *Rhizophora* where there are numerous pollen-sacs, fig. 187, F) and dehisce introrsely by longitudinal slits. According to the development of the floral axis the ovary is free or half- to completely inferior; it is generally formed by union of two to four carpels, and contains as many chambers, which may however be incompletely separated. There are usually two ovules in each chamber, pendulous from a central placenta, and anatropous with the micropyle directed upwards and outwards. There is generally a single style with a small lobed stigma. The fruit is a berry, more rarely a dry or somewhat fleshy capsule, and is sometimes one-seeded and indehiscent. There are one or more seeds, which are sometimes arillate. The generally straight embryo is often green, with two, sometimes three or four, cotyledons, which may be more or less united. Endosperm is usually present.

In the Mangrove-species the seeds germinate on the tree. By growth of the cotyledons, and later of the hypocotyl, the latter pushes through the widely open micropyle and the pericarp, reaching a length of 20 to 40 cm. or even a metre. In *Rhizophora, Ceriops* and others it becomes detached by separation from the cotyledons and falls from the tree; in *Bruguiera* the fruit falls with it. The root-end penetrates the mud and rapidly becomes attached by development of the main root or lateral roots. If

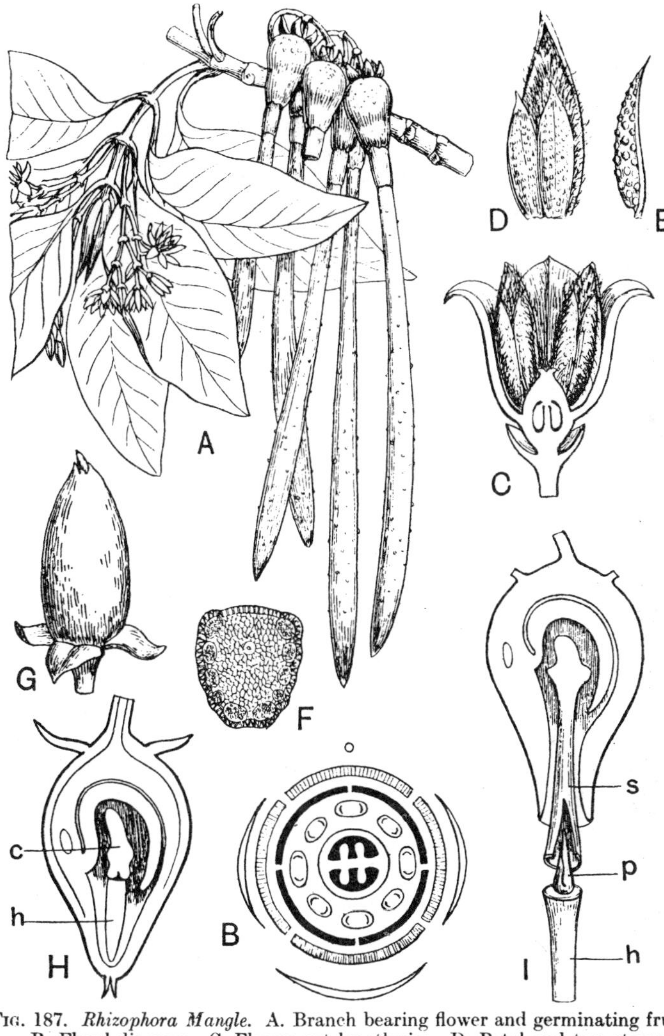

FIG. 187. *Rhizophora Mangle.* A. Branch bearing flower and germinating fruit. B. Floral diagram. C. Flower cut lengthwise. D. Petal and two stamens. E. Side view of stamen. F. Transverse section of anther shewing numerous pollen-sacs. G. Fruit; the tip of the hypocotyl protruding at the apex. H. Longitudinal section of a fruit; *c*, the cap-like cotyledon surrounded by endosperm; *h*, hypocotyl which has grown through the seed-coat into the apex of the pericarp. I. Longitudinal section at a later stage; the sheath, *s*, of the cotyledon, in which the plumule, *p*, has been enclosed, has pushed out the hypocotyl, *h*, bearing the plumule. C–F enlarged, A, G–I reduced. (A–C, G after Baillon; E, F after *Flor. Bras.*; H, I after Kerner.)

the mud is flooded the seedling may be carried on the tide and stranded later. In *Rhizophora* and *Ceriops* the endosperm also grows out of the micropyle and down over the wall of the seed.

The family is exclusively tropical and mainly Old World. The Mangrove-species have a wide range of distribution; *Rhizophora Mangle*, for example, is characteristic of the American and West African Mangrove vegetation; those growing inland occupy, on the contrary, a very restricted area.

Family VII. COMBRETACEAE

A tropical or subtropical family of trees or shrubs, sometimes climbing, with opposite entire leaves without stipules and flowers in racemes. The flowers are bisexual, or unisexual by reduction, regular with a tendency to zygomorphy, and 5-merous or sometimes 4-merous. The floral axis is carried above the ovary, forming a tube. Sepals with generally valvate aestivation. Petals small, alternating with the sepals, sometimes absent. Stamens bent inwards in the bud, 4, 5, 8 or 10, rarely ∞ ; generally twice as many as the sepals, in two series, the lower opposite the sepals, the upper alternating with them, the upper series sometimes reduced or absent. Ovary inferior and one-chambered, generally angled, the angles equal in number and alternating with the calyx-segments; ovules two to five, rarely more, anatropous, pendulous from the top of the ovary on long often united funicles; micropyle turned upwards and outwards. Style one, long, filiform, bearing a pointed, rarely capitate stigma. The receptacle tube bears a disc which is sometimes hairy. Fruit leathery or drupe-like, one-seeded, two- to five-angled, the angles often forming broad membranous wings. Seed without endosperm; cotyledons spirally rolled or irregularly folded, rarely planoconvex. Internal phloem is present in the stem.

There are 17 genera containing about 500 species, the greater number of which belong to *Terminalia* and *Combretum*.

The members of this family are rich in tannin, which occurs in bark, leaves and fruits; myrobalans (ripe fruits of *Terminalia Chebula*) are an important article of commerce in India. Species of *Terminalia* and others yield timber.

Family VIII. LECYTHIDACEAE

A small tropical family (about 150 species in 20 genera) of trees with often very large, alternately arranged, entire leaves, generally closely crowded at the ends of the branches. Stipules are wanting. An interesting anatomical character is the presence of a series of cortical bundles which also run up as separate strands into the leaf-stalk and remain distinct in the midrib and largest lateral veins. The bisexual flowers, which are often large, are solitary or in simple or compound racemes. The floral axis and ovary are completely united, and above the calyx the axis forms a flat disc to which the stamens and petals (when present) are attached, and which is further variously developed inside the androecium. There are generally four to six free sepals which persist in the fruit, and an equal number of strongly imbricate petals which fall with the stamens. In the remarkable west tropical African *Napoleona* there are no petals. The androecium shews great diversity of development; the numerous stamens are arranged in several whorls, often partly sterile, the members of which are more or less united at the base; in *Napoleona* the outer whorl forms a conspicuous corona-like structure. In the bud the stamens are bent inwards; their unequal development may render the flower zygomorphic, as in *Couroupita* or *Bertholletia*. The ovary has generally two to six chambers, with 1–∞ anatropous ovules in each; the position of the ovules is very various, being ascending, horizontal or pendulous. The simple style bears a capitate or lobed stigma. Pollination is presumably effected by insect-agency or in some cases by humming-birds.

The fruit is drupaceous, berry-like or capsular; often very large, as in the woody fruits of *Bertholletia* (Brazil-nut) and *Lecythis*. In the latter the intrastaminal disc persists and forms a lid-like cover which separates when the fruit is ripe. The embryo completely fills the seed, and is sometimes, as in the Brazil-nut, undifferentiated, consisting mainly of the much thickened hypocotyl. The oily seeds of *Bertholletia* are the Brazil-nuts of commerce; those of species of the allied genus *Lecythis* (Sapucaia-nut) are also eaten; both are natives of tropical America.

Family IX. MYRTACEAE

A large tropical and subtropical family containing 73 genera and about 2750 species of woody plants with opposite exstipulate generally entire leathery evergreen leaves. They are generally moderate-sized or small trees or shrubs, but shew every variety from small creeping undershrubs to giant trees, as the Eucalyptus, some species of which attain from 400 to 500 feet in height with a girth at the base of 90 feet. In Australia, which is one of the chief centres of development of the family, many of the species shew adaptations to the dry climate and strong sunlight in their needle-like leaves with a triangular or circular section, or in the assumption of a vertical position of the leaf-blade by twisting of the stalk. Constant anatomical characters are found in the bicollateral vascular bundles, and the numerous spherical glands containing ethereal oil which arise lysigenously in the cortical parenchyma of the young stem, beneath the epidermis of the leaf, as well as in the floral organs and the fruit.

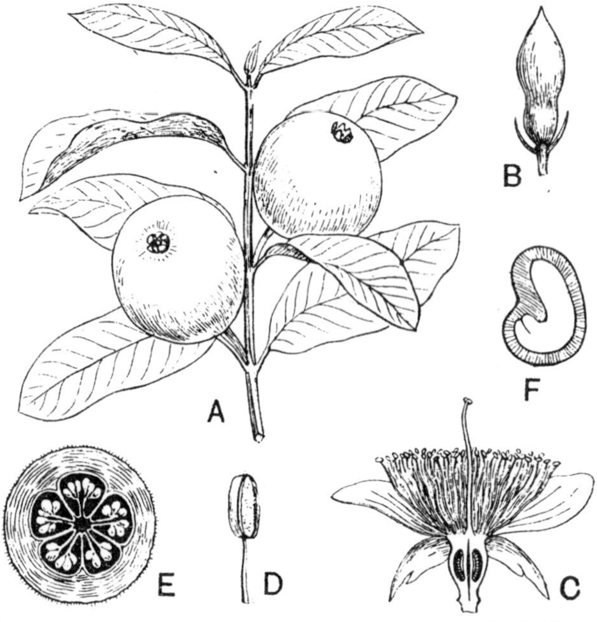

FIG. 188. *Psidium Guayava.* A. Branch with young fruit, × ½. B. Flower-bud, × 1½. C. Flower in vertical section. D. Stamen, × 15. E. Ovary cut across, × 8. F. Seed cut longitudinally shewing embryo, × 6. (After Niedenzu.)

The flowers are sometimes solitary in the leaf-axils, as in Myrtle (*Myrtus communis*) and Guava (*Psidium Guayava*); but generally in cymose, more rarely racemose, branched inflorescences. They are bisexual, regular, cyclic and generally epigynous (as in Myrtle, and *Psidium*, fig. 188), but union between receptacle and ovary is not always complete so that various degrees of perigyny occur. The most general arrangement of the flower is expressed by the formula S4–5,

P4–5, A∞, G(2–5). The sepals are generally free and imbricate with quincuncial arrangement; in some genera they are more or less united, and when the flower opens, may become regularly or irregularly torn off like a cap. In some cases they are much reduced, and in most species of *Eucalyptus* are inconspicuous or absent. The petals are generally free and more or less circular in form; in *Eucalyptus* they unite to form a cap which separates from the

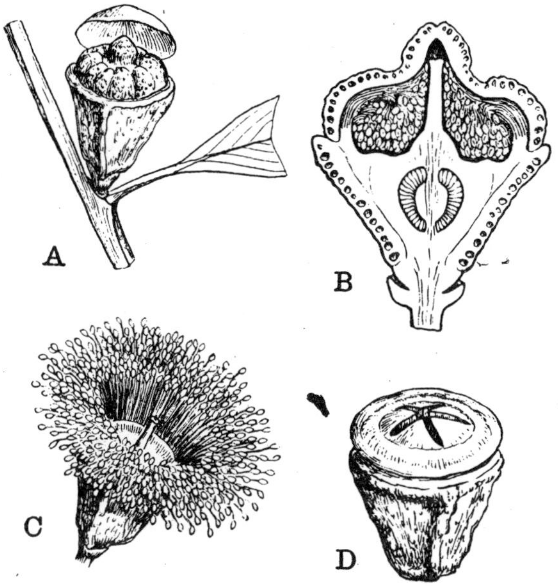

FIG. 189. *Eucalyptus globulus*. A. Flower-bud; the petals form a cap of two layers the outer of which has been raised. B. Flower-bud in vertical section, the outer petal-cap removed. C. Flower. D. Capsule splitting loculicidally in the upper portion. A, C, D slightly reduced; B somewhat enlarged.

receptacle when the flower opens; in *E. globulus* (fig. 189) the cap consists of an inner and an outer layer. The stamens are free and when numerous are arranged indefinitely in whorls on the edge of the receptacle. More rarely they are in obdiplostemonous whorls or groups isomerous with the corolla, or are reduced to a single whorl which is then generally antepetalous, as in *Melaleuca*. The anthers are usually dorsifixed and versatile, the cells dehiscing introrsely by longitudinal slits; the filaments are bent inwards in the bud (fig. 189, B). The simple style is long and generally kneed or flexuose in the bud; it ends in a simple capitate stigma. The ovary varies from one- to ∞-chambered, with 2–∞ generally somewhat obliquely pendulous anatropous or campylotropous ovules, situated on axile, rarely parietal, placentas.

The flowers are entomophilous. The conspicuous coloured androecium often serves as an organ of attraction; an excellent example is afforded by the Australian Bottle-brushes (*Callistemon*), where the numerous long filaments form a scarlet brush many times longer than the corolla. Several species are known in cultivation as greenhouse plants.

The fruit supplies valuable characters for subdivision of the family into subfamilies and tribes, as follows:

Subfamily I. MYRTOIDEAE. Fruit a berry or very rarely a drupe.

Tribe 1. *Myrteae*. Chief genera, *Myrtus*, *Psidium*, *Pimenta*, *Eugenia*.

Subfamily II. LEPTOSPERMOIDEAE. Fruit dry.

Tribe 2. *Leptospermeae*. Ovary several-chambered; fruit a loculicidal capsule. Chief genera, *Metrosideros*, *Eucalyptus*, *Callistemon*, *Melaleuca*.

Tribe 3. *Chamaelaucieae*. Floral axis produced above the ovary, which is unilocular, forming a dry, generally one-seeded fruit. Chief genera, *Calycothrix*, *Chamaelaucium*, *Darwinia*, *Verticordia*; mainly West Australian.

As a general rule only few or one of the ovules in the chamber or ovary develop into a seed. The testa is horny, as in *Myrtus* and its allies, or cartilaginous, leathery or membranous; in some species of *Eucalyptus* it is winged. The embryo which fills the seed is straight, or more or less bent or spirally rolled.

The family finds its chief development in two areas. Tropical America is the great centre of the berry-fruited *Myrteae*, and Australia of the other two tribes. They also occur generally throughout the warmer parts of the earth but fewer in number both of species and individuals. In S.W. and S.E. Australia Myrtaceae is the most prominent family, and the large forests of *Eucalyptus* represent a considerable portion of the wealth of West Australia. *Myrtus communis* (Myrtle) in the Mediterranean region is the only European representative; the genus (60 species) is one of the most widely distributed, occurring in the warmer parts of both hemispheres.

Eucalyptus is valuable on account of the hard wood of many of its species, their rapid growth, and the antipyretic property of the contained oil. They also yield a gum-resin. "Cloves" are the dried flower-buds of *Eugenia caryophyllata*, a native of the Moluccas. *Psidium Guayava* (Guava) and other species are generally cultivated in the tropics for their delicious succulent fruit.

Family X. MELASTOMACEAE

A considerable family (about 170 genera and 3000 species) occurring in the warmer parts of the earth, especially in South America. It is allied to Myrtaceae but shews striking distinctive characters, namely absence of oil-glands in the vegetative organs,

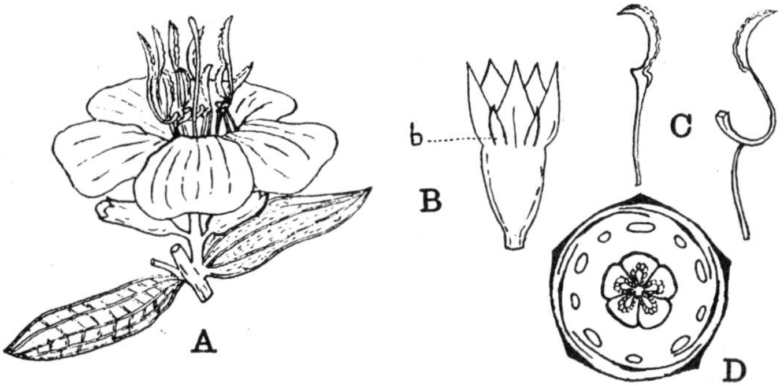

FIG. 190. A. Portion of inflorescence of *Dissotis incana*. B. Receptacle with calyx of same; *b*, stipular teeth of sepals. C. Two stamens of same. D. Floral diagram of *Tibouchina*. A slightly reduced; B, C enlarged. (A, B, C after Naudin; D after Eichler.)

prominent upwardly curving main leaf-nerves; and the anthers dehisce by apical pores and bear often remarkably developed appendages of the connective.

The plants shew great diversity of habit, including herbs, shrubs or trees, root-climbers, epiphytes and marsh or water plants. Aerial roots are formed in the root-climbers and epiphytes. On the roots of some of the latter (species of *Pachycentria*) are developed tuber-like swellings in which ants make a home. Loose air-containing tissue (aerenchyma) is formed in the submerged portions of the stem and roots of the marsh-inhabiting species. The vascular bundles have internal phloem, and concentric cortical and pith bundles also often occur. The leaves are opposite-decussate, rarely whorled and exstipulate. They are often large, and in some genera conspicuously spotted or variegated, as in species of *Sonerila* and *Bertolonia*. The two leaves of a pair are sometimes unequal (anisophylly), and this is frequently associated with the development in the larger leaf of sac-like outgrowths on the upper face which serve as shelters for ants; these *domatia* are entered from the lower leaf-face. They may occur also on the leaf-stalk. In return for shelter the ants protect the inflorescence from unbidden guests. The often showy flowers are arranged in cymose inflorescences. They are bisexual, regular or slightly zygomorphic, and generally 4- or 5-merous with two whorls of stamens; the sepals, petals and stamens are perigynous or epigynous on a tubular or bell-shaped receptacle. The sepals are sometimes reduced to a mere rim on the floral axis, or are united to form a cap; in other cases they are free and persistent. The petals are free, generally brightly coloured, and twisted to the right in the bud. The stamens are all alike, or alternately different in length and form, sometimes staminodial. The form of the appendage on the connective affords a useful character for distinguishing genera. The ovary is sometimes free, but generally more or less completely united with the sides of the hollowed receptacle; it bears a simple style and is generally chambered, with indefinite ovules.

Pollination is effected by aid of insects or in some cases presumably by humming-birds. The coloured bracteoles and red axis of the inflorescence serve in some cases as an attraction. The fruit is a capsule opening loculicidally, or a berry; the generally numerous small seeds are straight or bent and contain a very small embryo with fleshy cotyledons; endosperm is absent.

The berried fruits of epiphytic species are distributed by tree-loving birds, the seed being dropped in the excrement.

Species of *Medinilla*, *Centradenia* and others are grown as garden flowers in the tropics, or greenhouse plants in temperate climates; and species of *Gravesia*, *Sonerila* and *Bertolonia* for their variegated leaves.

Family XI. ONAGRACEAE

Flowers generally bisexual and regular, epigynous, the floral receptacle adhering completely to the ovary and being produced above it into the so-called calyx-tube which is often petaloid and sharply distinguished from the ovary from which it generally becomes separated after flowering; sepals, petals and stamens spring from its upper edge. Generally tetramerous with two whorls of stamens.

Generally herbs with alternate or opposite exstipulate leaves and axillary or racemose flowers.

About 500 species in 38 genera, in temperate and subtropical regions; a few in the tropics.

A few members of the family are annuals; such are species of *Epilobium*, *Clarkia*, *Godetia*. Others, such as *Oenothera biennis*, are biennial, forming in the first year a large taproot. The majority are, however, perennial. They are mostly herbaceous, sometimes becoming shrubby or arborescent (*Fuchsia*); *Fuchsia apetala* and allied species have a woody climbing stem.

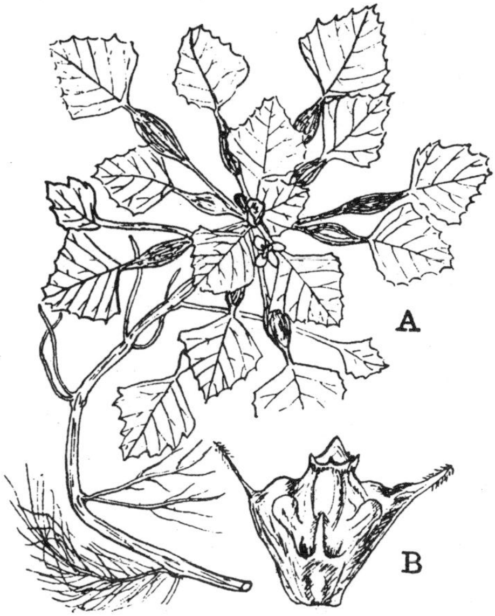

FIG. 191. *Trapa natans*. A. Upper part of plant. B. Fruit. Reduced. (After Wettstein.)

Jussieua and *Ludwigia* are herbaceous or shrubby marsh or water plants. The former, widely distributed in the tropics of both Worlds, contains aerenchyma in the cortex of the roots and submerged stems; frequently special respiratory roots are developed on the upper surface of the horizontal submerged stem and reach to the surface of the water.

Trapa (Water Chestnut) (three species, Central Europe to

Eastern Asia and tropical Africa) is a floating annual, with a rosette of leaves at the surface of the water supported by the inflated leaf-stalks; the submerged stem bears at the nodes pairs of finely-branched roots (fig. 191). The genus is anomalous in many respects in the family, and represents an ancient type which was more widely distributed in the Tertiary period and in later times had a wide range in North and Central Europe, indicated by partially fossilised fruits, as in the Cromer forest-beds.

In germination the cotyledons come above ground and function as the first green leaves of the plant. A number of species of different genera (*Clarkia*, *Oenothera*, *Epilobium* and others) are of interest from the fact that by intercalary growth, a large blade resembling the subsequently developed leaves in form, texture and hairiness is developed below the original cotyledon. The latter is then borne at the tip of the leaf thus formed, and is often delimited from the new growth by a constriction. A detailed account will be found in Lubbock's *Seedlings*, under the family Onagraceae. In *Trapa* the cotyledons are remarkably unequal, the larger, rich in starch, occupies the greater part of the large seed; the other is small and scale-like. In germination the larger remains in the seed while the smaller is carried up on the slender hypocotyl.

The leaves are simple and pinnately veined; they are generally undivided, with an entire or inconspicuously toothed margin; bundles of raphides occur in the tissue. They are alternate, opposite or whorled. *Circaea* has opposite leaves, but in *Epilobium* the same species may have them both alternate and opposite. They are generally exstipulate, but small caducous stipules occur in *Fuchsia*, *Circaea* and other genera. An anatomical feature is the presence of internal phloem in the stem. The flowers are often solitary in the leaf-axils, as in many Fuchsias, *Clarkia* and others; they are frequently associated, as in *Epilobium* and *Oenothera*, forming large showy terminal leafy spikes or racemes. The small white or red flowers of *Circaea* are borne in terminal and lateral racemes.

The flower is regular and formed on a tetramerous plan; S4, P4, A4 + 4, G$(\overline{4})$, may be regarded as the typical formula, of which *Epilobium* (Willow-herb), *Oenothera* (Evening Primrose) and *Fuchsia* are examples (fig. 192, A). In *Clarkia*, a

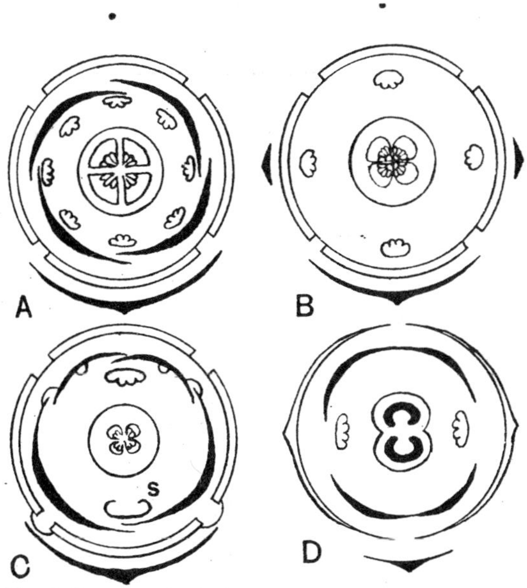

FIG. 192. Floral diagrams of: A, *Oenothera, Epilobium, Fuchsia*; B, *Ludwigia palustris*; C, *Lopezia*; *s*, staminode; D, *Circaea lutetiana*. (After Eichler.)

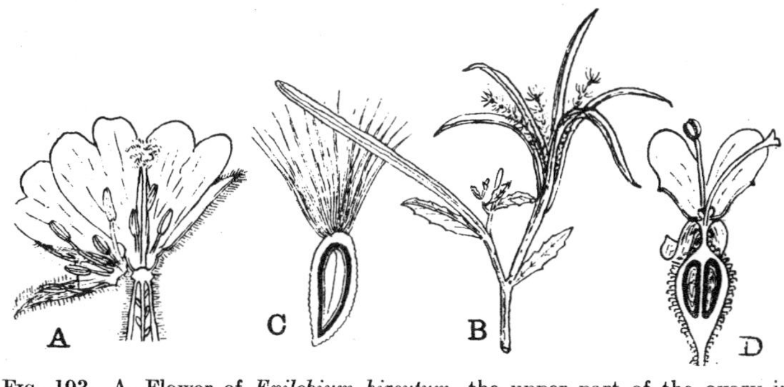

FIG. 193. A. Flower of *Epilobium hirsutum*, the upper part of the ovary in vertical section, the corolla cut open and turned back. B. Fruiting shoot of *E. parviflorum*. C. Seed of same in vertical section, enlarged. D. *Circaea lutetiana*, flower in vertical section, × 5. (After Raimann.)

western North American genus and a favourite herbaceous garden annual, the antepetalous stamens are often barren, while in the closely allied Californian *Eucharidium* they are absent, as also in *Trapa*. This is also the case in *Ludwigia* (fig. 192, B), a genus chiefly North American, but one species of which, *L. palustris*, is a very rare British plant, with minute green apetalous flowers, occurring in boggy pools in Sussex and Hants. *Circaea* (Enchanter's Nightshade) (fig. 192, D) shews still further reduction, having a dimerous flower, the parts of which spring from the top of a very slender stalk-like calyx-tube, which adheres to the base of the style (fig. 193, D). 3- and 5-merous flowers also occur in the family.

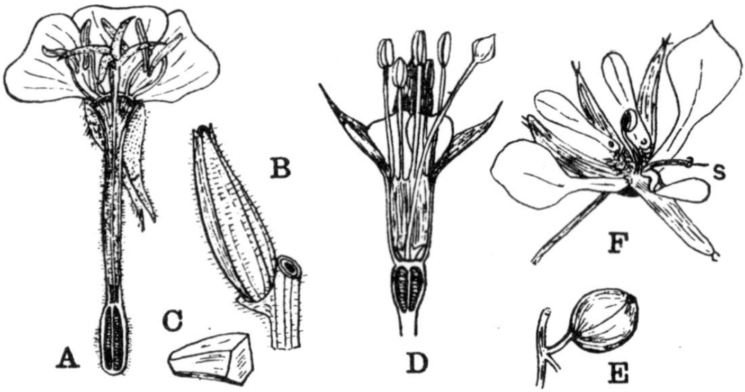

FIG. 194. A–C. *Oenothera biennis*. A. Flower in vertical section, × ½. B. capsule, × ½. C. Seed, enlarged. D. Flower of *Fuchsia magellanica* in vertical section, slightly reduced. E. Berry of *F. microphylla*, × ½. F. Flower of *Lopezia coronata*; *s*, staminode; × 3. (After Raimann.)

The sepals and petals are free; the former have a broad base, are valvate in aestivation and reflexed in the open flower; they may be petaloid in character, as in *Fuchsia*. The petals are generally convolute with a narrow insertion and are entire (many Fuchsias), or bilobed, as in most species of *Epilobium*, *Clarkia elegans* and others, or trilobed as in *Clarkia pulchella*. In some Fuchsias they are small and scale-like, in others absent (as in *F. apetala*). The stamens are free; those of the inner (antepetalous) whorl are generally shorter than those of the outer (antesepalous). The Central American genus *Lopezia* has only one fertile stamen, in the posterior plane, opposite which is a petaloid staminode (figs. 192, C, 194, F). The anthers are often divided into two

(*Circaea*) or several (*Clarkia*, etc.) stories by transverse plates of parenchymatous tissue. The pollen-grains are generally large and spherical, have three pores with protruding stoppers and are connected by viscid threads. In *Epilobium* they remain united in tetrads. The ovary is typically 4-celled with axile placentas and numerous ovules; the septa are often incomplete. *Circaea* has a 1–2-celled ovary with one ovule in each cell. The style is simple, generally long and filiform, with the stigma capitate (*Fuchsia*), or 4-rayed (species of *Epilobium*, *Oenothera*) or notched (*Circaea*). The ovules are anatropous, and ascending, horizontal or descending. *Trapa* is distinguished from the rest of the family by its half-inferior ovary, which is moreover bilocular with a pendulous anatropous ovule in each chamber.

The pendulous flowers of some Fuchsias are wind-pollinated, but most Onagraceae are adapted for insect-visits, chiefly of bees and Lepidoptera; they are often pollinated by night-fliers, in which case the flower is pale and opens towards evening (as Evening Primrose). Nectar is secreted by a swollen disc at the base of the style, or by nectaries situated on the lower part of the calyx-tube. The flowers are often proterandrous.

The fruit is generally a capsule splitting loculicidally, and leaving a central column bearing the seeds, as in *Oenothera* and *Epilobium*; in the Willow-herbs dissemination is effected by a long pencil of hairs at the broader (chalazal) end of the seed. *Fuchsia* has a berry, while in *Circaea* the indehiscent 1–2-seeded fruit is nut-like, and bears numerous recurved bristles. In *Trapa* it is a large somewhat top-shaped one-seeded drupe the fleshy layer of which soon disappears leaving the stony endocarp bearing two to four upwardly-projecting thorns (the persistent sepals) which are often barbed at the tip; the barbs represent strands of sclerenchyma which remain after disappearance of the soft tissue (fig. 191, B). The seeds are exendospermic.

The family finds its chief development in the temperate zone of the New World, especially on the Pacific side. The most widely distributed genus is *Epilobium* (Willow-herb), the seeds of which are eminently adapted for distribution by means of the long chalazal tuft of hairs; it contains nearly 200 species (12 British), and numerous hybrids, spread over the whole world with the exception of the tropics. The next largest genus is *Oenothera* with about 100 species in North and South America. *O. biennis* (Evening Primrose), a North American species, is less commonly naturalised in Europe than *O. Lamarckiana*, which is of special interest from its association with De Vries's work on Mutations. *Fuchsia* has 60 species in South and Central America and a few in New Zealand. *Circaea* is a small genus (seven species); *C. lutetiana* (Enchanter's Nightshade) is common in the north temperate zone in damp woods; a second species, *C. alpina*, which grows in hilly districts on the western side of Britain, extends to within the Arctic circle and to the mountains of Southern India. Several of the genera are of horticultural interest. *Fuchsia* (*F. coccinea*) was introduced from South America in 1788. The numerous garden forms are mostly hybrids in which *F. fulgens* plays an important part. The half-hardy shrubby species are varieties of the Chilean *F. macrostemma*, a shrub 6–12 ft. high with a scarlet calyx. The berries of some species are edible. *Oenothera, Clarkia* and *Godetia* are well-known border plants.

Family XII. HALORAGACEAE

A small but widely distributed family most nearly related to Onagraceae, but differing in anatomical characters (absence of internal phloem and raphides) and also in the one-celled ovary-chambers, free styles and copious endosperm in the seed. Adaptation to wind-pollination and a more or less aquatic life have been factors in its development. The plants are herbs with remarkable differences of habit more or less associated with the mode of life. They are chiefly water- or marsh-inhabiting, but land plants also occur. Our British genus *Myriophyllum* (Water Milfoil) comprises marsh or aquatic herbs, with generally whorled, but sometimes opposite or alternate, finely pinnatisect leaves. *Haloragis* consists of damp-loving herbs often reaching a considerable size (3 ft.), with alternate or opposite leaves which vary from linear to broadly cordate. Several species of *Gunnera* are enormous water-side herbs, with a subterranean or a short stout terrestrial stem, bearing large spirally arranged long-stalked leaves with a broad cordate or reniform blade. *G. chilensis* and other species grow on rocks; the former is a handsome plant known in cultivation as *G. scabra*. The leaves are generally exstipulate, but in *Gunnera* stipules are represented by intravaginal scales, several rows of which stand in the axil in front of the leaf-base; in *G. magellanica* they are ochreate. The stem of *Myriophyllum* shews characteristic aquatic structure—a large cortex traversed by numerous air-containing intercellular spaces, and a central axis limited by a well-marked endodermis in which the few vascular elements are arranged in bundles isomerous with the leaf-traces. The short stout stem of *Gunnera* is polystelic, consisting of a parenchymatous ground-tissue through which run numerous vascular strands. The stem affords an interesting case of symbiosis from the fact that colonies of a species of *Nostoc* (*N. Gunnerae*) penetrate and become enclosed within the outer layers of the cortex. The flowers are inconspicuous, with small free sepals; the petals when present are free and generally larger than the sepals, and have an imbricate

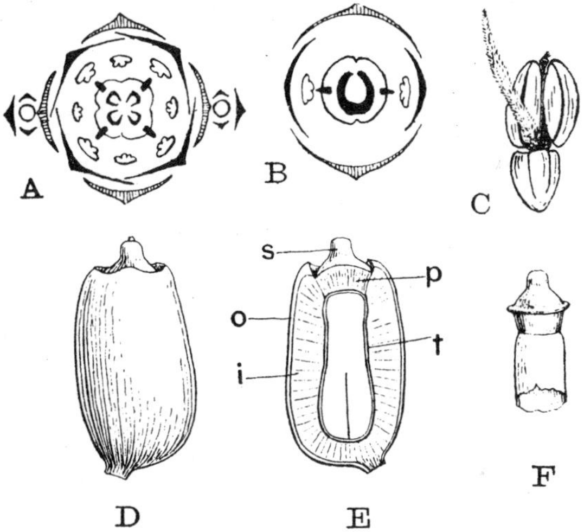

FIG. 195. A. Floral diagram of *Haloragis erecta*. B. Floral diagram of *Gunnera petaloidea*. C–F. *Hippuris vulgaris*. C. Flower, × 10. D. Fruit, × 15. E. Same cut lengthwise; *s*, remains of style; *o*, outer layer; *i*, hard inner layer of pericarp; *t*, testa surrounding the large embryo; *p*, plug in pericarp which becomes pushed out on germination, shewn in F with portion of membranous testa attached. (A, B after Eichler.)

or valvate aestivation; the stamens are free, with short filaments and normal anthers dehiscing longitudinally. Except in *Gunnera* the number of ovary-chambers corresponds with the number of the carpels; and there is a pendulous anatropous ovule, with two integuments, in each cell. In *Haloragis* and *Myriophyllum* the flowers are solitary or in dichasia in the leaf-axils, sometimes forming spikes, racemes or panicles by the reduction of the upper leaves to bracts. Each flower is subtended by a pair of bracteoles; in *Gunnera* they are ebracteate and crowded in large inflorescences, that of *G. insignis* from Costa Rica forming a pyramid 7 ft. high with a base 3 ft. broad. In *Haloragis* (fig. 195, A) the flower conforms to the common Onagraceae type, S4, P4, A4 + 4, G($\bar{4}$), the androecium being obdiplostemonous and the flowers bisexual or unisexual, while in *Myriophyllum* they are formed on the same plan but are generally monoecious or polygamous, the upper being male and the lower female, with the intermediate frequently bisexual. Suppression of a whorl of stamens also occurs in *Myriophyllum* and in the tropical genus *Serpicula*, while *Gunnera* (fig. 195, B) represents a further reduction, having a floral formula S2, P2, A2, G$(\bar{2})$, the ovary being moreover unilocular with a solitary ovule.

The inconspicuous flowers, with no nectar and with large often feathery stigmas, suggest wind-pollination. The fruit is a nut or drupe; the seed contains endosperm, often fleshy in character, surrounding the embryo, which, except in *Gunnera*, is straight and cylindrical. The embryo in *Gunnera* is minute, obcordate, and buried in endosperm close to the upper end of the seed.

The family contains seven genera, and about 160 species, and finds its chief development in Australasia to which two small genera are limited; *Haloragis*, the largest and richest in forms (60 species), is Australasian and Antarctic, a very few species passing into South-eastern Asia and one ascending into South America as far as Chile and Juan Fernandez. *Myriophyllum* (36 species) has a world-wide distribution; *Gunnera* (35 species) is almost confined to the southern hemisphere, especially South America. Of the two other genera, *Proserpinaca* is confined to North America, *Serpicula* is distributed through the tropics of both worlds.

Family XIII. HIPPURIDACEAE

This family comprises one genus, *Hippuris*, the affinity of which is very doubtful, represented by one species, *H. vulgaris* (Mare's-tail), widely distributed through Arctic and temperate regions and occurring also in Antarctic America. It has been included in Haloragaceae of which it was regarded as representing a very reduced floral type, but, if we except the aquatic habit, it has nothing in common with that family. In addition to the floral characters, a single stamen and carpel, Schindler(1) points out that the stem is a sympodium, in contrast with the monopodial development in the other family, and suggests an affinity with Santalaceae. Juel(2), who has recently worked out the development of the flower and fruit and the embryology, doubts whether the genus is rightly placed among the Archichlamydeae, but is unable to suggest any affinity with the Sympetalae.

The plant is an aquatic herb with a stout creeping submerged rhizome from which springs the stout erect many-jointed stem bearing rather closely set whorls of narrow entire sessile leaves. The minute solitary green flowers are sessile in the leaf-axils, and are borne on the upper portion of the stem, which projects from the water. They are typically bisexual, but sometimes unisexual by failure of the stamen or pistil to develop. The perianth is reduced to a rim round the top of the ovary; the single stamen is placed medianly in front of the long slender papillose style (fig. 195, C). The flowers are proterogynous and wind-pollinated. The one-celled ovary contains a pendulous anatropous ovule, which has been described as naked, but which Juel shews to have a single integument the micropyle in which becomes completely closed; above the micropyle, and lying upon the integument, is a development of the funicle which recalls the obturator of Euphorbiaceae and other families. The course of the pollen-tube is unusual (Juel), passing through the funicle and integument and penetrating the embryo-sac laterally.

The fruit has a thin outer coat and a thick hard inner coat; the portion beneath the persistent base of the style is scarcely thickened, but the gap becomes filled with a hard development of the funicle and seed-coat which forms a stopper and is ultimately pushed out by the radicle in germination (fig. 195). The seed-coat is otherwise a thin membrane. Juel points out that when the fruit is apparently ripe several layers of endosperm still remain around the large straight embryo but absorption of the endosperm continues after the fall of the fruit until a few layers only are left round the sides of the embryo which are with difficulty distinguished from the seed-coat.

REFERENCES

(1) SCHINDLER, A. K. "Die Abtrennung der Hippuridaceen von den Halorrhagaceen." Engler's *Bot. Jahrb.* XXXIV, Beiblatt 77 (1904).

(2) JUEL, H. O. "Studien über die Entwicklungsgeschichte von *Hippuris vulgaris*." *Nova Acta R. Soc. Sci. Upsaliensis*, ser. IV, 2, No. 11 (1911).

Order 16. *OPUNTIALES*

Family. CACTACEAE

Flowers bisexual, regular or becoming zygomorphic by curving of the perianth, stamens and pistil. Sepals and petals not clearly distinguished, generally indefinite and more or less united to form a tube, sometimes free. Stamens indefinite, united to the perianth-tube, rarely springing from the floral axis. Ovary inferior, one-celled with several parietal placentas bearing very many ovules on long funicles with two integuments; style simple, stigmas corresponding in number with the placentas. Fruit generally a fleshy one-celled berry with numerous seeds. Embryo straight or curved; endosperm present or absent. Succulent plants with fleshy columnar, conical or subglobose angular or cylindrical often ribbed and warted stems, which are simple or branched; more rarely with thinner cylindrical, sometimes rope-like, or broader leaf-like joints; very rarely leaf-bearing plants of normal appearance. Leaves usually reduced to scales, sometimes developed and flat or cylindrical in shape; generally falling soon; leaf-axils often with spines and more or less hairy. Flowers axillary or at the end of the tubercles; usually large and showy; solitary or few, rarely forming a panicle.

The affinity of the family is very doubtful. It has been classed with Aizoaceae (Ficoideae); Bentham and Hooker (*Genera Plantarum*) included the two families in one cohort Ficoidales, which was placed after Passiflorales and before Umbellales. Aizoaceae have however been regarded as more suitably placed among the Centrospermae and Wettstein also includes Cactaceae in that order. Engler, while retaining the order Opuntiales in the Dialypetalae near Myrtiflorae, admits an affinity, following K. Schumann, with Aizoaceae. The latter are mainly inhabitants of warm dry countries with a succulent habit in which the leaves are remarkably developed; the floral structure is, except in the large genus *Mesembryanthemum*, very different from that of Cactaceae. In *Mesembryanthemum* the ovary is inferior and the flower bears a strong superficial resemblance to that of a Cactus; but a comparison of the two shews important differences, as in the relation between the perianth and androecium (see Aizoaceae), the numerous styles of *Mesembryanthemum*, its many-celled generally capsular fruit, and the farinaceous endosperm in the seed. On the other hand, the general plan of structure of the flower of Cactaceae suggests a comparison with the Myrtiflorae.

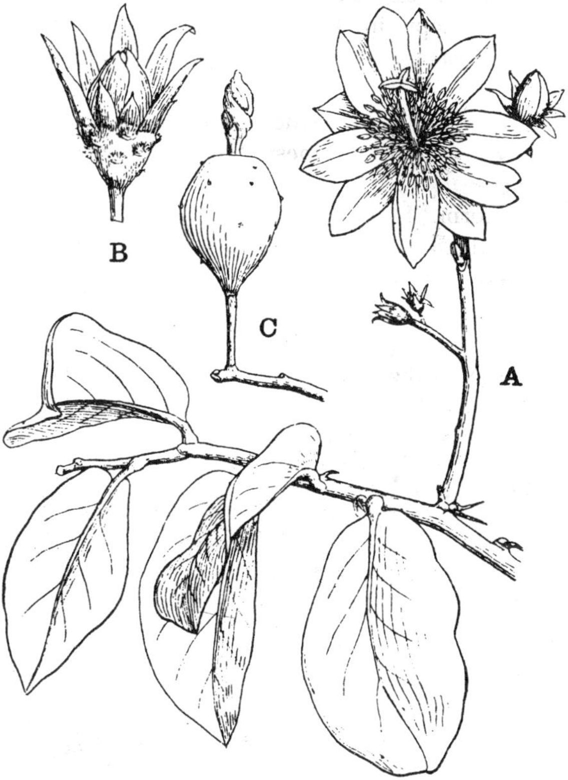

FIG. 196. *Pereskia aculeata.* A. Flowering branch, slightly reduced. B. Flower-bud, × 2½. C. Fruit, slightly enlarged. (A, C after Britton and Rose.)

The plants shew remarkable adaptation to desert conditions, which finds expression in a condensed form adapted to storage of water and protection from evaporation. The seeds germinate rapidly; germination may begin in the fruit, as in *Phyllocactus*. The seedling either bears fleshy cotyledons, as in *Opuntia*, the stem becoming fleshy and leafless immediately above them, or the cotyledons are undeveloped, as in *Mamillaria*.

Pereskia approaches a normal dicotyledonous growth, having a much-branched woody stem with broad, more or less fleshy leaves (fig. 196); in *Opuntia* the narrow fleshy leaves sometimes persist but generally fall early, and in the remaining genera the leaves are either very small or undeveloped. Except in the leafy Pereskias the work of assimilation is carried on by the simple or branched flattened or columnar stem, the surface of which is increased by the development of tubercles. One of the most striking features is the development of spines which are borne upon the tubercle, generally at the top, and shew remarkable variety in size and form. The spine-bearing area at the tip of the tubercle or along the ridges or edges of the columnar or flattened stem is known as the areole (figs. 197, 198, 199).

Various interpretations have been given as to the morphology of the tubercle and the spines; the general course of development suggests the following view. The lateral shoot originates soon after the leaf, not in the leaf-axil but upon the leaf-base; and leaf-base and axillary shoot form one structure, the tubercle. The tubercles may unite in vertical rows to form ribs, in the formation of which the stem may take part (figs. 198, 199). The variation in height and form of tubercle and rib determines the amount of assimilating surface. The spines correspond to leaves and are borne dorsiventrally at the growing point of the lateral shoot, which is sunken and protected by hairs and is either carried up entire by the growth of the tubercle and comes to stand finally at the tip (*Opuntia*, *Cereus*, *Echinopsis*) or may split into two parts (species of *Echinocactus*, *Mamillaria*), one part going up on the tubercle and producing spines, the other remaining behind, in or near the leaf-axil, to produce a branch or flower.

The presence of transitional structures between spines and leaves, found in *Opuntia* and *Echinopsis*, indicates the leaf-character of the former. Darbishire (2), however, working on *Mamillaria elongata*, regards the tubercle as representing only the leaf-base, and the spines as modified portions of the leaf-blade; and Rudolph (5) shews that the spines in *Opuntia missouriensis* arise in the leaf-axil and from the epidermis, so that in this case they are trichomes. K. Schumann (6), partly owing to the large numbers which are produced on the shoot, was unable to compare the spines with leaves, and regarded them as emergences.

In *Opuntia*, in addition to the spines, slender barbed bristles (glochidia) are formed on the inner side of the apex of the axillary shoot, which are homologous with the spines

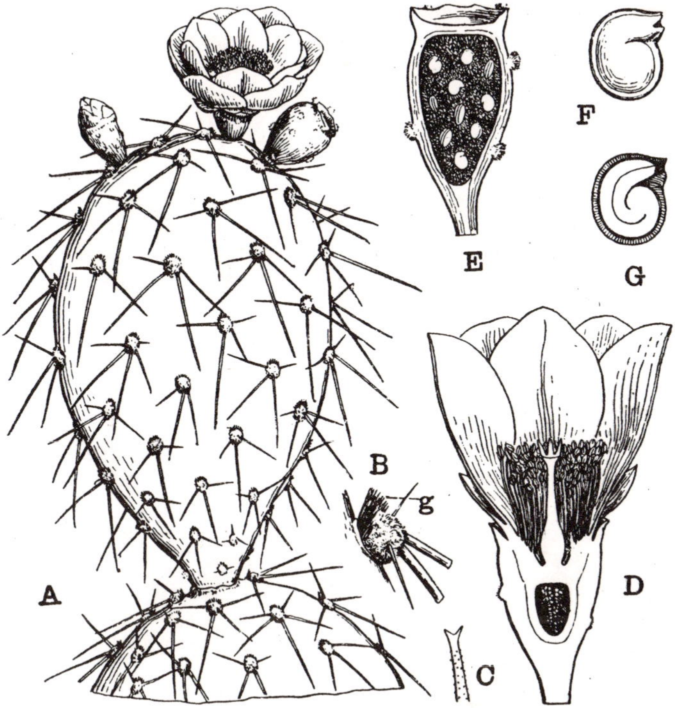

FIG. 197. *Opuntia Tuna.* A. Segment of stem bearing a flower-bud, open flower and young fruit, × ½. B. Areole shewing the spines embedded in wool and the glochidia, *g*, slightly enlarged. C. Upper part of a glochidium, enlarged. D. Flower in longitudinal section, slightly enlarged. E. Fruit in longitudinal section shewing the seeds embedded in pulp, slightly reduced. F. Seed, × 3. G. Seed in longitudinal section shewing the curved embryo.

(fig. 197, C). In the section *Cylindropuntiae* each spine is commonly covered with a thin sheath, formed by the union of a layer of hairs. In some cases the spines are weak or even hair-like (*Pilocereus*), and in species of *Mamillaria* the epidermis develops hairs so that the spine becomes feather-like. In others, as in *Leuchtenbergia*, they are flat and papery. In species of *Pereskia* some are modified to form hooks for climbing. In species of *Opuntia*, *Mamillaria*, *Cereus* and others, Ganong (3) found among the spines glands which exude nectar; these were shewn to be metamorphosed spines.

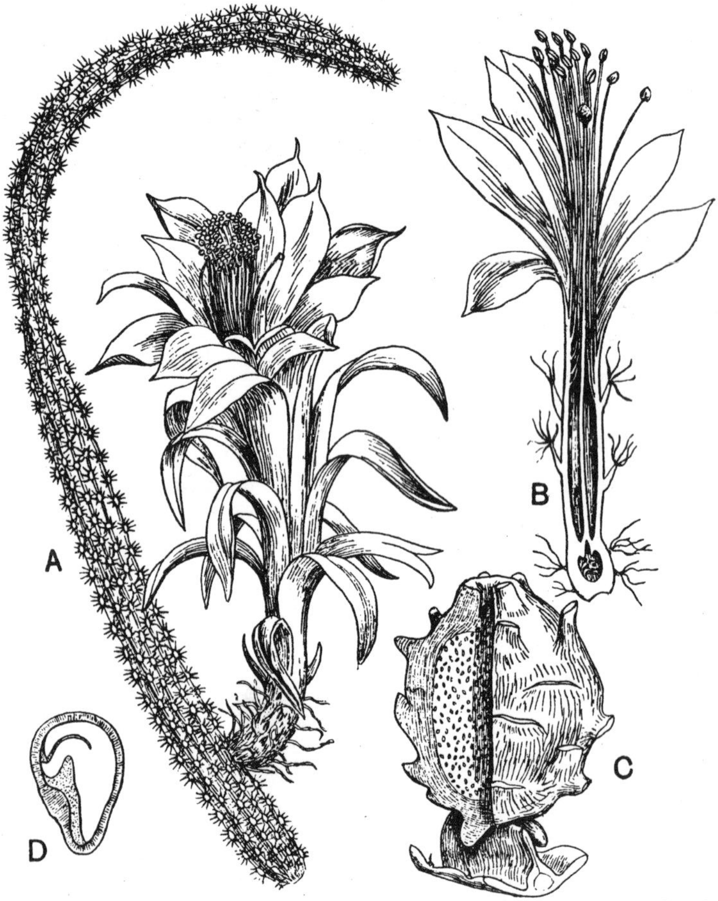

FIG. 198. *Cereus flagelliformis*. A. Shoot bearing a flower, slightly reduced. B. Flower in vertical section. C. Fruit of *C. triangularis*, cut open and shewing the numerous small seeds embedded in pulp, $\times \frac{1}{3}$. D. Seed of *C. grandiflorus*, enlarged. (A and B after Trew; C after K. Schumann.)

The great variety in the size, form and arrangement of the spines suggests a variety in function. Apart from a protective character and as a means for reducing transpiration, it has been shown (Darbishire(2)) that in *M. elongata* the set of spines by which the tubercles are crowned acts as a screen protecting the underlying tissues from the strong sunlight; a similar function is performed by the set of hairs at the top of the leaf of *Mesembryanthemum stellatum*.

The xerophilous character of the plants is also expressed in their anatomy. The thick epidermis has a strong cuticle with sunken stomata; beneath is a collenchymatous hypoderma, and below this the chlorophyll-containing tissue. There is a great development of pith and cortex, consisting of large round pitted water-storing cells, often containing mucilage. The fibro-vascular system conforms closely to the external form of the stem; the xylem consists of ringed and spiral tracheids, and there are no annual rings. All the tissues retain for a long time a capacity for growth, a fact which explains the ease with which the plants can be reproduced vegetatively or shoots be grafted on stems of another species.

The flowers are produced rapidly during or at the close of the rainy season, and shew no special adaptations to a dry climate. Branched inflorescences occur in *Pereskia*, where a terminal panicle is produced, and also frequently lateral panicles from the axils of the upper leaves. In the great majority of Cactaceae the flowers are solitary and spring from the middle of the upper part of an areole or behind it. In species of *Rhipsalis* several flowers develop successively from an areole. The ovary is inferior and more or less deeply sunken in the flower-bearing stem (fig. 197, D). It contains a large chamber lined by thickened placentas, which correspond in number with the stigmas. The surface often bears leaf-scales, which may be present in such quantity as to form a closely imbricating covering (e.g. in *Echinocactus concinnus*); as in the case of the leaves on the vegetative stem they often bear in their axils hairs, bristles, or even spines (figs. 196, 197, 198). The homology of the exterior covering of the ovary with the stem is especially plain in cases where, as in *Pereskia Bleo* and often in *Opuntia*, new flowers or leaf-shoots spring from the leaf-axil. D. S. Johnson(4) has studied *O. fulgida* with special reference to the perennation and proliferation of the fruits. Fruits of some Opuntias are known to remain attached to the plant and to grow actively for several years; in *O. fulgida* this condition is carried still farther, for not only does the fruit remain on the tree without shedding its seed, but both the ripe fruit and the unripe ovary may give rise to flowers and another crop of fruits. Four or five generations of fruits may thus be formed on a plant during one season. When a ripe fruit falls to the ground, it may put out adventitious roots and shoots, and so produce a new plant. The fleshy joints of the stem readily break off, sprout and form new plants, strongly resembling those produced from the end of the stem enclosing the sunken ovary. The numerous ovules are generally attached to a long funicle; several or many of the funicles become united below forming a much-branched bundle, which springs from the placenta. The ovules are anatropous and usually curved towards the funicle, which becomes broader near the micropyle and envelops it in a fold. The sides of the funicle often bear upwardly directed papillae, which probably serve as conductors of the pollen-tube. In *Opuntia* and the closely allied *Nopalea* the ovules are enclosed in a development of the micropylar fold of the funicle.

The floral envelope does not shew a distinction into calyx and corolla. It is radial in erect flowers, but tends to zygomorphy in oblique or horizontal flowers. In the latter the generally very elongated perianth-tube and the stamens become curved. An extreme case is found in *Epiphyllum*, where the flower is two-lipped. The number of the floral leaves is generally very large, and they usually unite below to form a long tube (fig. 198). In *Rhipsalideae* (fig. 200) and *Opuntieae* (fig. 197) the tube is very short and the floral leaves may be almost or quite free.

The lower floral leaves often resemble sepals, being green and having a thicker consistence than the upper more delicate and white or coloured leaves. In *Echinopsis* and many species of *Echinocactus* the perianth-tube is covered with overlapping scales, which may bear bundles of hairs and long spines in their axils as in the case of the ovary; the scales become larger above and pass into the petaloid leaves.

The larger sepaloid and petaloid leaves do not bear hairs or spines in the axils (see also, fig. 198). The stamens are always more than ten, generally the number is very large. They usually spring from the perianth-tube, arranged either in a

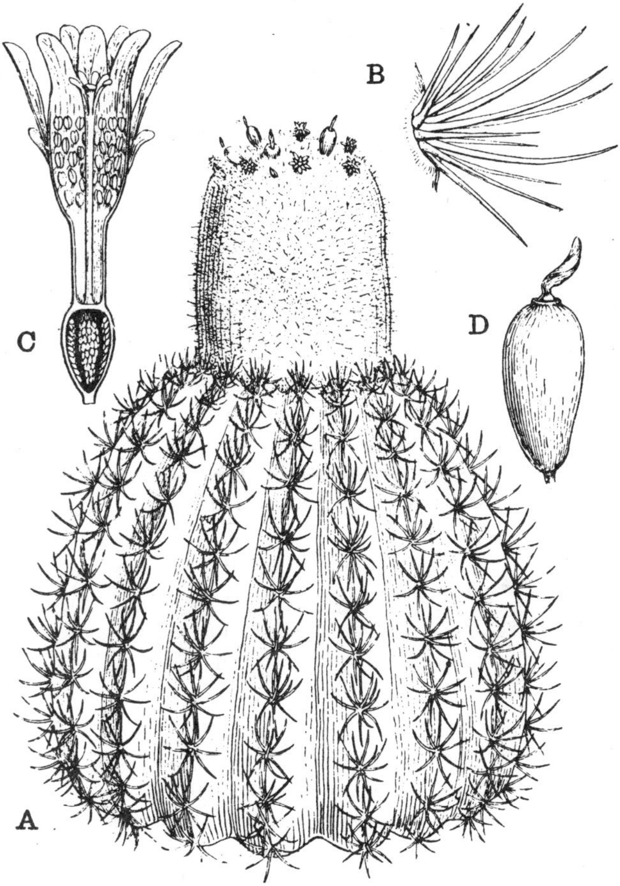

FIG. 199. *Melocactus communis.* A. Plant bearing flowers and fruit, × $\frac{1}{5}$. B. A single areole, slightly reduced. C. Flower in longitudinal section, × $2\frac{1}{2}$. D. Fruit, slightly enlarged.

regular spiral or in groups; often a ring of shorter stamens surrounds the throat of the tube. In *Epiphyllum truncatum* an inner series of about twenty stamens springs from the receptacle and these are united to form a tube, from the throat of which a membrane projects on the inside, forming a protection for the nectar. The anthers are attached at the base or on the back. When two groups of stamens are present the anthers of the lower are usually somewhat larger than those of the upper. The two-chambered anther opens longitudinally, either towards the inside or on the side. The pollen-grains are very small, yellow, spherical, with three small germ-pores, and usually smooth. The long slender style ends in two to many, usually thick, soft papillose stigmas.

We have little knowledge of the biology of the flower. The large, bright-coloured, often sweet-scented flowers suggest adaptation to the visits of insects. They are proterandrous; when the anthers open the stigmas are appressed, but separate later and become erect or spreading and overtop the stamens. Many hybrids have been formed by artificial pollination; as, for instance, in the genus *Phyllocactus*.

The fruit is usually a fleshy berry, except in *Echinocactus*, where it is often dry. The soft flesh is derived from the outer wall only, as in *Pereskia*, *Opuntia* (fig. 197, E) and *Rhipsalis*, or the funicles swell and form a considerable part of the soft flesh as in *Cereus* (fig. 198, C) and *Mamillaria*.

The fruits are often sweet and pleasant to the taste; a fact which doubtless conduces to their distribution by animals. Those of the epiphyte *Rhipsalis Cassytha* (fig. 200) resemble Mistletoe-berries in appearance, and contain a mucilage which facilitates the distribution of the seeds and their attachment to branches of trees.

The interior of the ripe fruit contains a homogeneous mass enclosing the usually numerous seeds (figs. 197, E, 198, C). These vary in form, being either flat or roundish, quite smooth or pitted or warted. The coat is usually black and shining; in *Opuntia* (fig. 197, F) and *Nopalea* pale or brownish with a more or less evident flattened margin. The embryo may be curved or spirally rolled and often associated with a greater or less amount of fleshy endosperm, especially in those species which have true leaves or flat stem-joints; in species with cylindrical or spherical stems it is generally straight, and often shews little or no differentiation and there is scarcely any endosperm.

The chief centre of distribution is in the dry districts of the warmer parts of America, especially of Mexico, including the temperate and colder part, as well as the neighbouring

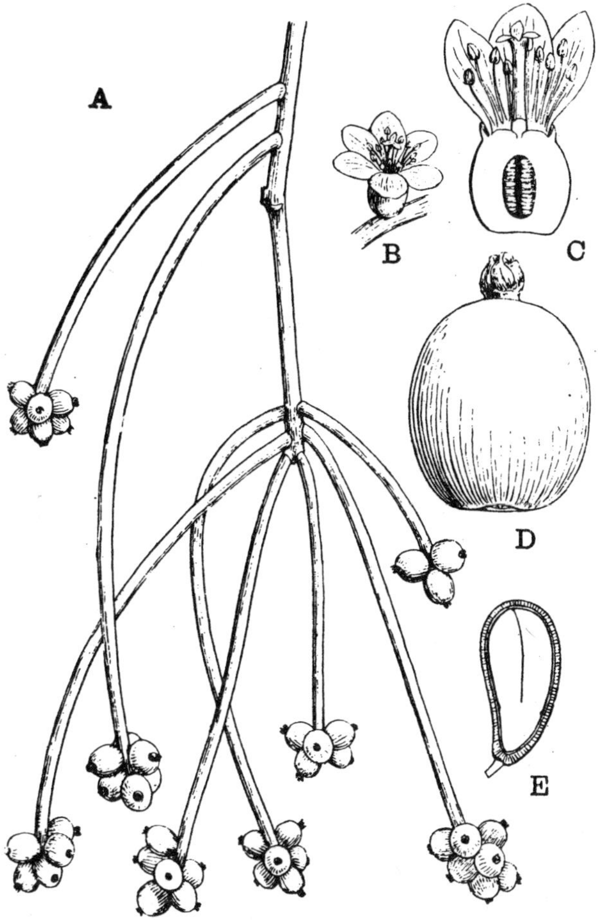

FIG. 200. *Rhipsalis Cassytha.* A. Portion of plant bearing fruits at the ends of branchlets, slightly reduced. B. Flower, × 2½. C. Flower in longitudinal section, × 5. D. Fruit, × 5. E. Seed in longitudinal section, × 20.

portion of the United States. The dry campos of the interior of Brazil are also the home of many species, and the Andine district contains some characteristic forms. In North America *Opuntia* penetrates as far north as 59° lat., and in South

America the family is represented as far south as Patagonia. They also reach a considerable elevation on the mountains, 3000 to 4000 metres on the Andes and the mountains of Mexico. On the mountains of Colorado species of *Opuntia* and *Echinocactus* are exposed during the whole year to frost. In cultivation in Europe many species last through the winter out-of-doors. Several genera occur in the West Indies. The epiphytic genera *Rhipsalis*, *Epiphyllum* and *Phyllocactus* occur in the tropical American forests, especially in Brazil. The family is represented in the Old World by a few species of *Rhipsalis* in tropical Africa, Madagascar and the Mascarene Islands, including *R. Cassytha*, which is widely distributed in tropical America. Species of *Opuntia* have long been naturalised in Southern Europe and the warmer parts of the Old World, and have spread so widely as often to give the appearance of true natives; in Australia and South Africa they have become a pest owing to their rapid vegetative propagation. Species of *Cereus* are grown in Mexico, and elsewhere in the warmer parts of America, for their sweet fleshy fruits, as is the prickly pear (*Opuntia Ficus-indica*) in the south of Europe. The latter is also often grown as a hedge. In the dry parts of Mexico, the southern United States and Brazil the large Echinocactuses and the Opuntias are used as water-containing fodder for cattle. The cultivation of the cochineal insect on *Nopalea coccinellifera*, *Opuntia Tuna* and other species was formerly of very considerable commercial importance.

Owing to lack of knowledge of the flower and fruit of many of the species and the difficulty of preserving the plants for comparative study in the herbarium, the systematic arrangement of the genera has presented much difficulty. It has been necessary to make more use of vegetative characters than is usually done in defining the various groups. A further difficulty is presented by the numerous transitional forms between the different types of vegetative structure. In recent years Britton and Rose (1) have made a detailed study of the family, based as far as possible on living material. While maintaining the same three main groups as those adopted by K. Schumann in the *Pflanzenfamilien* (6), they take a much narrower view as to the limitations of the genera, of which they recognise about 100 as compared with 23 recognised by Schumann. The genera fall into three subfamilies of very unequal size:

I. Pereskioideae. Plants with broad flat leaves and flowers stalked and generally panicled. One genus *Pereskia*, 19 species, from Mexico and the West Indies southwards to Paraguay and Argentina.

II. Opuntioideae. Succulent plants usually much branched, with round or flat leaf-like stem-joints; leaves small and generally fugacious; glochidia present. Flowers sessile, generally rotate. Cotyledons large. Britton and Rose recognise seven genera and at least 300 species, more than 250 of which belong to *Opuntia*. The subfamily is more closely related to the Pereskioideae than to the Cereoideae; it is widely distributed in America and the West Indies.

III. Cereoideae. Succulent plants, terrestrial or epiphytic; stems simple and one-jointed or much branched and many-jointed; joints globular to cylindrical, columnar or flattened, sometimes winged or leaf-like, often strongly ribbed or tubercled. Leaves usually wanting on the stem, sometimes small and scale-like, but usually present as scales on the ovary or perianth-tube. Spines usually present except in the epiphytes. Flowers sessile, generally with a perianth-tube. Cotyledons usually very small. This subfamily contains most of the genera and more than three-fourths of the species of the family and shews a much greater variety of structure in both stem and flower than the two preceding subfamilies; the genera are grouped in eight tribes. Principal genera: *Cereus*, *Phyllocactus*, *Epiphyllum*, *Echinopsis*, *Echinocereus*, *Echinocactus*, *Melocactus*, *Leuchtenbergia*, *Mamillaria*, *Rhipsalis*.

Pereskia is undoubtedly nearest to the original stem-form of the family, and *Opuntia* and the allied genera have diverged from it though still closely connected with it. Ganong(3) has depicted in tree-form the derivation of the Cereoideae from *Opuntia*, regarding the columnar species of *Cereus* as nearest to the primitive form of the family. His view on the phylogeny of the genera is based partly on a comparative study of the embryos and seedlings of the family, from which he concludes that there is on the whole a progressive condensation in bulk and diminution of surface effected by the increasing approach to a spherical form of the hypocotyl and diminution of the cotyledons, which is accompanied by a condensation in the adults due to increasing adaptation to a desert habitat. "There is however this important difference to be noted between the two series, that in the progressive condensation the embryos lag behind the adults, and this is plainest in the diminution of the cotyledons. Thus in *Opuntia* the leaves are very evident in both adults and embryos, but even here the cotyledons have diminished from those of *Pereskia* less than have the leaves of *Opuntia* diminished from those of *Pereskia*: in *Cereus* this is yet more evident, for here the cotyledons are still large and broad, while the leaves of the adults are reduced to small scales and often are but tiny rudiments. In *Echinocactus* the cotyledons are still somewhat prominent, but the leaves of the adults are represented only as microscopic rudiments soon merging into the stem in later growth, while in *Mamillaria* the leaves are as in *Echinocactus*, and the cotyledons themselves are nearing the vanishing-point" (Ganong).

REFERENCES

(1) Britton, N. L. and Rose, J. N. "The Cactaceae." *Carnegie Institute of Washington*, 1919–1923.

(2) Darbishire, O. V. "Observations on *Mamillaria elongata*." *Annals of Botany*, xviii, 375 (1904).

(3) Ganong, W. F. (*a*) "Beiträge z. Kenntniss d. Morphologie u. Biologie d. Cacteen." *Flora*, lxxix, 49 (1894).

—— (*b*) "Present problems of Cactaceae." *Botan. Gazette*, xx, 129, 213 (1895).

—— (*c*) "The Comparative morphology of the embryos and seedlings of the Cactaceae." *Annals of Botany*, xii, 423 (1898).

(4) Johnson, D. S. "The fruit of *Opuntia fulgida*." *Carnegie Institute of Washington*, 1918.

(5) Rudolph, K. "Beitr. z. Kenntniss d. Stachelbildung bei Cactaceen." *Österr. Botan. Zeitsch.* liii, 105 (1903).

(6) Schumann, K. "Cactaceae" in *Pflanzenfamilien*, iii, 6 *a* (1894).

Bluhende Kakteen (*Iconographia Cactacearum*), by K. Schumann and Max Gürke (continued by F. Vaupel), 1900–5, is a series of coloured plates, with descriptions, of species in cultivation.

Order 17. *UMBELLIFLORAE*

Flowers typically bisexual, regular, cyclic (polypetalous) and epigynous, generally 4- or 5-merous with one series of stamens (episepalous), and carpels often reduced to two. Calyx more or less reduced. Ovule solitary in each chamber of the ovary, pendulous, anatropous, with one integument and an outwardly directed micropyle. Endosperm copious; embryo small.

Plants woody or herbaceous with alternate (opposite in Cornaceae), often large leaves with a sheathing base and no stipules. Flowers generally umbellate. Oil- or resin-passages are often present.

The order is characterised by a simplification in the floral structure—great reduction in the calyx, a character frequently associated with epigyny, a single staminal whorl, reduction of the carpels to two, and a single ovule in each cell of the ovary; and a tendency to the aggregation of the flowers in a head, a biological factor of great importance in facilitating the visits of insects which finds its most complete expression in Compositae. Araliaceae and Umbelliferae are closely allied; Cornaceae stands somewhat apart, the secretory passages characteristic of the other two families are rarely present, the leaves are generally opposite, the number of carpels is less constant and the position of the micropyle varies; this family shews striking resemblances with the tribe *Sambuceae* of the sympetalous family Caprifoliaceae which is distinguished only by the union of the petals.

The first two families shew marked affinities with Celastrales and Rhamnales in the small size and general structure of the flower, as evidenced by the single staminal whorl, the conspicuous disc and the solitary ovule in the ovary-cell; the position of the ovule is however different, while in Rhamnales the stamens are opposite the petals. The order may be regarded as allied to these but representing a further development in the direction of epigyny accompanied by aggregation of the inconspicuous flowers.

Family I. ARALIACEAE

Generally trees or shrubs, the former often with a tall simple stem and a crown of large leaves giving a palm-like habit, as in *Tetrapanax papyrifer*, a native of Formosa, from the pith of which Chinese "rice-paper" is prepared. The shrubby species are sometimes climbers without special climbing organs but attaching themselves to stronger plants by short branches, or, as in the Ivy, (*Hedera*) by special climbing roots. A few are herbaceous. The leaves are often crowded at the ends of the branches, usually alternate, rarely opposite, generally with a broad sheathing base; they are often very large and palmately compound, or simple and entire or more or less palmately or pinnately divided. Different forms of leaf may occur on the same plant, as in Ivy, which has a lobed shade form, while leaves fully exposed to the light or borne on the inflorescence are entire; in other species the form of the leaf varies with the age of the branch. Stipular structures are often present in the form of a membranous border or a ligular structure above the sheath or as narrow lateral appendages of the leaf-base. Schizogenous passages containing resin or gum occur in root, stem and leaf. Stellate hairs are also of frequent occurrence.

The flowers are generally small, whitish, yellow or greenish, often in much branched showy inflorescences composed of umbels, heads, racemes or spikes. They are bisexual, often with a tendency to unisexuality, regular, and generally pentamerous. The calyx is generally inconspicuous and represented by a narrow seam which is quite simple or bears more or less conspicuous teeth; in *Meryta* it is absent. The petals are generally attached by a broad base; the usually pointed tip is sometimes incurved in the bud; they fall when the flower opens, either separately or several

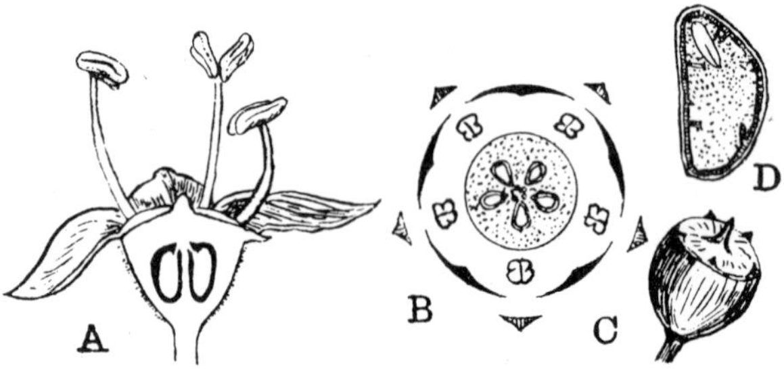

FIG. 201. *Hedera Helix*. A. Flower in vertical section, enlarged. B. Floral diagram. C. Fruit. D. Seed in vertical section, enlarged. (A, D after Strasburger; B, C after Le Maout and Decaisne.)

together or in the form of a cap. The stamens are equal in number to the petals, rarely more. The ovary is generally completely inferior, bears a flat or swollen nectar-secreting disc, and contains as many chambers as there are petals, sometimes fewer or only one, rarely more; the styles are free or united into a column, or absent, the stigmas then being sessile; there is a single pendulous, anatropous ovule in each chamber, and the micropyle is directed outwards. The fruit is generally a five-chambered berry or drupe; it sometimes separates into mericarps when ripe. The small embryo lies at the upper end of the seed; the endosperm is sometimes ruminate. The flowers of Ivy are visited by small flies, bees and moths which are attracted by the nectar freely exposed on the disc.

The family contains 63 genera with about 700 species and is mainly tropical with two great centres—the Indo-Malayan region and tropical America—each of which is characterised by distinct genera. It is represented in Europe by *Hedera Helix* (Ivy) (fig. 201)

Tobler[1], in a recent monograph, recognises six species of *Hedera*: *H. Helix* in Europe, and Asia Minor to Persia; *H. poetarum* in Greece; *H. canariensis* in North Africa and the West African Islands; *H. colchica* from the Black Sea to Persia; *H. himalaica*, Himalayas, with a variety in Southern China; and *H. japonica* in Japan and Korea. Many forms of Ivy are known in cultivation. The "Aralia," commonly grown as a room-plant, is *Fatsia japonica*, a monotypic Japanese genus. The Chinese "Ginseng" is the root of *Panax Ginseng*, a north-east Asiatic species.

Family II. UMBELLIFERAE

Flowers usually bisexual and regular, according to the plan S5, C5, A5, G($\overline{2}$). Sepals, petals and stamens free. Ovary bilocular, surmounted by an epigynous disc; styles two; ovules with a single integument, solitary and pendulous in each cell. Fruit of two indehiscent dry mericarps; seed with copious endosperm and minute embryo.

Herbs with usually hollow internodes; leaves alternate, compound, rarely simple, exstipulate; petiole with a broad sheathing base. Flowers small, in simple or compound umbels.

Genera about 270; species about 2700. Cosmopolitan but chiefly north temperate.

The family consists almost entirely of herbs, which may be small with a creeping filiform stem as in *Hydrocotyle vulgaris* (Penny-wort), but which have generally an erect stout hollow stem, often several feet high (as in species of *Heracleum*), or sometimes (as in species of *Angelica*) form gigantic plants which are the principal feature of the landscape. The leaves also shew great variety of form. They are generally pinnately compound, and branching of the lamina may be continued to the fifth or sixth order; but palmate division also occurs, and, though rarely, simple leaves. Leaf-characters are used to delimit the primary divisions of an artificial key to the genera. Thus the British species, representing 35 genera, fall into four very unequal groups, as follows:

1. Simple leaves: including only *Hydrocotyle* (fig. 202), with long-stalked peltate leaves, and *Bupleurum*, with entire sometimes perfoliate leaves.

[1] Tobler, F. *Die Gattung Hedera*. Jena, 1912.

2. Leaves palmate or simply ternately divided: *Sanicula* and *Astrantia*, leaves palmately lobed or partite; *Eryngium* (fig. 205), leaves spiny-toothed; *Peucedanum* (section *Imperatoria*), leaves trifoliolate.
3. Leaves simply pinnate, rarely compound at the base: eight genera.
4. Leaves bi- to tri-pinnate or bi- to tri-ternate: the remaining genera.

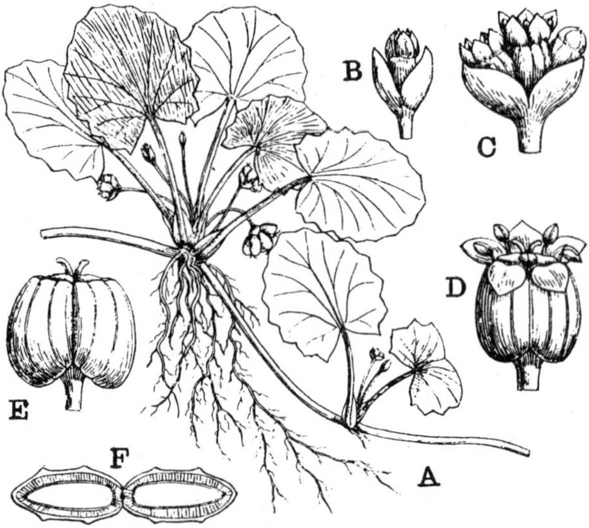

FIG. 202. *Hydrocotyle asiatica.* A. Portion of plant bearing flower and fruit. B. Young inflorescence, the central flower-bud only is seen emerging from the involucre of two bracts, ×2½. C. Older inflorescence, ×2½. D. Flower, ×4. E. Fruit, ×4. F. Fruit cut across, ×7.

Certain South American species of *Eryngium* and the Australasian *Aciphylla* are remarkable for their monocotyledon-like leaves, which with a narrow parallel-veined blade and broad sheathing base recall those of *Agave* and *Bromelia*. *Hydrocotyle* and allied genera have membranous stipules at the leaf-base. A genus, *Pseudocarum*, which climbs weakly by its petioles, occurs on Mt Ruwenzori in tropical Africa.

There is also a wide range in the duration of life. The plant may be monocarpic, either passing through its life-history in one season, as *Aethusa Cynapium* (Fool's Parsley), or accumulating strength for two or more seasons before flowering such are the biennials, as *Daucus Carota* (Carrot), while *Ferula* has a vegetation-period of several years. Polycarpic species persist by means of subterranean tubers or rhizomes or a thickened stem-base; as in species of *Heracleum*, *Peucedanum* and *Angelica*. The south temperate genus *Azorella* (fig. 203) has a caespitose habit, forming dense cushions which may be several feet in diameter and persist for many years. Shrubby species are rare, but occur in *Bupleurum* (in the Spanish *B. spinosum* the branches are short and spine-like), and several members of the Australian genus *Trachymene*.

Both the main axis and the lateral branches end in an inflorescence; the former is generally the more strongly

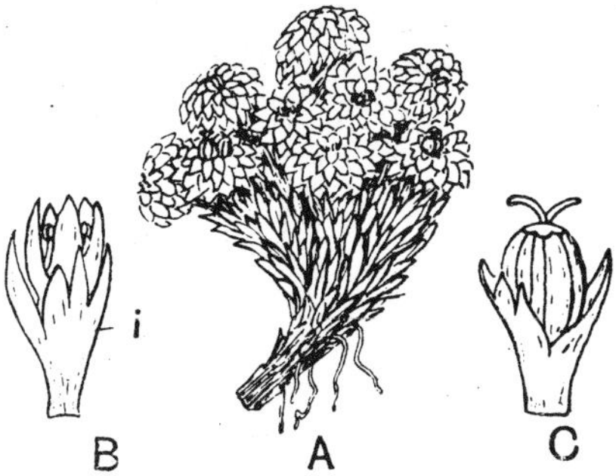

FIG. 203. *Azorella glabra*. A. Portion of plant, slightly reduced. B. Involucre, *i*, surrounding a single flower. C. Same in fruit. B and C enlarged. (After Weddell.)

developed, giving the characteristic habit of many of our common Umbellifers. In *Hydrocotyle*, and other genera with a creeping stem, inflorescences are borne only on lateral shoots.

An anatomical feature is the presence in all the organs of ethereal oil-, balsam- or resin-canals. Differences in the arrangement and structure of the vascular bundles and the bands of collenchyma afford characters which are more or less constant in individual tribes or smaller groups.

The umbellate inflorescence is a very constant characteristic. The arrangement and relative growth of the main axis and its branches afford characters which are often of generic or specific importance. The compound umbel is the commonest form, but simple umbels also occur (most species of

Hydrocotyle (fig. 202) and *Astrantia* (fig. 204)); very rarely is the umbel reduced to a single flower as in species of *Hydrocotyle*, and *Azorella* (fig. 203). In *Eryngium* (fig. 205) the flowers are crowded into dense heads surrounded by a whorl of rigid bracts; each flower is subtended by a bract. A terminal flower may be present, as in Carrot, where it differs from the rest in its form and purplish colour. The presence or

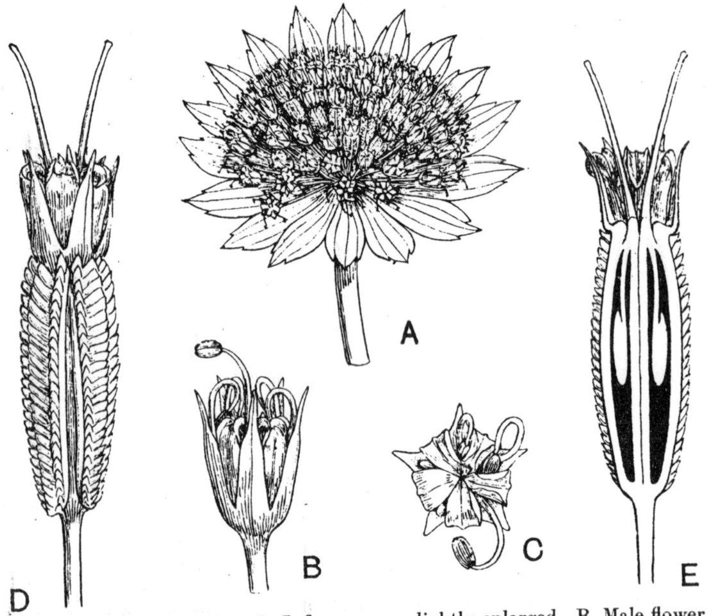

FIG. 204. *Astrantia major*. A. Inflorescence, slightly enlarged. B. Male flower, ×7. C. Same viewed from above; note the long sepals, the petals turned in towards the centre of the flower, and the stamens emerging in succession. D. Bisexual flower, after pollination, ×5. E. Same in vertical section.

absence of bracts, and their character, also afford useful distinctions. When present they form a whorl—the involucre—at the base of the rays of the compound umbel, or at the base of the partial umbel—the involucel. They are fewer in number than the rays or flower-pedicels, and represent the bracts subtending the outer pedicels. Either the involucre or the involucel or both may be absent. The simple umbel of *Astrantia* (fig. 204, A) is enveloped by a large, often coloured involucre.

The floral development is remarkable for the fact that the stamens appear first, followed generally by the petals, and then the sepals. Often, however, the calyx is only slightly indicated. The rudiments of the two carpels appear last; at first separate, they become united at their edges, while the roof forms the characteristic glandular disc or *stylopodium*,

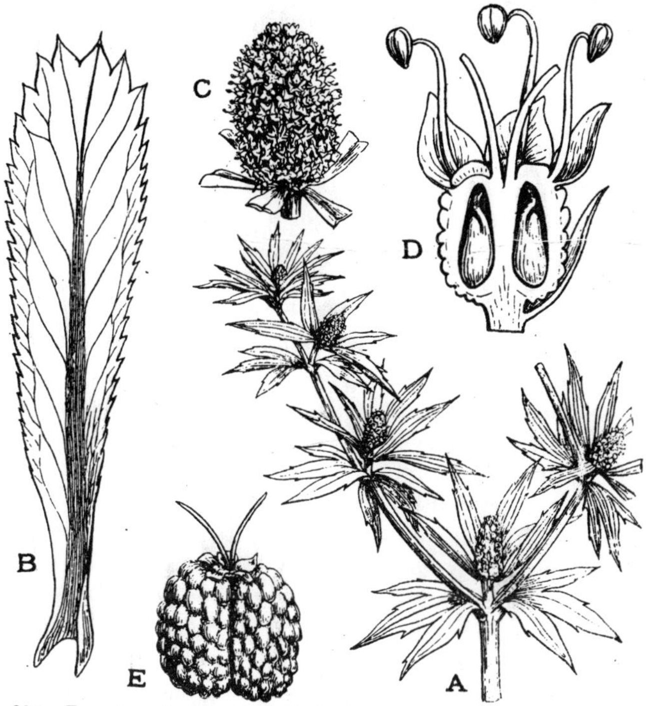

FIG. 205. *Eryngium foetidum*. A. End of flowering shoot. B. Leaf, slightly reduced. C. Single head, × 2½. D. Flower in vertical section, × 16. E. Fruit, × 10.

from which the styles grow out (fig. 206, F). The solitary ovule is pendulous and anatropous, the raphe being ventral, and has only one integument. During its development the ovary increases considerably, becomes inferior by intercalary growth of the receptacle and forms the most conspicuous organ of the whole flower; one of the carpels points towards the centre of the umbel, the other radially outwards. The calyx is much smaller than the corolla, and is often only indicated by inconspicuous teeth on the upper edge of the ovary; in *Saniculeae* (*Eryngium*, *Astrantia*, *Sanicula*), where they are best developed, the sepals are slightly imbricated with $\frac{2}{5}$ aestivation. Similarly in the corolla, the aestivation is open, valvate, or rarely imbricate with a $\frac{2}{5}$ arrangement, according to the breadth of the petals. The form of the apex of the petal, which is often pointed, the length of the point and its degree of inflexion in the bud afford characters for distinction of genera. The long slender filaments of the stamens are bent inwards in the bud; but are ultimately spreading. The anthers are basi- or dorsi-fixed, and the pollen is ellipsoidal with smooth poles and three equatorial stoppers, indicating points of egress for the pollen-tube.

The flowers are rendered conspicuous by the aggregation of large numbers into more or less dense inflorescences, the circumferential flowers of which are often sterile (male) and zygomorphic, the outer petals being much larger than the inner of the same flower and of the other flowers of the head. This arrangement recalls the development of ray-florets in Compositae. The flowers are generally white, sometimes pink or yellow (Parsnip, Fennel), very rarely blue; they are rarely scented; on the other hand, the whole plant has an ethereal odour which probably serves to attract insects. Nectar is secreted on the disc in the centre of the widely-open flower and is thus easily accessible to the smallest flies. Cross-pollination is necessitated by the very general well-marked proterandry; the stamens may have all dehisced and shed their pollen before the flowers of the same umbel have spread their styles and developed the stigmas. Drude (in the *Pflanzenfamilien*) distinguishes the following types:

1. *Flowers almost homogamous*, the maturing of the stigmas following quickly on that of the stamens, as in *Hydrocotyle vulgaris*. Self-pollination is rendered possible by the rapid spreading of the styles (which are bent inwards in the bud) and the development of stigmatic papillae before the fall of the stamens.
2. *Strongly marked proterandrous-dichogamy*, the most common type, alluded to above. The flowers of the last-developed lateral umbels are often male.
3. *Andromonoecism*. Both bisexual and male flowers are present. Examples are *Torilis Anthriscus*, *Astrantia* and *Scandix Pecten-Veneris*; the male are often distinguished by their longer stalks.

4. *Monoecism* occurs in the genus *Echinophora* (South Europe to West Asia) and allies, where a central female flower is surrounded by male umbels.
5. *Dioecism*, in *Arctopus* (South Africa) and species of *Aciphylla* (Antarctic).
6. Stamens more or less aborted in the flowers of the main umbels; flowers of the lateral umbels male; as in *Ferula*, where the main umbel of the stem and its primary branches bear numerous flowers in which the stamens are rudimentary as the flower opens, the anthers are often apparently fertile but contain few or no functional pollen-grains. The male lateral umbels develop later.

The fruit is crowned by the rim of the calyx and the remains of the style, and divides into two dry one-seeded mericarps in the plane of union of the two carpels. Before falling, the two mericarps often hang for a time on a thin simple or forked stalk, the carpophore. The form and structure of the mericarps shew great variety and supply characters for the distinction of genera and larger groups. A transverse section of the fruit shews five vascular bundles in each mericarp, corresponding with which are five longitudinal ridges on the surface—the primary ridges; between these five secondary ridges may occur. Between the primary ridges, or inside the secondary ridges when these are present, run oil-ducts (vittae—produced schizogenously). The line separating the mericarps is known as the commissure.

The endocarp is sometimes woody, forming a nut-like fruit as in *Hydrocotyle* (fig. 202, F); or more often of soft tissue. The pendulous seed has a membranous coat and usually adheres to the endocarp. The dense endosperm contains oil (not starch) and occurs in three different forms which supply useful systematic characters. In the greater number of the genera (*Orthospermae*) it is flat on the ventral side, as in *Petroselinum* (Parsley) (fig. 206, D); in others (*Campylospermae*) it has a longitudinal groove on the ventral side, the transverse section being more or less a crescent, as in *Conium* (fig. 206, B), *Chaerophyllum* (Chervil) or *Anthriscus*; while in others (*Coelospermae*) it is concave on the ventral side, as in *Coriandrum*.

Distribution of the mericarps is assisted by the wing-like

development of two or more of the ridges, as in *Heracleum* and *Angelica* (fig. 206, E); or by the development of spines or bristles, as on the secondary ridges of *Daucus* (Carrot) (fig. 206, G, H) or *Caucalis*, which render the mericarp liable to be dragged off and carried by birds or mammals. Occasionally the mericarps separate elastically from one another and from the top of the carpophore.

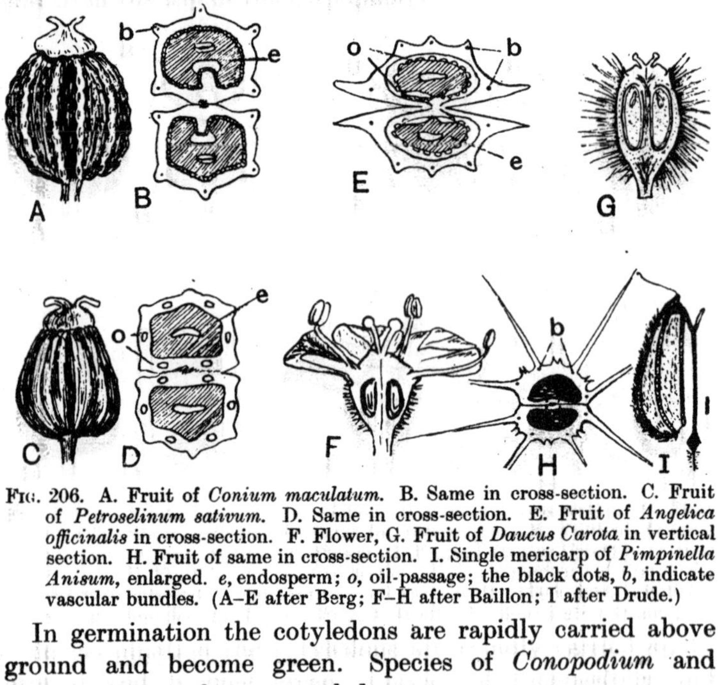

FIG. 206. A. Fruit of *Conium maculatum*. B. Same in cross-section. C. Fruit of *Petroselinum sativum*. D. Same in cross-section. E. Fruit of *Angelica officinalis* in cross-section. F. Flower, G. Fruit of *Daucus Carota* in vertical section. H. Fruit of same in cross-section. I. Single mericarp of *Pimpinella Anisum*, enlarged. *e*, endosperm; *o*, oil-passage; the black dots, *b*, indicate vascular bundles. (A–E after Berg; F–H after Baillon; I after Drude.)

In germination the cotyledons are rapidly carried above ground and become green. Species of *Conopodium* and *Bunium* have only one cotyledon.

The family occurs mainly in the northern hemisphere, from the subarctic to subtropical regions; in the tropics it occurs on the mountains, and reappears, but in smaller numbers than in the northern, in the southern hemisphere.

Characters for a natural division of the family are obtained from the fruit.

Three subfamilies are distinguished:

I. HYDROCOTYLOIDEAE. Fruit with a woody endocarp with no free carpophore. Vittae absent or sunk in the main ribs only. Mainly southern hemisphere. *Hydrocotyle*, with 78 species, almost cosmopolitan but mainly southern. *H. vulgaris* (British) occurs in Europe, West Asia and North Africa. *Azorella*, with 100 species mostly Andine, has a characteristic cushion-like habit.

II. SANICULOIDEAE. Mericarps with soft parenchymatous endocarp; exocarp rarely smooth. Style surrounded by a ring-like disc (stylopodium). Vittae various. *Eryngium*, 200 species in temperate and subtropical regions; *E. maritimum* (Sea-holly); *Astrantia*; *Sanicula*; *S. europaea* (Sanicle) is widely distributed in woods in Europe, North Africa and Western Asia.

III. APIOIDEAE. Mericarps with soft parenchymatous endocarp, sometimes with subepidermal layers of wood-fibres. Style on the apex of the stylopodium. Vittae various. Contains the great bulk of the genera and is subdivided into tribes by characters derived mainly from the form of the mericarp, the form and arrangement of the ribs and vittae, and the structure of the walls.

The following are the more important tribes of this subfamily and a few of the genera:

Scandiceae. *Myrrhis* (Europe to Central Asia, Chile), *M. odorata* (Cicely); *Scandix*; *Chaerophyllum* (Chervil); *Anthriscus*; mainly Europe to West or Central Asia.

Coriandreae. *Coriandrum*, *C. sativum* (Coriander), Mediterranean region.

Smyrnieae. *Conium*, *C. maculatum* (Hemlock), north temperate Old World.

Ammineae. *Bupleurum*, north temperate and subtropical (four British), one species South African, has simple entire leaves; *Apium*, world-wide, *A. graveolens* (Celery); *Petroselinum*, Central Europe and Mediterranean, *P. sativum* (Parsley); *Carum*, north temperate, *C. Carui* (Caraway); *Cicuta*, north temperate, *C. virosa* (Water-hemlock); *Pimpinella*, almost world-wide, but mainly Mediterranean and Western Asiatic; *Foeniculum*, mainly Mediterranean, *F. vulgare* (Fennel); *Aethusa Cynapium* (Fool's Parsley), Europe, North Asia.

Peucedaneae. *Angelica* (mainly north temperate and subarctic); *Ferula* (Mediterranean and Central Asia) affords various gum-resins, asa-foetida and galbanum; *Peucedanum*, north temperate, extending to Central America and tropical and South Africa; *Pastinaca*, Europe to Central and North Asia, *P. sativa* (Parsnip); *Heracleum*, north temperate, mountains of tropics, *H. Sphondylium* (Hog-weed).

Dauceae. *Daucus*, widely distributed, largely Mediterranean, *D. Carota* (Carrot).

Family III. CORNACEAE

Chiefly shrubs, sometimes arborescent, climbing in some South American species of *Griselinia*; *Cornus suecica* and *C. canadensis* are perennial herbs with a creeping underground rhizome. The leaves are opposite, more rarely alternate, simple, stalked and exstipulate. The small regular cyclic flowers are bisexual, or sometimes unisexual by reduction, and are arranged in cymose inflorescences of very various forms. Study of the development of the inflorescence shews that a terminal flower is always originally present on the main and lateral axes, but often does not develop, so that the mature inflorescence has the form of a panicle, umbel or head; sometimes it is corymbose, as in Dogwood (*Cornus sanguinea*). The head may be surrounded by a large leafy or

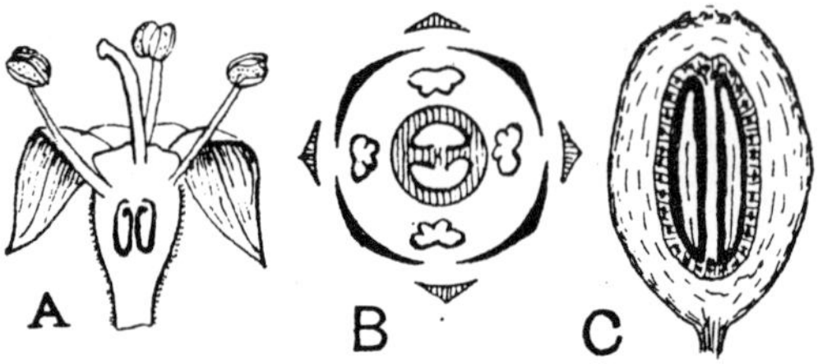

FIG. 207. *Cornus mas.* A. Flower in vertical section, ×4. B. Floral diagram. C. Fruit in vertical section, slightly enlarged. (A, C after Harms.)

petaloid involucre (as in *C. suecica* and *canadensis*). In *Helwingia* the few-flowered inflorescence springs from the middle of the leaf-blade; this, as Payer shewed, is due to the development of the leaf-base which carries up with it the axillary bud. We may regard as the typical floral arrangement that represented by the formula S4, P4, A4, G($\bar{2}$), which occurs in *Cornus* and other genera. The number of petals varies from three to five and the stamens equal them in number. The carpels vary from one to five; the female flower of *Aucuba* has one carpel only.

The calyx is generally but slightly developed, consisting of small teeth or lobes, or is reduced to a seam; in *Helwingia* it is quite undeveloped. The petals have, with few exceptions, a valvate aestivation. The dehiscence of the anthers is lateral or, more rarely, introrse. The top of the inferior ovary bears a disc; the simple style ends in a generally lobed stigma; the number of ovary-chambers corresponds with that of the carpels, and each contains one pendulous ovule with a dorsal or (in *Curtisia* and *Mastixia*) ventral raphe.

Pollination has been studied mainly in *Cornus*, most of the species of which are homogamous, the stamens and stigmas maturing simultaneously; the nectar lies free on the disc and in many species the smell of the flowers is attractive. The most frequent visitors are Hymenoptera, then flies and beetles. Cross-pollination is favoured by the position and difference in length of the style and the stamens which are touched by different parts of the insect's body; small creeping insects cause both self- and cross-pollination. The wide spreading of the stamens favours crossing between neighbouring flowers of an inflorescence. The flowers of *C. suecica* and *canadensis* are proterandrous; the capitate inflorescence is rendered conspicuous by the involucre. Insect-pollination is probably general in the family but *Griselinia littoralis* is said to be wind-pollinated.

The fruit is generally a drupe, but a berry in *Aucuba* and *Griselinia*; there are generally one to two, rarely to four seeds. In three closely allied species of *Cornus* (forming the section *Benthamia*) the ovaries are united and the fruits form a fleshy syncarp.

There are 10 genera and about 115 species, the great majority of which are northern extratropical. The largest genus *Cornus* has about 50 species, chiefly in north temperate regions. *Cornus sanguinea* (Dogwood), and *C. suecica*, a small perennial herb growing on alpine moors, are British. The umbel of minute purplish flowers of *C. suecica* is subtended by an involucre of four conspicuous white bracts. *C. mas* produces its heads of small yellow flowers in early spring before the appearance of the leaves; the drupes, which are about the size of an olive, are sometimes made into a preserve. *Aucuba* has three closely allied species in the Himalayas and Eastern Asia; *A. japonica*, a common garden shrub, is dioecious. *Kaliphora* has one, and *Melanophylla* three species in Madagascar. *Corokea* has three species in New Zealand; they are much-branched shrubs with branches often twisted and interlaced. *Griselinia* has six species in New Zealand, Chile and Brazil; they are trees or shrubs, often climbing, with thick leathery alternate leaves and minute flowers. *Helwingia* and *Torricellia*, each with three species, occur in the Himalayas, China and Japan.

The two genera *Curtisia* (one species in South Africa) and *Mastixia* (16 species of trees in the tropical forests of Further India and Malaya) are distinguished from the other genera by the ventral position of the raphe, which causes the micropyle to be on the side away from the placenta (extrorse).

A number of fossil species of *Cornus* have been described from leaf-remains; a few from the later Cretaceous, but the majority from Tertiary strata.

Cornaceae are closely allied to the Caprifoliaceae, in which they were included by Kunth; the resemblance between the flower of *Cornus* and that of *Viburnum* and *Sambucus* is very close, the former differing from the latter in having the petals free. In a tree-like arrangement which depicted the true relationships of families these two would stand near each other.

Several other genera were formerly included in this family (as in the *Pflanzenfamilien*), but in the light of more recent investigation have been removed. *Garrya* is now classed with the Monochlamydeae; *Nyssa*, *Camptotheca* and *Davidia*, forming the family Nyssaceae, and *Alangium*, constituting the family Alangiaceae, are regarded as more nearly allied to the Myrtiflorae.

REFERENCE

WANGERIN, W. *Das Pflanzenreich*, IV, 229 (1910); see also Engler's *Botan. Jahrb.* XXXVIII, Beiblatt 86 (1906).

See also HORNE, A. S. *A Contribution to the Study of the Evolution of the Flower, with special reference to the Hamamelidaceae, Caprifoliaceae and Cornaceae.* Transact. Linnean Society, (Bot.) VIII, 239 (1914).

GRADE C. SYMPETALAE

The following orders are characterised by greater or less union of the lower portion of the petals, either as the result of lateral union or by the development of their common base on which the stamens are generally also carried up; associated with this is the presence generally of only one integument to the ovule. The union of the petals to form a shallow or deep cup, funnel or tube, is obviously an adaptation to pollination by insect-aid. More efficient protection and concealment is thus ensured for the nectar, the pollen and the "essential" organs generally. The sympetalous families represent, generally speaking, a higher stage of entomophilous floral adaptation than the dialypetalous; and in several orders very great specialisation has been attained in the more advanced members; for instance, Labiatae in the Tubiflorae, Asclepiadaceae in the Contortae, Compositae in the Campanulales; the specialisation running on different lines in the different orders.

Associated with sympetaly is a progressive reduction in the number of whorls in the flower (the cyclic arrangement is general), and of parts in each whorl. The first four orders differ from the remainder in having a regular typically pentacyclic flower, that is with an outer and inner whorl of stamens, and characters recur that are associated with the Dialypetalae —freedom of petals, freedom of the stamens from the corolla, and a double integument in the ovule. These families are generally hypogynous but a greater or less tendency to epigyny occurs in each order. The remaining and greater number of orders have tetracyclic flowers, and in each a progressive adaptation to entomophily is worked out, often embodying reduction in number of parts in association with greater specialisation. The adaptation runs on different lines in different orders. Thus in Oleales and Contortae the flower remains regular, but in Tubiflorae the tendency to zygomorphy is the leading character associated with reduction in the staminal whorl. In the great majority the number of carpels is reduced to two, and there is also a greater or less tendency to the reduction of the number of ovules in each ovary-chamber. Two orders, Rubiales and Campanulales, are distinguished from the remainder by an inferior ovary and are also characterised in their more advanced representatives by a tendency to aggregation of the flowers. This culminates in the capitular inflorescence of Dipsacaceae and Compositae, where associated with the taking over of the protective function by the involucre of bracts is the reduction or late development of the calyx, which becomes functional only in the fruiting stage, and the progressive differentiation of the flowers of the head into an outer often purely attractive zygomorphic flower and an inner generally small and regular seed-producing flower. In the higher members of these two orders reduction of the ovary becomes complete, and it is a single cell containing one ovule. In Campanulales, which is regarded as representing the highest grade of adaptive floral development, the pollination-mechanism is associated with more or less complete lateral cohesion of the anthers.

The consideration of the sympetalous orders together after the dialypetalous is largely a matter of convenience and must not be regarded as implying that they form a distinct natural class. Sympetalous genera occur here and there widely distributed throughout the dialypetalous families, and the occurrence of dialypetaly has been noted among the typically sympetalous orders—especially those which we regard as representing a less advanced stage of adaptation. In a thoroughly natural system the sympetalous orders should follow those dialypetalous orders with which they are most nearly allied. But it must be borne in mind that the sympetalous orders can rarely, if ever, be traced from existing dialypetalous orders. At present all that we can say with any approach to truth is that certain orders of Sympetalae probably represent an advanced stage of the development characteristic of certain orders of Dialypetalae. The considerable differences of opinion as to phylogenetic relationships, among botanists who have carefully studied the orders of Seed-plants, and even the differences of opinion of one and the same worker at different periods, would suggest the advisability for the present of merely indicating possible lines of development.

A. PENTACYCLICAE.

Order 1. Ericales.
Family I. Clethraceae.
,, II. Ericaceae.
,, III. Pyrolaceae.
,, IV. Epacridaceae.
,, V. Diapensiaceae.

Order 2. Primulales.
Family I. Theophrastaceae.
,, II. Myrsinaceae.
,, III. Primulaceae.

Order 3. Plumbaginales.
Family Plumbaginaceae.

Order 4. Ebenales.
Family I. Sapotaceae.
,, II. Ebenaceae.
,, III. Styracaceae.
,, IV. Symplocaceae.

B. TETRACYCLICAE.

(*a*) *Superae.*

Order 1. Oleales.
Family I. Oleaceae.
,, II. Salvadoraceae.

Order 2. Contortae.
Family I. Loganiaceae.
,, II. Gentianaceae.
,, III. Apocynaceae.
,, IV. Asclepiadaceae.

Order 3. Convolvulales.
Family Convolvulaceae.

Order 4. Tubiflorae.
Family I. Polemoniaceae.
,, II. Hydrophyllaceae.
,, III. Boraginaceae.
,, IV. Verbenaceae.
,, V. Labiatae.
,, VI. Nolanaceae.
,, VII. Solanaceae.
,, VIII. Scrophulariaceae.
,, IX. Selaginaceae.
,, X. Globulariaceae.

Order 4. Tubiflorae (*contd.*).
Family XI. Orobanchaceae.
,, XII. Gesneriaceae.
,, XIII. Pedaliaceae.
,, XIV. Martyniaceae.
,, XV. Bignoniaceae.
,, XVI. Lentibulariaceae.
,, XVII. Acanthaceae.

Order 5. Plantaginales.
Family Plantaginaceae.

(*b*) *Inferae.*

Order 1. Rubiales.
Family I. Rubiaceae.
,, II. Caprifoliaceae.
,, III. Adoxaceae.
,, IV. Valerianaceae.
,, V. Dipsacaceae.

Order 2. Campanulales.
Family I. Campanulaceae.
,, II. Goodeniaceae.
,, III. Stylidiaceae.
,, IV. Compositae.

A. PENTACYCLICAE

Flowers regular, with typically five isomerous whorls, conforming to the formula Sn, Pn, An + n, G(n), n often being 5; one of the staminal whorls may be rudimentary or suppressed, and in other cases additional whorls are formed. The pentacyclic orders may be regarded as those most closely allied to the Dialypetalae; the petals are sometimes free, in Ericales the stamens are generally free from the corolla with a hypogynous insertion, and in Primulales the ovule has two integuments.

Order 1. *ERICALES*

Flowers bisexual and regular conforming to the formula Sn, Pn, An[+ n], G(n), where n is 4 or 5; generally obdiplostemonous. Stamens generally free from the petals, and inserted like the latter at the edge of a usually hypogynous, nectar-secreting disc; petals sometimes free. Carpels generally opposite the petals; ovary superior to inferior, generally multilocular with axile placentas bearing numerous ovules; ovule with one integument; style simple. Seeds small containing a straight embryo surrounded by endosperm; embryo undifferentiated in Pyrolaceae. Usually shrubby plants with simple often leathery persistent leaves, chiefly in cold and temperate climates.

The order represents a comparatively unadvanced state of sympetaly still retaining characters of the Dialypetalae. Thus the petals are sometimes free, as in the small family Clethraceae (containing one genus *Clethra*, widely distributed in the warmer parts of both hemispheres), frequently in Pyrolaceae, and occasionally in Ericaceae. The stamens are generally free from the corolla and in two series; Epacridaceae represents a more advanced type with petals constantly united to form a tube and a single whorl of generally epipetalous stamens, and the same obtains in Diapensiaceae, where however a second whorl of stamens is sometimes indicated by staminodes. Diapensiaceae is also of interest from the reduction in the number of carpels to three. In the more recent editions of his *System* Engler separates this family as a distinct order, Diapensiales.

The subfamily Vaccinioideae of Ericaceae is exceptional in having the ovary inferior, but its close resemblance in other characters with the subfamily Arbutoideae forbids its separation from the Ericaceae, and it must therefore be regarded as representing an advance, in the direction of simplification of the flower, on the other members of the order.

Family I. CLETHRACEAE

Contains the single genus *Clethra*, with about 30 species widely distributed in America from the Atlantic United States, through Central America and the West Indies to Brazil, and occurring also in Eastern Asia and Madeira. They are shrubs or small trees with alternate leaves and flowers in terminal racemes or panicles. The flowers are bisexual, regular, with five persistent sepals, five free deciduous petals, and obdiplostemonous with two whorls of five hypogynous stamens. There is no disc. The anthers dehisce by a pair of apical pores. The three carpels form a superior three-celled ovary containing in each chamber numerous anatropous ovules on club-shaped outgrowths of the central placenta. The style divides above into three stigma-bearing branches. The fruit is a three-valved capsule opening loculicidally, the septa separating from the central column. The small seeds have a loose testa and a fleshy endosperm round a short cylindrical embryo.

Family II. ERICACEAE

Flowers bisexual, generally regular, 5- to 4-merous, obdiplostemonous. Petals sometimes free, situated with the stamens on the outer edge of a hypogynous or epigynous nectar-secreting disc. Anthers introrse, the halves free and often spreading above, often appendaged, dehiscing by apical pores or slits, pollen in tetrads. Carpels opposite the petals, united to form a 5- to 4-chambered ovary; style central, simple, elongated, with a capitate stigma; placentas axile, bearing each 1–∞ anatropous or obliquely amphitropous ovules. Fruit generally a many-seeded capsule, sometimes a berry or a few-seeded drupe; seeds small or minute, containing generally copious endosperm and a cylindrical, often very short embryo.

Woody plants, shrubby or sometimes arborescent, with alternate, opposite or whorled leaves, which are often needle-like, and frequently stiff, leathery and evergreen. Flowers solitary in the leaf-axils, or in lateral or terminal racemes or panicles.

Genera 80; species about 1500. Temperate and cold regions; mountains in the tropics.

Ericaceae are generally woody plants, sometimes with a slender creeping stem as in Bilberry (*Vaccinium*), or forming low bushes like our native Heaths, or of larger size sometimes even becoming arborescent, as in some Rhododendrons.

The tendency for the strongest shoots to end in an inflorescence causes vigorous lateral branching, as in the Rhododendrons, from just beneath the flowering shoots, often resulting in a sympodial growth, or, as in many Heaths, from the lower branches; sometimes, as in *Vaccinium* and *Andromeda*, new shoots break through the ground from a widely-creeping subterranean rhizome (compare the mode of growth in *Pyrola*). The nature and development of the terminal bud also lead to differences in habit. In the true Heaths (Ericoideae) the new season's growth on a leafy shoot follows without a break on that of the past season, the closely arranged leaf-whorls succeeding without the intervention of bud-scales; in the Rhododendrons, and generally in the other subfamilies, the foliage-leaves in the terminal bud are protected by bud-scales and the more rapid growth of the lower part of the shoot results in the separation of the foliage of successive seasons by a number of long naked internodes, above which the spirally-arranged leaves form a more or less dense rosette. Endotrophic mycorrhiza in association with the roots is of widespread occurrence in the family, and in *Calluna vulgaris* (Ling)[1] the fungus has been traced from the roots through the plant to the ovary and fruit, where it is associated with the seeds and thus serves to infect the next generation; the fungus has been found to occur in the fruit very generally in the family.

The leaves shew well-marked xerophytic characters; in the Heaths, which flourish in open situations in temperate and colder climates, they are needle-like, and often deeply grooved, forming an almost closed chamber on the under side; in the other subfamilies they are generally stalked and more or less elliptical in shape, with an entire margin, often leathery and evergreen, with a strongly cuticularised upper epidermis, and often containing water-storing tissue between it and the green assimilating layers; characteristic hairs frequently occur on the lower epidermis.

[1] Rayner, M. C. "Obligate symbiosis in *Calluna vulgaris*." *Annals of Botany*, XXIX, 97 (1915).

The leaf-anatomy affords characters which are found to be of value in distinguishing genera and sometimes even tribes. They are founded on the arrangement of the chlorophyll parenchyma, the presence of water-storing tissue, the arrangement of the stomata on the lower face, and the structure of the hairs.

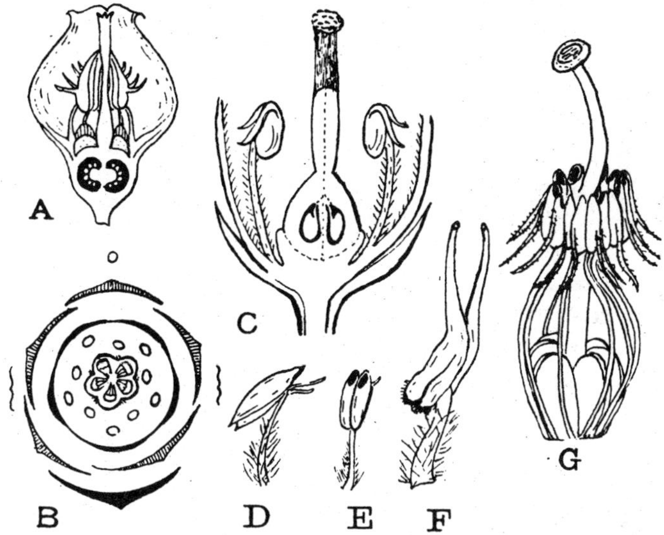

FIG. 208. A. *Vaccinium Myrtillus*, flower cut vertically. B. *V. Vitis-idaea*, diagram of a pentamerous flower. C. *Arctostaphylos Uva-ursi*, flower in vertical section (upper part of corolla not shewn). D, E. Side and front views of stamen of *Arbutus Unedo*. F. Stamen of *Vaccinium Vitis-idaea*. G. Stamens and pistil of *Erica caffra*. All enlarged. (A after Engler; B after Eichler; C after Warming; D–G after Drude.)

The flowers are sometimes solitary and axillary or terminal, but are generally arranged in few- to many-flowered racemose inflorescences at the end of the branches, as in *Rhododendron* (an umbel), *Arbutus* (a panicle), or on small lateral shoots (*Erica*); a bract and pair of bracteoles are normally present. They are regular, or in *Rhododendron* (fig. 209) slightly zygomorphic with stamens of unequal length and bent downwards towards the mouth of the flower (declinate), and conform with few exceptions to the formula Sn, Pn, An + n, G(n), with four or five members in each whorl; the androecium is obdiplostemonous, and the carpels are opposite the petals. The corolla is usually more or less bell-shaped; in the Ericoideae it persists, becoming dry after fertilisation.

A characteristic of the flower is the nectar-secreting disc, on the edge of which the corolla and stamens are separately attached; it varies in form in different genera. There is great variety in the form of the anthers, according to the degree of separation of the halves; the shape and position of the apical pore or slit and the absence or presence and form of the horn-like appendages affording useful systematic characters.

The conspicuous bright-coloured corollas, the presence of nectar, and the scent, render the flowers attractive to insects, and the position of the sticky stigma projecting beyond the introrsely dehiscent anthers favours crossing. Our British Heaths are good examples of bee-flowers; the stigma which protrudes at the mouth of the pendulous flower is first touched by the insect, which in probing for nectar at the base of the flower touches the anthers or their appendages and shakes out a shower of pollen from the apical pores.

There are four subfamilies; the distinctions are based chiefly on the position of the ovary and the character of the fruit and seed.

Subfamily I. Rhododendroideae. Fruit a septicidal capsule; seeds with a strongly-ribbed loose coat, often winged. Petals deciduous, sometimes free (*Ledum* and others); anthers without appendages. Includes 17 genera with 850 or more species. The largest genus, *Rhododendron* (including *Azalea*), with about 800 species, has its chief centre of distribution in the mountains of Eastern Asia, many species occurring on the Himalayas and the mountains of South-west China; there is a less important centre in North America, and a few species occur in the mountains of Central and Southern Europe and Asia Minor; the flowers are often large and showy. *Dabeocia* (St Dabeoc's heath), a monotypic genus of Western Europe, occurs in Ireland. The British representatives are *Loiseleuria procumbens* and *Phyllodoce coerulea*, both north circumpolar species found in Scotland.

Subfamily II. Arbutoideae. Fruit a berry or loculicidal capsule seeds triangular-ovate, not winged. Petals united, deciduous; anthers often with bristle-like appendages; ovary superior. 20 genera with about 250 species, chiefly north temperate to arctic. The largest genus, *Gaultheria* (about 100 species), is chiefly American, extending from the north along the Andes through central and tropical South America to Chile; a few species occur in the Himalayas and Eastern Asia, extending south to Australia and New Zealand. *Arbutus Unedo* (Strawberry-tree), so called from its orange-scarlet berry, is a South European species which extends into South Ireland; *Arctostaphylos alpina*, *A. Uva-ursi* and *Andromeda polifolia* are arctic and alpine species occurring in Britain. *Epigaea repens* is the Trailing Arbutus or Mayflower of Atlantic America.

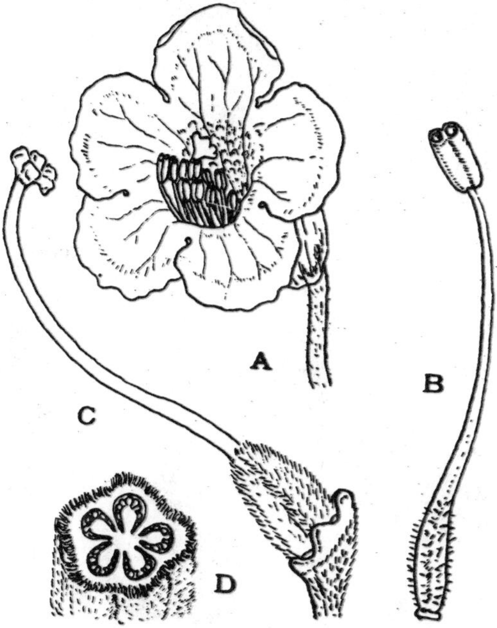

FIG. 209. *Rhododendron lanatum*. A. Flower. B. Stamen. C. Pistil. D. Ovary cut across. B, C, D enlarged. (After Hooker.)

Subfamily III. VACCINIOIDEAE. Ovary inferior, fruit a berry. Petals united, corolla often globose, bell-shaped or tubular, deciduous. 23 genera with about 330 species, extending from the north temperate zone, in which several species have a wide circumpolar distribution, to the mountains of the tropics. *Vaccinium*, the largest genus, has about 100 species, four of which are British; *V. Myrtillus* is the Bilberry and *V. Oxycoccos* the Cranberry. Epiphytes are frequent in tropical mountain genera, such as *Thibaudia* (South American Andes) and others.

Subfamily IV. ERICOIDEAE. Fruit usually a loculicidal capsule; seeds round, not winged; corolla persistent; anthers often appendaged. 17 genera with about 600 species, 500 of which belong to *Erica*. The great majority are confined to the Cape; others occur on the mountains of tropical Africa (species of *Erica*, *Blaeria*) and the Malagasy Islands, and in Europe and North Africa, especially the Mediterranean region. *Calluna vulgaris* (Ling) is a wide-spread heath-plant in Europe, spreading into North Asia and occurring also in Greenland and in a few places in the Atlantic States of America. *Erica Tetralix* and *E. cinerea* are our common British Heaths; two other species in S.W. England and one in West Ireland (*E. mediterranea*) are representative of a South-west European flora.

Family III. PYROLACEAE

Petals free or united; stamens free, hypogynous. Ovary incompletely chambered; placentas thick fleshy, bearing very numerous small anatropous ovules. Fruit a loculicidal capsule. Seeds minute with a loose testa; embryo small, undifferentiated within the endosperm. Perennial evergreen herbs, or saprophytes without chlorophyll.

Genera 10; species about 30. Chiefly woods of the north temperate zone.

Pyrola and *Chimaphila*, forming the subfamily Pyroloideae, have a widely-branching underground rhizome, the branches forming above ground leafy shoots, often a dense leaf-rosette, and ending in one, or several racemose flowers. The leaves are leathery and evergreen, hence the name Winter-green (*Pyrola*). *Monotropa* (Indian pipe) and allied genera (*Pterospora*, *Sarcodes*) forming the subfamily Monotropoideae are leafless saprophytes, living in the moist humus of woods and thickets; they have a richly branched fleshy root-system in which flower-shoots originate as endogenous buds, and breaking through the cortex appear above ground as succulent scapes, generally yellowish in colour, bearing brownish, crimson, pinkish, yellow or white scale-leaves and ending in a single flower or a raceme; the roots are associated with a symbiotic fungus forming a mycorrhiza, which is also present in less quantity in the Pyroloideae.

The flowers stand in the axil of a scale-like bract, bracteoles are wanting; they are bisexual, regular, 4- to 5-merous, and conform to the typical formula, with obdiplostemonous androecium. The petals are generally free but united to form an urnshaped or campanulate corolla in the monotypic American genera *Pterospora* and *Sarcodes*. The anthers in Pyroloideae dehisce by apical pores and the pollen remains united in tetrads, except in *P. secunda*; in

Monotropoideae the dehiscence is by slits and the pollen is simple.

The corolla and stamens are often inserted at the edge of a nectar-secreting disc, which is sometimes represented by separate nectaries or is rarely completely absent.

The fleshy ovary bears a simple columnar style; in some genera of Monotropoideae it is unilocular except at the extreme base, but generally the septa are wanting only in the upper part of the chamber. The ovary-cavity is almost filled by the thick fleshy placenta, the surface of which is crowded with ovules. The minute seeds have a delicate loose testa enclosing a central nucleus consisting of an oily endosperm and a few-celled undifferentiated embryo. *Pyrola*, the largest genus (15 species), extends into the

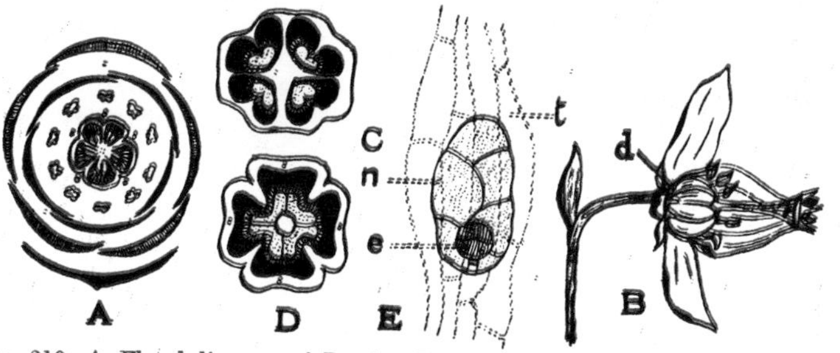

FIG. 210. A. Floral diagram of *Pyrola*. B. Flower of *P. uniflora*, enlarged; two petals have been removed to shew the glandular disc, *d*. C. Transverse section of base of ovary of *Monotropa Hypopitys*. D. Similar section above the middle; both enlarged. E. Median portion of the seed of *M. Hypopitys*, much enlarged; *e*, embryo; *n*, endosperm; *t*, loose seed-coat. (A after Eichler; B–E after Drude.)

Arctic zone, and southwards into the Himalayas and the mountains of Mexico; four species are British. *Monotropa*, with three species, has almost as wide a distribution; the British species, *M. Hypopitys*, occurs in Europe, Siberia, Japan and in the New World from Canada to Mexico; of the allied genera seven are North American and one is Himalayan; *Sarcodes*, a monotypic genus in the Sierra Nevada of California, sends up a fleshy spike of brilliant red flowers through the melting snow[1].

Pyrolaceae are closely allied to Ericaceae from which they are distinguished by incomplete septation of the ovary and the undifferentiated embryo. They may be derived from an Ericaceous stock with free petals (such as are found in Ericaceae in *Ledum* and allied genera) in which the shrubby habit has been lost and in which the herbaceous habit has gradually become saprophytic.

[1] Oliver, F. W. "On *Sarcodes sanguinea*," *Annals of Botany*, IV, 303 (1889).

Family IV. EPACRIDACEAE

Flowers bisexual, regular, conforming to the formula Sn, Pn, An, G(n), with mostly pentamerous whorls. Sepals free; petals united; stamens generally epipetalous; anthers bilocular, dehiscing by a common longitudinal slit, and without appendages. Ovary superior, with 1–∞ anatropous, generally pendulous ovules on an axile placenta in each of the five chambers; style simple with a capitate stigma. Fruit a five-valved capsule or a drupe with one to five stones. Embryo straight, cylindrical, surrounded by a copious endosperm rich in proteid.

Shrubby plants with a heath-like habit, rarely arborescent.

Genera 23; species about 350. Mainly Australasian.

The Epacridaceae replace in Australia and Tasmania the heaths (Ericoideae) of South Africa, which they resemble in habit, being generally small shrubs, somewhat sparingly branched, with closely arranged generally alternate small stiff narrow entire leaves. The flowers are racemose, each with a bract and a pair of bracteoles, or often a number of bracteoles passing into the calyx, the sepals of which are free and resemble the bracteoles. The corolla is tubular with five lobes, which are often spreading or reflexed; it is inserted on the edge of a disc of varying form. Two monotypic genera, *Prionotes* (Tasmania) and *Lebetanthus* (Patagonia), more nearly approach Ericaceae in the hypogynous insertion of the stamens and the separate dehiscence of the distinct anther-halves. In the great majority of cases the five stamens, which alternate regularly with the segments of the corolla, are attached to the corolla-tube and dehisce by a longitudinal slit common to the two cells of the anther, which becomes at the same time one-celled by the disappearance of the partition. The pollen is simple or in tetrads. The carpels are opposite the petals and the simple style generally springs from a central depression. In about half the genera, including *Epacris*, the fruit is a several- to many-seeded capsule with a loculicidal dehiscence; in the remaining genera (*Styphelia* and allies) where there is only one ovule in each ovary-chamber the fruit is a berry or a drupe containing one to five stones.

The great majority of the species are Australasian; the family extends to New Caledonia and the Pacific Islands, and in *Styphelia*, to Malaya and India (one species), and is represented in antarctic South America by the monotypic genus *Lebetanthus*. *Epacris* and *Styphelia* are commonly grown as greenhouse plants.

It is distinguished from Ericaceae, to the subfamily Ericoideae of which it is most nearly allied, by the single whorl of generally epipetalous stamens with their characteristic anther-dehiscence.

Family V. DIAPENSIACEAE

A small family with six genera and 12 species, chiefly alpine and arctic, in the northern hemisphere. The plants are low-growing evergreen shrubs or perennial herbs, with leaves arranged in dense or loose rosettes and solitary or racemose flowers. The flowers are bisexual and regular, with the formula S5, P5, A5 + 0, G(3), and preceded by a pair of bracteoles. The sepals are often free and the corolla is sometimes divided almost to the base; there is no hypogynous disc, and the five stamens are epipetalous, alternating with the corolla-segments; an alternating whorl of staminodes is sometimes present. Each anther-half opens by a longitudinal slit, and the pollen is simple. The three carpels form a trilocular ovary which bears a simple style with a three-lobed terminal stigma, and contains indefinite anatropous or amphitropous ovules on a central placenta in each chamber. The fruit is a three-valved loculicidally dehiscing capsule; the seeds are numerous and contain a cylindrical, straight or slightly bent embryo surrounded by a copious fleshy endosperm.

Diapensiaceae is distinguished from the other families of the order by the reduction in the number of the carpels; also by the absence of the disc. It resembles Epacridaceae in the simple whorl of epipetalous stamens.

Diapensia lapponica, which is found in Scandinavia and is the only European species, has a circumpolar distribution; a second species occurs in the Himalayas. Two monotypic genera occur in the Atlantic States of North America; *Schizocodon* has two species in Japan; *Berneuxia* one in Eastern Tibet; and *Shortia* three species in Eastern Asia and one (*S. galacifolia*) on the mountains of North Carolina.

Order 2. *PRIMULALES*

Flowers regular, bisexual or unisexual by abortion, conforming to the formula Sn, Pn | An, G(n), where n is generally 5. Stamens antepetalous and generally inserted on the petals, which are very rarely free; an outer antesepalous series is present in Theophrastaceae. Ovary superior, rarely inferior, unilocular, with indefinite ovules on a basal or free-central placenta; ovule with two integuments. Style simple. Seeds small, embryo straight, surrounded by fleshy or horny endosperm. Herbs, shrubs or trees with simple, entire, exstipulate leaves.

The pentamerous plan of the flower, the unilocular ovary and central placentation of the ovules (which moreover resemble the characteristic type of Dialypetalae in having two integuments) have suggested a connection of the order with the Centrospermae. The evidence is less clear than in the case of the next order, Plumbaginales; but it seems possible that the order is derived from a Dialypetalous stock allied to that from which sprang the Centrospermae. Theophrastaceae and Myrsinaceae are woody plants; Primulaceae are herbs. In the first diplostemony is suggested by an outer series of staminodes alternating with the petals; in the other two families, which are more closely allied, a tendency to epigyny occurs.

Family I. THEOPHRASTACEAE

Flowers bisexual or dioecious, regular, generally pentamerous. Sepals free or slightly connate (*Clavija*); corolla rotate, urceolate or funnel-shaped; an outer series of five antesepalous staminodes alternates with an inner series of five antepetalous stamens, the filaments of which in *Clavija* are united in a tube. Ovary superior, unilocular, with indefinite anatropous ovules with two integuments, immersed in mucilage on a central placenta. Style simple; stigma entire or irregularly lobed. Fruit a drupe or berry containing many to few seeds (very rarely only one). Embryo well developed, surrounded by endosperm.

Trees or shrubs with alternate simple, entire or serrate, exstipulate leaves, which are generally crowded at the ends of the branches, and terminal, or sometimes lateral, racemose inflorescences of often showy flowers. An important anatomical character is the presence of long strands of sclerenchymatous tissue beneath the epidermis of the leaves; a distinguishing feature from Myrsinaceae is the absence of resin-passages.

The family contains four genera with about 70 species in the warmer parts of America, from Florida, through Central America and the West Indies, to North Paraguay.

Family II. MYRSINACEAE

Flowers bisexual, or more often unisexual by abortion, regular, 4- to 5-merous, resembling those of Primulaceae, but carpels sometimes fewer. Ovules numerous or few, generally sunk in the tissue of the free-central placenta. Fruit generally a one- to few-seeded drupe. Embryo surrounded by a fleshy or horny endosperm.

Shrubs or trees, with alternate, entire, exstipulate, often leathery leaves, and axillary or terminal many-flowered simple or compound racemose inflorescences. Bracteoles are typically present.

Genera 32; species about 1000. Widely distributed mainly in tropical and subtropical countries.

Schizogenous resin-passages are present in the leaves, and occur also in the pith and cortex.

The bracteoles when present stand right and left of the flower, in the usual position; when absent they may be regarded as suppressed, their presence being implied in the normal arrangement of the sepals, thus contrasting with Primulaceae where the arrangement of the sepals is co-ordinated with the typical absence of bracteoles. The corolla-tube is generally short; in the large genus *Embelia* (60 species) the petals are free. The filaments of the stamens are sometimes united into a tube; in the monotypic genus *Amblyanthus*, from the Khasya Hills, N.E. India, the anthers are united. In *Aegiceras* the anther-chambers contain numerous transverse septa. *Aegiceras* (two species) are Mangrove plants growing on the coasts of the Old World tropics; as in other Mangroves the seed germinates before leaving the fruit.

The genus *Maesa* forms a distinct subfamily (Maesoideae) characterised by the inferior or half-inferior ovary; it has 102 species in the tropics of the Old World. In the remaining genera the ovary is superior. The ovules, which have two integuments, are semianatropous to semicampylotropous and are almost always buried in the placenta; generally only one or a few seeds develop, and these become surrounded at the base by the dry cup-like remains of the placenta. The style is short and simple; the stigma simple or sometimes lobed; when the lobes are equal in number to that of the members of the other whorls they are opposite the sepals, indicating a similar position for the carpels. The fruits are generally small and round with a crustaceous or hard endocarp and a fleshy exocarp. The cylindrical embryo is straight or slightly bent and completely surrounded by the endosperm. Polyembryony occurs in species of *Ardisia* (as *A. japonica*), and the germination in this genus is also of interest; *A. crenata* recalls that of *Aegiceras* in the early growth of the embryo, the radicle having ruptured the fruit-coat before the fruit has fallen. In *A. crenulata* and *A. japonica* the cotyledons

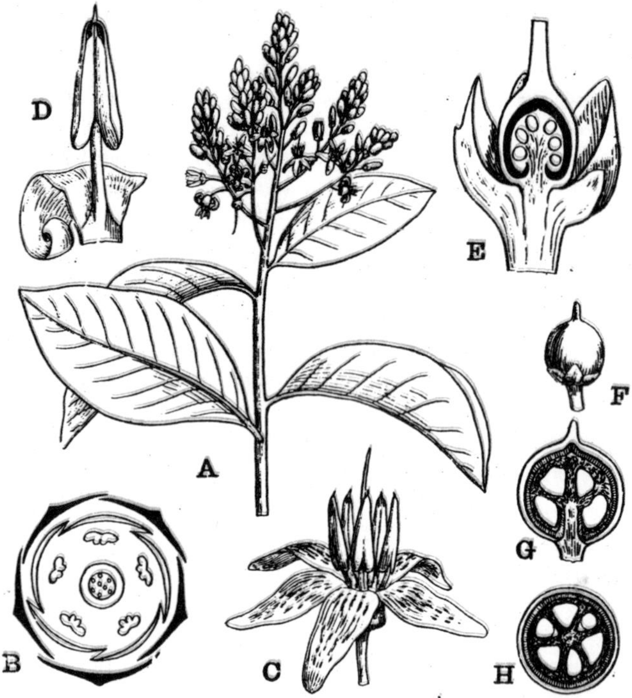

FIG. 211. *Ardisia tinifolia.* A. Flowering shoot, × ⅓. B. Floral diagram. C. Flower, × 2. D. Single stamen and petal, × 4. E. Ovary and calyx in longitudinal section, × 8. F. Fruit, nat. size. G, H. Fruit in vertical and transverse section, × 2.

never leave the seed, whereas in other species, as *A. polycephala*, they emerge and form broad long-persistent coriaceous leaves.

The family is widely distributed in tropical and subtropical countries, extending southwards with a few species at the Cape and in New Zealand, and northwards to Japan, Mexico and Florida. The largest genus *Ardisia* has 240 species in the warmer parts of both hemispheres; *Rapanea* has 140 species and *Embelia*

60 species; both genera are widely distributed. *Myrsine africana* is cultivated in gardens.

Myrsinaceae are closely allied to Primulaceae, from which they are distinguished by their woody habit, and the one- to few-seeded drupaceous fruit.

Family III. PRIMULACEAE

Flowers regular, bisexual, typically pentamerous, conforming to the formula S5, P5, | A5, G(5), the stamens situated on the corolla-tube and opposite the segments. Ovary superior, rarely half-inferior, unilocular with a free-central placenta bearing indefinite semianatropous ovules with two integuments. Fruit a capsule. Seeds containing a small embryo surrounded by fleshy endosperm.

Herbs with simple exstipulate leaves and flowers solitary or arranged in umbels, racemes or panicles; bracteoles wanting.

About 600 species in 22 genera. Cosmopolitan, but mainly in the temperate and colder parts of the northern hemisphere; many arctic and alpine.

The plants are herbs, sometimes annual but generally perennial, as in *Primula*, persisting by a sympodial rhizome, or in *Cyclamen*, by a tuber formed from the swollen hypocotyl. The aerial stem-internodes are suppressed, the leaves forming a dense radical rosette, as in *Primula* and others, or there is a well-developed aerial stem, erect, as in some species of *Lysimachia* (*L. vulgaris*), or creeping, as in *L. Nummularia* (Moneywort, Creeping Jenny) and *L. nemorum*.

The leaves are generally simple, often with a toothed margin; the submerged leaves of *Hottonia* are cut into fine linear segments; the arrangement is alternate, opposite or whorled, and all three arrangements occur in the genus *Lysimachia*. The flowers are solitary and axillary, as in *Anagallis* (Pimpernel), *Lysimachia Nummularia* and others, or racemose on terminal or lateral axes, forming an umbel, as in species of *Primula* (either sessile, as in Primrose, or stalked, as in Cowslip and Oxlip), or racemes or spikes as in species of *Lysimachia*.

Each flower stands in the axil of a bract, but there are no

bracteoles, and corresponding with their absence the two first developed leaves of the quincuncial calyx stand right and left towards the posterior aspect of the flower (fig. 212, A), while the odd sepal is posterior. Exceptions from the typical pentamerous arrangement occur, as in *Trientalis*, 5- to 9-mery, *Lysimachia*, 5- to 6-mery, *Centunculus*, 4- to 5-mery. The calyx is leafy and usually persists till the fruit is ripe. The corolla is generally differentiated into a longer (*Primula*) or shorter (*Anagallis*) tube, and a limb which is more or less spreading (Primrose, Pimpernel, etc.) or connivent into a cup (Cowslip) or reflexed (*Cyclamen*); in *Soldanella* it is bell-shaped; in *Lysimachia* and others the tube is often very short and in the small genus *Pelletiera*, which is remarkable for its trimerous flowers, it is absent, the petals being free; in *Glaux* there are no petals. In *Coris* the flower is zygomorphic.

The single whorl of stamens is situated on the corolla-tube with its members opposite the petals; the position is explained by assuming the disappearance of an outer antesepalous whorl which is occasionally represented by scales, as in *Samolus* (fig. 212) and *Soldanella*. Another explanation is based on the fact that the petals arise late in the development of the flower (fig. 212, F) and spring from the backs of the staminal primordia; this assumes the presence of only three whorls, regularly alternating, namely calyx, stamens with petal-like dorsal outgrowths, and carpels; that is to say, the apetaly, which is obvious in *Glaux*, is general in the family. The former theory is the one now generally accepted for the reason given and because it applies also to the closely allied family Myrsinaceae, where, as in Primulaceae, the vascular bundle-traces of an outer staminal whorl are found in the receptacle. The anthers dehisce introrsely by longitudinal slits. The ovary is superior, or half-inferior in *Samolus*; the simple terminal style ends in a capitate undivided stigma. The free-central placenta sometimes bears only five ovules, opposite the petals, but most frequently there are a large number developed spirally and in basipetal succession. The ovary arises as a ring on the floral axis, so that its development affords as little evidence as does the adult structure of the pistil for the existence of five carpels. The fruit, however, forms a capsule which dehisces in most genera by five teeth or valves opposite the sepals; in *Hottonia*, *Cyclamen* and some species of *Primula* they are opposite the petals, while in other species there are ten valves. In *Anagallis* and other genera the capsule is a pyxidium, dehiscing transversely by a sharply-defined lid (fig. 212, H).

The flowers are always bisexual, but cross-pollination is favoured by a very general and well-marked dimorphism in the flower. The two forms are characterised by long and short styles respectively, the stamens occupying corresponding positions half-way down and at the mouth of the corolla-tube, differ ences which also entail slight differences in the shape of the tube and the width of the mouth; the short-styled form has smaller stigmatic papillae corresponding with the smaller pollen-grains of the long-styled. Experiments shew that if the visits of insects be prevented there is little or no fertility, and that the best results are obtained by a legitimate cross, that is by the pollination of a long-styled flower from a short-styled, or vice versa.

The ovules are generally semianatropous, so that the seed is peltate, having the hilum in the centre of one side (ventral). In *Hottonia* and *Samolus* the ovules are anatropous and the hilum is in consequence basal.

The seeds of *Primula vulgaris*, according to Sernander, differ from those of other species of the genus, which are distributed by wind, in being carried by ants; in this species the funicle has been converted into an oil-body (elaiosome). Hildebrand[1] states that the seeds of *Cyclamen* are also myrmecochorous, ants invading the capsule as soon as it opens and carrying off the seeds.

The family is divided into the following tribes according to the position of the ovary, the aestivation of the corolla and the symmetry of the flowers.

Tribe 1. *Primuleae.* Ovary superior, capsule valvular, corolla-lobes imbricate or quincuncial in the bud. *Primula*, the largest genus in the family, contains more than 300 species, native of

[1] Hildebrand, F. *Die Gattung* Cyclamen *L., eine systemat. u. biologische Monographie*, 1898.

the north temperate, subarctic and arctic regions, especially alpine and subalpine; a great number occur in the mountains of Northern India and China; five are British. *Androsace* has about 90 species in the north temperate zone; they are often more or less caespitose in habit. *Soldanella* has six species in the European Alps. *Dodecatheon*, characterised by the sharply reflexed petal-limbs, has 30 species in North America, chiefly on the Pacific side. *Hottonia* are water-plants with much-divided swimming leaves and an erect stem bearing the white or lilac racemose flowers above the water; *H. palustris* (Water Violet) occurs in Europe and North Asia; *H. inflata* in Atlantic North America.

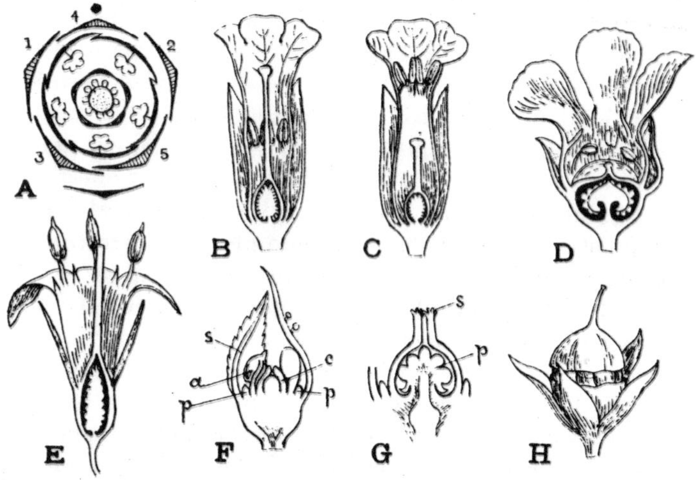

FIG. 212. A. Diagram of flower of *Primula*. B, C. Long- and short-styled flowers of *P. elatior*, slightly enlarged. D. Flower of *Samolus Valerandi* in longitudinal section, × 2½. E. Flower of *S. repens* in longitudinal section, × 2½. F. Longitudinal section of young flower-bud of *Anagallis arvensis*, much enlarged; *s*, sepal; *p*, petal; *a*, anther; *c*, carpel. G. An older stage of the same shewing the pistil, much enlarged; *s*, stigma; *p*, central placenta. H. Fruit of *Anagallis* shewing transverse dehiscence, × 2½. (A after Eichler; B, C after Prantl; D, E, H after Pax and Knuth; F, G after Sachs.)

Tribe 2. *Cyclamineae*. Ovary superior, capsule valvular. Tuberous plants; flowers with sharply reflexed petal-limbs. One genus, *Cyclamen*, with 16 species chiefly in the Mediterranean region: one (*C. europaeum*) in Central Europe.

Tribe 3. *Lysimachieae*. Ovary superior, capsule valvular (subtribe *Lysimachiinae*) or a pyxidium (subtribe *Anagallidinae*), corolla-lobes contorted in bud. *Lysimachia* has 110 species in temperate regions and mountains in the tropics. *Trientalis*, three species, in the temperate and cold regions of the northern hemisphere; *T. europaea* (British) is a small plant with white nectarless flowers. *Centunculus*, three species of very small annual herbs widely distributed in temperate and warm countries; *C. minimus* is British. *Anagallis* (Pimpernel), 24 species (two British), is almost cosmopolitan. *Glaux*, one species, is a small low-growing maritime herb with opposite fleshy leaves widely distributed in the north temperate zone (British).

Tribe 4. *Samoleae*. Ovary half-inferior. One genus, *Samolus*, with nine species, one of which, *S. Valerandi* (Brook-weed), is cosmopolitan (British), growing in damp ground, often near the sea. The other species have a wide distribution in temperate climates.

Tribe 5. *Corideae* is distinguished by its medianly zygomorphic flowers and spiny calyx. One genus, *Coris*, with two species in South Europe; small perennial herbs with a thyme-like habit.

Order 3. *PLUMBAGINALES*

Family. PLUMBAGINACEAE

Flowers bisexual, regular, with the formula S5, P5, | A5, G(5); bracteoles present. Calyx generally membranous, persistent; stamens opposite the petals. Ovary superior, unilocular, with a solitary basal anatropous ovule with two integuments borne at the top of a long filiform funicle with the micropyle pointing upwards; styles five, opposite the sepals. Embryo straight, endosperm mealy.

Perennial herbs or shrubs with narrow leaves, with or without stipules, bearing epidermal water- or chalk-glands.

One family with 10 genera and about 260 species; cosmopolitan.

The stem is erect or sometimes climbing, as in the well-known greenhouse plant *Plumbago capensis*; in *Acantholimon* the plants are low-growing cushion-forming undershrubs. The leaves are often radical, as in the two British genera *Armeria* (Thrift) and *Limonium* (*Statice*) (Sea-lavender); in *Plumbago* and allied genera they have a well-developed flat blade, but frequently they are very narrow and grass-like (*Armeria*), or needle-like, as in most species of *Acantholimon*. This reduction of leaf-surface is associated with the habitat, many members being salt-steppe, maritime or alpine plants, as are our British species of *Armeria* and *Limonium*. In species living in dry situations the water-pores on the leaf-surface are sunk below the level of the other epidermal cells or otherwise protected; the addition of chalk to the secretion is characteristic of xerophilous species, the chalk-scales covering the glands or even the whole leaf-surface and acting as a very efficient check to transpiration. Another anatomical character of interest is the presence in individual genera of vascular bundles in the pith and cortex, of phloem in the wood, and of strands of sclerenchyma in the primary cortex and in the pith.

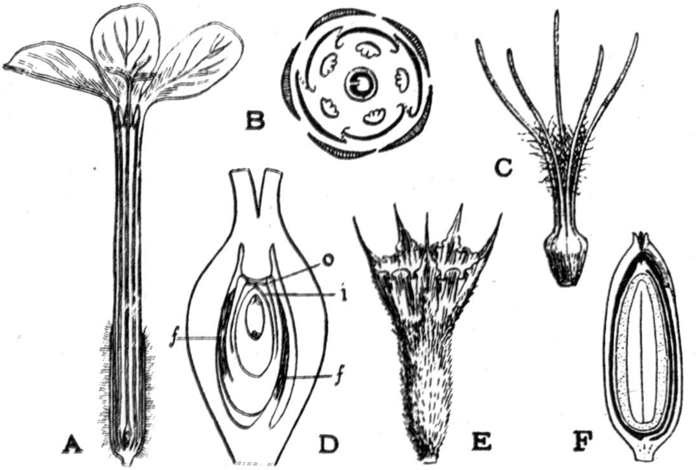

FIG. 213. A. Flower of *Plumbago rosea* in vertical section. B. Floral diagram of *P. europaea*. C. Pistil of *Armeria maritima*. D. Vertical section of ovary of *Armeria alpina*; *f*, funicle; *o*, *i*, outer and inner integuments of ovule. E. Persistent calyx enclosing fruit in *Limonium Thouini*. F. Vertical section of fruit of same. C–F enlarged. (A, C, E, F after Baillon; B after Eichler; D after Wettstein.)

The inflorescence in *Plumbago* and allied genera is a raceme, each flower having two lateral bracteoles which always remain sterile. In the *Staticeae*, on the contrary, branching occurs in the axil of one of the bracteoles, the other being sometimes suppressed (*Armeria*) so that in place of the single flower in *Plumbago* a few-flowered cyme is developed. The whole inflorescence shews widely-differing forms. In *Armeria* the cymes are closely aggregated into a head, the bases of the outer scarious bracts being prolonged down the scape and uniting to form a sheathing tube. In

Limonium the cymes are generally three-flowered and are arranged sometimes in a secund manner, in often widely-branching panicles. The calyx is tubular with a five-cleft limb and plaited between the segments; in some species of *Armeria* it is spurred; in *Plumbago* it is covered with glandular hairs; it is often scarious and coloured, as in *Limonium*. The petals, which are often free almost to the base, have a contorted aestivation. The stamens are hypogynous in *Plumbagineae*; in *Staticeae* they are inserted on the petals. They dehisce introrsely. There is no evidence of an outer antesepalous whorl of stamens; even the vascular bundle-traces which occur in Primulaceae are absent. The ovary is always superior; it is often five-ribbed. A peg-like mass of conducting tissue projects from the base of the style and spreads over the micropyle (fig. 213, D). Pollination is effected by pollen-eating flies, short-tongued bees, nectar-sucking butterflies and the like, but self-pollination also occurs. Dimorphic flowers occur in *Limonium vulgare* and *Plumbago*. The fruit is enveloped in the calyx, the light membranous character or the sticky glandular hairs (*Plumbago*) of which aid in distribution.

The pericarp is thin and papery, it may become hardened at the apex and separate as a lid, or tear irregularly round the base, or remain closed. The seed completely fills it; a thin seed coat surrounds the mealy endosperm.

The family is cosmopolitan, but favours especially salt-steppes and sea-coasts. *Plumbago* (10 species) occurs in the tropics of both Old and New Worlds; one species, *P. europaea*, is endemic in the Mediterranean region. *Acantholimon* (80 species) has its main development in Persia, spreading eastward to Tibet and westward as far as Greece. *Armeria* (about 50 species) is an alpine, arctic and maritime genus of the north temperate zone, reappearing on the Andes of Chile. *Limonium* (more than 120 species), the largest genus of the family, is cosmopolitan but occurs chiefly in maritime districts and salt-steppes in the Old World.

The ten genera contain about 260 species, the number in some genera, such as *Limonium* and *Armeria*, varying widely according to estimates of different authors. They fall into two tribes.

Tribe 1. *Plumbagineae* (including *Plumbago* and three small genera). Characterised by a simple inflorescence, both bracteoles being sterile, hypogynous stamens, and styles free only in the upper part.

Tribe 2. *Staticeae*. Inflorescence compound, one of the bracteoles being fertile, stamens epipetalous, styles free almost to the base. *Acantholimon*, *Armeria*, *Limonium* and three smaller genera.

K. V. O. Dahlgren[1] has indicated a striking difference in the embryogeny of the two tribes. In *Staticeae* there is the usual eight-nuclear embryo-sac, but in *Plumbagineae* only four nuclei are formed; the uppermost becomes the nucleus of the egg-cell, the lowest the antipodal nucleus, and the remaining two the polar nuclei which fuse very early. The antipodal cell soon disappears. The author comments on this extreme reduction of the gameto-phyte generation.

The Plumbaginaceae were formerly classed with the Primulales with which they have a common flower-plan, with antepetalous stamens, a unilocular ovary, and an ovule with two integuments. They shew, however, important differences, especially in the structure of the ovary, which has five styles, and the attachment of the solitary ovule, the mealy nature of the endosperm, and the absence of any trace of an outer staminal whorl.

An affinity has been suggested with the Centrospermae which Plumbaginaceae resemble in the ovary formed of several carpels but unilocular, with a basal ovule borne on a long funicle, and the embryo surrounded by a mealy endosperm; also in the presence of bracteoles and the persistent calyx. Analogous deviations from the normal arrangement of the vascular bundles in the stem occur in each order. The small tropical family Basellaceae (Centro-spermae) has flowers recalling those of Plumbaginaceae in having the five petals united below, and one series of antepetalous stamens. The characteristic campylotropous ovule and curved embryo of Centrospermae contrast with the anatropous ovule and straight embryo of Plumbaginaceae, but as the latter also occur in the Centrospermae this does not preclude the view that Plumbaginaceae are derived from the Centrospermae or from a form closely allied to that order.

Order 4. *EBENALES*

Flowers bisexual and regular, often pentamerous, diplo-stemonous, or triplostemonous, or sometimes haplostemonous

[1] *Arkiv Bot.* XIV, No. 8, 1–8 (Stockholm, 1915).

by abortion; petals united, stamens inserted on the corolla-tube; ovary superior, rarely inferior, multilocular; placentation axile with one to few ovules in each chamber. Seeds with a straight or slightly bent embryo surrounded by endosperm.

Woody plants, mainly tropical, with simple, often leathery, leaves.

The families fall into two suborders. (1) Sapotineae, comprising Sapotaceae, with a superior completely septate ovary, and a single ascending ovule with one integument in each chamber, and laticiferous sacs in stem and leaves; and (2) Diospyrineae with a superior or more or less inferior ovary, completely chambered or not chambered above, and ovules with two integuments; latex is absent. Diospyrineae contains three families which are distinguished by the following characters:

Ovary superior rarely half-inferior:

- Flowers usually dioecious, ovary completely septate ... Ebenaceae
- Flowers bisexual, ovary septate only in the lower portion ... ... Styracaceae

Ovary almost completely inferior; flowers bisexual, ovary completely septate ... ... Symplocaceae

The order is isolated, with no close affinities with the other Pentacyclic orders, and may be considered to have had a distinct origin from the Dialypetalae. There is a tendency to an indefinite number of stamens and to epigyny, both finding their highest expression in the Symplocaceae. The order represents two distinct lines of development, one represented by Sapotaceae and the other by the Diospyrineae. in which the ovule still retains the two integuments and in which also the epigynous tendency is strongly marked. Its phylogeny is extremely doubtful; Wettstein has suggested a possible affinity with the Malvales, Geraniales and Rutales, among which occur the diplostemonous androecium with a tendency to indefinite stamens by branching, and the septate isomerous gynoecium with axile placentation.

Family I. SAPOTACEAE

Flowers generally bisexual; sepals 4–8 in two isomerous whorls, or five in one whorl; petals more or less united, as many as the sepals but in one whorl, more rarely double as many in two whorls, sometimes with lateral or dorsal appendages. Stamens situated on the petals, typically in two or three whorls, the outer antesepalous whorl often reduced to staminodes or absent. Carpels as many or double as many as the number of stamens in a whorl; ovary superior with as many chambers, each with one anatropous ovule having a single integument at the base of the axile placenta; micropyle pointing downwards. Style simple, stigma inconspicuous. Fruit a berry, the outer layer sometimes sclerenchymatous; seeds few or one, with a broad hilum; testa hard, shiny; endosperm oily, sometimes absent.

Mostly trees with alternate simple entire penninerved leathery leaves, sometimes stipulate. Rows of laticiferous sacs are found in the pith and cortex and in the leaves. Flowers small, solitary or in cymose clusters in the leaf-axils, or above the scars of fallen leaves, sometimes on old stems.

Genera 40; species about 600; widely spread in the tropics.

The leaves, flowers and fruit are often clothed with unicellular two-armed hairs. The bracteoles when present are small and caducous. The sepals are slightly united at the base or free, they shew considerable variation in number and arrangement, thus 2 + 2, 3 + 3, 4 + 4, or 5. The petals follow in a single whorl, isomerous and alternating with the sepals. Occasionally a second whorl is present; they are united below, the segments having an imbricate aestivation. In *Mimusops* the number is apparently increased by the development from the base of each segment of a pair of dorsal outgrowths which may resemble the segment or be much cut. In other genera each segment bears a pair of lateral outgrowths. The stamens or staminodes follow in regular alternating whorls. When the number of carpels is doubled, the ovary-chambers alternate with the members of the two preceding staminal whorls in the same way as the petals alternate with the sepals in the two-whorled calyx.

The flowers are proterogynous; the tip of the style with the sticky stigma appearing above the corolla before the expan-

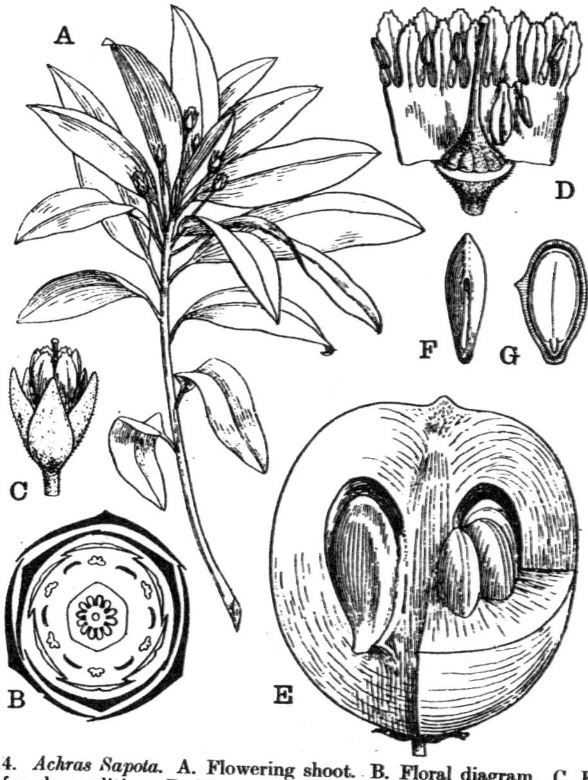

FIG. 214. *Achras Sapota*. A. Flowering shoot. B. Floral diagram. C. Flower in female condition. D. Flower with corolla opened out; a whorl of petal-like staminodes alternates with the corolla-lobes. E. Fruit; the left side in longitudinal section; the right shewing transverse and longitudinal sections. F. Seed, seen from the back, shewing the long hilum. G. Seed in longitudinal section shewing embryo. C, D enlarged; A reduced; E–G slightly reduced. (After Engler and *Flor. Bras.*)

sion of the latter; at this stage the flowers are erect, but they usually become pendulous by bending of the stalk during dehiscence of the anthers, which is generally extrorse by longitudinal slits.

The inner pulp of the berry is traversed by numerous latex-sacs; in the larger fruits the epidermis is replaced by brown cork-layers, and sclerenchymatous tissue is also found in the periphery. Sapotaceae occur throughout the tropics; they are absent from Europe and extra-tropical Asia, but in America *Bumelia* extends northwards to Illinois and southwards to Argentina. *Sideroxylum* extends as far south as New Zealand.

The sticky nature of the berries favours their distribution by birds, though in many cases the size of the fruit is a bar to this means of dispersal. Engler notes that several species with fruits of more than 2 cm. diameter occur in the West Indies and the Sunda Islands, and also in the islands of the Malagasy group, suggesting a transport of the seeds by sea-currents.

The family is a useful one. The fruits are edible, and the larger and more juicy, such as *Achras Sapota*, the Sapodilla of the West Indies, *Chrysophyllum* and *Mimusops*, are cultivated in the tropics; the endosperm is also often a source of oil, as in *Butyrospermum*. The wood generally is very hard, affording useful timber often known as iron-wood, e.g. *Sideroxylum*; the bark is frequently bitter and astringent. The latex of many species, especially of *Mimusops*, *Payena* (Malaya) and *Palaquium* (India, Malaya), is a source of gutta percha.

The largest genera are *Sideroxylum*, with 100 species in tropical and subtropical parts of the Old World, *Chrysophyllum*, 70 species in the tropics of both Worlds but chiefly of America, and *Mimusops*, with 160 species widely distributed through the tropics.

Family II. EBENACEAE

Generally dioecious; flowers rarely bisexual or polygamous, 3- to 7-merous, regular. Calyx persistent, more or less deeply cut. Corolla gamopetalous; segments with generally contorted aestivation, equal in number to those of the calyx. Stamens attached to the corolla-tube and isomerous with its segments, or twice as many or still more numerous; staminodes generally present in the female flower. Pistil rudimentary or absent in the male flowers, superior in the female and bisexual; ovary with 2 to 16 chambers, containing one or two anatropous pendulous ovules in each chamber; styles two to eight, free or united at the base; stigmas small, entire or two-lobed. Fruit fleshy or leathery, generally indehiscent, often few-chambered and one- to few-seeded by abortion. Endosperm copious, cartilaginous, often ruminate; embryo axile straight or slightly curved. Woody plants; leaves generally alternate, more rarely opposite or whorled, entire, usually leathery, exstipulate. Flowers axillary, solitary or

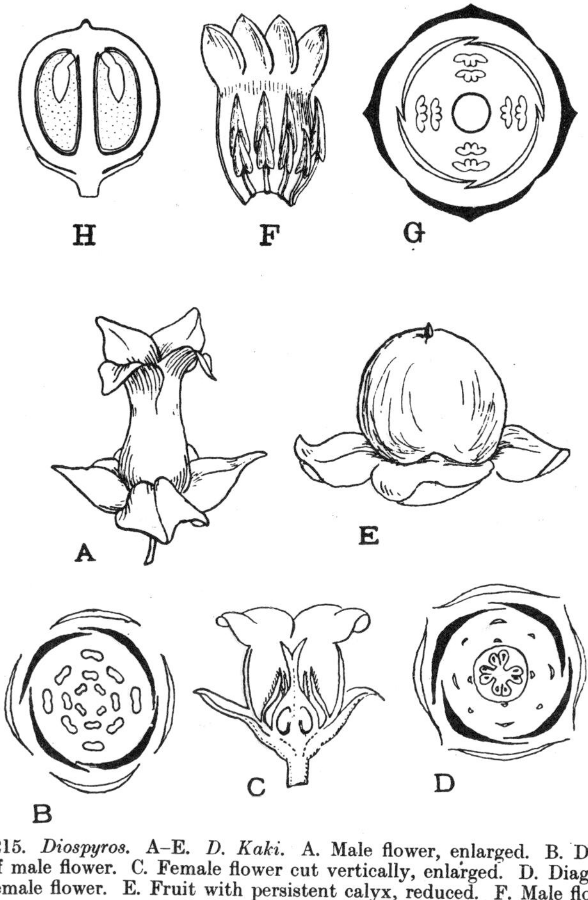

FIG. 215. *Diospyros.* A–E. *D. Kaki.* A. Male flower, enlarged. B. Diagram of male flower. C. Female flower cut vertically, enlarged. D. Diagram of female flower. E. Fruit with persistent calyx, reduced. F. Male flower of *D. tetrasperma*; corolla cut open shewing stamens; × 3. G. Diagram of male flower. H. Fruit of same cut vertically, shewing seeds with copious endosperm and slightly curved apical embryo. (A–E after Le Maout and Decaisne.)

in small cymose inflorescences, of fewer flowers in the female plants; bracteoles typically present.

Genera 5; species 320. Tropical and subtropical.

The adult stem is characterised by an extraordinarily hard and heavy heart-wood, generally black in colour, more rarely green or red; ebony is the heart-wood of various species of *Diospyros*. Cauliflorous inflorescences occur in some species of *Diospyros* and *Maba*.

The corolla is tubular, bell-shaped or rounded; the stamens, all of which are fertile, are inserted at the base of the corolla-tube or near it on the receptacle, they are usually in two whorls, but the number of stamens may be considerably increased by branching. The anther-dehiscence is longitudinal by lateral slits, more rarely, as in a section of the genus *Diospyros*, by apical pores.

The carpels are typically isomerous with the sepals and petals and opposite the former; the ovary-chambers are often divided into two by a false septum, in which case the number of carpels is apparently doubled and there is a solitary ovule pendulous from the central angle of each chamber.

The fruit is generally indehiscent forming a fleshy berry, but in a few species has a valvular dehiscence. *Diospyros Kaki*, the Kaki or Persimmon, a native of China and Japan, has a luscious fruit as big as an orange; the fruit of other species is also edible, as *D. virginiana* (United States) and *D. Lotus*, the Date-plum of the eastern Mediterranean region.

There are five genera with 320 species; the family finds its chief development in India and the Malay Archipelago in the two genera *Diospyros* and *Maba* with more than 200 species. *Royena* (20 species) and *Euclea* (17 species) are exclusively African, and *Tetraclis* is a monotypic genus from Madagascar. Both *Diospyros* and *Maba* are widely spread in the warmer parts of the world.

Ebenaceae are most nearly allied to Styracaceae but are distinguished by the septate ovary and generally unisexual flowers. From Sapotaceae they are distinguished by absence of latex and the presence of two ovules (each with two integuments) in each carpel.

Family III. STYRACACEAE

Flowers bisexual, regular, generally conforming to the formula S (5–4), P (5–4), A (10–8), G (3–5). Calyx tubular, with segments generally reduced to short or inconspicuous teeth, and petals free nearly to the base with imbricate or valvate aestivation. Stamens in one series but rarely isomerous with the petals, epipetalous, very rarely free from the petals; filaments united below; anthers generally dehiscing by lateral slits. Ovary superior or slightly inferior, rarely almost completely inferior (*Halesia*, *Pterostyrax*), generally three- to five-chambered in the lower part only, unilocular above; ovules anatropous, with one or two integuments, pendulous or erect, one to few in each chamber. Style cylindrical or filiform, with a capitate or inconspicuously two- to five-lobed stigma. Fruit generally a drupe, with a fleshy or dry pericarp often splitting by three valves; seeds one to few, with a thin or leathery coat and a broad hilum. Embryo generally straight in the axis of the endosperm.

Trees or shrubs, with alternate simple, entire or serrate, membranous or leathery leaves; and small or moderate-sized flowers, generally arranged in simple or compound racemose inflorescences; bracteoles small or absent.

In the majority of the members of the family the stem, leaves, inflorescence, and the outside of the calyx and corolla are covered with yellowish or brownish stellate hairs.

There are eight genera, with about 120 species, of which 100 belong to *Styrax*, the greater number in the warmer parts of America extending northwards into the southern United States; others are found in tropical Asia and northwards to China and Japan. Two small genera occur in west tropical Africa. The family does not occur in Australasia. *Styrax officinalis* (Mediterranean region) yielded the Styrax of the Ancients; *Styrax Benzoïn* is the source of gum-benzoïn, a fragrant resin which exudes from cuts in the bark; it is a native of Sumatra.

Family IV. SYMPLOCACEAE

Contains one genus, *Symplocos*, included by Bentham and Hooker in Styracaceae, from which however it is distinguished by its inferior, completely septate ovary, yielding a drupaceous fruit which is crowned by the persistent calyx-teeth. The flowers are typically pentamerous; the number of petals is however very varied (3 to 11), they are gamopetalous but divided almost to the base. The number and arrangement of the epipetalous stamens also shew great variety; they are generally numerous in two or more series but often shew evidence of an arrangement in five bundles; the filaments are often more or less united. There are about 300 species of trees or shrubs with leathery leaves. The chief centre is in India and the Malay Archipelago, a few species extending to Japan; a smaller centre is in tropical America, one species, *S. tinctoria*, extending into North America. They are absent from Africa; two species are found in Australia and as many as 13 are endemic in New Caledonia.

B. TETRACYCLICAE

Flowers with four whorls, that is with only one whorl of stamens which alternates with the petals; the carpels when equal in number with the stamens alternate with them, but in the great majority of families the number is reduced to two. The flowers are regular or zygomorphic; with few exceptions the petals are united and the stamens spring from the corolla-tube. The ovule has rarely more than one integument.

(*a*) Superae

Flowers with a superior ovary (in a few families there is a tendency to epigyny); regular, with generally isomerous stamens, as in Contortae and Convolvulales; or, as in Tubiflorae, regular or zygomorphic, in the latter case with reduction in the number of stamens. In the first order, Oleales, the petals are often free, and in the family Salvadoraceae the ovule has two integuments.

Five orders: Oleales, Contortae, Convolvulales, Tubiflorae, Plantaginales.

Order 1. *OLEALES*

Flowers bisexual or unisexual by abortion; regular, generally tetramerous, with united or sometimes free petals, and stamens generally reduced to two. Ovary superior, of two united carpels, bilocular with one or two pendulous or basal anatropous ovules with one or two (Salvadoraceae) integuments.

Woody plants with opposite simple or pinnately compound leaves; stipules absent or small (Salvadoraceae); flowers often small in racemose inflorescences.

The families of this order are included in the next in the Systems of Bentham and Hooker, and of Engler, the joint orders being termed Gentianales and Contortae in the respective Systems. But, as Wettstein points out, there are essential differences between the two families Oleaceae and Salvadoraceae (Oleales) on the one hand and the families included here in the Contortae on the other; such are the placentation of the ovules, the absence of internal phloem in Oleales, the dimerous androecium in Oleaceae which constitutes by far the greater part of the order (an advanced character which is associated with the apparently primitive character of frequent polypetaly) and (in Salvadoraceae) an ovule with two integuments. The frequent presence of compound leaves (in Oleaceae), an exceptional character in sympetalous families, is also worthy of note. It is difficult to suggest a point of origin for the Oleales; Wettstein derives them from Celastrales, which are woody plants with simple or pinnately compound (Staphyleaceae) leaves, tetracyclic flowers with generally isomerous stamens, and ovary-chambers containing one or two pendulous or erect ovules; indications of sympetaly also occur.

Family I. OLEACEAE

Flowers bisexual, rarely unisexual, regular, conforming to or derived from the formula S4(or 5), P4(or 5), A2, G(2). Calyx small, rarely absent; petals united, sometimes free or absent, aestivation imbricate or valvate, rarely contorted; stamens epipetalous. Ovary bilocular with generally two pendulous or ascending anatropous ovules in each chamber; style simple, stigma bilobed. Fruit various, containing one or a few seeds. Endosperm present or absent; embryo straight.

Woody plants with generally opposite simple or imparipinnate leaves without stipules, and terminal or axillary compound racemose inflorescences.

Genera 22; species about 400. Temperate, subtropical and tropical regions.

The plants are shrubs, as *Syringa* (Lilac) and *Ligustrum* (Privet); sometimes climbing as *Jasminum*, or trees, as *Olea* (Olive) or *Fraxinus* (Ash). Several buds are sometimes found one above the other in a single leaf-axil (accessory buds) in floral as well as in vegetative shoots, as in *Syringa* and *Fraxinus*.

The scales of the winter-buds, where these are found, represent the whole leaf, as in *Syringa*, or the leaf-base only, as in *Fraxinus*; in *Syringa* additional protection is afforded by a resinous excretion.

Extrafloral nectaries are described in various species. Peltate hairs are very common in the family, as is also the occurrence of small crystals of calcium oxalate. The flowers are small, generally in a compound raceme as in *Syringa*, *Ligustrum*, *Fraxinus*; in *Fraxinus* the dense inflorescences of small polygamous flowers appear before the leaves from lateral or terminal buds on last year's shoots. In the Chinese genus *Forsythia*, cultivated in shrubberies, and in *Jasminum nudiflorum* (China), common in gardens, the bright yellow flowers, which also appear before the leaves, stand one to three together at the end of short scale-bearing shoots. Each

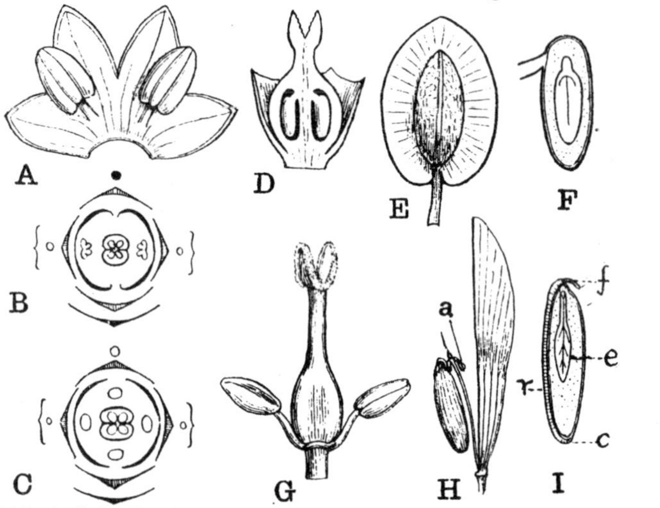

FIG. 216. A, B, D–F. *Olea europaea*. A. Corolla with stamens opened out, × 5. B. Floral diagram. D. Calyx and ovary in vertical section, × 8. E. Drupe, the fleshy coat partly cut away to shew the stone, slightly reduced. F. Seed, slightly enlarged, in vertical section. C. Floral diagram of *Tessarandra fluminensis*. G–I. *Fraxinus excelsior*. G. Bisexual flower, enlarged. H. Samara, with half removed and the seed pulled out. The seed is borne on the long placental axis, on the top of which are seen the aborted ovules, *a*. I. Seed in vertical section, × about 1½; *c*, chalaza; *r*, raphe; *f*, funicle; *e*, embryo embedded in endosperm. (A, D–F after Knoblauch; B, C after Eichler.)

flower stands in the axil of a bract and a pair of bracteoles are typically present; but both bracts and bracteoles are often very caducous. The flowers are generally bisexual, but *Fraxinus* is polygamous or dioecious, as also are species of *Olea*.

The calyx is absent in *Fraxinus excelsior* (fig. 216, G) and other species of the genus; when present it is usually tetramerous, the sepals being median and transverse; when a fifth sepal is present, as in species of *Jasminum*, it is either anterior or posterior; the aestivation is valvate. The petals are isomerous and alternate with the sepals; aestivation may be valvate (*Syringa*) or imbricate (*Jasminum*); in *Fraxinus* the petals are free or united in pairs by the insertion of the stamens at their base. Sections of *Fraxinus* and *Olea* are characterised by apetaly, as, for instance, in *F. excelsior*. Occasionally more than five sepals and petals are present, as in *Jasminum* (*J. nudiflorum* has six); the number rarely exceeds six. The most general

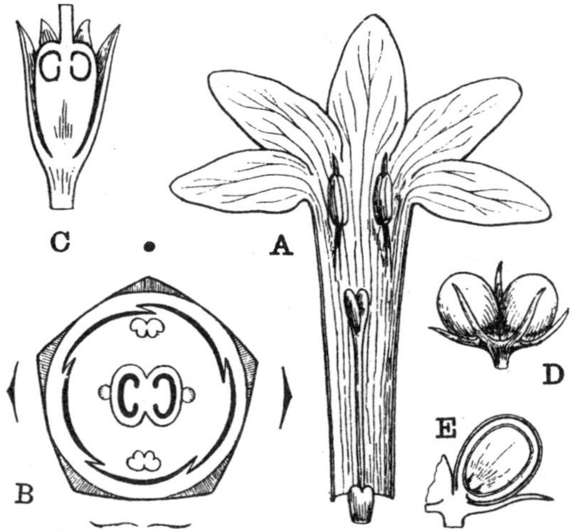

FIG. 217. A–C. *Jasminum azoricum*. A. Corolla cut open and pistil. B. Floral diagram, the stamens are median. C. Calyx and ovary in longitudinal section. D. Fruit of *J. floribundum*. E. One half of same cut longitudinally to shew seed and embryo. A and C enlarged. (After Eichler and Engler.)

form of corolla is a tube with more or less spreading limbs (*Syringa, Ligustrum, Olea, Jasminum*). The two stamens are generally transverse, but sometimes median (fig. 217, B); both positions are found in the same genus; occasionally there are four, as in the monotypic American genera *Hesperella* and *Tessarandra* (fig. 216, C), where they alternate with the petals; three or five may also occur. The anthers dehisce laterally by a longitudinal slit. The two carpels alternate with the stamens and are therefore usually median; a three- or four-chambered ovary sometimes occurs. There are generally two ovules in each chamber, pendulous or ascending from the axile placenta; sometimes only one is present, while in *Forsythia* and other genera there are from four to ten.

Many species are visited by insects, which are attracted by the colour and smell of the flower and presence of nectar, and various adaptations occur favourable to cross-pollination. Thus long- and short-styled flowers are known in species of *Jasminum* and in *Forsythia*. Both Lilac and Privet are homogamous and, in the absence of insect-visits, self-pollinated. Ash is wind-pollinated. In *Fraxinus* the fruit is a dry, indehiscent one-seeded (rarely two-seeded) samara, only one (or two) of the four ovules developing to seed. In *Forsythia* and *Syringa* it is a few-seeded capsule dehiscing loculicidally into two valves; in *Olea* a drupe with a thick mostly one-seeded endocarp; in *Ligustrum* a few-seeded berry; in *Jasminum* a berry which is often divided into two one-seeded lobes by a vertical constriction (fig. 217, D). Endosperm when present is oily.

The greater number of species are found in the Indo-Malayan and Chinese-Japanese areas. Six genera (13 species) are European, namely, *Fraxinus*, *Syringa*, *Phillyrea* (South), *Ligustrum* (one species), *Forsythia* (one species in the Balkan peninsula) and *Jasminum* (one species in the Mediterranean region). *Fraxinus excelsior* (Ash) and *Ligustrum vulgare* (Privet) are British. Olive (*Olea europaea*), the most important member of the family, an evergreen tree which reaches a great age, is a native of Western Asia, whence it was introduced into Southern Europe before the Christian era; olive oil is expressed from the fleshy drupe.

The family is divided into two subfamilies and three tribes depending on the nature of the fruit and position of the seed.

Subfamily I. OLEOIDEAE. Seeds pendulous; fruits not vertically constricted.

Tribe 1. *Fraxineae*. Fruit a samara. *Fraxinus*, 60 species chiefly in North America, Eastern Asia and South Europe. *F. Ornus* (Manna Ash), a native of South Europe, is a small tree with whitish flowers, which yields a sweet exudation from cuts in the bark.

Tribe 2. *Syringeae*. Fruit a loculicidal capsule. *Syringa*, 20 species, Europe and Asia; *S. vulgaris* is the Lilac. *Forsythia*, several species in E. Asia, one in the Balkan Peninsula.

Tribe 3. *Oleeae*. Fruit a drupe or berry. *Olea*, 35 species, mainly in South Africa, India, Australia and Polynesia. *Ligustrum*, 50 species in Europe, Asia and Australia.

Subfamily II. JASMINOIDEAE. Seeds generally erect; fruits divided into two parts by a vertical constriction except when one carpel fails to develop. *Jasminum*, 200 species chiefly in the warmer parts of the western hemisphere.

Family II. SALVADORACEAE

A small family of woody plants with nine species contained in three genera in the tropical regions of Asia and Africa. The small bisexual or unisexual flowers are 4- or 5- merous, with two carpels forming a one- or two-chambered ovary, each chamber containing one or two erect anatropous ovules. The petals are free or united, the stamens hypogynous or epipetalous and the ovule has two integuments. The fruit is a berry or drupe, generally one-seeded, and the seed is without endosperm.

The plants are shrubs or trees with opposite entire leaves with very small stipules and racemes or spikes of small flowers.

A family of doubtful affinity placed by Baillon, followed recently by Engler, among the Dialypetalae near Celastraceae owing to the free petals and hypogynous stamens in two out of the three genera and ovule with two integuments. Among sympetalous families it is most nearly allied to Oleaceae, in which free petals occasionally occur (*Fraxinus*), and the two families are regarded by Wettstein as representing an order derived from the Celastrales, which are woody plants with a haplostemonous flower, a frequently bilocular ovary with a single anatropous pendulous or erect ovule, occasionally with only one integument, and a tendency to union of the petals.

Order 2. *CONTORTAE*

Flowers bisexual, regular, hypogynous, gamopetalous, conforming to the formula S4 to 5, P4 to 5, A4 to 5, G2. Corolla-segments usually with twisted aestivation. Stamens epipetalous; a nectar-secreting disc generally round the base of the ovary. Carpels united, or free below with united styles; ovules few or numerous on parietal or axile placentas, with one integument; embryo usually surrounded by more or less endosperm. Herbs, shrubs or trees with generally opposite, exstipulate, often simple leaves. An important anatomical character is the presence of internal phloem.

Of the four families Loganiaceae and Gentianaceae stand somewhat apart from the Apocynaceae and Asclepiadaceae, which are very closely allied and shew a remarkable specialisation of androecium and stigma in relation to insect-pollination.

Wettstein suggests a common origin for this order and the Tubiflorae, regarding them as somewhat parallel series, adaptation to insect-pollination having progressed in the one in association with a regular flower, in the other with the development of zygomorphy. Hallier goes further and combines the two orders under the name Tubiflorae and seeks their origin in the Linaceae among Dialypetalae. Wettstein, on the other hand, would derive the two orders from Rosales.

A. Stamens four to five; carpels completely united, style not expanded above; fruit a capsule, berry or drupe; endosperm copious, embryo small. Laticiferous tubes absent.

Generally woody plants, stipules obvious or reduced, ovary bilocular ... ... I. Loganiaceae.

Herbs with exstipulate leaves, ovary typically unilocular with parietal placentation ... II. Gentianaceae.

B. Stamens five; carpels generally united only by the styles; style swollen into a stigma-head which bears the stigmatic surface below the apex; fruit generally a pair of follicles; embryo large. Laticiferous tubes present.

Pollen transferred directly from the anthers III. Apocynaceae.

Pollen transferred by means of specialised translators IV. Asclepiadaceae.

Family I. LOGANIACEAE

Flowers usually bisexual, regular, 4- or 5-merous (with rare pleiomery in corolla and androecium). Calyx-segments imbricate; corolla gamopetalous, segments valvate, imbricate or contorted. Stamens as many as the corolla-segments, inserted at the mouth of the corolla or on the tube; disc absent or small. Carpels two, completely united; ovary superior, generally completely bilocular; style usually simple with capitate or bilobed stigma; ovules usually indefinite, amphitropous or anatropous. Fruit a capsule, berry or drupe; seeds with endosperm.

Herbs, shrubs or trees with simple opposite stipulate leaves, and generally cymose inflorescence. Latex absent.

Genera 33; species about 600. Mostly tropical.

Many are woody climbing plants (lianes), either twining, as *Fagraea*, or tendril-climbers, as in many species of *Strychnos*, the tendrils of which are hook-like axillary shoots, which on twining round a support become thickened and lignified; the subtending leaf is reduced to a scale. In other species of this genus axillary or terminal shoots become modified into straight or curved thorns. The stipules are variously developed; they may be conspicuous interpetiolar structures, or reduced to a stipular line connecting the petioles (*Strychnos*). In *Fagraea* and others they form intrapetiolar structures by which the growing point is

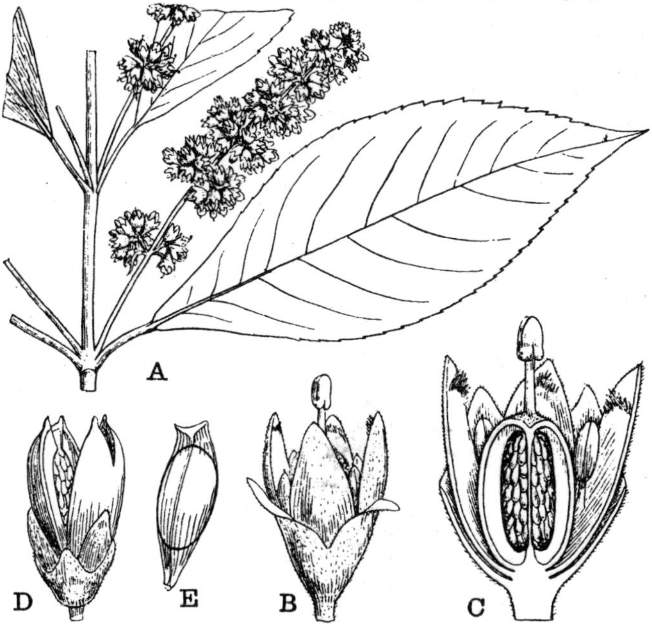

FIG. 218. *Buddleia americana*. A. Leaf and inflorescence, × ¾. B. Flower, × 6. C. Flower in longitudinal section, × 8. D. Fruit dehiscing, × 4. E. Seed, × 22.

originally protected. Certain anatomical characters are constant in larger or smaller groups. The larger subfamily Loganioideae is characterised by presence of internal phloem and absence of glandular hairs, and the smaller Buddleioideae by absence of internal phloem and presence of glandular hairs. Laticiferous tissue is always absent. *Strychnos* and a few allied genera are characterised by isolated groups of soft bast in the wood (phloem-islands). The cymose inflorescence shews various degrees of development; bract and bracteoles are generally present and dichasial or monochasial developments according as branching occurs in both bracteoles or in one only. Sepals, petals and stamens generally follow in regularly alternating 4- or 5-merous whorls, pleiomery of petals and stamens (8–16 members) occurs in two small genera; the aestivation of the corolla-segments affords good tribal characters. The mouth of the corolla-tube often bears a crown of unicellular hairs. The anthers dehisce by two longitudinal slits. The two carpels are generally median; rarely oblique or transverse. In some species of *Strychnos* the ovary is unilocular, and in *Fagraea* it is more or less completely bilocular; but in the great majority it is completely bilocular.

The dehiscence of the capsular fruit is generally septicidal, the valves breaking away from the central placenta. The berry of *Strychnos* may reach a considerable size.

The seeds are generally small and flattened or ellipsoidal; in the dry fruits they are often winged. In *Strychnos* they attain a considerable size, and are disc-like with a central hilum; the seed-coat contains powerful alkaloids (strychnine, brucine). The endosperm, which is present in greater or less quantity, is fleshy or cartilaginous, and contains oil and proteid. The embryo is straight; *Strychnos* has leaf-like cotyledons.

The family is poorly represented outside the tropics; *Gelsemium* extends to North America and China; *Logania* and two other genera occur in New Zealand. *Strychnos* (about 150 species) and *Buddleia* (100 species) are generally distributed in the tropics, and the other genera have restricted areas, mainly tropical.

There are two well-marked subfamilies.

I. Loganioideae. Internal phloem and no gland-hairs; leaves entire; flowers 4- or 5-merous. The 21 genera are subdivided into six tribes, chiefly according to the nature of the fruit, and the aestivation of the corolla. Chief genera: *Gelsemium*, two species; *G. sempervirens*, a twining shrub with axillary yellow funnel-shaped flowers, a native of the South Atlantic United States, contains the alkaloid gelsemin in its rootstock; *Logania*, 21 species; *Strychnos*, 150 species; *Spigelia*, 30 species, American; *Fagraea* (30 species, Indo-Malaya to Pacific Islands and Australia) has large flowers.

II. Buddleioideae. No internal phloem, gland-hairs present; leaves generally toothed or lobed and the stipules generally reduced to a line connecting the leaf-bases; flowers generally tetramerous. 10 genera. *Nuxia* (30 species, Africa, Madagascar and Mascarene Islands); *Buddleia* (fig. 218) (100 species).

The seeds of the tree *Strychnos Nux-vomica* (tropical India and Malaya), yield the alkaloids strychnine and brucine. The bark of *S. toxifera* yields Curare poison, used by the South American Indians for their arrows. The seeds of *S. Ignatia* are the Ignatius Beans, used in India as a remedy for cholera. Other members of the family are used in various parts of the world for the medicinal and poisonous alkaloids contained in bark, seed or elsewhere. Several species of *Buddleia* are grown as ornamental plants.

Loganiaceae, especially the subfamily Loganioideae, shew points of resemblance with the Rubiaceae (q.v.) in the opposite stipulate leaves and regular isostemonous flower, but are distinguished by the superior ovary and the presence of internal phloem; the presence of genera with half inferior ovaries in each family (e.g. *Mitreola* in Loganiaceae) makes the connection still closer. The Buddleioideae, on the other hand, with their toothed leaves and falsely tetramerous flower, suggest an affinity with Scrophulariaceae.

Family II. GENTIANACEAE

Flowers bisexual, regular, generally 4- to 5-merous, with reduction to two in the pistil. Calyx variously divided, aestivation generally imbricate. Petals united below into a tube varying in form, segments erect or spreading with generally contorted aestivation. Stamens as many as and alternating with the corolla-segments, variously placed on the corolla; anthers attached at the base or the back, generally dehiscing introrsely by two longitudinal slits. Ovary superior (half-inferior in *Villarsia*), generally unilocular with two variously developed parietal placentas which rarely meet in the centre forming two chambers; ovules generally very numerous, anatropous or half-anatropous; style simple with undivided or bilobed or bipartite stigma; glandular disc-developments are frequently found at the base of the ovary. Fruit generally a membranous or leathery capsule splitting septicidally into two valves. Seeds small and numerous, testa various; embryo small, endosperm copious.

Generally annual or perennial herbs with opposite entire exstipulate leaves and cymose flowers.

Genera 70; species about 800. Distribution world-wide, but mainly temperate.

The plants are herbs, either annual, or perennial by an underground rhizome, rarely shrubs. They generally grow erect and have a marked dichotomous branching; the Asiatic *Crawfurdia* has a climbing stem. Others are low-growing and caespitose, often forming large clusters, as in the alpine Gentians.

The leaves are opposite and decussate, except in *Menyanthes* and a few allied aquatic or marsh genera, where they are radical or alternate.

A saprophytic habit has originated in several parts of the family in single or small groups of closely allied genera, chiefly American, but also Asiatic and African. The plants are small, slender, low-growing herbs containing little or no chlorophyll with leaves reduced to scales. *Voyria*, with

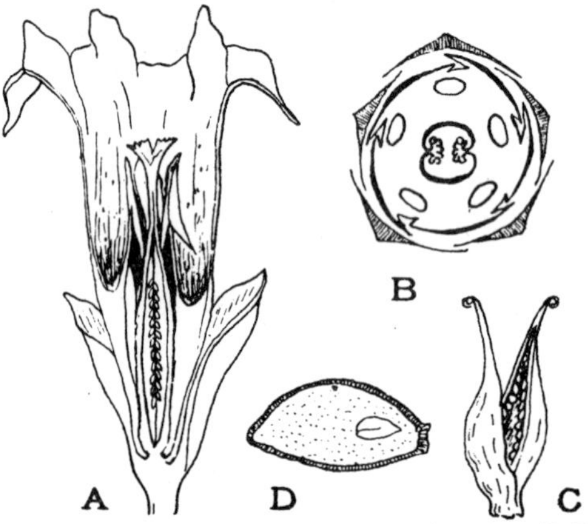

FIG. 219. *Gentiana acaulis.* A. Flower in longitudinal section. B. Floral diagram (of *G. verna*). C. Fruit. D. Seed, enlarged. (After Le Maout and Decaisne; B after Eichler.)

brilliant-coloured scale-leaves, has three species in Guiana, and the nearly allied *Leiphaimos* (fig. 220) 20 species, chiefly in the tropical forests of America and the West Indies, with two in tropical Africa. *Cotylanthera* is Asiatic, while *Bartonia* and *Obolaria*, both of which contain chlorophyll, are Atlantic North American.

Menyanthes and the allied genera with alternate leaves have been separated under the name Menyanthoideae as a distinct subfamily from the rest of the genera, Gentianoideae, and anatomical characters support this distinction, the Gentianoideae having bicollateral bundles in the stem, while in Menyanthoideae they are collateral; in the latter also intercellular hairs occur comparable with those in Nymphaeaceae.

The inflorescence is generally cymose, often dichasial, resembling that of Caryophyllaceae, the lateral branches frequently passing into monochasia; sometimes it is reduced to a few flowers or a single axillary or terminal one, as in some Gentians; true spikes or racemes are rare. Bracts and bracteoles may be present or absent. There are rarely more than four or five members in each whorl; *Blackstonia* (*Chlora*) (Yellow Wort) has six to eight. The calyx is sometimes divided almost to the base, generally however it forms a tube with teeth or segments which are imbricate in the bud, more rarely valvate or open. The corolla shews great variety in form; for instance among our British genera it is rotate in *Blackstonia*, salver-shaped in *Cicendia*, funnel-shaped in *Erythraea*, and rotate-cylindrical, bell-shaped, funnel-shaped or salver-shaped in *Gentiana* (fig. 219). In the subfamily Gentianoideae, which includes the great majority of the genera, the corolla-lobes are with few exceptions twisted to the right in the bud, in the Menyanthoideae they are induplicate and valvate. The throat is often fimbriate or scaly, and in *Swertia* nectar-secreting pits, often with fringed margins, are found near the base of the segments.

The stamens, which are generally equal in length, are inserted at very different heights on the corolla-tube; the filaments are slender, rarely broadening at the base. The anthers are most frequently dorsally attached and versatile, sometimes basifixed; *Exacum* (tropical Asia) is exceptional in having dehiscence by two apical pores; after escape of the pollen they may become spirally twisted (as in *Erythraea*); in a few genera they bear glandular structures; occasionally, as in many species of the saprophytic genera *Leiphaimos* and *Voyria* they unite laterally, forming a tube. Dimorphic flowers are frequent, as in the Bog-bean (*Menyanthes*), and pleomorphy occurs in the monotypic Brazilian genus *Hockinia*. There is considerable variation in the size and shape of the pollen-grain, the sculpture of its extine and the number and position of the germ-pores, and the division of the family into tribes and subtribes by Gilg, in the *Pflanzenfamilien*, is based primarily on pollen-characters. There is also much variety in the form of the nectar-secreting disc-developments, which are usually situated at the base of the ovary; in the length of the style, which may conspicuously exceed the corolla or be altogether absent; and the stigma may be entire with great diversity of form, or more or less variously bilobed. Characters derived from the stigma are largely used by Bentham and Hooker for the discrimination of genera.

The two carpels have generally a median, more rarely a transverse position; there is great variety in the development of the parietal placentas, which are very slightly developed or protrude so far as to meet in the centre and form an axile structure, the ovary becoming bilocular, a characteristic of the tribe *Exaceae*; occasionally it grows out again to form an apparently 4-locular ovary. The ovules have a very short funicle and are often sunk in the tissue of the placenta. Rarely, as in *Pleurogyne*, more or less of the inner surface of the carpels is ovule-bearing (superficial placentation).

The flowers shew many adaptations for pollination by insect-visits; such are the brilliant colour and the association of the plants in large numbers, as in the alpine Gentians; the presence of nectaries at the base of the calyx, on the corolla, or very generally at the base of the ovary, and the frequency of dimorphy, proterandry and proterogyny. In *Gentiana* the flowers are adapted to very different types of visitor—*G. lutea* with a rotate corolla, yellow in colour, and the nectar freely exposed, is adapted to short-tongued insect-visitors; *G. Pneumonanthe* has a bright blue corolla with a long tube and is visited by humble-bees; *G. verna* has a still longer and narrower tube and is visited by Lepidoptera.

The numerous small seeds are borne on the incurved edges of the pair of valves resulting from the septicidal dehiscence of the capsule; in the Menyanthoideae the dehiscence is more or less irregular. The seeds are variously shaped, with a smooth, ribbed, warty, reticulately or variously marked testa. Investigation of some saprophytic species reveals marked exceptions to the copious endosperm and small embryo which characterise the family; in *Leiphaimos* (fig. 220) the ovules are straight and naked, many remain sterile, and the fertile develop seeds which contain only a few endosperm cells and an undifferentiated embryo consisting of a row of cells (1–4). Guerin has observed the frequent occurrence of numerous antipodal cells in the embryo-sac.

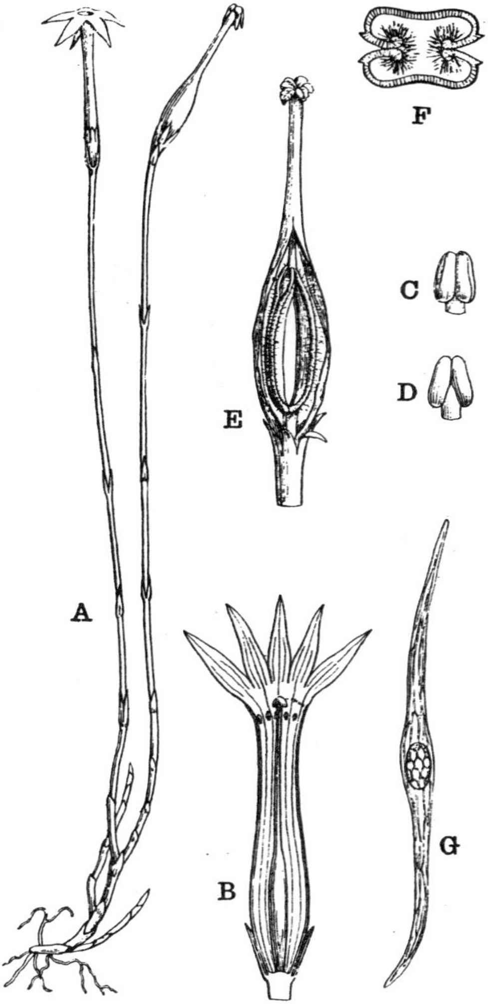

FIG. 220. *Leiphaimos aphylla*. A. Plant shewing flower and fruit, × ¾. B. Flower in longitudinal section, × 1½. C, D. Front and back views of stamen, × 15. E. Fruit surrounded by withered corolla, cut open shewing placentas bearing the thread-like seeds, × 1½. F. Transverse section of fruit, × 7. G. Seed, × 45. (A, C, D, F, G after *Flor. Bras.* and Gilg.)

The family has a world-wide distribution, and representatives adapted to very various conditions; for instance, among our British species dry sand-loving plants, as *Cicendia filiformis* (rare in Britain), marsh-plants, as *Menyanthes* (Bog-bean), floating water-plants, as *Limnanthemum peltatum*, grass-land plants, as *Blackstonia* (*Chlora*) *perfoliata* or *Erythraea Centaurium*, or alpines, as *Gentiana*. The largest genus *Gentiana* (400 species) is distributed over the whole of Europe (including arctic but excluding South Italy and Greece), the mountains of Asia, South-east Australia and New Zealand, the whole of North America and along the Andes to Cape Horn; it is absent from Africa. Some genera, on the other hand, have a remarkably local distribution; several, for instance, are confined to limited areas in tropical America.

The family falls naturally into two subfamilies.

I. GENTIANOIDEAE. Leaves opposite and decussate; corolla-lobes twisted in bud. 65 genera.

II. MENYANTHOIDEAE. Leaves alternate; corolla-lobes induplicate and valvate in bud. Five genera of aquatic or marsh plants.

The Gentianoideae are variously subdivided into tribes by characters based on the length of style, the size and importance of the placentas in the subdivision of the ovary, the development of the anther-connective, and, in more recent arrangements, the nature of the pollen.

The very general presence of bitter principles in all the vegetative parts, especially in the rhizomes and roots, has given a medicinal value to many species, as *Gentiana lutea* (and other species) and *Menyanthes trifoliolata*.

Gentianaceae are allied to Loganiaceae, from which they differ in their usually unilocular ovary with parietal placentation, their exstipulate leaves, and the presence of a bitter principle.

Family III. APOCYNACEAE

Flowers bisexual, regular, generally pentamerous (sometimes tetramerous) with reduction to two in the pistil. Calyx generally divided almost to the base, segments with quincuncial arrangement. Corolla generally salver- or funnel-shaped, aestivation of the segments contorted, the throat or interior of the tube often with hairs, scales or other outgrowths. Stamens on the tube or throat of the corolla and alternate with its segments; filaments short, anthers introrse, free or connate and often adhering to the stigma, the cells sometimes empty below and prolonged into spines; pollen granular; disc rarely absent, entire or lobed or of 2–5 scales. Carpels rarely exceeding two, free or connate; superior or sometimes partly inferior (as in *Plumiera*); style simple, often with a thickened stigma-head; ovules ∞, anatropous, generally pendulous. Fruit various; seeds often flat, winged or comose; endosperm cartilaginous or fleshy, sometimes absent.

Generally woody climbers, more rarely erect, or rarely perennial herbs; latex present in unsegmented tubes; stem-bundles bicollateral. Leaves simple, usually opposite, more rarely whorled or alternate, entire, rarely with small inter-petiolar stipules. Flowers solitary or in racemose inflorescences passing into cymose; bracts and bracteoles present.

Genera 155; species about 1000. Mainly in the warmer parts of the world.

The perennial herbaceous or suffruticose habit of our British representative *Vinca* (Periwinkle) obtains only in a few genera; erect shrubs and trees are not very numerous, such are the Oleander (*Nerium*), a large shrub of the Mediterranean region growing on riverbanks, or the east Asiatic *Alstonia*, which comprises shrubs or trees reaching 100 ft. in height and with generally whorled leaves. In some genera the erect stem has a tendency to become succulent by development of a thick parenchymatous cortex; an extreme case is seen in *Adenium*, a small genus from tropical Africa and Arabia, with a thick tuber-like stem with fleshy branches and spirally arranged more or less fleshy leaves. The commonest type is the climbing shrub, many species forming lianes in the tropics of both hemispheres, having generally a twining stem, but occasionally, as in species of *Landolphia* and *Clitandra*, the stems of the inflorescences are more or less tendril-like. The flowers are very rarely solitary, as in *Vinca*, generally they are in several- or many-flowered inflorescences which correspond with the leaf-arrangement, being opposite-decussate when the leaves have the usual arrangement, or umbellate when, as in *Alstonia*, the leaves are whorled; in the ultimate branches dichasial or monochasial cymes are developed.

The arrangement of the sepals, which are often united only at the base, is quincuncial, the odd member, the second in the order of development, being posterior. Gland-structures varying in form and number are frequently found at the base of the calyx. While the salver-shape (as *Vinca*) and funnel-shape (as *Nerium*) are the commonest forms of corolla, other shapes also occur, as for instance in *Allamanda*, a well-known greenhouse climber with handsome bell-shaped flowers. The hairy or other outgrowths frequently occurring at the throat or on the tube may be compared with the definite corona of the Asclepiads. The right or left direction of the twisting of the corolla-segments is constant in individual genera; a few genera are characterised by valvate aestivation.

The position of the stamens, at the throat, or deeper in the corolla-tube, varies, but the filaments are always short, and the anthers as a rule included in the tube. Each half of the anther has two cells; the inner and outer cells are similar or the former are shorter than the latter which are more or less drawn out into tails empty of pollen and becoming thickened to form spine-like processes which may be compared with the more definite appendage in the Asclepiads. The pollen is generally of simple form, spherical with a fine granulation and three equatorial bands; in the genus *Condylocarpus* the grains remain united in tetrads.

The nectar-secreting hypogynous disc shews great variation in form, passing from a bowl-shape through lobed structures to a gradual separation into five scales. Very rarely are there more than two carpels, a few genera having 3–5; they are most frequently free, passing above into a common simple style; where they are united the ovary is one- or two-chambered. A characteristic of the family is the cylindrical, ellipsoidal, disc-like, or otherwise-shaped enlargement of the stigma-head with which the anthers are often closely related. The stigma is often bilobed above but the receptive portion is below the division, on the median line or at the base of the head. Owing to the position of the anthers self-pollination is rendered almost impossible, and insect-visits, for which the flower is eminently adapted, are necessitated (see *Vinca minor*, fig. 221). The generally numerous ovules are situated on parietal placentas or, in two-chambered ovaries, on the partition wall.

In the most common form of fruit each carpel ripens to a follicle dehiscing by the ventral suture to allow the escape of the seeds which are generally winged (fig. 222) or crowned with a tuft of hair which assists distribution.

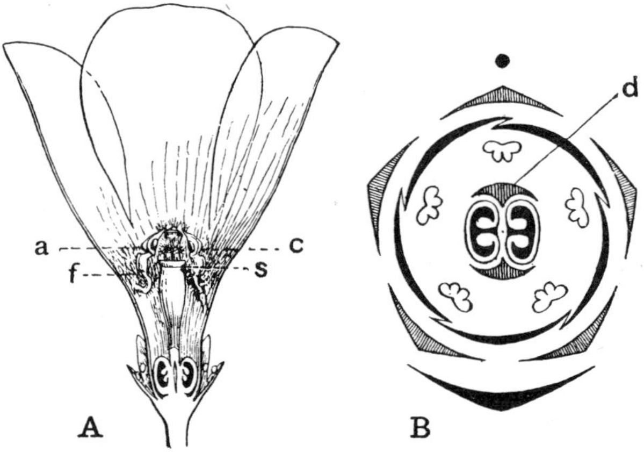

FIG. 221. *Vinca minor*. A. Flower in vertical section shewing the entire style and the stigmatic apparatus; *s*, disc with viscid edge which forms the receptive surface for the pollen; above the disc is a ring of hairs, *c*, on which pollen collects as shed from the anthers, *a*. A slender proboscis on entering the tube of the flower becomes smeared with viscid matter, which will draw pollen from the collecting hairs as the proboscis is withdrawn; this pollen will become attached to the stigmatic disc of the next flower visited. *f*, kneed filament of the stamen; the barren outer cell of each anther is drawn out into a point which bears a small tuft of hairs and comes in close contact with the extreme top of the stigma. The corolla-lobes are more spreading than shewn in the drawing. B. Floral diagram; *d*, pair of nectaries at the base of the ovary. (B after Eichler.)

Apocarpous fruits are sometimes fleshy and indehiscent, or may be one-seeded, forming sometimes a pair of winged mericarps, as in the West Indian *Cameraria*. The tropical South American genus *Condylocarpus* takes its name from the separation of the ripe fruit into one-seeded joints. In *Rauvolfia*, a large genus in the tropics of both Worlds, the fruits are free or united at the base.

In genera with a syncarpous ovary the fruit is generally indehiscent, fleshy and berry-like, as in *Landolphia* (a tropical and South African rubber-yielding genus), where the fruit has a leathery exocarp enclosing an acid pulp derived from the hairy seed-coat. In *Cerbera Manghas* the fruit is drupaceous, a fleshy exocarp surrounding a tough fibrous mesocarp and hard woody endocarp containing 1–2 seeds; it is adapted for transport by sea-currents, the fibre forming a float, as in the Coco-nut, and has a wide distribution, growing on the shores between the tides from Madagascar and further India to China, N.W. Australia and the Pacific Islands. Finally the syncarpous ovary may produce a two-valved capsule, as in *Aspidosperma* and *Allamanda*; in the latter it is prickly.

The seeds in the dry fruits are generally winged (as in *Plumiera, Allamanda*) or bear a tuft of hairs, usually developed at the micropyle, but sometimes at the base of the seed (*Kickxia*), or sometimes, as in *Strophanthus*, at both ends.

The embryo has large cotyledons which are thin, and flat, or more or less bent or rolled and enveloped in a copious endosperm, or fleshy when the endosperm is absent.

The great majority are tropical, the number of species being fairly divided between Old and New Worlds. A few genera are temperate, as *Apocynum* (five species from South Europe to China and the Eastern United States), *Nerium* (three species from the Mediterranean to Arabia, Persia and India) and *Vinca* (five species in Europe and Western Asia); in *Vinca minor*, our only British representative, the family reaches its most northerly limit in Europe. Representatives of the family are also found in New Zealand, at the Cape (as *Landolphia capensis*) and in temperate South America.

The caoutchouc-containing latex renders many species of value commercially; such are *Landolphia* (Africa), *Kickxia* (Africa and tropical Asia) and others. The latex is frequently extremely poisonous. Seeds of species of *Strophanthus* (Africa and Asia) yield a powerful drug. Some of the pulpy fruits are eaten. Many species are cultivated for their showy flowers; such are *Allamanda, Dipladenia* and *Nerium* (Oleander).

In the exposition of the family by Schumann in the *Pflanzenfamilien* two subfamilies are recognised, each containing about half the number of genera.

I. Plumieroideae. Stamens free or only slightly attached to the stigma-head; anthers generally without tails; seeds usually without a coma; corolla-lobes often twisting to the left. Includes *Plumiera* (fig. 222) and *Vinca* (fig. 221).

II. ECHITOIDEAE. Stamens closely attached to the stigma-head; anthers tailed, and the outer cells empty at the base; seeds generally with a coma; corolla-lobes generally twisting to the right.

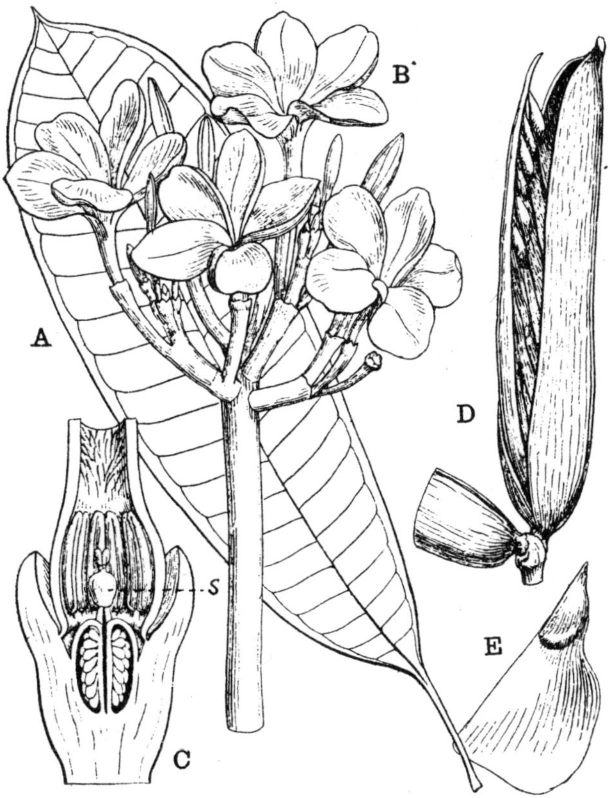

FIG. 222. *Plumiera rubra.* A. Leaf. B. Inflorescence. C. Base of flower in longitudinal section, × 12; note the half-inferior ovary; *s*, swollen receptive part of the stigma. D. Fruit; the right-hand follicle is dehiscing, the placenta bearing the seeds projects through the separated ventral edges of the carpel. The lower portion only of the left-hand follicle is seen. E. Winged seed, nat. size. A, B, D about $\frac{2}{3}$ nat. size.

The Echitoideae shew in the anther-characters, the relation of the anthers to the stigma and other points, a strong affinity with the next family Asclepiadaceae.

The subfamilies are subdivided into tribes based on characters derived from the greater or less union of the carpels, the number of the seeds, and the inclusion or not of the anthers in the corolla-tube.

Family IV. ASCLEPIADACEAE

Flowers bisexual, regular, conforming to the plan S5, P5, A5, G2. Calyx very deeply divided, arrangement quincuncial, with the odd sepal posterior. Corolla gamopetalous, often rotate, segments with contorted, more rarely valvate, aestivation; either corolla or stamens or both may bear appendages of various forms forming a single or double corona. Filaments of stamens very short or none; anthers generally very closely attached to the pistil forming a gynostegium, rarely free; pollen sometimes united in tetrads and granular, but generally the contents of each half-anther cohere in waxy pollinia; disc absent. Carpels superior, free, with numerous anatropous pendulous ovules on the ventral placentas; styles two, cohering above and dilating to form a flat five-angular or -lobed stigma-head, on the underside of which are five receptive surfaces. Fruit of two follicles; seeds generally bearing an apical tuft of hair; endosperm usually sparse, cartilaginous.

Perennial erect herbs, or shrubby or woody climbers; rarely erect shrubs or trees; sometimes succulent. Leaves opposite-decussate, rarely whorled or alternate, simple, generally entire. Flowers cymose or racemose.

Genera 280; species over 1800. Mainly tropical.

As in Apocynaceae the common habit is that of the shrubby climber; perennial herbs also occur, especially in the dry districts of South and Central Africa, the plant persisting by an underground tuber-like rhizome; trees and large shrubs are rare.

A succulent habit characterises some genera; *Hoya*, a well-known greenhouse plant, has fleshy stem and leaves; extreme cases are found in the South African *Stapelia* and allied genera which are eminently xerophytic plants with a Cactus-like habit, having thick fleshy stems and leaves reduced to thorns or scales. The Indo-Malayan *Dischidia* is epiphytic, climbing by means of adventitious roots; the leaves are fleshy with a covering of wax; *D. Rafflesiana* has modified pitcher-leaves in which debris and water collect and into which an adventitious root-system grows.

Busich[1] finds endotrophic mycorrhiza to be of frequent occurrence in the roots, especially in succulent species.

The Asclepiadaceae closely resemble Apocynaceae in anatomical characters, and have, like the latter, laticiferous tubes and bicollateral vascular bundles.

There are two types of inflorescence. In one the axis ends in a flower, while branching occurs in the axils of the two bracteoles producing a dichasial cyme; as usual, one branch grows more strongly than the other and in the ultimate branching the inflorescence becomes monochasial. The other type is racemose, either a raceme or umbel (fig. 223). When the inflorescence is axillary there is usually only one at each node, the other leaf remaining sterile or subtending a vegetative shoot. In some relatively large-leaved species, as *Asclepias Cornuti* of gardens, the inflorescence arises, not in the leaf-axil but in the space between two leaves, and often at a higher level. This may be due to a sympodial development, the inflorescence-axis being terminal, or to subsequent displacement and adhesion of the flowering axis to the shoot. The flowers are generally small; relatively large in *Ceropegia, Stapelia* and *Stephanotis*.

The plan of the flower is remarkably uniform, with three regularly alternating pentamerous whorls of sepals, petals, and stamens respectively, and a bicarpellary pistil crowning the axis. The corolla is generally deeply divided with spreading segments (rotate), but the tube may be longer, forming a salver-shaped corolla, as in *Stephanotis*, or the remarkable pitcher-like structures of *Ceropegia* which are often zygomorphic. Great variations occur in the form of the corona-like petaloid appendages springing from the backs of the stamens, or occasionally from the corolla. The five stamens and two carpels arise separately on the floral axis. The ovaries remain free, but the styles unite to form a common swollen stigma-head which may be flattened or more or less conical

[1] *Verh. Zool.-Bot. Gesell. Wien*, LXIII, 240–64 (1913).

or even beaked; the receptive surfaces are on the edge or under side.

The anthers are united laterally to form a five-sided blunt cone which, except in the subfamily Periplocoideae, is attached on the inside to the stigma-head; this union of anthers and pistil forms the gynostegium. As in Apocynaceae we may distinguish two subfamilies which differ in staminal characters, the Periplocoideae containing 46 closely allied genera, representing a less specialised type, with granular pollen associated in tetrads and stamens often stalked, and the Cynanchoideae comprising the rest of the family and representing a more specialised type with the pollen in each anther-half united into a single wax-like pollinium, and the anthers almost or quite sessile.

The most marked distinction between Asclepiadaceae and Apocynaceae is the presence in the former of the so-called translator by aid of which the pollen is transferred. In the simpler group (Periplocoideae) it consists of a somewhat spoon- or funnel-shaped structure between each anther, ending below in a sticky disc. The granular pollen from one half of each of the two adjacent anthers is poured into its concave receptacle. In the more elaborate typical members, the translator consists of two parts, the corpusculum which is attached to the margin of the stigma-head between the anthers, and a pair of arms by which the corpusculum is attached to the pollinia of the adjacent anther-halves. The length and form of the arms vary widely, affording characters for the distinction of genera.

Pollination in the simpler group recalls that of the *Ophrydeae* among the orchids; in the male stage of the flower the sticky basal disc is turned outwards and will adhere to the head of an insect-visitor, which on retiring will carry off the whole translator with its pollen-contents, and on visiting another flower may deposit pollen-grains on the stigmatic surfaces on the underside of the stigma-head. In the Cynanchoideae an insect visiting the flower for nectar catches its leg in the slit left between the cells of adjacent anthers and in drawing it up comes in contact with the notched base of the corpusculum and drags the latter off, bearing with it the pair of pollinia. The arms of the translator are hygroscopic, and as they dry bring the pollinia together, increasing the hold on the insect's leg. The stigmatic surface lies beneath

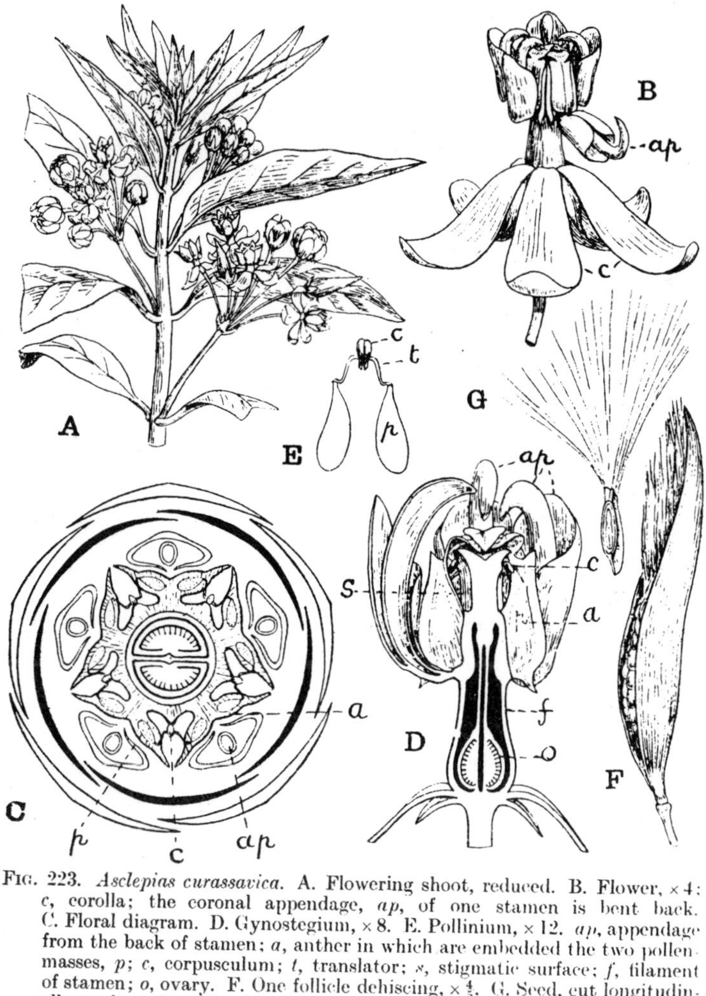

FIG. 223. *Asclepias curassavica.* A. Flowering shoot, reduced. B. Flower, × 4; *c*, corolla; the coronal appendage, *ap*, of one stamen is bent back. C. Floral diagram. D. Gynostegium, × 8. E. Pollinium, × 12. *ap*, appendage from the back of stamen; *a*, anther in which are embedded the two pollen-masses, *p*; *c*, corpusculum; *t*, translator; *s*, stigmatic surface; *f*, filament of stamen; *o*, ovary. F. One follicle dehiscing, × $\frac{4}{5}$. G. Seed, cut longitudinally to shew the embryo, × $2\frac{1}{2}$.

the anther-slit, so that in visiting a flower in the female stage the act of catching the leg in the slit will cause the pollinia to become attached to the receptive surface. The corona-like appendages are often associated with the secretion and storing of nectar (fig. 223).

There is a great uniformity in the fruit, which consists of a pair of follicles containing numerous ovate-oblong to oblong seeds which with few exceptions are crowned by a tuft of hairs facilitating their dispersal by aid of air-currents.

The family is almost exclusively tropical, only a few genera occurring in temperate regions. The Old World is richer in generic types, containing two-thirds of the total number of genera as against one-third in America. In the Old World the northern limit is represented in *Vincetoxicum*, which reaches 61° latitude in Scandinavia; in North America *Asclepias* is Canadian, one species marking the northern limit at 54° latitude on the Saskatchewan. In the southern hemisphere the family is sparingly represented in temperate South America and Australia, but South Africa is very rich, containing many genera and more than 400 species.

As already indicated there are two well-marked subfamilies.

I. PERIPLOCOIDEAE. Pollen granular, in tetrads; translators spoon-shaped with a basal adhesive disc. 46 genera.

II. CYNANCHOIDEAE. Pollen aggregated in two, or in *Secamone* four, wax-like pollinia; translators differentiated into a corpusculum and a pair of arms. 230 genera. This is subdivided into tribes, according to the number or position of the pollinia with regard to the arms, as follows:

Asclepiadeae. Pollinia pendulous on the arms. *Asclepias* (fig. 223), *Cynanchum*.

Secamoneae. Pollinia four in each anther. *Secamone*.

Tylophoreae. Pollinia erect. *Ceropegia, Stapelia, Stephanotis, Hoya*.

Gonolobeae. Pollinia horizontal. *Gonolobus*.

Asclepiadaceae are closely related to Apocynaceae, the only important difference being the great specialisation in the stamens and pistil to ensure transference of the pollen in the Asclepiads.

Order 3. *CONVOLVULALES*

Family. CONVOLVULACEAE

Flowers regular, bisexual, rarely dioecious by abortion, hypogynous, conforming to the formula S5, P5, A5, G(2). Sepals generally free, with quincuncial aestivation, the outer often larger than the inner, persistent. Corolla gamopetalous, generally entire or slightly five-lobed, and induplicate-valvate and twisted in the bud. Stamens inserted on the base of the corolla-tube and alternating with the petals; anthers two-celled, generally introrse, each cell dehiscing by a longitudinal slit; intrastaminal disc generally present and ring- or cup-shaped. Ovary superior, rarely of three or five carpels, syncarpous, with a broad base, 1–2-locular, rarely 3-locular (or by development of a spurious septum 4-locular), with two erect anatropous sessile ovules springing from the inner angle of each chamber, the micropyle pointing downwards; style generally simple, filiform or more or less bipartite, or two distinct styles; stigma terminal and capitate, or two distinct stigmas of various shape. Fruit generally a capsule with loculicidal dehiscence, rarely indehiscent and fleshy or nut-like; seeds as many as the ovules or fewer, as usually in the indehiscent fruit; endosperm cartilaginous, surrounding the embryo which has generally broad folded or bent, often emarginate or bilobed cotyledons, and the radicle directed towards the hilum.

Annual or perennial herbs, often twining, more rarely shrubby, very rarely arborescent; leaves simple, alternate, exstipulate. Flowers often showy, solitary and axillary, or in few- to many-flowered axillary dichasia; bracteoles generally small. Important anatomical characters are the presence of latex and bicollateral vascular bundles.

The family Convolvulaceae is generally included in the next order with which it shews many resemblances, especially to the earlier families, but is distinguished by the placentation of the ovules and the presence of latex; the large embryo with folded cotyledons is also a notable feature.

Wettstein suggests that the Convolvulales may have had a separate origin from the dialypetalous Dicotyledons near that which gave rise to the Tubiflorae, and regards them as allied phylogenetically to the Malvales or Geraniales.

Genera 47; species about 1100. Mainly in the warmer parts of the world.

Cuscuta (Dodder) is a genus of leafless parasites with

slender pink, yellow or whitish twining stems; the seeds germinate in the ground, producing a filiform leafless stem, the continued existence of which depends on its attachment to a host-plant, from which it henceforth derives nourishment by suckers; by withering at the base it loses all connection with the soil.

The most highly developed members of the family are twining plants with generally cordate leaves, and large showy white or purplish flowers; such, for instance, are many

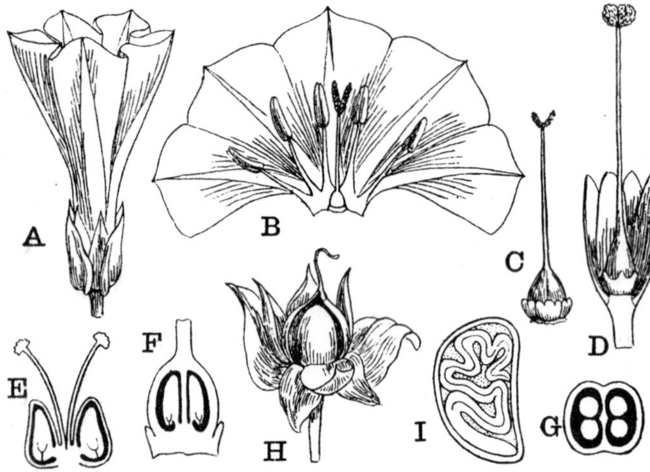

FIG. 224. A. Flower of *Calystegia sepium.* B. Flower of *Convolvulus Scammonia,* opened out. C. Pistil of *Calystegia sepium.* D. Flower of *Argyreia pomacea,* the corolla and stamens removed. E. Longitudinal section of pistil of *Dichondra.* F. Same of *Convolvulus.* G. Ovary of *Convolvulus* in transverse section. H. Capsule of *Calystegia,* with persistent bracts and sepals. One valve of the capsule turned back to shew one of the seeds. I. Seed of *Pharbitis hispida,* cut lengthwise shewing folded embryo. C–G and I enlarged. (A, C, H after Nees; D, E after Peter; I after Wettstein.)

species of the largest genus *Ipomoea*, and those of the allied genera *Calonyction*, *Argyreia* and others, or our native *Calystegia sepium* (Bindweed). But among the less highly developed groups of genera other types prevail. In the small tribe *Dichondreae*, the ovary of which recalls the more characteristic Boraginaceae in its ultimate division into two or four portions and its two distinct gynobasic styles (fig. 224, E), the plants are small prostrate or creeping herbs with very small axillary solitary flowers. In the larger tribe

Dicranostyleae, in which the styles are more or less completely distinct, the plants are generally low-growing or undershrubs with small flowers, solitary or in few-flowered cymes, as in the widely-spread *Cressa cretica*, or the common tropical weed *Evolvulus alsinoides*. Others of this tribe developed under desert conditions, such as *Hildebrandtia*, endemic in the arid regions of north-east tropical Africa, have a much-branched shrubby habit with the slender rigid woody branches often ending in spines and bearing the small leaves and small solitary flowers on short dwarf shoots. Another, *Seddera* (chiefly African and Arabian), comprises undershrubs with spreading rigid branches bearing small sessile or subsessile leaves, generally protected by hairs, and very small solitary flowers. To the same tribe belong a few small genera comprising high-climbing shrubs, nearly glabrous, with large elliptical leathery leaves and a densely hairy inflorescence of very small flowers arranged in racemes or panicles. Such are the tropical Old World *Neuropeltis*, and the tropical American *Dicranostyles*. Of similar habit are the *Erycibeae*, in which however the style is simple or suppressed; *Humbertia* from Madagascar is exceptional in forming a large tree.

The two largest genera, *Convolvulus* and *Ipomoea*, representing the tribes *Convolvuleae* and *Ipomoeae*, shew great variety in habit and include erect or prostrate herbs or undershrubs, sometimes annual but more often perennial from a tough woody rhizome by means of which, under desert conditions, the plant is enabled to persist from season to season. In *Convolvulus Scammonia* (Scammony) the rhizome is thick and tuberous. In *Ipomoea Batatas* (Sweet Potato) and others the lateral roots become fleshy. In the dry districts of the eastern Mediterranean area and Arabia, species of *Convolvulus* become spiny shrubs, with great reduction in the size of the leaves, or assume a broom-like habit. Under more favourable conditions long trailing leafy stems are developed, as in our common Small Bindweed (*C. arvensis*), which often tend more or less to climb, and suggest the development of the climbing habit which characterises higher members of the group, and is generally associated with an increase in the size and number of the flowers. Another adaptation of the creeping stem is found in the sand-binders, such as our native *Calystegia Soldanella* or the tropical *Ipomoea pes-caprae* (fig. 225).

There is great diversity in the form of the leaves, which are generally stalked and without exception exstipulate. Scale-leaves are absent from the aerial shoot; this, like the

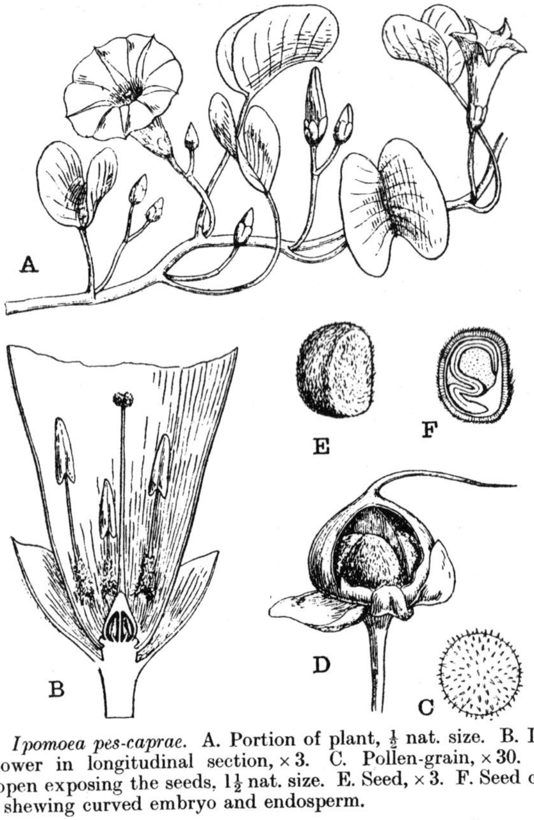

FIG. 225. *Ipomoea pes-caprae.* A. Portion of plant, $\frac{1}{2}$ nat. size. B. Lower part of flower in longitudinal section, × 3. C. Pollen-grain, × 30. D. Fruit, cut open exposing the seeds, $1\frac{1}{2}$ nat. size. E. Seed, × 3. F. Seed cut lengthwise shewing curved embryo and endosperm.

absence of stipules, whose function is frequently protective, is perhaps associated with the fact that the family is mainly tropical and subtropical, and those species which are exposed to rigorous conditions at certain seasons of the year remain dormant during the unfavourable period. The buds on subterranean shoots are protected by scale-leaves. The leaf-margin is entire, often with a hastate or cordate base, or the blade is more or less deeply digitately or pinnately (*Quamoclit vulgaris*) lobed. The shoots are glabrous, or variously covered with hairs.

The flowers are solitary in the leaf-axils, with the two bracteoles scale-like and at some distance below the flower (Bindweed) or, as in *Calystegia*, large and enclosing the calyx; in *Neuropeltis* one of the bracteoles becomes much enlarged in fruit, forming a membranous wing to which the fruit is attached, and serving as a float. Generally branching occurs forming compound axillary dichasia, which may become capitate through suppression of the stalks of the second and following orders, as in *Ipomoea pes-tigridis* and others; in the higher branchings the inflorescence often becomes monochasial. By diminution of the foliage-leaves, in the axils of which the solitary flowers or dichasial inflorescences are borne, a terminal raceme or raceme of dichasia may arise, a more or less complete separation occurring between the flowering and vegetative portions of the shoot. In the small tribes, *Poraneae* and *Erycibeae*, the members of which are chiefly shrubby climbers with large leathery leaves, the flowers are in true racemes or panicles.

The flowers are regularly pentamerous, with reduction to two in the gynoecium; in some species of *Ipomoea* there are three carpels and in the tropical Asiatic *Erycibe* the gynoecium is pentamerous; *Hildebrandtia*, an East African xerophyte with small leaves and stiff thorn-like branches, is dioecious and has generally tetramerous calyx, corolla and androecium. The sepals, which are usually free, rarely more or less united, shew considerable variation in size, shape and covering, affording valuable systematic characters; they are smooth, warty or variously hairy, orbicular or more or less elongated, determining the shape of the bud, and the two or three outer ones are often strikingly larger than the inner, a difference which sometimes becomes exaggerated in the fruiting stage.

The corolla is somewhat uniform in shape, generally funnel-shaped, with entire or slightly lobed margin, more rarely bell-shaped or tubular; a characteristic feature, variation in details of which afford useful generic characters, is the more or less evident demarcation into longitudinal areas, five stronger areas tapering from base to apex, corresponding to the middle of the petals and characterised by well-marked longitudinal vascular bundles, alternating with five non-striated weaker triangular areas, which in the bud are folded inwards, the stronger areas only being exposed and shewing a twist to the right. The length of the slender filaments of the stamens varies widely, often in the same flower; the ovate to linear anthers are attached dorsally, they are two-celled and dehisce more or less introrsely by longitudinal slits.

In Convolvulaceae, as in some other families, considerable importance has been attached to the character of the pollen in the distinction of genera. Two principal forms are distinguished, an ellipsoidal form with longitudinal bands, characteristic of *Convolvulus* and many other genera, forming the *Convolvulus*-type, and a spherical form with a spiny extine, characterising a smaller number of genera, including *Ipomoea*, and known as the *Ipomoea*-type.

The two carpels are median. The ovary has often a dense covering of hairs; it is occasionally more or less unilocular, a median partition being however generally present in the lower part (as in *Calystegia*); generally a complete median wall is present; sometimes a secondary wall appears growing from the primary and making the ovary more or less completely four-chambered. In most species of *Pharbitis* there are three chambers, but there may be three or two in the same species. In the typically bilocular ovary and in the trilocular there are two collateral anatropous ovules at the inner angle of each chamber, the micropyle pointing downwards and outwards; the ovules are solitary in the four-chambered ovaries; in the unilocular there are four or two. The shape of the stigma, which terminates the style or its branches, shews some variation, linear, capitate, globose, etc., affording useful systematic characters. *Erycibe* is exceptional in its sessile stigma.

The flowers, which are often large and generally brightly coloured, with red or violet, blue or white corollas, are visited by insects for the nectar excreted on the hypogynous disc.

Species of *Quamoclit* and *Ipomoea* with narrow tubes are adapted for visits of nectar-seeking birds. Many of the large-flowered species (such as *Ipomoea Batatas*) have extra-floral nectaries on the leaf-stalk near the blade. Self-pollination also occurs; in *Convolvulus arvensis* flowers are produced late in the season which are short-styled and adapted for self-pollination. Cleistogamic flowers have been observed in *Dichondra repens*. In most species of the family the flowers remain open for one day only or for a few hours, after which the edges of the corolla roll up inwards, closing the tube with the ovary.

The fruit is generally a four- or six-valved capsule, but sometimes dehisces irregularly from base or apex. It is rarely indehiscent with a membranous (*Porana*) or strongly lignified (*Maripa, Erycibe*) pericarp, and the seeds generally reduced to one. In the typical dehiscent fruit all the ovules become seeds, each being shaped like the quadrant of a sphere, but one or more may abort, the remainder becoming ellipsoidal or spherical. The testa is generally smooth but sometimes warty or hairy (as in *Ipomoea pes-caprae* (fig. 225) and others); its character is of importance for distinction of species. Thus the section *Eriospermum* of *Ipomoea* takes its name from the long-haired seeds.

The embryo of *Cuscuta* differs from that typical of the family in being filiform and spirally twisted with generally no trace of cotyledons; according to some observers there is a layer of nutrient tissue outside the true endosperm, derived from the nucellus (perisperm).

The family occurs in all parts of the world except the coldest, and is especially numerous in tropical Asia and tropical America. *Ipomoea*, the largest genus, has about 400 species in the warmer parts of the earth; two only are found in Southern Europe. *Convolvulus* has 200 species in temperate, more rarely in tropical climates, and finds its principal development in the Mediterranean region and Western Asia; it is represented in Britain by the small Bindweed (*C. arvensis*). The large Bindweed, *Calystegia sepium*, and *C. Soldanella*, a sandy seashore plant with procumbent stem, are distinguished from *Convolvulus* by the large bracteoles enveloping the calyx.

The Dodders, *Cuscuta*, of which there are 100 species in the warmer and temperate regions, have been separated as a distinct family on account of their parasitic habit, imbricate corolla, more copious endosperm and filiform spiral embryo; their floral structure is however that of Convolvulaceae (fig. 226) and it is preferable to regard the genus as a reduced parasitic form of that family, a simplification of the embryo being commonly associated with a parasitic habit. Another point of interest in *Cuscuta* is the presence of a whorl of scales on the inside of the corolla-tube below the

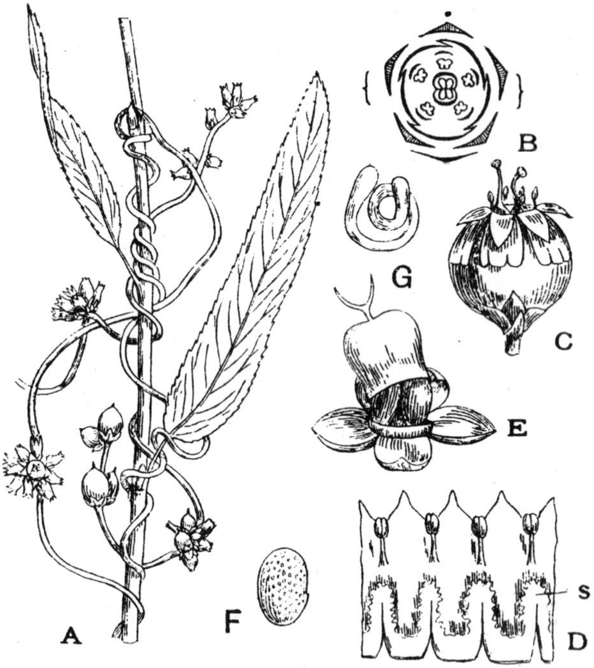

FIG. 226. *Cuscuta*. A. *C. monogyna*, on a species of *Salix*. B. Floral diagram of *C. epithymum*. C. Fruit of *C. reflexa* covered above by the withered flower, ×4. D. Part of corolla of *C. Trifolii* spread out; *s*, scale. E. Fruit of same, opening. F. Seed of *C europaea*. G. Embryo of *C. epilinum*. D–G enlarged. (A, E, F after Peter; B after Eichler; C after Hooker; D after Reichenbach; G after Nees.)

stamens, structures which are more general in Hydrophyllaceae and recall the well-known corona of the Boraginaceae.

Convolvulaceae have much in common with Polemoniaceae—contorted aestivation, frequently the structure of the inflorescence, and occasionally a trimerous pistil—but are distinguished by habit, the definite number of ovules with basal placentation, the characteristic corolla, and also by the presence of latex.

They are also allied to Solanaceae and Boraginaceae; from the former they are distinguished by the definite number of ovules and the inferior micropyle towards which the radicle is directed; the position of micropyle and radicle distinguish them also from Boraginaceae; the latter family, and also Hydrophyllaceae, are further distinguished by their unilateral scorpioid inflorescence.

Ipomoea Batatas is one of the most widely distributed food-plants in tropical and subtropical regions, the tuberous lateral roots, rich in starch and sugar, forming the Sweet Potato. The rootstock of *Calystegia sepium* is also cooked and eaten in New Zealand. Several species are used in medicine on account of the strong purging properties of the latex; Scammony is the dried latex from the rootstock of the oriental *Convolvulus Scammonia*, and Jalap is obtained from the tubercles of *Exogonium Purga*, and derives its name from Jalapa in West Mexico, the home of the species. *Ipomoea pes-caprae* is planted on the Indian coast to bind the sand, and species of *Ipomoea*, *Calystegia* and *Convolvulus* are widely grown as ornamental plants. *Convolvulus arvensis*, a native of Europe and Western Asia and introduced elsewhere, is a pest in fields and gardens from its deep-growing branched rhizome. Many of the Dodders are the cause of much damage, from their parasitic habit; such are *Cuscuta Trifolii* on Clover, and *C. epilinum* on Flax.

Order 4. *TUBIFLORAE*

Flowers typically bisexual, regular or more often zygomorphic, hypogynous, rarely epigynous, tetracyclic. Corolla rarely induplicate-valvate in bud. Stamens epipetalous, generally four, or two by reduction. Ovary usually bicarpellary and two- to one-celled, rarely three- to five-carpellary with as many cells; ovules many to few in each cell, with typically marginal placentation.

Generally herbs, but woody plants occur; with alternate or opposite, generally simple leaves.

The Tubiflorae form a natural group illustrating the development, from a regular tetracyclic isostemonous flower, of a highly specialised zygomorphic type with reduction in the androecium to four or two members.

Several lines of development may be traced from the regular isostemonous-flowered families. In Polemoniaceae the tendency to zygomorphy is slight, the ovary is generally tricarpellary, but occasionally two or five carpels occur, with ovules numerous to solitary in each chamber; the position of the micropyle resembles that in Convolvulaceae. It is mainly an American family, especially Western American. Hydrophyllaceae recall Polemoniaceae in the plan of the regular flower, but the ovary is bicarpellary and bilocular, or unilocular from the failure of the large marginal placentas to unite in the centre; the number of ovules varies from numerous to two. On the other hand they approach Boraginaceae in the tendency to arrange the flowers in a scorpioid cyme, the frequent presence of a ring of scales on the corolla-tube and the position of the ovule with an upwardly-directed micropyle. It is a small family, widely distributed, but especially North American. In Boraginaceae there is a tendency to zygomorphic development of the corolla, but the chief point of interest is the great and constant reduction and specialisation in the pistil; the originally bilocular ovary is divided by a later-formed partition into four one-ovuled portions, and in the minority of genera forms a four- to one-seeded drupe, but in the majority divides into four distinct one-seeded nutlets.

The two closely allied families Verbenaceae and Labiatae illustrate the perfection of the zygomorphic type. The flower is, with few exceptions, zygomorphic. In Verbenaceae zygomorphy does not extend to the calyx but the tubular corolla is generally more or less two-lipped; in Labiatae the calyx, though sometimes regular, is often markedly two-lipped, and the corolla is rarely regular but generally conspicuously two-lipped. The four stamens are generally didynamous and reduction to two occurs. In Verbenaceae the two carpels (very rarely four or even five) unite to form an originally bilocular (very rarely 4- or 5-locular) ovary; each chamber has two ovules, and later is divided by a septum into one-ovuled chambers; the fruit is a chambered drupe or splits along the septa to form two-chambered or one-chambered mericarps. The style is terminal or sometimes situated in an apical depression of the ovary. In Labiatae the reduction and specialisation of the ovary has become complete. The bilocular ovary (with a gynobasic style) becomes divided into four one-ovuled chambers, forming in the fruit four one-seeded nutlets.

Another line of development within the series may be traced from Solanaceae with generally regular isostemonous flowers and a bicarpellary, generally bilocular, ovary with numerous ovules. The tribe *Salpiglossideae* with four or two fertile stamens and an often more or less zygomorphic flower forms a link with Scrophulariaceae where zygomorphy of corolla and androecium is generally strongly marked. Along this line of development there is comparatively little tendency towards that specialisation of the ovary and fruit associated with reduction in number of seeds which we have noted as culminating in Labiatae. Several families are closely allied to Scrophulariaceae. Gesneriaceae, a mainly tropical family, has a unilocular ovary with indefinite ovules, and shews a marked tendency to epigyny; the closely allied Orobanchaceae, also with a unilocular many-ovuled ovary, represents the development of the parasitic tendency evidenced in several genera of Scrophulariaceae; while the Lentibulariaceae, with generally a markedly two-lipped corolla and reduction to two stamens, and also a free-central placentation in the ovary, are a more or less aquatic family specialised in the direction of the insectivorous habit. Pedaliaceae shew specialisation of the fruit. Bignoniaceae are evidently allied; the plan of the flower is that of Scrophulariaceae, but the woody habit, often climbing, and the two-valved capsular fruit with flattened winged seeds suggest an affinity with Apocynaceae.

Acanthaceae, while shewing a similar floral plan, have a highly specialised flower and fruit, as indicated by the great variety in the development of the stamens, and the remarkable mechanism for the dispersion of the seeds in the more typical genera.

It is difficult to suggest a point of origin among polypetalous dicotyledons for the Tubiflorae. They were presumably derived from a group characterised by pentamerous flowers with oligomerous gynoecium, numerous marginal ovules and a disc-development beneath the ovary. Wettstein suggests a type akin to Rosales, in which sympetaly occasionally occurs, for instance in Crassulaceae and Saxifragaceae (*Roussea* has a campanulate corolla), the latter family is typically bicarpellary, and the bilocular ovary with numerous anatropous ovules on the large placentas resembles that of Solanaceae or Scrophulariaceae (the oblique position of the ovary characteristic of Solanaceae occurs also in *Saxifraga*); there is also a marked tendency to epigyny.

Suborder 1. POLEMONINEAE. Flowers generally regular. Ovary generally trilocular with ∞ to 1 ascending anatropous ovules with inferior micropyle. Fruit generally a loculicidal capsule.

Family I. POLEMONIACEAE

Flowers bisexual, regular or slightly zygomorphic, conforming to the formula S5, P5, A5, G(3). Calyx generally bell-shaped or tubular; segments imbricate or valvate. Corolla hypogynous, with generally a well-developed tube and more or less spreading limb, aestivation contorted. Stamens inserted at various, often unequal heights on the corolla-tube; anthers two-celled with longitudinal dehiscence; an intrastaminal disc is generally present. Ovary superior, 3-locular (rarely of two or five carpels); style terminal, filiform, dividing at the top into three arms which are stigmatic on the upper face; ovules 1–∞, sessile in the inner angle of each chamber, anatropous. Fruit usually a loculicidal capsule; seeds 1–∞ in each chamber; embryo straight, in the axis of a generally copious fleshy or cartilaginous endosperm.

Annual, biennial or perennial herbs, rarely shrubby; leaves usually alternate or the lower or all opposite, entire or divided, exstipulate. Flowers generally showy and in a complicated cymose inflorescence, rarely solitary and axillary.

Genera 12; species about 280. Distribution America, chiefly western North American; very few in the Old World.

Cobaea, a tropical American genus, has pinnately compound leaves ending in much branched tendrils by means of which the plant climbs. The other genera comprise mostly erect or more or less spreading perennial herbs, as in *Phlox*, or *Polemonium*, the former with simple entire mostly opposite leaves; the latter with alternate pinnately divided leaves. The large bell-shaped flowers of *Cobaea* stand singly in the leaf-axils, but usually a number of flowers are arranged in an inflorescence formed on a dichasial plan, each shoot ending in a flower, and the branching at first dichasial becoming dorsiventral from the stronger development in the lower bracteole of each pair. The flowers are often numerous, forming, for instance in *Phlox*, large involucrate heads; bracteoles may be present or absent.

There is great variety in the form of the corolla, which is bell-shaped in *Cobaea*, from funnel-shaped to almost rotate in *Polemonium*, salver-shaped in *Phlox*, while in some species of *Gilia* the tube is almost filiform. The often unequally spreading lobes in *Loeselia* (Western America) give the corolla a zygomorphic appearance, while in the small Mexican genus *Bonplandia* the lobes are unequal and the corolla is bilabiate.

The white or brightly coloured (often blue) flowers are strongly proterandrous and adapted for pollination by bees or Lepidoptera.

Cobaea has a septicidal capsule; in the other genera it is loculicidal. In *Collomia* and species of *Gilia* the outer layer of the seed-coat is mucilaginous, swelling when wetted and fixing the seed to the soil.

The great majority of the species are North American, especially western. *Cobaea* (nine species) ranges from Central America to Venezuela and North Chile; *Cantua* (six species) occurs on the Andes southwards to Chile; and a few species of other genera (*Gilia*, *Polemonium*, *Collomia*) occur in Andine South America. The family is represented in the eastern hemisphere by three species, *Phlox sibirica*, and *Polemonium coeruleum* and *lanatum*. *Polemonium* contains about 30 species; *P. coeruleum* (Jacob's Ladder), native in Central and North England, is a widespread north temperate plant in both hemispheres; *P. humile* is a widely distributed Arctic species. Other species occur in the Rocky Mountains and the mountains of California; *P. antarcticum* at the Strait of Magellan. *Phlox* has about 50 species in North America and one in North Asia. *Gilia* has about 100 species in extratropical and subtropical North and South

America; many grow at considerable altitudes on the western mountain ranges. Species of *Phlox, Gilia, Polemonium, Cobaea scandens* and others are favourite garden plants.

Polemoniaceae shew some resemblance to Convolvulaceae, with which they agree in general plan of floral structure. They differ in the gamosepalous calyx, the simpler aestivation of the corolla, the typically trimerous pistil, and the varying number of seeds; also in the absence of latex.

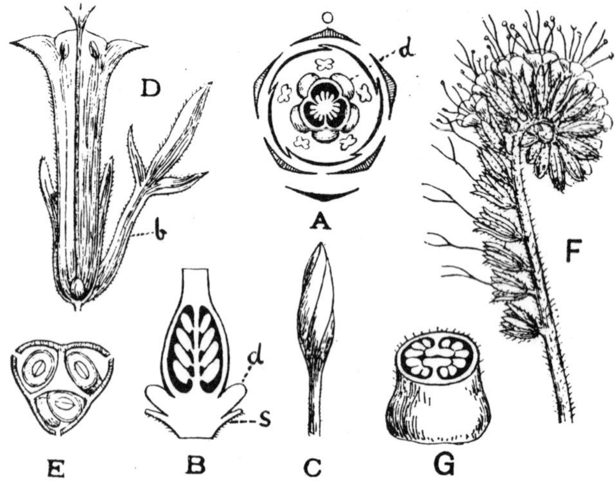

FIG. 227. A–E. Polemoniaceae. A. Floral diagram of *Polemonium coeruleum*; *d*, disc. B. Median longitudinal section shewing ovary, disc (*d*) and base of sepals (*s*). C. Upper portion of corolla of *Gilia androsacea* shewing contorted aestivation. D. Flower of *G. Brandegei* cut vertically, with subtending bract *b*. E. Cross-section of fruit of *G. capitata* shewing loculicidal dehiscence. F, G. Hydrophyllaceae. F. Cyme of *Phacelia integrifolia*. G. Ovary of *P. Whitlavia* cut across. B–E and G enlarged. (A after Eichler; B after Engler; C–E after Peter; F, G after Brand.)

Suborder 2. BORAGININEAE. Flowers generally regular in unilateral cymes. Carpels two; ovules ∞ to two in each carpel, anatropous with upwardly directed micropyle. Fruit a capsule or drupe or splitting into four one-seeded mericarps.

Family II. HYDROPHYLLACEAE

Flowers bisexual, regular, generally pentamerous with reduction to two in the pistil. Sepals united at the base, aestivation imbricate. Corolla rotate, bell- or funnel-shaped, aestivation imbricate, seldom contorted. Stamens epipetalous, generally inserted on the base of the corolla, alternating with the petals; filaments equal or unequal; anthers versatile, two-celled, dehiscing introrsely by longitudinal slits; scale-like appendages are often present at the base of or alternating with the stamens. Ovary superior, generally one- more rarely two-chambered; ovules ∞–2, sessile or pendulous, anatropous or amphitropous with micropyle directed upwards and outwards; styles one and generally more or less divided, or two; stigma terminal, generally capitate. Fruit usually a loculicidal capsule; seeds as many as the ovules or fewer; embryo small in the axis of a copious or sparse fleshy or cartilaginous endosperm.

Annual, biennial or perennial herbs or undershrubs; leaves entire or pinnately, rarely palmately divided, exstipulate, alternate or opposite, often forming a basal rosette. Flowers small or showy, solitary or in scorpioid cymes usually without bracteoles.

Genera 18; species about 170; mainly North American.

Mostly annual or perennial herbs of various habit, with simple or variously pinnately divided leaves. There is a wide variation in the nature of the hairy covering, which may be softly hairy, stiff, thorny, or glandular. The white, blue or purple flowers may be solitary, as generally in *Nemophila*, but are mostly in scorpioid cymes without bracteoles; often two or more cymes are arranged in dichasial or umbel-like compound inflorescences. The South African genus *Codon* (two species) is distinguished from the rest of the family by its pleiomerous flowers, with 6–10- or 12-merous whorls. In *Nemophila* and others stipule-like appendages occur between the sepals; more frequent are the fold- or scale-like appendages on the corolla-tube which stand in pairs between or in front of the stamens or attached to the base of the filament, and recall the corona-like structures of Boraginaceae. They are often of obvious significance in relation to insect-visits, forming, for instance in *Hydrophyllum*, nectar-conducting tubes.

There is great variety in the number and position of the ovules, which are sometimes inserted on large parietal placentas, which may grow out into the ovary-space, or the dividing wall being complete the placentation becomes axile; corresponding differences occur in the details of the capsular dehiscence.

The family is chiefly North American, extending north-westwards to Alaska and north-east Asia, and southwards through Mexico and Central America along the Andes to Argentina and the Strait of Magellan. The genus *Hydrolea* (19 species) has the widest distribution, occurring in the tropics of Asia and Africa, including Madagascar, as well as of the New World. The largest genus, *Phacelia*, has 100 species, chiefly in western North America and Mexico, with a few in the Andes of Chile and Peru.

Hydrophyllaceae stand between Polemoniaceae on the one hand and Boraginaceae on the other.

The structure of the flower recalls Polemoniaceae, though the aestivation is generally imbricate; the frequently large number of ovules indicates an affinity with Polemoniaceae, while in the reduction to two ovules there is an approach to Boraginaceae, to which also an affinity is shewn in the occurrence of scales on the corolla-tube and more especially in the arrangement of the flowers in scorpioid cymes.

Species of *Nemophila*, *Phacelia* and *Wigandia* are cultivated in gardens and greenhouses. *Hydrophyllum canadense* is used as a remedy for *Rhus Toxicodendron* poisoning and various skin-diseases.

Family III. BORAGINACEAE

Flowers bisexual, regular, rarely zygomorphic, hypogynous; floral formula S5, P5, A5, G(2). Sepals free or united below into a longer or shorter tube which is generally bell-shaped; aestivation imbricate or open. Corolla tubular or funnel-shaped, with generally a spreading limb; lobes imbricate, rarely contorted, seldom unequal, the throat often closed by scale-outgrowths. Stamens alternate with the corolla-lobes, inserted at the mouth or on the tube, equal or unequal, anthers introrse; a ring-like nectar-secreting disc beneath the ovary. Ovary of two median carpels, originally bilocular but soon separating into four one-ovuled portions with a gynobasic style; style more rarely terminal, generally simple with a simple stigma; ovule inserted at the inner angle of each of the four chambers; more or less erect, with upwardly directed micropyle. Fruit of four nutlets; more rarely a 4- to 1-seeded drupe. Endosperm generally absent, sometimes scanty.

Hispid annual or perennial herbs, more rarely shrubs or trees, with generally alternate leaves, which are simple and exstipulate, and flowers arranged in simple or double scorpioid cymes which are spirally coiled in the bud.

Genera 88; species about 1600.

Generally herbs, annual, as in some species of *Myosotis*, or perennial by a thickened creeping rhizome, as in *Pulmonaria* or *Symphytum*. Many species have a strong tap-root, which, as in *Alkanna*, may contain a purple dye, Alkanet.

When perennial they become woody below, while the genus *Tournefortia* and the subfamilies Cordioideae and Ehretioideae, which are almost exclusively tropical and subtropical, comprise trees and shrubs; some species of *Tournefortia* are climbers. A characteristic of the family is the covering of stiff hairs which often spring from a hardened tuberous base, as in Borage, *Echium* and others.

The leaves are generally alternate, the lower in some genera opposite; they are mostly entire and usually narrow, lanceolate or linear. Stipules are absent. The radical leaves in many genera, for instance *Pulmonaria*, *Cynoglossum* and others, are strikingly different from the cauline, generally broader and stalked, sometimes cordate, the cauline being narrow and sessile.

The characteristic dorsiventral inflorescence is very complicated. It consists of one or sometimes a pair of scorpioid cymes, resulting from development in the higher only of the two bracteoles; in *Myosotis* both the bracteoles are suppressed, in other cases only the lower infertile one of each pair; the developed bracteoles are arranged on the lower side of the cyme while the flowers are crowded towards the upper. The flower-stalks may be adnate to the sympodium for a greater or less distance above the bracteole in the axil of which they arise and may thus appear extra-axillary. The cyme is at first closely coiled, becoming uncoiled as the flowers open. At the same time a change in colour may occur, the flowers being red in the bud and becoming blue as they expand, as in *Myosotis*, *Echium*, *Symphytum* and others.

A departure from the usual radial symmetry of the flower occurs in *Echium* and a few allied genera, where the corolla is oblique and the stamens are unequal in size. *Lycopsis* also is zygomorphic from the bending of the corolla-tube and inequality of the lobes. The plane of symmetry in the zygomorphic flowers is an oblique one.

The aestivation of the sepals is open or imbricate or more rarely valvate; they shew very various degrees of union and may be free almost to the base, as in *Lithospermum*, or united to form a longer or shorter tube, five-toothed or cleft above, as in *Myosotis*. The form of the corolla also varies widely.

Thus in the genera found in Britain the corolla is rotate in Borage, tubular in Comfrey (*Symphytum*), salver-shaped in *Anchusa*, funnel-shaped in *Cynoglossum*, and in each case the throat is closed by scale-like outgrowths from the middle line of each petal. In *Lithospermum* the throat of the funnel- or salver-shaped corolla may be naked or closed with five tumid folds, while in *Pulmonaria* the funnel-shaped corolla bears five pencils of hairs alternating with the stamens.

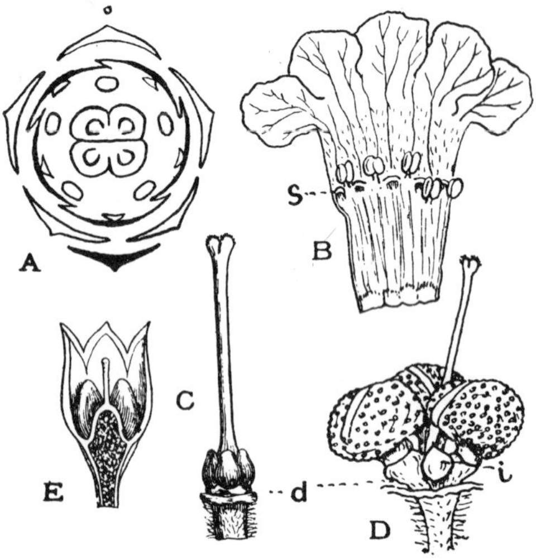

FIG. 228. A. *Anchusa officinalis*, floral diagram (after Eichler). B–D. *Alkanna tinctoria*. B. Corolla cut open, two stamens are at a lower level than the remaining three; *s*, scales at throat of corolla; × 3. C. Pistil, × 6. D. Fruit, with three nutlets, × 3; *i*, fourth carpel which has not formed fruit; *d*, disc. E. Fruit of *Myosotis* enclosed in the calyx, in longitudinal section, enlarged. (B–D after Berg and Schmidt.)

The anthers are generally attached to the filament on the back near their base, are more or less linear in shape and dehisce introrsely by longitudinal slits. The ovary is seated on a ring-like nectar-secreting disc. The two median carpels become early divided by a median constriction or division into four distinct one-ovuled portions, the style springing from the centre of the group. In Cordioideae (fig. 229) and Ehretioideae the ovary retains its original form and bears the style on its apex. One ovule is attached at the inner angle of each ovary-segment; it is anatropous with an upwardly directed micropyle, and generally erect but sometimes more or less oblique or almost horizontal; in Heliotropioideae it is descending with a downwardly directed micropyle. The simple style generally ends in a simple stigma, as in *Borago, Symphytum, Myosotis* (fig. 228, E), more rarely the stigma is bilobed, as in *Anchusa* (fig. 228, C) and *Pulmonaria*, while in *Echium* and others the style is bipartite. In *Cordia* each stigma is again divided (fig. 229, B).

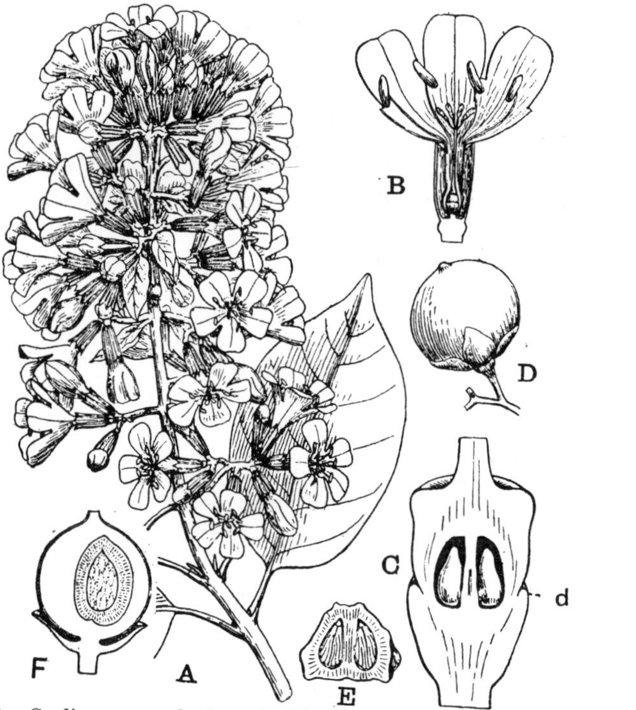

FIG. 229. *Cordia gerascanthoides*. A. Flowering shoot somewhat reduced. B. Flower, the calyx and corolla partly cut away exposing the pistil, × 2. C. Ovary and disc, *d*, cut longitudinally, × 12. D. Drupe, × 2. E. Endocarp cut across shewing two seeds, × 3. F. Drupe of *C. Myxa* in longitudinal section.

The flowers shew well-marked adaptations to insect-visits. Their aggregation on the upper face of the dorsiventral inflorescence, the colour and the presence of nectar serve to attract. The corona which is frequently formed by the corolla-scales protects the pollen and nectar from wet or undesirable visitors. It may also indicate the position of the nectar to the insect-visitor, as in the yellow eye of the blue Forget-me-not, one of the first points noted by Sprengel, which led him to his investigations into the relations between flowers and insects.

In *Pulmonaria* the nectar is protected from the great majority of insects except humble-bees by the length of the corolla-tube; heterostyly occurs here and in other genera. In *Anchusa* the narrow entrance to the tube reserves the nectar for bees; the drooping flowers of *Borago* allow the visits only of insects which (like bees) can hang below the flower, while *Symphytum* has in addition a tube of sufficient length to exclude all except insects with a long proboscis. The remarkable change of colour during the life of the flower has been described as a recapitulation of the stages of evolution—thus the rosy and blue in some species of *Myosotis*, yellow, bluish and violet in *M. versicolor*, and red, violet and blue in *Pulmonaria*, *Echium* and others. Here yellow seems to be a primitive colour and, at least in many cases, violet and blue seem to have been preceded by red—an assumption which is strengthened by the fact that many blue and violet species (as of *Myosotis*, *Anchusa*, *Symphytum*) give rose-red varieties apparently by reversion to more primitive characters.

In Cordioideae, Ehretioideae and Heliotropioideae the fruit is drupaceous, containing typically four seeds, though all may not be developed. In the greater number of genera, comprising the Boraginoideae, the fruit consists of separate nutlets, generally four in number, one or more of which may become aborted. The form of the nut and the nature of the pericarp are extremely varied; thus in *Lithospermum* they are stony, in *Mertensia* rather fleshy, in *Myosotis* usually highly polished, in *Cynoglossum* covered with hooked or barbed bristles. Species of *Symphytum*, *Borago*, *Anchusa*, *Pulmonaria* and *Myosotis* have been described as myrmecochorous. A small portion of the receptacle remains attached to the base of the fruit on dispersal and forms an elaiosome, by which ants are attracted. The elaiosome is often white, thus contrasting with the dark fruit. The seed has a membranous coat and contains a sparsely developed endosperm or, more often, endosperm is absent. In Heliotropioideae endosperm is often copious. The embryo is straight or curved, with flat or fleshy cotyledons and a short upwardly directed radicle.

The family is widely spread throughout the temperate and tropical parts of the world. Its principal centre is in the Mediterranean region, whence it extends over Central Europe and Asia, becoming less frequent as we approach the cold temperate zone. There is a smaller centre on the Pacific side of North America. It is less frequent in the south temperate zone; one genus only, *Halgania*, occurs in Australia where it is endemic. The Boraginoideae are relatively scarce in the tropics, to which however the Cordioideae and Ehretioideae, and, generally speaking, the woody members of the family, are almost restricted.

The following division into subfamilies depends on the characters of fruit and style.

(*a*) Style terminal; fruit a drupe.

Subfamily I. CORDIOIDEAE. Style twice bilobed, drupe with a typically four-chambered stone. *Cordia*, a large genus with 250 species especially in tropical America (fig. 229), and two small genera.

Subfamily II. EHRETIOIDEAE. Style simple, bilobed, or two distinct styles. Drupe with two two-seeded, or four one-seeded stones. 10 genera. Warmer parts of both hemispheres.

Subfamily III. HELIOTROPIOIDEAE. Style simple or bilobed, the stigmatic surface in the form of a swollen hairy ring below the apex of the style. Ovules anatropous and descending. Seed with endosperm. Three genera: *Tournefortia*; *Heliotropium* with 220 species in the temperate and warmer parts of both Old and New Worlds.

(*b*) Style gynobasic; fruit of separate (generally four) nutlets.

Subfamily IV. BORAGINOIDEAE. 73 genera which are distributed among seven tribes according to the number of achenes, regularity or dorsiventrality (*Echieae*) of the flower, and the mode of insertion of the nutlets. Our eight British genera are included in this subfamily. Of world-wide distribution, but the principal centre lies in the Mediterranean region, the numbers diminishing through Central Europe and Asia. A second but less important centre occurs in Pacific North America.

Van Tieghem draws attention to the marked differences between Heliotropioideae and the three remaining subfamilies, in the position of the stigma below the apex of the style, the anatropous descending ovules, as contrasted with the generally more or less erect ascending character, and the presence of endosperm frequently in considerable quantity. He considers them as forming a distinct family, but also regards the remaining three subfamilies as distinct families.

Suborder 3. VERBENINEAE. Flowers generally medianly zygomorphic. Carpels generally two, each with two anatropous ovules with a downwardly directed micropyle. Fruit drupaceous or dividing into four one-seeded mericarps. Leaves generally opposite or whorled.

Family IV. VERBENACEAE

Flowers zygomorphic, rarely regular, bisexual or polygamous by abortion; typical formula S5, P5, A4, G(2). Calyx inferior, persistent, gamosepalous, bell-shaped or tubular, bearing five, four, or more rarely six to eight teeth, lobes or segments. Corolla tubular, often curved, with a spreading limb, more rarely bell-shaped, limb 5- to 4-fid, more rarely multifid, lobes equal or more or less two-lipped, aestivation imbricate. Fertile stamens four didynamous, or two, rarely isomerous with the corolla-lobes and equal, inserted on the corolla-tube and alternate with its lobes, the fifth (posterior) stamen, or, in diandrous flowers, three stamens, reduced to staminodes or absent; filaments free, anthers dorsifixed, two-celled, dehiscence longitudinal and introrse; hypogynous disc often inconspicuous. Carpels two; ovary entire or four-furrowed or rarely shortly four-lobed, two- to four-celled; very rarely four to five carpels with an eight- or ten-celled ovary; style terminal, simple, entire or dividing above into as many stigma-bearing lobes as there are carpels; position of ovules various, but the micropyle always inferior. Fruit usually drupaceous; divisions generally two or four, one-seeded; more rarely dividing into two-celled or one-celled mericarps; testa thin, endosperm fleshy when present, usually absent; embryo about as long as the seed, straight; radicle short, inferior.

Herbs, shrubs or trees, with generally opposite or whorled

leaves, which are simple or more rarely palmately or pinnately compound; stipules absent. Inflorescence racemose or cymose.

Genera 80; species about 800. Almost exclusively tropical or subtropical.

The plants shew a great variety in habit. They are mostly shrubs or undershrubs; but many are trees, as *Tectona grandis* (Teak) and others. The genus *Verbena*, to which belongs the only British representative of the family, *V. officinalis* (Vervain), contains, besides herbs or shrubs with well-developed leaves, also species with much reduced leaves, the function of assimilation being more or less relegated to the stem. Many are shrubby climbers (lianes), sprawling, twining or climbing by means of thorns or spines, as for instance species of *Lantana*, *Clerodendron*, *Vitex* and others. *Avicennia* is a mangrove shrub inhabiting tropical shores in both hemispheres; the stem increases in thickness by repeated production of new cambiums, outside and concentric with the original one.

A few species have alternate leaves. The leaves shew great variety in form, but are generally simple, more rarely compound as in *Vitex*, where they are 3- to 7-foliolate. There is also a wide diversity in the inflorescence. The indefinite include racemes, spikes, or heads, or the flowers are axillary; the bracts are often well-developed, sometimes coloured, forming, as in species of *Lantana*, a conspicuous involucre. Both bracteoles are present, or one or both are suppressed. Cymose inflorescences arise by dichasial branching in the axils of the bracteoles, often passing into monochasial cymes as in *Vitex* and *Clerodendron*. The cymes are axillary, or variously aggregated in paniculate, often showy inflorescences.

The flowers are generally pentamerous. Reduction in the calyx is rare, as in *Physopsis*, which has a regular tetramerous flower. A reduction in the number of the teeth or lobes occasionally occurs by union or abortion in the formation of a two-lipped calyx, as in species of *Vitex*; but the tendency to zygomorphy is much less than in the corolla, where union of the two upper lobes is frequent in the formation of an upper lip, as in *Lantana*, *Lippia* and others. In the fruiting stage the calyx is often much altered and enlarged, sometimes bladder-like and more or less enclosing the fruit and sometimes brightly coloured. The aestivation of the sepals is generally open; that of the petals imbricate with a varying arrange-

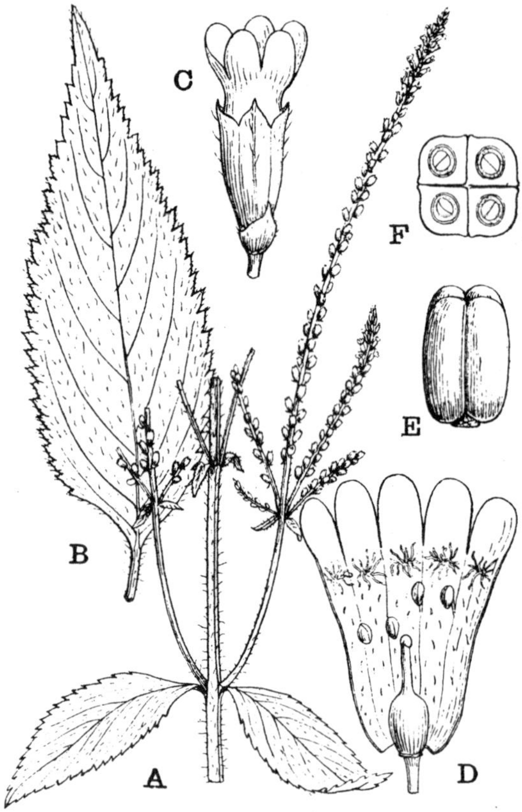

FIG. 230. *Verbena urticifolia*. A, B. Flowering stem and leaf, slightly reduced. C. Flower, × 8. D. Flower with corolla cut open and rolled back, × 13. E. Fruit, × 7. F. Fruit in transverse section shewing four one-seeded drupes, × 7.

ment. A pentamerous androecium is very rare, occurring in *Tectona* and in the Malayan *Geunsia*, where the flower is pentamerous throughout. The fifth stamen is sometimes represented by a staminode, but is generally absent; diandry may arise by reduction to staminodes of the two posterior or more rarely the two anterior stamens; where four stamens are present and fertile they are rarely equal, one pair (usually the anterior) being longer or having a higher insertion on the corolla-tube (figs. 230, D, 231, C).

There are usually two carpels which are median, more rarely there are four (as in *Duranta*, fig. 231) or five (*Geunsia*);

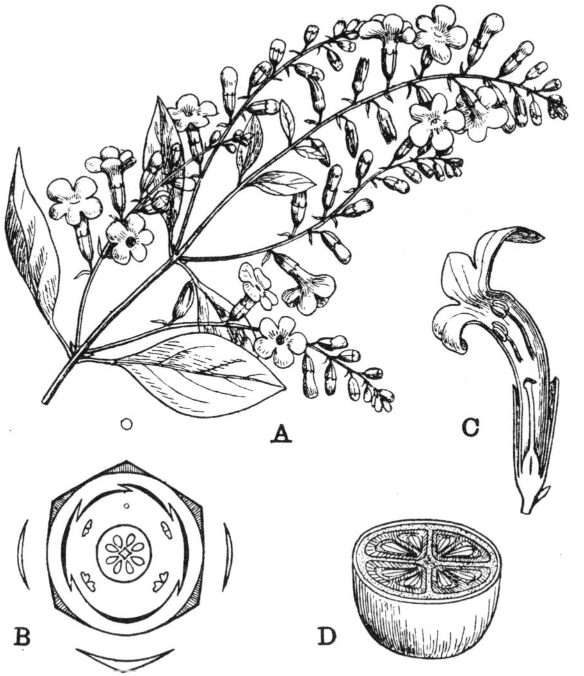

FIG. 231. *Duranta Plumieri*. A. Flowering shoot, × ¾. B. Floral diagram. C. Flower cut lengthwise, × 3. D. Drupe cut across; each cell of the ovary has been again divided forming in all eight one-seeded chambers, × 3.

in *Lantana*, *Lippia* and allied genera the posterior carpel becomes aborted. Placentation in the young ovary is parietal, the placentas subsequently uniting more or less completely in the centre of the ovary, each becoming rolled back right and left and bearing a single ovule, sometimes at the base or higher, or sometimes at the apex of the ovary-chamber, but always with a downward-pointing micropyle. A secondary ("false") septum frequently grows from the middle of each carpel inwards, separating the placenta-lobes and dividing the ovary into one-ovuled chambers (figs. 230, 231). The ovary is sometimes rounded, but generally more or less lobed according to the number of the chambers; thus two-lobed in *Lantana*, four-lobed in *Clerodendron* and *Verbena* (fig. 230, E).

From the resemblance of the floral structure to that of Labiatae and the occurrence of proterandry it is probable that pollination is effected as in the Labiatae by the agency of bees and flies. The flowers of *Aegiphila* are diclinous by reduction or abortion.

The fruit is generally drupaceous. The exocarp is fleshy or dry; the endocarp is almost always hard. The arrangement of the stones affords useful characters for the grouping of the genera; the endocarp forms a four-chambered stone in *Tectona*, *Vitex* and allied genera; two two-chambered stones in *Priva* and others; four two-chambered stones in *Duranta* (fig. 231); two undivided stones in *Lantana* and others; and four distinct stones in *Clerodendron* and others. In *Verbena* the exocarp is dry and thin, and the fruit consists of four nutlets (fig. 230). Two- or four-valved capsular fruits also occur, as in *Avicennia*. The seeds are very uniform; they are ovate in shape, sessile and have a basal, lateral or apical attachment according to the attachment of the ovule. There is normally one for each cell or stone in the fruit, occasionally, by abortion, only one for the whole fruit as in *Avicennia*. The seed-coat is thin and membranous. Two of the smaller groups of genera are characterised by presence of endosperm, which is absent in the majority of genera. The embryo is straight with a small radicle directed towards the micropyle and the base of the seed, and a pair of larger parallel cotyledons.

Avicennia has a remarkable embryology, details of which have been worked out by Treub[1]. Early in its history the embryo is carried in the endosperm out through the micropyle into the nucellus, where one cell (cotyloid cell) of the endosperm grows enormously and forms a multinuclear branched hypha-like sucker spreading in all directions through the nucellus and finally reaching the placenta. By its means the embryo is nourished. The fruit-chamber ultimately becomes filled by the highly-developed embryo,

[1] "Notes sur l'embryon, le sac embryonnaire et l'ovule." *Annales Jardin Botan. Buitenzorg*, III, 1883.

which has large green folded cotyledons and a hairy radicle, the latter striking root in the mud on the fall of the fruit.

Several Malayan species of *Clerodendron* are myrmecophilous, as *C. fistulosum*, where ants inhabit the hollow stem-internodes and are attracted by a series of nectaries on the midrib on the lower face of the leaf.

The family is almost exclusively tropical or subtropical. *Verbena*, a very polymorphic genus, has 80 species mainly in tropical and extratropical America, a very few occurring in the Old World; *V. officinalis* (British) grows in Central and Southern Europe, and extends from the Mediterranean region to the Himalayas. Several species are familiar garden plants. *Vitex* has 100 species in the warmer parts of both hemispheres, a few extending into temperate Europe and Asia; *Vitex Agnus-Castus* of the Mediterranean region was esteemed by the ancients as an anaphrodisiac. *Lantana* has 50 species, mainly in the warmer parts of America, with a few in Asia and Africa; the flowers shew a great variety of colour, several species are known as garden plants. *Tectona grandis* (Teak) is a large East Indian tree noted for its hard wood. *Clerodendron* is a large tropical and subtropical genus (90 species), including the well-known greenhouse-plant with red calyx and white corolla.

Verbenaceae are most nearly allied to Labiatae from which they are distinguished by the terminal, not gynobasic style and the undivided ovary, also by the habit.

Those Boraginaceae with a terminal style and regular flowers have a strong resemblance to Verbenaceae, but are distinguished by the ascending ovule with superior micropyle and radicle.

The family is divided into seven tribes depending on the racemose or cymose nature of the inflorescence, the presence or absence of endosperm, the structure of the fruit and the position of the ovule.

Family V. LABIATAE

Flowers usually bisexual, zygomorphic; typical formula S5, P5, A4, G(2). Calyx inferior, persistent, bell-shaped or tubular, with free teeth or lobes, sometimes two-lipped. Corolla tubular, limb variously bilabiate, aestivation imbricate.

Stamens epipetalous, alternating with the corolla-lobes, typically four, didynamous, sometimes reduced to two; anthers typically two-celled, cells dehiscing introrsely by a longitudinal slit; hypogynous disc usually fleshy. Carpels median; ovary soon four-partite by a secondary division with a single anatropous erect basal ovule attached to an axile placenta in each segment; micropyle directed downwards; style generally gynobasic and simple with bifid apex. Fruit of four one-seeded nutlets included within the persistent calyx; seed with thin testa; endosperm absent or scanty; embryo with generally flat cotyledons parallel to the fruit-axis and a short inferior radicle.

Herbs or shrubs, very rarely small trees or climbers, with generally a four-angled stem and opposite, rarely alternate, ternate or verticillate, simple exstipulate leaves; often with a covering of glandular hairs. Inflorescence generally cymose, the pair of cymes in opposite leaf-axils forming a pseudo-whorl; bracteoles usually present.

Genera 170; species about 3000; warm and temperate regions.

The great majority are annual or perennial herbs inhabiting the temperate zone, more rarely becoming shrubby, especially in warmer climates; trees are extremely rare, as in a few species of *Hyptis* (Brazil), and in the Indian genus *Leucosceptrum*. A few American species of *Scutellaria* are climbers, but a climbing habit is very rare. Propagation of the individual may be effected by new shoots which either remain attached to the parent-plant, as in *Salvia*, *Ballota* and others, or the latter dies off, as in *Mentha*, *Lycopus* or *Stachys palustris*. The new shoots form aerial runners with somewhat reduced leaves, as in *Ajuga reptans*, or subterranean stolons with shortened internodes and colourless scale-leaves, as in *Stachys palustris* or *Lycopus*. In many species, as *Mentha aquatica*, both forms occur. The young shoot-axes are four-sided. The simple leaves shew all possible variations from an entire blade to toothed, lobed, cut, or finely multi-sected, as in many Salvias. A whorled leaf-arrangement (three to eight) is characteristic of some genera. Stem, leaves and inflorescence are frequently more or less hairy, often with glandular hairs, the secretion of which has a scent characteristic of the genus or species; sessile glands on the epidermis are also frequent.

The flowers are rarely solitary in the axils of the leaves or bracts, as in *Scutellaria*; generally they form an apparent whorl (verticillaster) at the node, consisting of a pair of cymose inflorescences, each forming a simple three-flowered dichasium, as in species of *Salvia* (fig. 234), *Prunella* and others, or more generally branching and passing over into a pair of monochasial cymes, as in *Lamium*, *Ballota*, *Nepeta* and others; the axes are either all reduced, forming a dense sessile inflorescence, or the main axis or lateral axes or both are more or less developed. A number of whorls may be crowded at the apex of the stem, and the subtending leaves being reduced to a bract-form, a raceme- or spike-like inflorescence results, as in *Prunella*, or *Teucrium* (fig. 233), more rarely a head, as in *Hyptis* (fig. 232); the bracts are simple, or highly differentiated, as in *Monarda* and others. Dorsiventral inflorescences arise by crowding of the flowers towards one side of the main axis, as in species of *Scutellaria*, *Teucrium* and others.

There is usually a pair of bracteoles above the bract, only one of which is developed in the monochasial cymes, and both are often suppressed, as in *Teucrium*, *Prunella* and others. The plan of construction of the flowers is remarkably uniform, but there are wide variations in detail. The calyx-tube may, for instance, be tubular, bell-shaped or spherical, straight or bent, and the form of the teeth or lobes is equally varied; the number of nerves affords useful systematic characters; there are normally five main nerves between which simple or forking or doubled secondary nerves are developed to greater or less extent; the calyx is equally toothed, as in *Origanum*, or two-lipped, as in *Thymus* (fig. 235, E) or *Salvia* (fig. 234, E); it is persistent, and often variously accrescent, sometimes fleshy (*Hoslundia*), in the fruit. The aestivation is generally open, more rarely quincuncial.

The corolla is differentiated into tube and limb; the former is straight or very variously bent, and often widens towards the mouth. The limb is seldom equally five-toothed, or becomes by union of the two upper teeth an almost regular tetramerous corolla, as in *Mentha*; usually there is a well-marked division into upper and lower lip, which arises in three ways. The most general is the 2/3 arrangement, the posterior pair of petals forming the upper lip which is flat, as in *Thymus* (235, D), etc., or concave, as in *Lamium*, *Stachys*, *Salvia* (fig. 234), etc., and shewing very great variety in form;

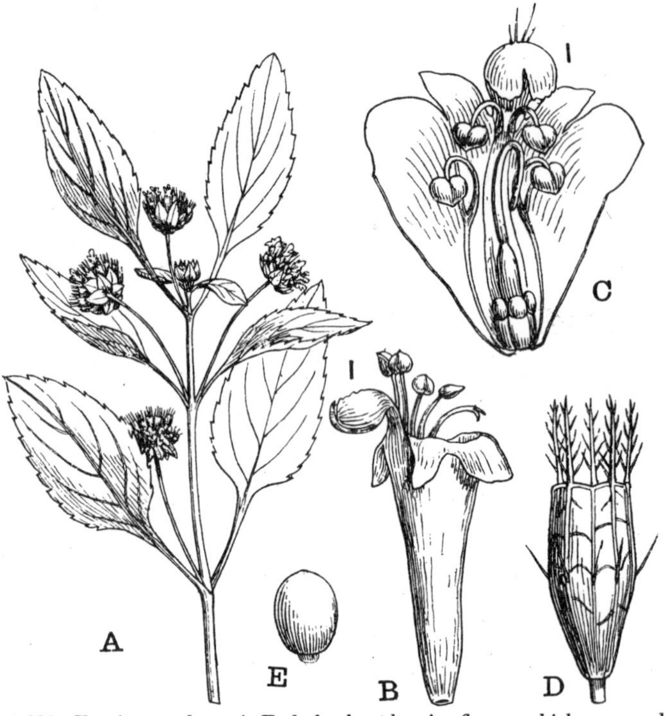

FIG. 232. *Hyptis atrorubens*. A. End of a shoot bearing flowers which are crowded into roundish heads subtended by reduced leaves which form an involucre, reduced. B. Young flower removed from calyx, × 5; *l*, lower lip of corolla. C. Flower in bud, the corolla cut open and folded back, × 13. D. Calyx, equally 5-toothed, enclosing fruit, × 5. E. Nutlet, × 8.

the three lower petals form the lower lip, the median lobe of which is generally most developed. A 4/1 arrangement characterises the subfamily Ocimoideae, the four upper petals forming a slightly developed upper lip, while the anterior forms a very variously differentiated lower lip, as in *Hyptis* (fig. 232). The 0/5 formed by pushing all the lobes forward is characteristic of *Teucrium* (fig. 233).

The fifth, posterior, stamen is very rarely developed; it is occasionally present as a staminode, but generally is completely suppressed. The two upper stamens are often reduced to staminodes or more or less completely suppressed (*Lycopus*, *Salvia*, etc.), rarely are these developed and the anterior pair reduced. *Coleus* has monadelphous stamens. The anterior pair are generally the longer, rarely, as in *Nepeta* and allied genera, the posterior pair. The anthers are two-celled, each

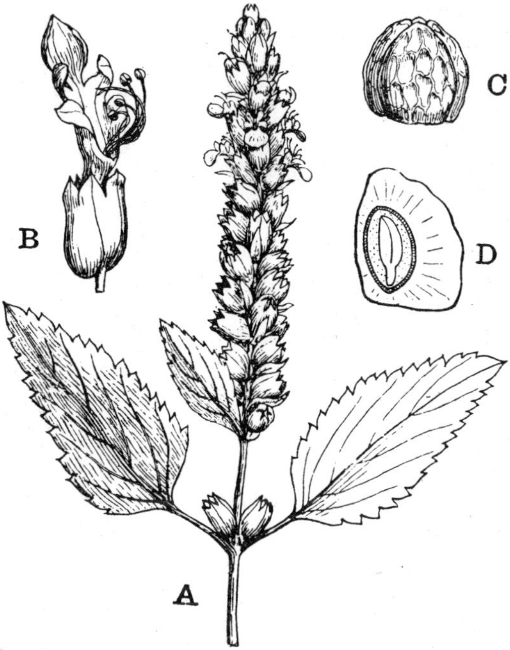

FIG. 233. *Teucrium inflatum*. A. Shoot with a terminal spike of flower and fruit, slightly reduced. B. Flower, × 2½. C. The four nutlets, × 7. D. One nutlet cut longitudinally to shew seed and embryo, × 8.

cell dehiscing longitudinally; the cells may be separated by development of the connective, which in *Salvieae* becomes filiform and articulated with the filament, the anterior cell being reduced or sterile and modified (fig. 234, C). A hypogynous disc is common, generally four-lobed, the anterior pair of lobes serving to secrete nectar.

The two median carpels are very early divided by a constriction into four one-ovuled segments. The stigmatic papillae are usually only at the tip of the style-arms. The anatropous ovules are attached at the inner corner of each ovary-segment; the raphe is on the side next the axis and the micropyle is inferior.

Two types of flower-arrangement are distinguished in relation to insect-pollination. In the less common the anterior part of the flower is more protruded, the stamens and style lie on the under lip, and nectar is secreted on the upper side of the disc; the insect in probing the flower gets its belly and legs smeared with pollen (see *Hyptis*, fig. 232). In *Coleus* and *Plectranthus* the stamens and style are included in the boat-like lower lip, while the upper forms a more or less erect attracting mechanism. The flowers may be resupinate by torsion of the pedicel, or, in species of *Ajuga* and *Teucrium*, by torsion of the corolla-tube.

In the commoner type the stamens and stigma are protected by the arching upper lip, as in *Lamium* and many of our British species, and nectar is secreted on the lower side of the disc. The lower lip affords a resting place for the insect, which in probing the flower collects the pollen on its back. Different genera and species shew very numerous variations in detail. The Salvias, for instance (fig. 234), are characterised by the well-known lever arrangement, the barren half of each anther forming a knob at the end of the short arm, which, when touched by the proboscis, brings the pollen-containing half at the end of the longer arm down upon the back of the insect.

In both types flowers occur with longer tubes and brightly coloured, adapted to visits of butterflies and moths; as, for instance, the bright-red-flowered *Monarda*, species of *Salvia*, *Stachys*, etc. Most British species have shorter tubes and are bee-flowers; some South American Salvias are pollinated by humming-birds. In *Mentha*, *Thymus* (fig. 235, D), *Origanum* and allied genera, the corolla is nearly regular with stamens ultimately spreading beyond it. The bracts sometimes form organs of attraction; as, for instance, in species of *Salvia*, where the upper are sterile and coloured, or the fertile bracts are large and coloured or white.

Gynodioecism occurs in widely different groups; the reduction of the stamens is generally associated with reduction of the corolla, as in *Mentha*, *Thymus*, *Salvia*, *Nepeta* and others. Heteranthy, or occurrence of large-flowered and small-flowered specimens, has been observed in species of *Satureia* and *Galeopsis*. Cleistogamy occurs in genera of different groups (*Salvia*, *Lamium*, *Ajuga*) and in *Lycopus virginicus* Meehan has described the production of autumn flowers which are buried beneath the earth where self-pollination and fruit-development take place.

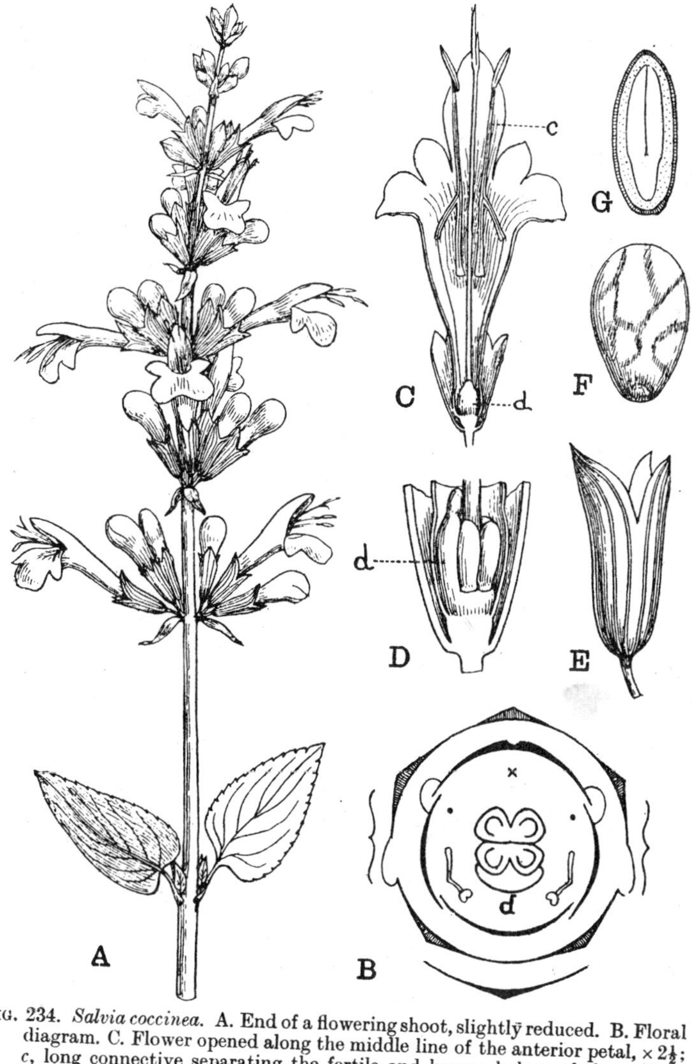

FIG. 234. *Salvia coccinea*. A. End of a flowering shoot, slightly reduced. B. Floral diagram. C. Flower opened along the middle line of the anterior petal, × 2½; *c*, long connective separating the fertile and barren halves of the anther. D. Base of flower, × 7; *d*, large nectary developed from the hypogynous disc. E. Calyx in fruiting stage, × 2½. F. Nutlet, × 8. G. Same in longitudinal section. (B after Eichler.)

The fruit is very uniform; all four nutlets are generally equally developed; rarely is the pericarp fleshy or produced into a membranous wing.

The testa is thin and often more or less absorbed by growth of the embryo which fills the seed; or a scanty endosperm is present. The embryo lies generally parallel to the fruit-axis with short inferior radicle and generally flat cotyledons. The persistent calyx envelops the nutlets, which it often protects by closing above them, and may also assist distribution, forming occasionally a swollen bladder, or having winged, thorny, barbed or hairy lobes or teeth (figs. 232, D, 235, A, C). In species of *Ajuga*, *Lamium* and others

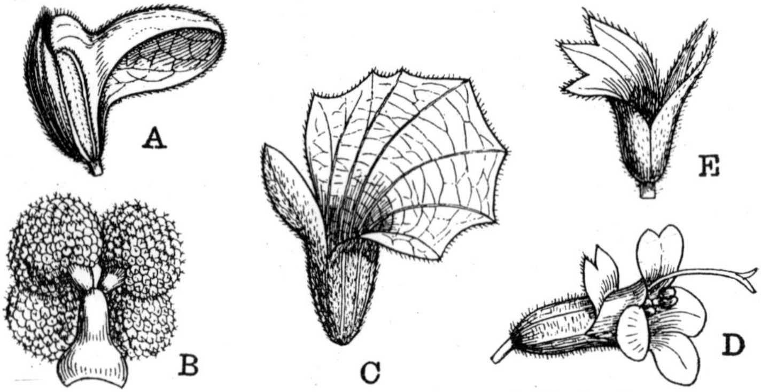

FIG. 235. A. Calyx of *Scutellaria alpina*, the posterior lip bearing a concave shield-like process, × 3. B. Nutlets of *S. splendens*, borne on a columnar development of the disc (gynophore), × 5. C. Calyx of *Otostegia*, the upper lip small, the lower lip (of four sepals) large and spreading, enlarged. D. Flower of *Thymus vulgaris*, × 5; the third tooth on the upper lip of the calyx is not indicated (cf. fig. E). E. Calyx of same, × 6. (After Briquet.)

a small portion of the receptacle becomes separated with the nutlets, and forms an elaiosome, in the same way as noted in *Borago* and other genera of the family Boraginaceae.

The chief centre of distribution of the family is the Mediterranean region, both as regards numbers of genera and species and their importance in the general vegetation, such genera as *Lavandula*, *Thymus*, *Rosmarinus* and others often occurring in large quantity and forming a characteristic feature of the vegetation. The subfamily Ocimoideae is almost exclusively tropical and subtropical and occurs in both hemispheres. The large subfamilies have a very wide distribution. Prostantheroideae, a small group of six genera, is confined to Australia and Tasmania, and Prasioideae has also a limited distribution, chiefly in tropical Asia and the Sandwich Islands.

The family is subdivided by Briquet (in the *Pflanzenfamilien*) as follows.

A. Style not gynobasic. Nutlets with lateral-ventral attachment, the usually large surface of contact often more than half the height of the ovary.

Subfamily I. AJUGOIDEAE. Seeds without endosperm. 11 genera; two, *Teucrium* (fig. 233) and *Ajuga*, are British; *Rosmarinus* is Mediterranean.

Subfamily II. PROSTANTHEROIDEAE. Seeds with endosperm. Six genera; Shrubs or undershrubs; Australian.

B. Style completely gynobasic. Nutlets with basal attachment and small surface of contact.

Subfamily III. PRASIOIDEAE. Nutlets drupaceous with fleshy or strongly thickened exocarp and hard crustaceous endocarp. Seven genera, mainly tropical; *Prasium* is Mediterranean.

Subfamily IV. SCUTELLARIOIDEAE. Nutlets dry, pericarp thin; seeds more or less transverse; embryo with a bent radicle lying on one cotyledon. Two genera; *Salazaria*, monotypic, in Western America and Mexico; *Scutellaria* (British), 180 species, almost cosmopolitan (wanting in South Africa, rare in tropical Africa) (fig. 235, A, B).

Subfamily V. LAVANDULOIDEAE. Nutlets dry; seeds erect; embryo with short straight superior radicle; lobes of disc opposite the ovary-lobes; anthers becoming unilocular by fusion of the cells at the top. One genus, *Lavandula*, with 26 species, spreading from the Canary Islands through the Mediterranean region to further India.

Subfamily VI. STACHYOIDEAE. Nutlets dry; seeds erect; embryo with short straight inferior radicle; disc-lobes when distinct alternate with ovary-lobes; stamens ascending or spreading and projecting straight forwards. 110 genera subdivided into 12 tribes depending on the shapes of the corolla and calyx, and on the number and relative length of the stamens. The following are British: *Marrubieae* (*Marrubium*), *Nepeteae* (*Nepeta*), *Stachyeae* (*Prunella*, *Galeopsis*, *Melittis*, *Lamium*, *Ballota*, *Stachys*), *Salvieae* (*Salvia*) (fig. 234), *Satureieae* (*Calamintha*, *Origanum*, *Thymus* (fig. 235, D, E), *Mentha*, *Lycopus*).

Subfamily VII. OCIMOIDEAE. As VI, but stamens descending, lying upon or enclosed in the upper lip. 30 genera; tropical and subtropical; *Hyptis* (fig. 232), *Ocimum*, *Plectranthus*, *Coleus*.

Subfamily VIII. CATOPHENOIDEAE. Nutlet dry; seed erect; embryo with curved radicle lying against the cotyledons. One genus with three species in tropical America.

Labiatae are readily distinguished from all other families except Verbenaceae, with which they are closely united by the first two subfamilies in which the style approaches more or less a terminal position. Several genera of Verbenaceae, on the other hand, have the style more or less deeply sunk between the four ovary-lobes, and the relation becomes closer from their cymose inflorescence, and the fact that several of the verbenoid genera of Labiatae have the laterally attached hemianatropous ovule which also characterises the Verbenaceae in question. It is on this account impossible to draw a sharp line between the two families.

The separation of the ripe fruit into four nutlets has led to the association of Labiatae with Boraginaceae in one order Nuculiferae by some botanists; but the Boraginaceae are readily distinguished by their superior micropyle, outwardly directed raphe and the superior radicle of the embryo.

The resemblance with Scrophulariaceae and Acanthaceae based on the occasional occurrence in these families of opposite leaves and a four-angled stem is superficial; the fruit structure completely separates them from Labiatae.

Many genera are of economic use on account of the volatile oils which they contain. Such are Thyme, Marjoram (*Origanum*), Sage (*Salvia*), Rosemary (*Rosmarinus*), Lavender (*Lavandula*), Patchouly (*Pogostemon*). Tubers of *Stachys* and *Coleus* are eaten.

Suborder 4. SOLANINEAE. Flowers sometimes regular and isostemonous, but usually medianly zygomorphic, with four or two fertile stamens. Carpels generally two (five in Nolanaceae), with usually indefinite ovules. Fruit generally a capsule.

Family VI. NOLANACEAE

A small family of about 40 species in three genera, on the western side of South America, mainly strand-plants. They are herbs or small shrubs with alternate, often fleshy, simple leaves and regular bisexual pentamerous flowers, solitary or associated in leafy racemes. The corolla is bell- or funnel-shaped, with a five-lobed limb folded in bud; the stamens are all fertile and of equal length. The typically five-carpellary ovary is many-ovuled and is divided longitudinally into a series of 5 to 10 segments, or transversely into 10–30 segments standing in two to three superposed series; each segment forms a one- to seven-seeded mericarp, usually with a stony endocarp. The embryo is curved and surrounded by endosperm.

The family has been variously placed by different authorities. Bentham and Hooker included it in Convolvulaceae, an affinity with which is suggested by the form and aestivation of the corolla, and the pentamerous ovary. The separation of the fruit into mericarps recalls Boraginaceae. It is however probably most nearly allied with Solanaceae, with which it was united by Dunal and Baillon, but from which it is distinguished by the remarkable structure of the fruit.

Family VII. SOLANACEAE

Flowers bisexual, regular or sometimes zygomorphic, hypogynous, conforming to the formula S5, P5, A5, G(2). Sepals persistent, often enlarging in the fruit. Corolla gamopetalous, shape various, rarely two-lipped; aestivation various, generally plicate, sometimes also convolute. Stamens inserted on the corolla-tube and alternating with the lobes, in zygomorphic flowers often unequal, sometimes one is rudimentary; anthers bilocular, dehiscing by slits or pores; hypogynous disc generally obvious. Ovary bilocular, or falsely 3- to 5-locular, the carpels generally obliquely placed; ovules generally numerous on swollen axile placentas, anatropous or slightly amphitropous; style simple, stigma bilobed. Fruit a many-seeded berry or capsule; seeds often with a pitted testa; embryo bent or straight, embedded in endosperm.

Herbs, shrubs, or small trees, leaves various. Flowers terminal on the main and lateral axes, solitary or in cymes.

Genera 85; species about 2000; tropical to temperate regions.

The plants are small annual herbs, as in *Solanum nigrum*, a common weed in waste places, or biennial, or, generally, perennial, as *Atropa Belladonna* (Deadly Nightshade), a large branching herb which is indigenous on chalk and limestone in Britain. Climbing plants occur, as *Solanum Dulcamara*, an irregularly trailing twiner. Plants of very various habit, including herbs, shrubs, and trees, creeping, erect and climbing plants, are found in the single genus *Solanum*, which comprises more than half the number of species in the family.

The leaves are generally alternate in the vegetative parts, in the flowering parts often in pairs, an arrangement which, like the extra-axillary position of the flowers or cymes, is the result of congenital union of axes.

Thus in *Datura* (fig. 236, A), where branching is dichasial, the leaf at any given node belongs really to the node below,

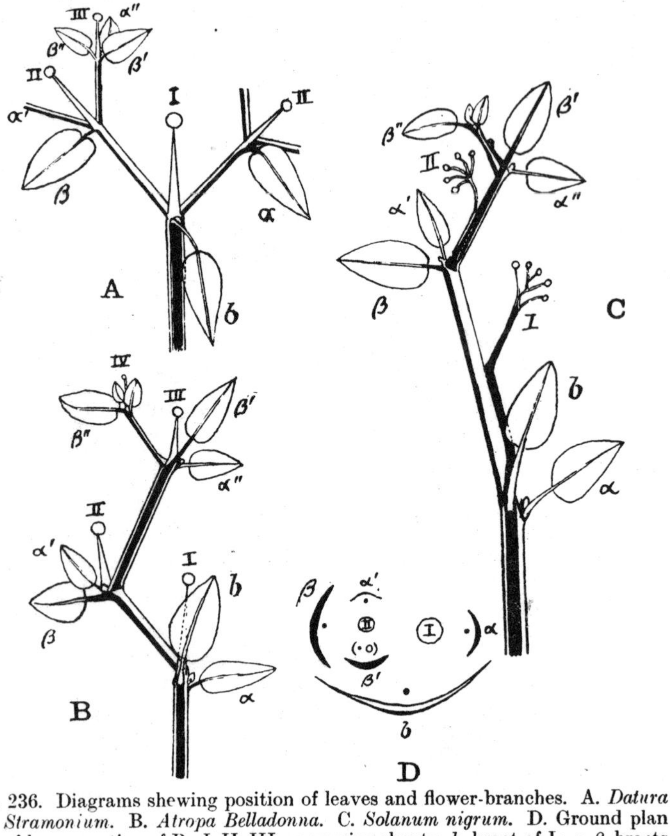

FIG. 236. Diagrams shewing position of leaves and flower-branches. A. *Datura Stramonium*. B. *Atropa Belladonna*. C. *Solanum nigrum*. D. Ground plan of lower portion of B. I, II, III, successive shoots; *b*, bract of I; *α*, *β*, bracts of II; *α′*, *β′*, bracts of III, etc. (After Eichler.)

but has become adnate to, or raised up on, its axillary shoot as far as the next node; this union is indicated in black in the figure. In *Atropa Belladonna* (fig. 236, B) one of the two branches at each node remains undeveloped, and there are a pair of leaves, a smaller (*α*, *α′*, *α″*, etc.) which subtends the undeveloped branch at the node in question and a larger (b, β, β', etc.) which has been carried up from the node below. *Solanum nigrum* (fig. 236, C) illustrates a further complication owing to the union of the flower-shoot for some distance above the position of its bract with the axis of the next higher degree; the flower-shoot thus appearing to spring from the middle of an internode (extra-axillary).

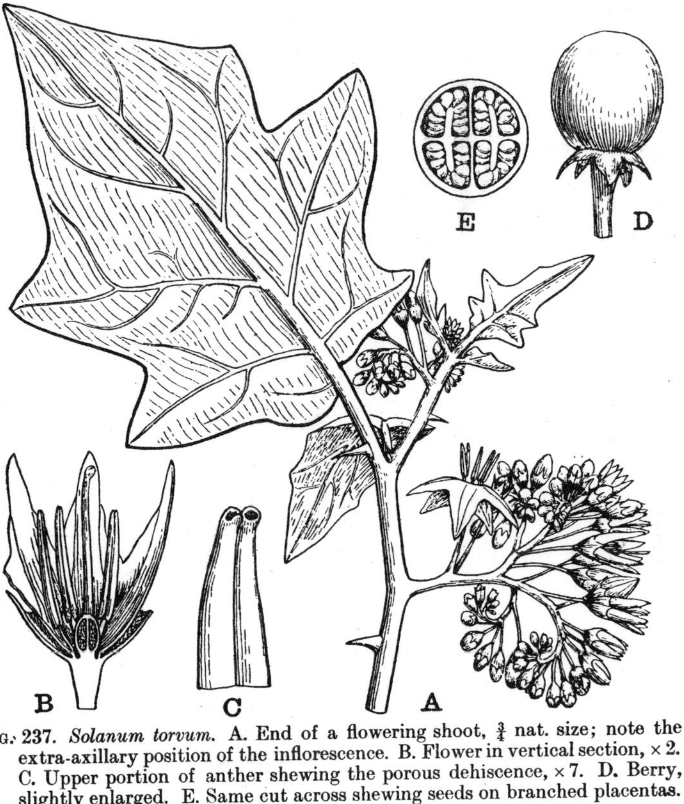

FIG. 237. *Solanum torvum.* A. End of a flowering shoot, $\frac{3}{4}$ nat. size; note the extra-axillary position of the inflorescence. B. Flower in vertical section, × 2. C. Upper portion of anther shewing the porous dehiscence, × 7. D. Berry, slightly enlarged. E. Same cut across shewing seeds on branched placentas.

A constant anatomical character is the occurrence in the stem of bicollateral bundles, affording a distinction from the simple collateral bundles of Scrophulariaceae.

The flower, speaking strictly, is rarely actinomorphic owing to the oblique position of the ovary as indicated in the floral diagram (fig. 239, D). The corolla is regular, for instance the rotate corolla of *Solanum* (fig. 237), or the bell-shaped of *Atropa*, or, as in *Hyoscyamus* (Henbane), somewhat irregular. Zygomorphy is more strongly marked in the tribe *Salpiglossideae*, where only two or four fertile stamens are present (fig. 241), and the corolla may become strongly two-lipped, as in *Schizanthus*. The stamens are generally more or less unequal in length. Exceptions from the typical dimerous ovary occur. *Henoonia*, a monotypic Cuban genus, has a unilocular one-ovuled ovary. In *Capsicum* the ovary becomes unilocular in the upper part. On the other hand, several genera are characterised by an ingrowth of false dissepiments forming, as in *Datura* (Thorn-apple) (fig. 239, F), a regularly four-celled or, as in *Nicandra*, an irregularly three- to five-celled chamber. The ovules are generally numerous but sometimes few, as in *Cestrum*, a large American genus with tubular flowers, species of which are greenhouse plants. The great majority are bisexual, with conspicuous flowers adapted to insect-pollination; nectar is secreted at the base of the ovary or at the bottom of the corolla-tube between the stamens. The flowers are homogamous or proterogynous. Cleistogamic flowers occur in *Salpiglossis*.

The fruit is a berry or capsule, the two forms being connected by dry indehiscent and irregularly dehiscent berry-like fruits. The greater or less curvature of the embryo affords characters for the separation of the genera into tribes. The persistent calyx may serve to protect or distribute the fruit; as, for instance, the prickly calyx of some species of *Solanum*, or the bladdery structure enveloping the fruit of *Physalis* (Winter Cherry, Cape Gooseberry) (fig. 238).

The main distribution centre of the family lies in Central and South America; 38 out of 85 genera are endemic in Central and South America and the West Indies.

The genera are arranged as follows by Wettstein (in the *Pflanzenfamilien*).

Series A. Embryo curved through more than a semicircle. All five stamens fertile, equal or slightly unequal.

Tribe 1. *Nicandreae*. Ovary three- to five-chambered, placenta irregularly three- to five-lobed. Contains a single monotypic genus *Nicandra*, a native of Peru, but widely cultivated as an ornamental plant, the calyx becoming enlarged in the fruit, as in *Physalis*.

Tribe 2. *Solaneae*. Ovary two-chambered. Contains 40 genera, which are grouped into subtribes according to the form of the corolla, the character of the fruit, and the insertion of the anther. *Lycium* has 100 species in the extratropical parts of both hemispheres, comprising small trees or shrubs, often thorny with a long cylindrical or narrowly bell-shaped

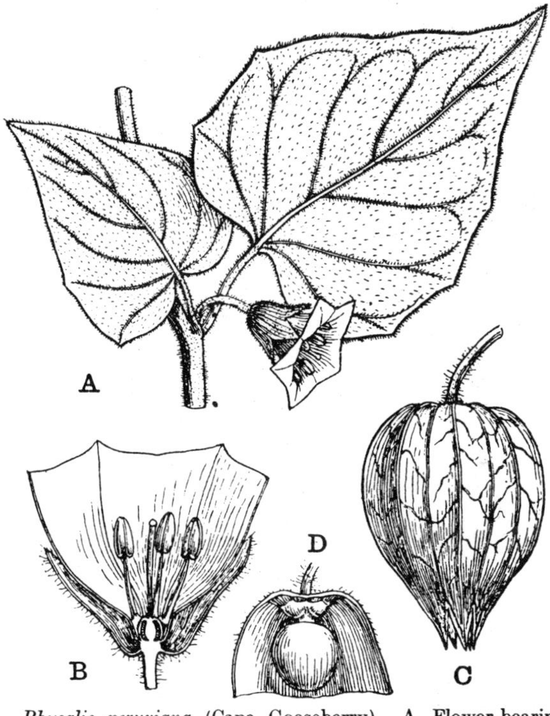

FIG. 238. *Physalis peruviana* (Cape Gooseberry). A. Flower-bearing node, $\frac{3}{4}$ nat. size. B. Flower in vertical section, × $1\frac{1}{2}$. C. Enlarged calyx surrounding the fruit, $\frac{3}{4}$ nat. size. D. Lower part of calyx cut open exposing the fruit, slightly enlarged.

corolla-tube and a juicy berry. *Lycium chinense* is a straggling climber often cultivated under the name Tea-plant.

Atropa, two species in Europe, North Africa and West and Central Asia; *A. Belladonna* (British) yields the powerful drug atropin; corolla campanulate, fruit a fleshy berry.

Hyoscyamus, 11 species from the Canary Islands through Europe and North Africa to Asia; corolla irregularly bell-

or funnel-shaped; fruit a many-seeded capsule, enveloped by the calyx-tube, with circumscissile dehiscence; *H. niger* (Henbane) is British.

Physalis, 45 species mostly in the warmer parts of North and South America. The calyx becomes large and bladder-

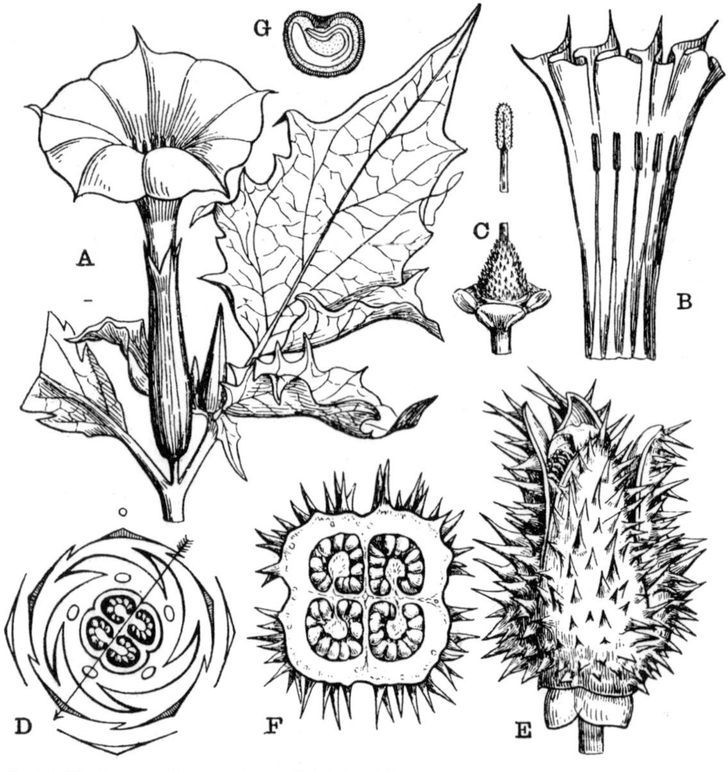

FIG. 239. *Datura Stramonium.* A. End of flowering shoot. B. Corolla cut open. C. Pistil, the rest of the flower has been cut away. D. Floral diagram, the arrow indicates the plane of symmetry. E. Capsule opening. F. Capsule in transverse section. G. Seed in transverse section shewing curved embryo. All slightly reduced except G which is enlarged.

like, enclosing the ripe, sometimes edible, berry. *P. Alkekengi* (Europe and Asia), Winter Cherry, and *P. peruviana*, Cape Gooseberry (fig. 238).

Capsicum, 30 species in Central and South America, one in Japan; widely cultivated for the sake of the spherical or elongated more or less fleshy berried fruit (Chillies).

Solanum, 1200 species of herbs, shrubs or rarely trees, of very various habit, with a rotate corolla and many-seeded juicy berry; the anthers form a cone in the centre of the flower and dehisce by apical pores (fig. 237, C). *S. tuberosum* (Potato), *S. Lycopersicum* (Tomato); *S. Dulcamara* (Bittersweet) and *S. nigrum* are British.

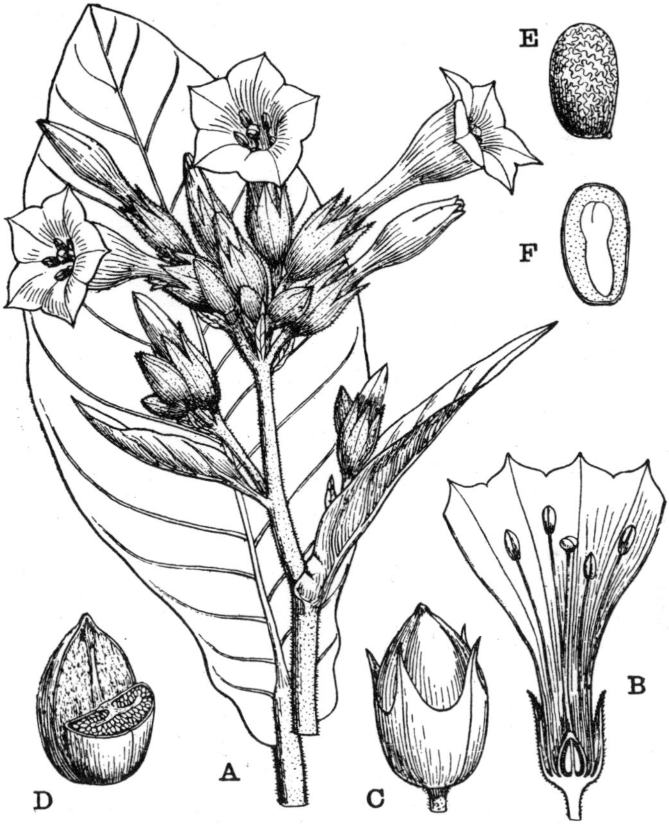

FIG. 240. *Nicotiana tabacum*. A. Leaf and flowers. B. Flower in vertical section. C. Fruit. D. Fruit opening, one half cut transversely shewing the T-shaped placenta. E. Seed. F. Seed cut longitudinally shewing the straight embryo. A, B, slightly reduced. C, D, × 1½. E, F, × 18. (A after Bentley and Trimen.)

Mandragora (Mandrake), three or four species in the Mediterranean and Himalayan areas, are acaulescent herbs with large forked swollen roots.

Tribe 3. *Datureae*. Ovary four-chambered. Three genera. *Datura*, 15 species of shrubs, trees or herbs in the warmer parts of the earth; corolla large funnel-shaped; fruit a capsule or berry with numerous flat seeds. *D. Stramonium* (Thorn-apple), sometimes found as an escape in Britain, is officinal (fig. 239).

Series B. Embryo straight or only slightly bent.

Tribe 4. *Cestreae*. All five stamens fertile. 24 genera, almost exclusively American. *Nicotiana*, 50 species; *N. tabacum* (Tobacco) (fig. 240) and other species are cultivated. *Cestrum*, 140 species in tropical and subtropical America. *Petunia*, 14 species, chiefly tropical American.

Tribe 5. *Salpiglossideae*. Two to four stamens fertile, always unequal; flowers obviously zygomorphic. 10 genera, tropical American and Australasian. *Salpiglossis* (fig. 241) (eight species in subtropical South America), *Schizanthus* (11 species in Chile) and *Brunfelsia* (22 species, tropical America) are known in cultivation as ornamental plants.

FIG. 241. A. Flower of *Salpiglossis*, slightly reduced. B. Corolla-tube opened shewing the two pairs of fertile stamens and a rudimentary fifth stamen, enlarged.

Solanaceae are most nearly allied to Scrophulariaceae, from which the majority, or what we may regard as the typical members (Series A in the above arrangement), are separated by the actinomorphic flower, and the complete staminal whorl, as well as by the bicollateral vascular bundles of the stem; while the tribes *Cestreae* and *Salpiglossideae* approach this family much more nearly, being distinguished mainly by the anatomical character, but having also a plicate aestivation in the corolla, while in genera which in zygomorphy of the corolla closely resemble Scrophulariaceae, the oblique position of the carpels is a distinguishing character.

Several genera are of economic importance as foods, as *Solanum tuberosum* (Potato), *S. Lycopersicum* (Tomato), *S. Melongena* (Aubergine, Egg-plant), *Capsicum* and others. The Potato was introduced into Europe from South America by the Spaniards between 1560 and 1570, whence its cultivation spread to Italy, the Netherlands and Central Europe. It was brought to England, from South America, perhaps by Drake, in 1586, and probably first cultivated by Raleigh at Youghal in Ireland[1]. Some genera, as *Atropa*, *Datura*, *Solanum*, *Hyoscyamus*, containing strong alkaloids, are of use medicinally. Many are garden plants, such as species of *Solanum*,

[1] See R. N. Salaman, "The Potato in its early home," *Journ. Roy. Hort. Soc.* 1937, 253.

Petunia, Nicotiana, Salpiglossis, Schizanthus, Datura, or greenhouse, as *Cestrum* and *Brunfelsia.* Tobacco is derived mainly from *Nicotiana tabacum,* Virginian tobacco, a native of Central or South America and now widely cultivated in temperate and warmer countries; it is the source of the tobaccos of Cuba, the United States, the Philippine Islands, and the Latakia of Turkey; East Indian or Green Tobacco is the product of *N. rustica,* a native of Mexico and widely cultivated in Southern Germany, Hungary and the East Indies.

Family VIII. SCROPHULARIACEAE

Flowers bisexual, zygomorphic, hypogynous conforming to or derived from the formula S5, P5, A5, G(2), a reduction to four or sometimes two occurring in the androecium. Calyx more or less deeply 4- to 5-partite. Corolla medianly zygomorphic, variously imbricate, not plicate, shape very various. Stamens rarely all fertile, the posterior generally sterile or 0, rarely only two fertile; hypogynous disc annular or unilateral. Ovary bilocular, rarely unilocular, carpels median; ovules usually numerous on a thick axile placenta and anatropous; style simple or bilobed. Fruit a capsule or sometimes a berry; seeds numerous and small, or few and larger; embryo straight or slightly curved, surrounded by endosperm.

Herbs (sometimes hemiparasites or parasites) and undershrubs, rarely shrubs or trees, with alternate, opposite or whorled exstipulate leaves, and racemose or cymose inflorescences.

Genera 205; species about 2600; cosmopolitan.

According to Wettstein 64 per cent. of the species are perennial herbs and undershrubs, 30 per cent. annual herbs, while shrubs and trees (as *Paulownia*) are rare. There are wide differences in habit. The stem may be prostrate and creeping, as in some of our Veronicas and *Linaria Cymbalaria* (Ivy-leaved Toad-flax); generally it is erect, as in Fox-glove, Figwort, Mullein, etc. Climbing plants occur, for instance twiners as *Rhodochiton* (Mexico), leaf-stalk climbers as *Antirrhinum cirrhosum, Maurandia* (Mexico), and others. The South African *Harveya* and *Hyobanche* are parasites almost devoid of chlorophyll, with scale-like leaves, while among British plants *Lathraea* (Tooth-wort) is a leafless root-parasite without chlorophyll; and several genera are hemi-parasites, such as *Euphrasia*, *Odontites*, *Melampyrum* and others, which though provided with green-leaved shoots attach themselves by suckers, which are reduced lateral roots, to roots of grasses. Adaptation to life in water occurs in *Hydrotriche* and *Ambulia* (fig. 242). In anatomical structure, and the frequently developed hairiness of stem and leaf, Scrophulariaceae resemble Labiatae.

The leaves are alternate throughout, as in *Verbascum*, or the lower opposite and the upper alternate, as in *Antirrhinum*, or all opposite, as in *Mimulus*, or whorled, as in some Veronicas. All varieties of leaf-arrangement occur in the genus *Veronica*. Heterophylly obtains in certain New Zealand species of this genus, shoots with scale-like appressed leaves (recalling the habit of a Cypress or *Lycopodium*), and also shoots with flattened spreading leaves, occurring on the same plant. Certain aquatic genera, *Ambulia* (South Asia, East Africa, Australia) (fig. 242), *Hydrotriche* (Madagascar) and others have entire aerial leaves and much divided submerged leaves.

The flowers terminate axes of the secondary or some higher degree. They are solitary, as in *Linaria*, *Mimulus* and others, or form spikes or racemes, either terminal, as in *Digitalis*, species of *Veronica* (*V. arvensis*) and others, or axillary, as in other Veronicas (section *Chamaedrys*). In other cases cymose inflorescences occur, forming dichasia arranged in terminal or axillary racemes, spikes or panicles. In *Verbascum* the arrangement becomes highly complicated by the formation below each three- to seven-flowered dichasium of two to five accessory dichasia with gradually decreasing numbers of flowers. Bracts and bracteoles are generally present. When a terminal flower is present it shews a radial structure, becoming peloric, as in Toad-flax, where symmetry is restored by production of a spur to each petal; all the flowers are sometimes peloric, as in the peloric form of the garden Snapdragon (*Antirrhinum*). Departure from the typically pentamerous calyx is seen in *Euphrasia*

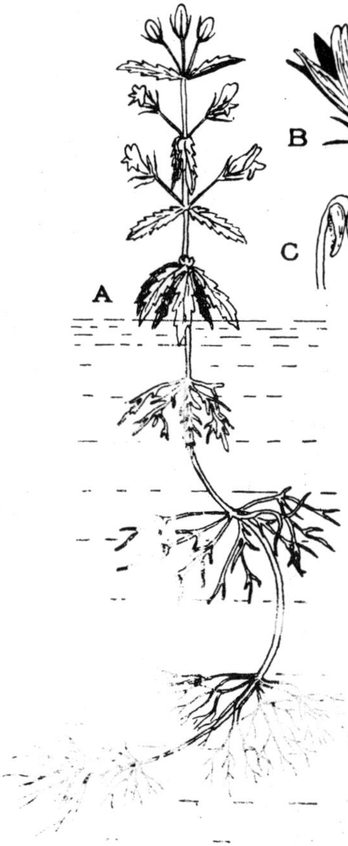

FIG. 242. *Ambulia hottonioides*. A. Plant, × $\frac{1}{3}$. B. Flower, with pair of bracteoles, nat. size. C. Stamen, enlarged. (After Wettstein.)

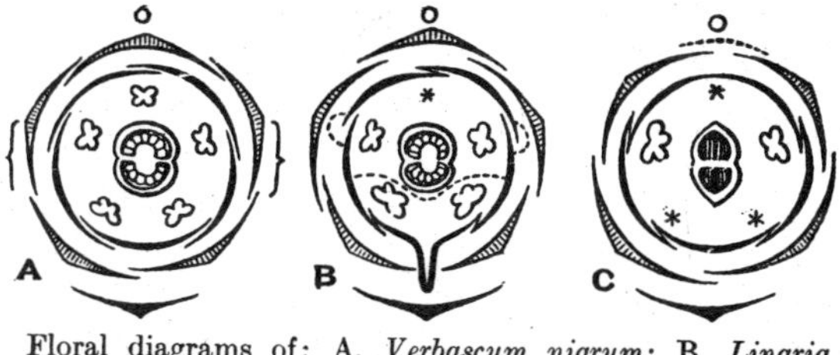

FIG. 243. Floral diagrams of: A, *Verbascum nigrum*; B, *Linaria vulgaris*; C, *Veronica Chamaedrys*. (After Eichler.)

and many Veronicas, where the posterior sepal is suppressed; in *Pedicularis* two to five teeth are present, and in *Calceolaria* the two anterior sepals are completely united.

The typical medianly zygomorphic corolla shews great diversity in form, depending on the length and breadth of the tube, which may be almost obsolete, as in *Veronica* (fig. 244, C), or large and bell-shaped, as in Foxglove (fig. 244, D), and on the union and development of the limbs, which are spreading, as in *Veronica*, small and suberect, as in *Scrophularia* (Figwort) (fig. 244, G), or form a pair of closed lips, as in *Linaria* or *Antirrhinum*. *Linaria* is characterised by a spurred anterior petal, *Diascia* (South Africa) by a spur on each of the lateral anterior petals. In *Calceolaria* (fig. 244, B) the tube is very short and the limb forms two lips, an upper smaller, formed by the union of the posterior pair, and a lower larger, formed from the three anterior petals. A regular corolla occurs in a few genera; *Verbascum* has five nearly equal segments. In *Veronica*, *Gratiola* and others the corolla becomes four-lobed by a complete union of the two posterior petals (figs. 243 and 244, C). Of the five stamens all may be present and equal, as in some species of *Verbascum*; generally however (as usual in *Verbascum*) the three posterior are longer than the anterior (fig. 244, A). In the great majority the posterior stamen is more or less aborted (243, B), while of the four fertile the anterior pair are generally longer than the posterior, as in *Linaria*, *Digitalis*, etc. In *Veronica*, *Calceolaria* and others two only are present. The form of the anthers is very various; the two chambers are sometimes unequal, they may remain distinct or become united above; appendages or hair-developments are sometimes present. Dehiscence is introrse, generally by a longitudinal slit. The two carpels are generally equal; in *Antirrhinum* the anterior is the larger.

The flowers are adapted for pollination by insects. Nectar is secreted by the disc around the base of the ovary, or by specialised nectaries on the under side. The differences in the form of the corolla are associated with the visits of different classes of insects. We may contrast, for instance, the open flower with short tube of *Verbascum* or *Veronica*, where the nectar is more exposed, the long wide tube of *Digitalis*, where the stigmas and stamens are placed so as to touch the back of the visiting insect (a large bee), the closed flower of Snapdragon requiring a certain amount of strength to separate the lips, and *Euphrasia* and others with loose powdery pollen and appendaged anthers, which when disturbed shake out the pollen on the visitor's head. In species of *Mimulus* and others the stigmas are sensitive to contact in such a way as to favour cross-pollination by the visiting insect. There is

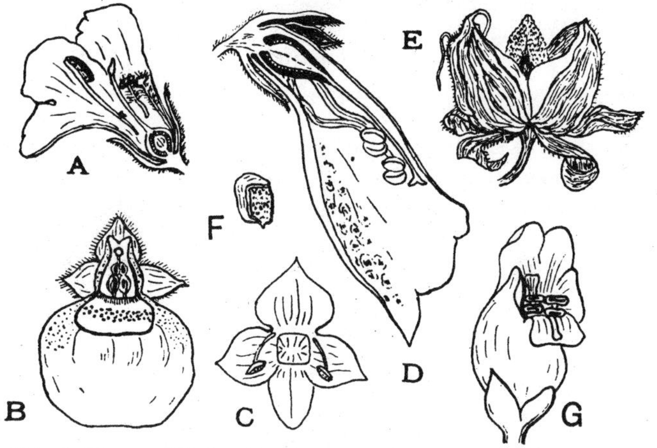

FIG. 244. A. Flower of *Verbascum Thapsus* cut longitudinally, slightly reduced. B. Flower of *Calceolaria*, front view, the two stamens and the style with capitate stigma project above the large lower lip of the corolla. C. Corolla with stamens of *Veronica officinalis*, enlarged. D–F. *Digitalis purpurea*. D. Flower cut longitudinally, slightly reduced. E. Fruit, opening along the septa into two valves, the placenta covered with the minute seeds remains in the centre. F. Seed, much enlarged. G. Flower of *Scrophularia nodosa*, a proterogynous wasp-flower, in second stage; the style has bent over the lip of the corolla, the four fertile anthers stand above the lip and the sterile stamen stands erect behind them, enlarged. (G after Warming; the rest after Wettstein.)

also generally means for ensuring self- in default of cross-pollination. Cleistogamic flowers occasionally occur, as in *Linaria vulgaris*.

The capsule is surrounded at the base by the persistent calyx; sometimes, as in *Rhinanthus* (Yellow-rattle), completely enveloped by it. Dehiscence is valvular, by two or four valves, either loculicidal or septicidal (fig. 244, E), or, as in *Antirrhinum*, by pores. Occasionally the fruit is a berry. Seeds are generally small and numerous; rarely few and large, as in *Veronica*. The testa is smooth or variously granular, angled or winged. In *Linaria Cymbalaria* the plant buries its seeds by the downward bending of the fruit-stalks. Species of *Veronica*, *Lathraea* and *Melampyrum* are classed as myrmecochorous. An elaiosome is formed from the raphe in the first two genera and from the chalaza in the last.

Scrophulariaceae find their chief development in the temperate regions of both hemispheres, diminishing rapidly towards the tropics and colder regions.

The following are the subfamilies and principal tribes adopted by Wettstein in the *Pflanzenfamilien*.

A. The two posterior corolla-lobes covering the lateral lobes in the bud.

Subfamily I. PSEUDOSOLANOIDEAE. Leaves alternate, five stamens often present.

Tribe *Verbasceae*. *Verbascum*, five stamens (British); *Celsia*, four stamens.

Subfamily II. ANTIRRHINOIDEAE. At least the lower leaves opposite; the posterior stamen absent or barren.

Tribe *Calceolarieae*. Corolla two-lipped with a large concave lower lip. *Calceolaria*, 200 species, chiefly temperate South American.

Tribe *Antirrhineae*. Corolla two-lipped, spurred or saccate at base. *Linaria* (British), 95 species, and *Antirrhinum* (British), 32 species, northern hemisphere; *Nemesia* (South Africa).

Tribe *Cheloneae*. Corolla two-lipped, not spurred or saccate; fruit a capsule or a berry; inflorescence cymose. *Collinsia*, North America, cultivated in gardens; *Scrophularia* (British); *Pentstemon*, chiefly North American (gardens); *Paulownia*, an arborescent Japanese genus, grown in European gardens.

Tribe *Gratioleae*. Corolla two-lipped, not spurred or saccate; flowers solitary or racemose. *Mimulus* (Musk), chiefly extratropical American, *M. guttatus* (*Langsdorfii*) has become widely naturalised by the side of streams in Britain; *Gratiola*; *Limosella* (British).

B. The posterior corolla-lobes covered in the bud by one or both of the lateral lobes.

Subfamily III. RHINANTHOIDEAE.

Tribe *Digitaleae*. Corolla-lobes flat and divergent, or the two upper erect; not parasitic. *Veronica* (British), 200 species mostly north temperate, but also in New Zealand and Australia; *Digitalis* (British), 25 species, Europe and West Asia; *Sibthorpia* (British).

Tribe *Gerardieae*. Similar to *Digitaleae*, but one of the anther-lobes is often reduced. Hemiparasites, or, as in the South African *Harveya* and *Hyobanche*, parasitic.

Tribe *Rhinantheae*. Upper corolla-lobes forming a helmet-like lip. Hemiparasites as *Odontites, Bartsia, Euphrasia, Rhinanthus, Pedicularis, Melampyrum*, all of which have British representatives; or parasitic as *Lathraea*, which includes a few species of root-parasites without chlorophyll and bearing scale-like leaves (*L. Squamaria*, Toothwort, is British).

Heinricher[1] has studied in detail the development and life-history of the *Rhinantheae* and indicates an interesting gradation in parasitism from the hemiparasitic annual forms (such as *Euphrasia, Odontites* and *Melampyrum*) to the truly parasitic *Lathraea*, as found in the perennial forms, *Pedicularis, Bartsia* and *Tozzia*; the last named will only germinate in association with the host-plant.

The family is most nearly allied to Solanaceae, the affinity being most marked in the tribes *Verbasceae* and *Salpiglossideae* respectively. Scrophulariaceae are distinguished by the median position of the carpels, the aestivation of the corolla, and also in most cases by the zygomorphic corolla, and the reduction of the posterior stamen; an anatomical distinction is afforded by the collateral, not bicollateral, structure of the vascular bundles in the stem. Of the other nearly allied families, Orobanchaceae and Gesneriaceae differ in their unilocular ovary with parietal placentation, and Bignoniaceae and Pedaliaceae in the absence of endosperm. Acanthaceae and Verbenaceae, which have often a similar habit, are distinguished by the structure of the fruit.

The uses of Scrophulariaceae are limited. *Digitalis* and other genera are or have been used in medicine, and many genera are cultivated in garden or greenhouse for their showy flowers.

The small family (IX) SELAGINACEAE is regarded by Wettstein as a tribe of the subfamily Antirrhinoideae of Scrophulariaceae, with which it is certainly closely allied. It consists of about 120 species of herbaceous or shrubby plants of heath-like habit, mainly South African, but found also in Madagascar and on the mountains of tropical Africa. The small strongly zygomorphic hypogynous flowers are aggregated in heads or spikes;

[1] Heinricher, E. "Die grunen Halbschmarotzer." *Jahrbucher f. Wissenschaftliche Botanik*, XXXI, 77 (1897); XXXII, 389 (1898); XXXVI, 665 (1901); XXXVII, 264 (1902).

the plan of structure is that of Scrophulariaceae (fertile stamens four or two, anthers becoming one-celled), but there is a single pendulous ovule from the top of each cell of the ovary and the fruit consists of two achenes which become free when ripe; the achenes are often unequal and one may be sterile or obsolete. Principal genera, *Hebenstreitia* and *Selago*; species of the former are grown in the greenhouse.

The closely allied family (X) GLOBULARIACEAE, with 20 species, mainly South European and Mediterranean, comprises perennial herbs or shrubs, with simple alternate exstipulate leaves and small hypogynous zygomorphic flowers crowded in a head surrounded by an involucre of bracts. The individual flower with its tubular calyx, markedly two-lipped corolla, and didynamous stamens suggests Labiatae, but the ovary is unilocular, with a single anatropous ovule pendulous from the apex; the style is terminal with a capitate or short two-lobed stigma. The fruit is a one-seeded nutlet enclosed in the persistent calyx. Most of the species belong to the alpine and Mediterranean genus *Globularia*. Bentham and Hooker included the family in Selaginaceae; it may be regarded as a further development of the tendency expressed in that family to aggregation of the flowers and reduction in the gynoecium.

Family XI. OROBANCHACEAE

Flowers bisexual, medianly zygomorphic. Calyx hypogynous, gamosepalous, two- to five-partite. Petals five, imbricate, united to form a two-lipped corolla. Stamens four, didynamous, epipetalous. Carpels two, median, rarely three, ovary superior, unilocular, placentas (two to six) parietal, often T-shaped in section and variously branched; ovules indefinite, anatropous; style single; stigma terminal, often two- to four-lobed. Fruit a capsule dehiscing loculicidally; seeds minute, containing a few-celled undifferentiated embryo in an oily endosperm; testa pitted or rough.

Annual or perennial parasitic herbs containing little or no chlorophyll, with scale-like leaves and flowers in a terminal raceme, rarely solitary and terminal.

The members of this family live parasitically upon the roots of other plants to which they attach themselves by root-suckers. The seeds of *Orobanche* will only germinate when in contact with the root of the host-plant. In *Phelipaea*, a genus with two species in Asia Minor, parasitic on *Centaurea*, the scarlet flowers are solitary; usually however the aerial upright stem bears a raceme or spike of flowers, as in our British Broomrape (*Orobanche*); in *Orobanche ramosa* it is branched. Each flower stands in the axil of a bract and a pair of bracteoles are also often present on the pedicel. The form of the calyx shews considerable variation affording useful systematic characters. It is typically pentamerous, but reduction or suppression of one or more members may occur; thus the anterior lateral sepals are absent in some species of *Orobanche*. The posterior stamen is always suppressed; the anthers dehisce lengthwise; occasionally one half-anther is barren, as in

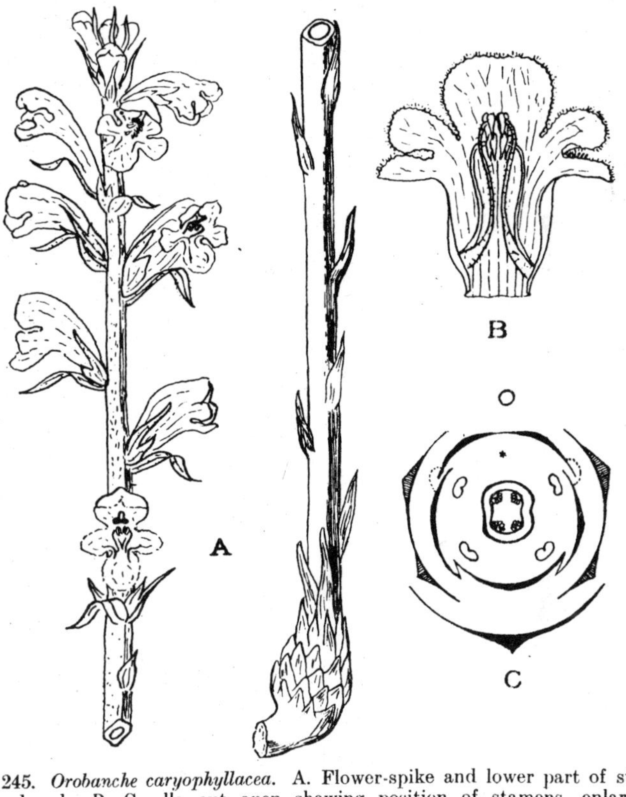

FIG. 245. *Orobanche caryophyllacea*. A. Flower-spike and lower part of stem, reduced. B. Corolla cut open shewing position of stamens, enlarged. C. Floral diagram. (A, B after Reichenbach; C after Eichler.)

the tropical Asiatic *Christisonia*. The two carpels are median; where three are present, as in three monotypic Asiatic genera, the odd one is posterior. Each carpel bears two parietal placentas, which may unite in pairs at the ventral edge or become displaced towards the middle line of the carpel; they are often branched. In *Christisonia* the ovary becomes two-celled below by union of the placentas.

The flowers are proterogynous and pollinated by insects, which are attracted by the scent and the nectar secreted at the base of the stamens or ovary.

The capsule has a leathery pericarp splitting loculicidally in drying; the minute seeds, which are produced in great numbers, have a honeycombed testa and are readily carried by the wind.

The 14 genera contain about 130 species, which are almost confined to the northern hemisphere, reaching their greatest development in the warm temperate zone of the Old World. *Orobanche* contains 100 species, nine of which are British; some occur only on definite host-plants, for instance, *O. Hederae* on Ivy; others are confined to single genera, or families, or occur, like *O. minor* or *O. ramosa*, on species of various families. They may become dangerous pests to cultivation, as *O. ramosa* in Hemp and Tobacco fields, or *O. minor* on Clover.

The Orobanchaceae have been regarded (as by Baillon and Warming) owing to the unilocular ovary as a parasitic offshoot of the Gesneriaceae. The genera of Gesneriaceae are however strictly autotrophic. The Scrophulariaceae, on the other hand, include various degrees of parasitism in the tribes *Gerardieae* and *Rhinantheae*; thus in *Gerardia* some species are tall leafy plants with slight root-parasitism, as *G. flava*, and from these, transitions, shewing increased parasitism, lead to *G. aphylla*, which has very small leaves and is almost completely parasitic. The South African genera *Harveya* and *Hyobanche* are completely parasitic, the latter having a short fleshy stem covered with scale-leaves and a dense spike of flowers suggesting in its habit the Orobanchaceae. Heinricher's intensive morphological and physiological studies on the *Rhinantheae* indicate a very close relationship between such genera as the hemiparasitic *Bartsia*, and *Tozzia*, which is successively holoparasitic and hemiparasitic, and the completely parasitic *Lathraea*, which on account of its incompletely two-celled ovary has been included in Orobanchaceae. Boeshore[1] has recently reviewed the evidence for regarding the Orobanchaceae as a direct continuity of the line of parasitism developed in the Scrophulariaceae.

Family XII. GESNERIACEAE

Flowers bisexual, zygomorphic, with the formula S5, P5, A5 or 4 or 2, G(2). Calyx gamosepalous, teeth generally short, with valvate aestivation. Corolla gamopetalous, tube generally well-developed, often two-lipped, aestivation im-

[1] *Contributions from the Botanical Laboratory of the University of Pennsylvania*, v, 139 (1920).

bricate. Stamens generally four, didynamous, epipetalous; disc obvious. Ovary superior or more or less inferior, unilocular, sometimes imperfectly bilocular by the ingrowth of the parietal placentas; ovules indefinite, anatropous; style simple, stigma usually bilobed. Fruit a two- or four-valved capsule, sometimes indehiscent and fleshy; seeds numerous, small; endosperm present or absent; embryo straight.

Generally herbaceous or slightly woody plants, with opposite entire exstipulate leaves and solitary or cymose showy flowers.

Genera about 100; species about 1100; mainly tropical and subtropical.

The members of this family are herbs or more or less woody undershrubs, rarely arborescent; some are tree-climbers, attaching themselves by adventitious roots produced at the nodes. *Sinningia speciosa* (the Gloxinia of greenhouses) is an example of a tuberous habit which is met with in the subfamily Gesnerioideae, the tuberous thickened rhizome producing annual aerial shoots. These species are propagated by leaf-cuttings. A more frequent habit in this subfamily is associated with the production of generally subterranean runners which are thickly covered with scale-leaves.

In the other subfamily, Cyrtandroideae, a remarkable reduction of the axis is found in species of *Streptocarpus* and other genera. In some of these the adult plant has but one leaf, which is a much developed cotyledon, the second cotyledon having perished early in the life of the plant (fig. 247). Attachment and nutrition are obtained by adventitious roots from the base of the leaf from which spring also the flower-bearing shoot and adventitious leaf-shoots. The adult plant is thus comparable with an individual which, as in *Sinningia*, has been developed vegetatively from a leaf.

To the same subfamily belongs the large Indo-Malayan genus *Aeschynanthus*, comprising subshrubby epiphytes with fleshy or leathery leaves.

The leaves are simple with an entire or toothed margin; they often have a soft thick hairy covering. They are generally decussate, sometimes in whorls of three or four, rarely alternate. Stipules are absent.

The flowers may be solitary in the leaf-axils but are generally arranged in denser or looser cymose inflorescences; they are showy and markedly zygomorphic (except *Ramondia*). The calyx is usually tubular with five more or less equal teeth. The form of the corolla shews considerable variation from the rotate almost regular form characteristic of *Ramondia*, a small South European alpine genus, to the bell-shaped zygomorphic Gloxinia type. Associated with the more regular corolla is the presence of five stamens (*Ramondia*), while in the majority of the genera the zygomorphy is expressed by four didynamous stamens; a further reduction to two fertile stamens is frequent in the Cyrtandroideae. The form of the anther, the development of the connective and the position of the cells afford good generic characters, as also does the structure of the disc, which may be ring-like or lobed or form five or fewer distinct nectaries. The position of the ovary allows a separation into two distinct subfamilies, Cyrtandroideae, having a superior, and Gesnerioideae, having a more or less inferior ovary (fig. 246).

The large bright-coloured flowers, their zygomorphy, and the general arrangement of the stamens and style indicate adaptation to insect-pollination.

The fruit is generally a two- or four-valved capsule dehiscing loculicidally; in *Ramondia* and allied genera it is septicidal. Fleshy berry-like fruits are characteristic of *Cyrtandra*, a large genus found chiefly in the islands of the Indian and Pacific Oceans.

The seeds are small and numerous, with a considerable amount of endosperm in the Gesnerioideae, or none or only a trace in Cyrtandroideae.

The species inhabit mainly the tropical and subtropical portions of both hemispheres. The family is represented in Europe by two endemic genera, *Ramondia* (four species, including *Jankaea*), and *Haberlea* (one species), found only in the Pyrenees and the mountains of the Balkan peninsula. Allied to these is the small tropical East African genus *Saintpaulia*; one species, *S. ionantha*, is well-known in cultivation.

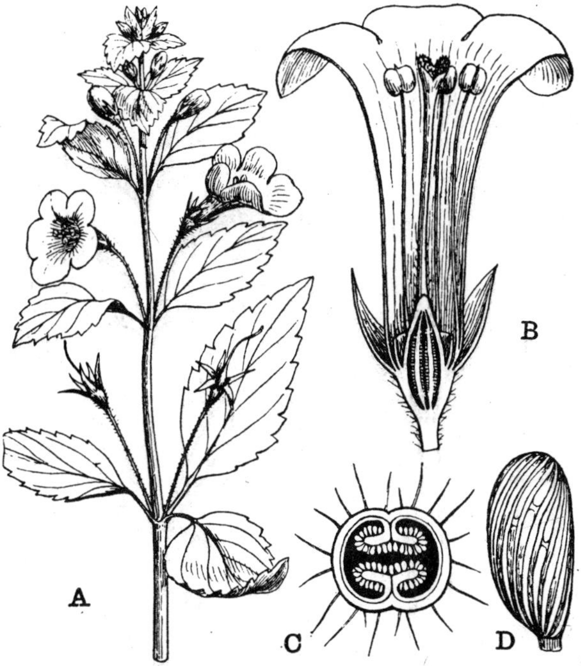

FIG. 246. *Achimenes coccinea.* A. Flowering shoot, reduced. B. Flower in vertical section, × 3. C. Ovary in transverse section, × 8. D. Seed, much enlarged.

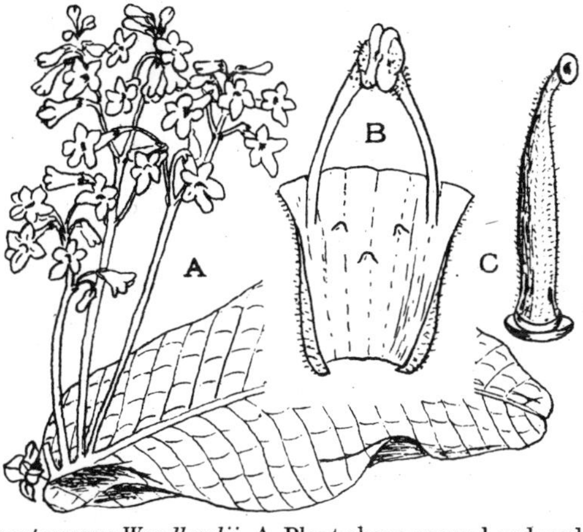

FIG. 247. *Streptocarpus Wendlandii.* A. Plant above-ground, reduced. B. Portion of corolla cut open, shewing the two fertile stamens, enlarged. C. Pistil, enlarged. (After *Bot. Mag.*)

The principal genera in the two subfamilies are:

CYRTANDROIDEAE. *Ramondia; Didymocarpus*, 100 species, mostly Indo-Malayan; *Streptocarpus*, 50 species, mainly South Africa and Madagascar; *S. polyanthus* and others are cultivated; *Aeschynanthus*, 70 species, Indo-Malaya and South China; *Besleria*, 50 species, tropical America, the fruit is generally a berry; *Cyrtandra*, 180 species, mainly in the islands of the Indian and Pacific Oceans, also known in cultivation; *Columnea*, 100 species, tropical America.

GESNERIOIDEAE. *Achimenes*, 25 species, tropical America, cultivated; *Sinningia*, 20 species, Brazil; *S. speciosa* is the well-known Gloxinia of greenhouses; *Gesneria*, 35 species mainly West Indian.

The Gesneriaceae are closely allied to the Scrophulariaceae, Orobanchaceae and Bignoniaceae, and it is difficult to draw sharp distinctions between the four families. Bignoniaceae are separated by the siliqua-like woody fruit and large generally winged seeds; frequently also by the divided leaves, the habit and the anomalous structure of the wood. The best distinction from Scrophulariaceae is the unilocular ovary. Orobanchaceae are sometimes regarded as a parasitic degenerated group of Gesneriaceae.

Family XIII. PEDALIACEAE

A small family of annual or perennial herbs, more rarely shrubs, with opposite leaves, or the upper leaves spirally arranged, and generally axillary flowers. Mucilage-containing glandular hairs are common.

The flowers are bisexual and zygomorphic, with five united sepals and petals and four didynamous stamens, the fifth stamen being represented by a small staminode. The two carpels form a superior two- to four-chambered ovary (inferior in *Trapella*); the chambers may be more or less completely divided by the development of false septa. There are one to many ovules on an axile placenta in each chamber. The fruit is a capsule or nut, often winged or provided with thorns or hooks. The seeds contain a straight embryo surrounded by a thin layer of endosperm.

There are about 60 species in 16 genera, mostly shore- and desert-plants in tropical and South Africa, Madagascar, the Indo-Malayan region and tropical Australia. *Sesamum*, the largest genus, contains 16 species in tropical Africa and two in further India; *Sesamum indicum* (Sesame) is widely cultivated in the warmer parts of both hemispheres for its oily seeds; it was known to the old Greeks and Romans. *Harpagophytum* (two species in

South Africa) has large woody fruits which bear stiff barbed processes.

Trapella[1], an aquatic herb with broad floating leaves and narrow submerged leaves, a genus with two species in China and Japan, forms a distinct subtribe characterised by its inferior ovary and only two fertile stamens.

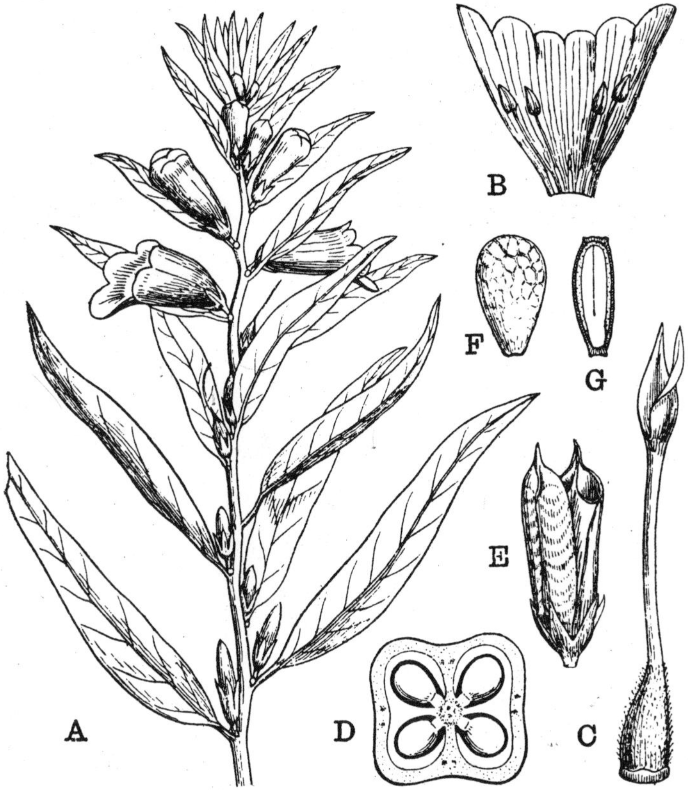

Fig. 248. *Sesamum indicum.* A. Portion of plant shewing flowers and fruit, slightly reduced. B. Corolla opened from the front. C. Pistil, enlarged. D. Transverse section of ovary, much enlarged. E. Fruit dehiscing, slightly reduced. F. Seed. G. Seed cut lengthwise, × 5. (A, C, D, E after Stapf.)

Martyniaceae (XIV) form a small tropical and subtropical American family with three genera and nine species which are often included in Pedaliaceae but may be distinguished by their

[1] Oliver, F. W. "On the structure, development and affinities of *Trapella*." *Annals of Botany*, II, 75 (1888).

unilocular ovary with parietal placentation. The fruit is a horned capsule, the soft outer layers of the pericarp falling and leaving a woody endocarp. The T-shaped placentas becoming united with each other and the woody endocarp tend to form a four-chambered interior.

The monotypic genus *Martynia* is an annual herb, probably a native of Mexico but now found in the tropics of both hemispheres. The stigmas are sensitive, and the fruit (fig. 249) bears two long curved horns, an admirable adaptation for distribution by attachment to the coats of large animals.

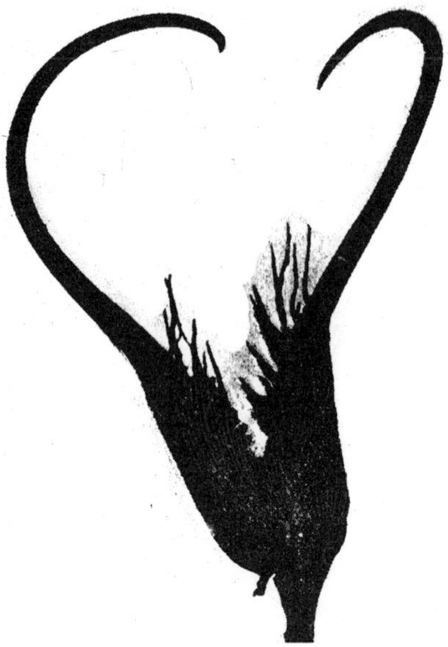

Fig. 249. Fruit of *Martynia lutea*, $\times \frac{3}{8}$.

Family XV. BIGNONIACEAE

Flowers bisexual, zygomorphic, with the formula S5, P5, A4, G(2). Calyx gamosepalous. Corolla usually bell- or funnel-shaped with descending imbricate aestivation. Stamens springing from the lower part of the corolla-tube, didynamous, the posterior represented by a staminode; anthers two-celled, the cells usually placed one above the other, dehiscence longitudinal; hypogynous disc present. Ovary superior, bilocular, more rarely unilocular; ovules numerous, anatropous and usually erect on axile placentas;

style simple, filiform; stigma bilobed. Fruit generally a two-valved capsule containing numerous large flattened membranous-winged seeds; more rarely fleshy and indehiscent; endosperm absent.

Trees or shrubs, often climbers, with opposite-decussate generally compound exstipulate leaves, and showy flowers.

Genera 110; species about 550; mainly tropical.

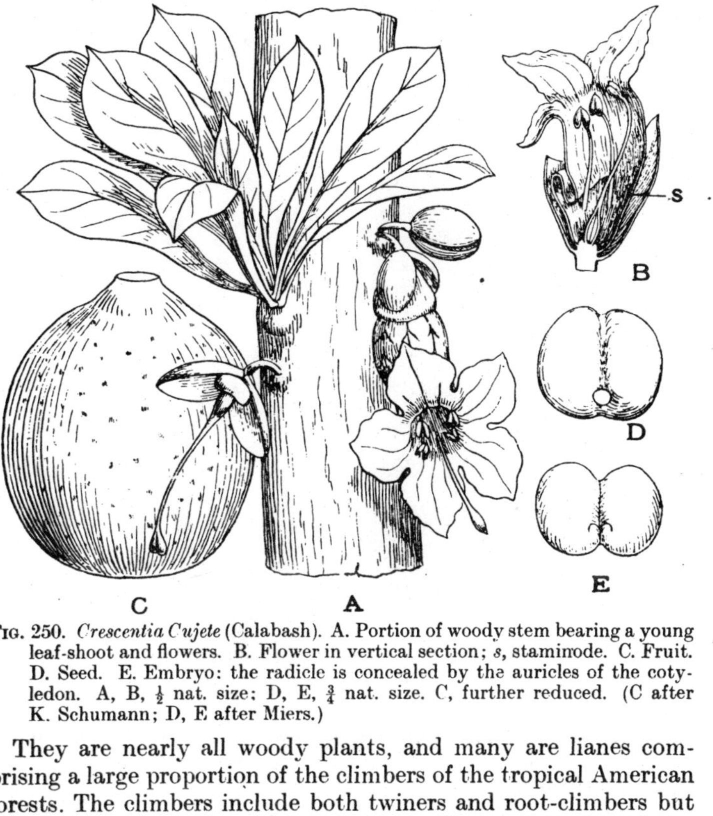

FIG. 250. *Crescentia Cujete* (Calabash). A. Portion of woody stem bearing a young leaf-shoot and flowers. B. Flower in vertical section; *s*, staminode. C. Fruit. D. Seed. E. Embryo: the radicle is concealed by the auricles of the cotyledon. A, B, ½ nat. size; D, E, ¾ nat. size. C, further reduced. (C after K. Schumann; D, E after Miers.)

They are nearly all woody plants, and many are lianes comprising a large proportion of the climbers of the tropical American forests. The climbers include both twiners and root-climbers but the great majority are tendril-climbers. The tendrils are at the ends of the leaves representing modified leaflets; they are simple or branched and in some cases end in adhesive discs or in hooks. Their stems shew anomalous secondary thickening, the wood being often more or less divided into wedge-shaped areas from the failure of the cambium to form wood in the intervening spaces.

Another type, examples of which are found in the Brazilian campos, shews a xerophytic habit with reduced stems.

Crescentia (including the Calabash), *Jacaranda* and *Catalpa* include many fine trees; *Catalpa bignonioides* is often grown in our parks and gardens.

The flowers are generally arranged, following the phyllotaxy, in opposite-decussate compound inflorescences, which are usually dichasia passing into monochasial cymes; both bracts and bracteoles are present. They are sometimes produced on the old wood (cauliflory), as in the Calabash (fig. 250). The calyx shews

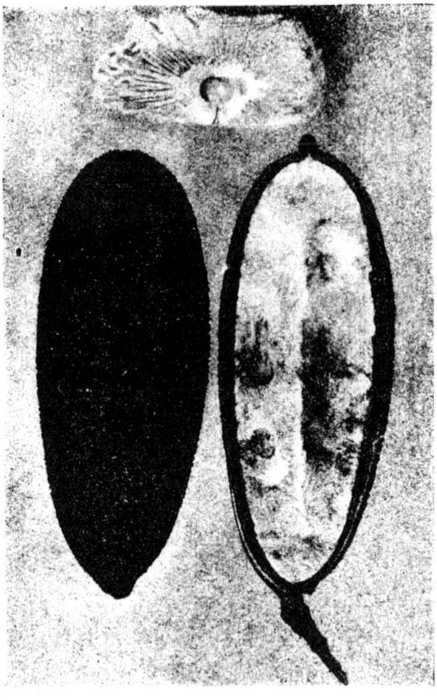

FIG. 251. Fruit of *Pithecoctenium muricatum*, × $\frac{3}{8}$. One valve has been removed exposing the other valve on which the membranous winged seeds are densely packed. Above is a single seed.

considerable variation in form and structure and the manner of opening, the differences affording useful generic characters. The showy corolla is generally bell- or funnel-shaped, rarely two-lipped; the posterior stamen is generally present as a small filiform staminode (fig. 250, B, *s*). The pollen is remarkably uniform, the grains being spherical with three equatorial bands in which the germ-pores are situated. A unilocular ovary is found in *Eccremocarpus* and in the *Crescentieae*, but the latter include all stages between unilocular and completely bilocular. The ovules are almost always anatropous with the micropyle directed downwards; the placentas in the unilocular ovaries are parietal, in the bilocular the numerous ovules are arranged in two or more rows on the vertical septum. The shape of the capsule and the method of dehiscence afford the best characters for the distinction of genera; in the most frequent form the two woody valves separate siliqua-like from the septum on which the flattened winged seeds are densely packed (fig. 251).

The *Crescentieae* are characterised by an indehiscent fruit, the frequently hard tough exocarp containing a juicy pulp in which the seeds are embedded. They are often large and gourd-like, as in the Calabash (*Crescentia Cujete*) (fig. 250).

The chief centre of distribution is found in tropical South America, especially Brazil. *Catalpa* is common to the east and west hemispheres; *C. bignonioides* is a native of the Eastern United States; two closely allied species are natives of Japan and North China, and two others occur in Cuba. The small genus *Campsis* has a similar distribution with two species in North America and one in Japan; *C. radicans* (often known as *Tecoma radicans*), a native of Atlantic North America, is a handsome root-climber. The chief distinction from the Scrophulariaceae lies in the structure of the fruit and the large often winged seeds and absence of endosperm. They also differ in their habit and geographical distribution, being mainly tropical lianes or trees.

Family XVI. LENTIBULARIACEAE

Flowers bisexual, zygomorphic, with the formula S5, P5, A2, G(2). Calyx often two-lipped. Corolla two-lipped, lower lip saccate or spurred. Stamens attached to the base of the corolla, anthers unilocular. Carpels forming a unilocular ovary with free-central placenta bearing numerous anatropous ovules somewhat sunk in its tissue, or only two ovules; stigma sessile, two-lobed, with the posterior lobe much reduced. Fruit a many-seeded capsule opening by two to four valves or by a ring-like or irregular slit, or indehiscent and one-seeded; seeds generally small, without endosperm; embryo varying in form.

Insectivorous herbs of very different habit, often aquatic or marsh-plants; flowers solitary or generally in a raceme or spike.

Genera five; species about 250; tropical and temperate.

About 200 of the species are included in *Utricularia*, a genus widely spread in the tropical and temperate parts of the earth. Many of its species are aquatic, as our native Bladderwort; these are submerged rootless water-plants with finely-divided leaves on which are borne bladders (fig. 252, F, G); the mouth of the bladder is closed by a valve (*v*), opening inwards in response to a stimulus conveyed by four sensory hairs (*s*), and immediately closing again, thus forming a trap for small aquatic animals, the products of decay of which are absorbed by hairs lining the wall of the bladder[1]. The flowers are borne on long racemes projecting above the water; the flowering shoots are sometimes provided with inflated swimming leaves. Other species are land-plants, often growing among moss, with long runner-like shoots on which the bladders are borne; while others are epiphytic (as *U. montana*) and have a much-branched rhizome with tuber-like thickening containing water-storing tissue. The terrestrial and epiphytic species have simple entire leaves.

The embryo consists of a cell-mass bearing only a few leaf-rudiments; in germination it forms a rosette of bristle-like primary leaves, and there is no primary root. Goebel has shewn that the ordinary distinctions between stem and leaf do not hold in *Utricularia*, precisely similar rudiments may form leaves, bladders or shoots.

The other British genus, *Pinguicula* (Butterwort), contains about 30 species, chiefly northern extratropical. Their habit is very uniform; they are perennial herbs growing in damp places attached by true roots, often alpine, with a rosette of simple entire leaves and one-flowered axillary leafless scapes. The upper surfaces of the leaf bear glandular hairs, the sticky secretion on which, often aided by inrolling of the leaf-margins, captures small insects and renders soluble the proteid material of their bodies which is then absorbed. The embryo has generally a single cotyledon.

Genlisea, a tropical American and African genus, is a land-plant with densely crowded rosettes of simple foliage-leaves and peculiar pitcher-like leaves appressed to the soil; the lowest portion of the leaf forms the pitcher, above which is a long tubular neck ending in a pair of twisted spreading limbs. The limbs are lined like the neck with bristles pointing towards the pitcher, an entry to which is facilitated while an exit is prevented.

Polypompholyx is a small genus occurring in tropical Australia and South America; the plants resemble the land forms of *Utricularia*. *Biovularia* (two species), a West Indian genus of small swimming water-plants, resembles *Utricularia* in habit, but is distinguished by the ovary containing only two ovules, which are united, and the one-seeded indehiscent fruit.

[1] See Withycombe, C. L. "On the function of the Bladders in *Utricularia vulgaris*." *Journ. Linn. Soc.* (*Bot.*) XLVI, 401 (1924).

Pollination has been studied in European species of *Utricularia* and *Pinguicula*, which are entomophilous. The broad lower lip of the flower gives a resting-place for the insect which, in probing for nectar in the spur, comes first in contact with the stigma-lobe, which in *Utricularia* is sensitive, closing on being touched. In

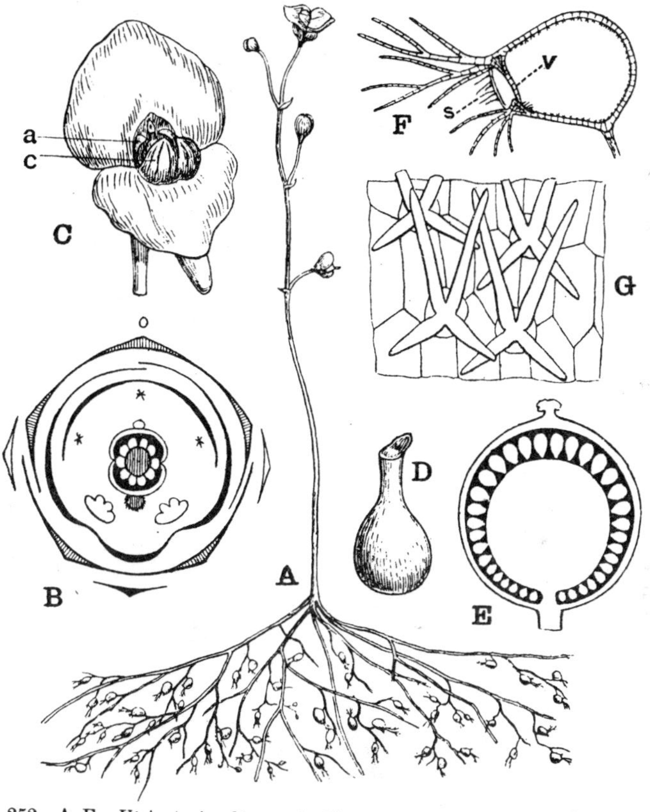

FIG. 252. A–E. *Utricularia obtusa*. A. Plant, $\frac{3}{4}$ nat. size. B. Floral diagram. C. Flower, ×4; *c*, crest on the lip, behind which are seen the two anthers, *a*, and the upper part of the pistil. D. Pistil, ×5. E. Fruit in vertical section, enlarged. F. Pitcher of *U. intermedia* in section, ×8; *s*, sensory hairs, *v*, valve. G. Stellate hairs on inner surface of pitcher, much enlarged. (B after Eichler; F, G after Withycombe.)

withdrawing its head the insect carries off pollen from the anthers Cleistogamic flowers occur in species of *Utricularia*.

Lentibulariaceae are most nearly allied to Scrophulariaceae, which they resemble in the zygomorphic flower, the reduction in the androecium, the median position of the carpels and the capsular fruit. They are distinguished by the unilocular ovary with the free-central placentation. Their insectivorous habit is also distinctive.

Family XVII. ACANTHACEAE

Flowers bisexual, medianly zygomorphic, hypogynous, conforming to or derived from the formula S5, P5, A5, G(2), with generally a reduction to 4 or 2 in the androecium. Calyx more or less deeply 5-, sometimes 4- (rarely 3-) partite. Corolla often bilabiate. Stamens when four generally didynamous; one or two staminodes often present; anthers often spurred and hairy, the halves frequently inserted at different heights, dehiscence generally longitudinal; pollen very various in form; hypogynous nectar-secreting disc variously developed. Carpels median; ovary bilocular with 2 (rarely 1)–∞ anatropous or amphitropous ovules in the middle line of the septum; style simple, generally long; stigma shortly two-lobed, the posterior lobe often reduced. Fruit a loculicidal capsule (rarely a drupe) with generally 4–∞ seeds supported on a development (often hook-like) of the funicle (retinaculum or jaculator); endosperm absent (except in the small subfamily Nelsonioideae).

Herbs or shrubs, more rarely trees, with opposite-decussate exstipulate leaves and cymose or racemose inflorescence. Cystoliths are often present in the epidermis or parenchyma of the stems and leaves.

Genera 220; species over 2000; in the warmer parts of the world.

Acanthaceae are mostly inhabitants of tropical to subtropical forests, growing especially in damp or marshy places. Some are however xerophytic, as *Barleria*, *Blepharis*, or *Acanthus*, the leaves and bracts of which are often more or less spiny. Climbing plants are represented by *Mendoncia*, *Thunbergia* and allies, the stems of which also shew abnormal secondary thickening. Arborescent species are rare.

In the usual mesophytic type the leaves are thin and delicate with entire margin, while in the xerophytes the lamina is more or less reduced and spiny. The cystoliths are a very characteristic feature, absent only in a few groups of genera; they are found generally in the epidermal cells of the leaf and stem and their form is characteristic of single genera or tribes.

The most frequent type of inflorescence is the cymose, namely dichasial passing into monochasial in the higher branching; it is often in the form of short axillary clusters. Spicate and racemose inflorescences are also frequent. Bracts and bracteoles are generally well developed and often brightly coloured, adding to the attractiveness of the inflorescence.

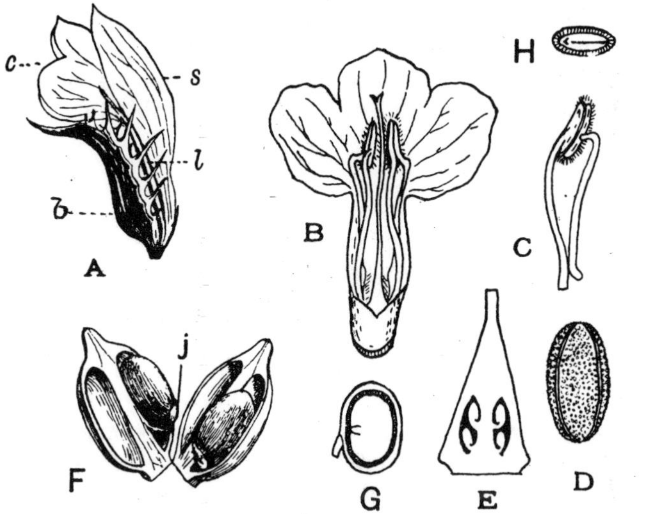

FIG. 253. *Acanthus mollis*. A. Flower, side view, × ⅔; *b*, bract; *s*, posterior sepal; *l*, lateral sepal; *c*, corolla. B. Flower opened out, three petals form the lower lip, the upper lip is absent. C. One pair of stamens. D. A pollen-grain, × 450. E. Ovary in longitudinal section, × 2½. F. Capsule shewing the two valves; *j*, the jaculator. G. Seed in longitudinal section. H. Seed in transverse section. (After Lindau, except F.)

In the Mendoncioideae and Thunbergioideae the bracteoles are large, forming an involucre round the tube of the flower.

The structure of the flower is very uniform. In *Thunbergia* the calyx is often reduced to a narrow seam, its protective function being transferred to the large bracteoles. The corolla has a longer or shorter often slender tube which passes above into an almost equally five-lobed limb, as in *Thunbergia*, the large genus *Ruellia* and many other genera, or the limb is more or less deeply two-lipped. The upper lip is generally straight and erect and bifid at the apex, but is completely absent in *Acanthus* and allied genera, the corolla being cut away almost to the base of the tube (fig. 253, B). The lower lip is sometimes rolled up but generally more or less horizontal, forming a platform for insect-visitors, and three-lobed; the inner side of the lip is often more or less densely hairy, the hairs often extending to the mouth of the corolla. The subfamily Acanthoideae, which includes the great majority of the genera, is subdivided according to the aestivation of the corolla into the *Contortae* where the lobes are contorted, and the *Imbricatae* where they are imbricate.

There are rarely five fertile stamens, as in the small Brazilian genus *Pentstemonacanthus*; more often there are four, didynamous, while in the *Imbricatae* there are generally

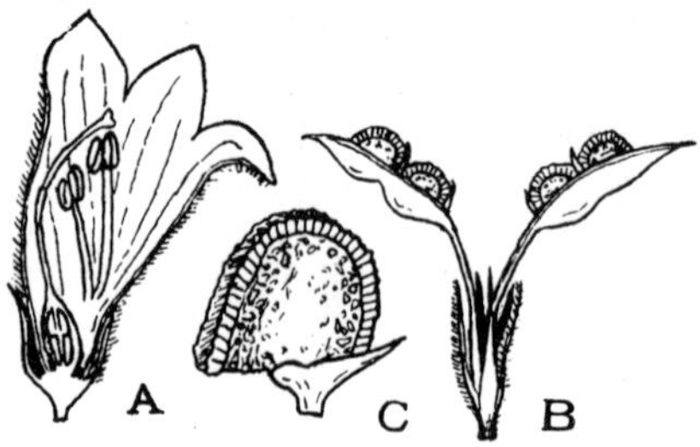

FIG. 254. *Asystasia*. A. Flower in vertical section, × 2. B. Fruit open, shewing the two seeds in each valve, nat size. C. A seed with its jaculator, × 3. (After Lindau.)

two. The posterior stamens are the first to be reduced to staminodes or to disappear completely; *Brillantaisia* is exceptional in that the anterior pair are staminodial. The filaments are generally quite free and project from the mouth of the flower; in *Thunbergia* the stamens are short and included in the tube. The anthers shew great variety in form and position in different genera; they are two- or one-celled; in the latter case a rudiment of the second cell may be present. In the two-celled, the cells are equal and approximated at the same level, or more or less separated by a development of the connective, when they are often different in size and at different levels; the lower is often spurred. The form and sculpturing of the pollen-grains shew a remarkable diversity and afford useful characters for generic distinction.

The ovary is more or less elongated and passes above into a long narrow style which projects from the mouth of the flower, ending in two small variously-shaped stigmas; the posterior stigma is often reduced. The species are generally markedly proterandrous.

Except in the small subfamily Mendoncioideae, where it is a one- or two-seeded drupe, the fruit is always a bilocular capsule splitting loculicidally to the base and bearing 2–10 seeds, arranged in a double row, on each valve. In the small subfamily Nelsonioideae the seeds are small and numerous. In the latter and in the subfamily Thunbergioideae the funicle is enlarged to form a papilla below the hilum and the seeds are spherical; the efficient seed-ejecting mechanism of the Acanthoideae is absent but the pointed tips of the capsule-valves, as in *Nelsonia*, may be effective in clinging to an animal's fur and causing the small seeds to be jerked out.

In the Acanthoideae the funicle forms a hook-like projection, the jaculator, in which the seed rests; the woody capsule opens elastically, the two valves separating to the base and springing backwards, causing at the same time an oscillation of the longer or shorter stiff stalk. As a result the seeds are thrown out, the jaculator, which is usually slightly twisted to the side, directing the flat lens-shaped seeds in a lateral direction so that the seeds from the four rows are sent cross-wise in four directions. In the section *Contortae*, where the capsule is only shortly stalked and the distributing mechanism therefore less effective, special adaptations of the seed-coat occur, as, for instance, in the form of closely lying toothed scales which become erect and mucilaginous when damp, thus readily clinging to a suitable object (as in *Crossandra*), or more generally there is a covering of long unicellular hairs which when dry lie close and form a smooth coat but when wetted stand up and become mucilaginous; these occur also in *Nelsonia*, where they are barbed at the tip.

Though mainly tropical, the family spreads beyond the tropics in most parts of the world. Thus in the Mediterranean region and South Europe, several species of *Acanthus* occur as steppe and desert plants. Similarly it is represented in the southern United States, extratropical Australia, South Africa, China and Japan.

Many genera are favourite ornamental, mainly greenhouse or hothouse plants, by reason of their brightly coloured flowers, which are sometimes large or massed in conspicuous inflorescences in which the bracts may also play an attractive part, or their variegated spiny leaves. Among these are *Barleria*, *Ruellia*, *Justicia*, *Jacobinia*, *Thunbergia* (climbing) and others.

Order 5. *PLANTAGINALES*

Family PLANTAGINACEAE

Flowers generally bisexual, regular, conforming to the formula S4, P4, A4, G(2); ovary four- to one-chambered; ovules several to one on a central placenta, anatropous. Fruit a transversely dehiscing capsule or a nut. Embryo straight, surrounded by fleshy endosperm.

Generally herbs, sometimes shrubby with woody stems; leaves generally alternate and narrow. Flowers small, crowded in heads or spikes on axillary scapes.

The family contains the genus *Plantago*, with 200 species, world-wide but mainly in the temperate zones; *Litorella*, two species, Europe and Antarctic, and the monotypic genus *Bougueria* on the high Andes.

The British species of *Plantago* (Plantain) are small herbs with radical generally narrow leaves and parallel nerves, and this habit is the most general one, though erect branched sometimes woody species also occur. They are often xerophytic, inhabiting dry waste places, or maritime or alpine localities. Each flower stands in the axil of a bract; bracteoles are absent and the four teeth of the tubular calyx are placed diagonally. The membranous corolla has a four-toothed limb; the stamens have large versatile anthers containing smooth dry pollen, and long slender variously coloured filaments projecting from the mouth of the corolla. The ovary is generally two-celled, or four-celled through development of false septa; the filiform style bears stigmatic hairs.

Pollination is generally effected by wind. The flowers are proterogynous; the somewhat feathery stigmas project from the unopened flower but remain receptive while the stamens mature so that self-pollination is possible. The upper flowers in the female stage may be pollinated by pollen from the lower more advanced flowers on the same spike (fig. 255, B). Departures occur from the normal type of flower; female (or more rarely male) as well as bisexual flowers are found sometimes, either on the same or different plants, and dimorphic flowers have also been observed. Indications of entomophily are associated with these. The spikes are rendered more or less conspicuous by the colour of the filaments, especially in *P. media*, where they are violet and the flower has a sweet scent and attracts bees in addition to the flies which occasionally

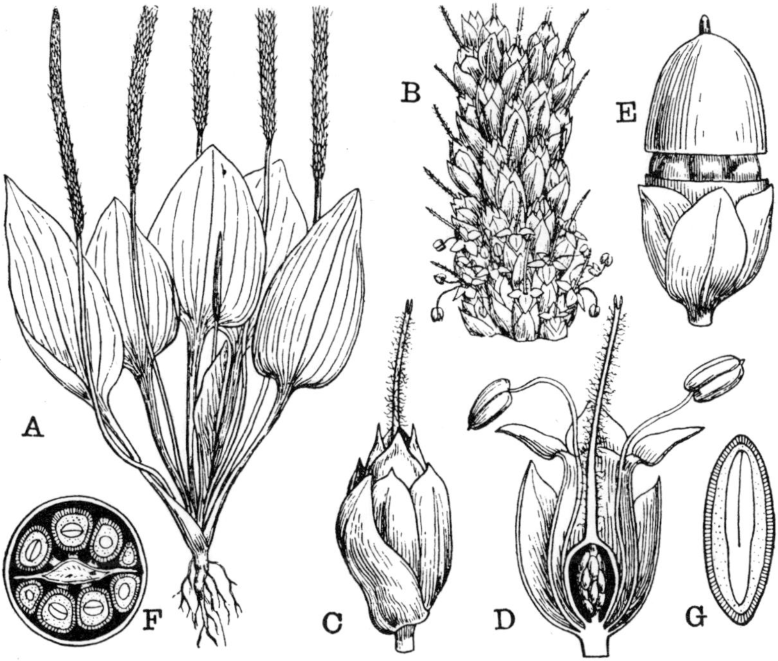

FIG. 255. *Plantago major*. A. Plant, the upper portion of the four longer spikes cut off, × ¼. B. Portion of spike, × 3; the upper flowers in the female condition, the lower (older) in the male. C, D. Flowers in the two successive states further enlarged, in D in vertical section, × 8. E. Fruit with circumscissile dehiscence, × 8. F. Fruit cut across. G. Seed cut longitudinally, × 12.

visit this and other species. In this species H. Müller distinguished an anemophilous and an entomophilous form, the latter having shorter scapes, shorter reddish filaments, more adhesive pollen and only slightly projecting stigmas. *P. lanceolata* (Ribwort) is also visited by the honey-bee, which renders the pollen adhesive by pouring honey upon it; Delpino distinguishes a dwarf entomophilous mountain form with short spikes and stamens.

Litorella is a small creeping perennial herb growing on edges of lakes and ponds, with tufts of radical awl-shaped leaves and unisexual flowers in groups of three, a stalked male with versatile anthers on long filaments, and on either side a sessile female with a long feathery style; the female mature earlier. The fruit is a one-seeded nut. Submerged plants do not flower but spread by means of long runners. *L. uniflora* occurs in North and Central Europe (British); a second species is a native of Antarctic South America.

Plantaginaceae are an anomalous family of doubtful affinity. They are probably reduced from a sympetalous group with entomophilous flowers and are perhaps placed most naturally near Tubiflorae—the position generally assigned to them in recent systems of classification.

(*b*) **Inferae**

Flowers tetracyclic with inferior ovary; generally regular with isomerous stamens but with an increasing tendency to zygomorphy of the flower in the order Rubiales (compare especially Caprifoliaceae, and Valerianaceae, and also the subfamily Lobelioideae of Campanulaceae in the order Campanulales); also with a development of zygomorphy associated with the aggregation of the flowers in heads, the outer series of flowers of which tend to become zygomorphic. There is an increasing tendency in both orders to reduction of the number of carpels and ovules culminating in an unilocular ovary with a single ovule, and also to reduction or late development of the calyx, which in the most advanced families becomes merely an organ for dissemination of the dry one-seeded fruit.

Two orders, Rubiales and Campanulales.

Order 1. *RUBIALES*

Flowers typically bisexual, regular to zygomorphic, tetramerous or pentamerous; stamens generally equal in number to the corolla-lobes, sometimes fewer; carpels generally fewer. Ovary inferior, rarely half-inferior, one- to several-chambered with $\infty - 1$ anatropous ovules.

Woody plants or herbs with opposite generally simple leaves, and usually cymose inflorescences.

The group is a natural one and the families illustrate a gradual reduction in the number of parts of the flower associated with an increasing tendency to zygomorphy in adaptation to insect-pollination. Rubiaceae and Caprifoliaceae are mainly woody plants, but the herbaceous development is indicated, especially in the former, the plants of which include a wide range of habit. In Valerianaceae and Dipsacaceae the herbaceous habit prevails. The inflorescence is typically cymose, and may be derived from a dichasium which in its higher branches passes into monochasia. In Rubiaceae, Caprifoliaceae and Valerianaceae there is a tendency to attain conspicuousness by aggregation of the flowers, sometimes even into dense heads; this tendency reaches an extreme development in Dipsacaceae, where the compact head is subtended by an involucre of bracts and the outer flowers may become enlarged and less regular, suggesting the ray-florets of Compositae. Associated with the aggregation of the florets is the loss of the protective function of the calyx and its development as a means of distribution of the fruit, the pappus of many Valerianaceae being homologous with that of Compositae.

The flower in Rubiaceae is, with rare exceptions, regular and the stamens are isomerous with the corolla; Caprifoliaceae include both regular and medianly zygomorphic types; the androecium is generally isomerous, very rarely is the posterior stamen reduced or suppressed. In Valerianaceae zygomorphy of the flower is a striking character and is associated with progressive suppression of the stamens, which are never isomerous. The flower of Dipsacaceae is medianly zygomorphic and the suppression of the posterior stamen has become a fixed character.

The gynoecium in Rubiaceae is in the great majority of genera bicarpellary, but varies to isomery with the corolla; the number ot ovules is also very variable, from many to one in each chamber; a similar range of variation occurs in Caprifoliaceae, but the most general number of carpels in this family is three; in Caprifoliaceae there is a tendency to the degradation of all but one of the ovary-chambers, resulting in the development of a few- or one-seeded fruit. This tendency has become accentuated in Valerianaceae where the gynoecium is typically three-carpellary, but only one chamber is fertile and contains a solitary ovule; the various genera shew a progressive series in the degree of abortion of the two sterile chambers. In Dipsacaceae the ovary is unilocular and one-ovuled and there is no trace in the developed flower of a second carpel except in the occasional forking of the stigma.

The origin of the Rubiales may be sought among the epigynous tetracyclic Dialypetalae. They may be regarded as a sympetalous development of a stock allied to Umbelliflorae which shew considerable resemblance to Rubiales in their tetramerous to pentamerous tetracyclic epigynous flowers, more or less suppression of the calyx, the fleshy disc crowning the ovary, the anatropous ovules with a single integument and the embryo surrounded with endosperm. The resemblance between some genera of Rubiaceae (e.g. *Knoxia*) and Caprifoliaceae (e.g. *Viburnum*) on the one hand and of Cornaceae on the other is very close, there being little beyond the polypetalous character of the latter to distinguish them.

Family I. RUBIACEAE

Flowers bisexual (rarely unisexual), regular (rarely zygomorphic), 5- to 4-merous, with isomerous stamens and generally fewer, usually two, carpels. Calyx-segments generally open in aestivation, sometimes one or more enlarged. Corolla generally funnel-shaped, hypocrateriform or rotate; segments with valvate, imbricate or contorted aestivation. Stamens inserted on the corolla-tube; anthers introrse, two-celled, with longitudinal dehiscence. Ovary crowned by a more or less developed fleshy disc, very rarely semi-inferior, one- to several-, usually two-chambered, with $\infty - 1$ anatropous ovules in each chamber; style filiform, sometimes bifid or multifid. Fruit a capsule, berry or drupe. Seeds with a straight, more rarely curved, embryo, generally in the base or axis of a fleshy, cartilaginous or horny endosperm.

Shrubs or trees or sometimes herbs, with opposite or whorled entire leaves and interpetiolar or intrapetiolar, sometimes foliaceous stipules. Flowers usually in decussate panicles or cymes, sometimes aggregated into heads.

Genera 380; species about 4600; mainly in the warmer parts of the earth but extending into the temperate and even represented in the frigid zones.

The plants show much diversity in habit. The great majority are tropical trees and shrubs; herbaceous forms though not uncommon are mainly confined to certain tribes, but the *Galieae*, the somewhat anomalous tribe to which our British representatives belong, are almost exclusively herbaceous. The tendency to the production of herbaceous forms is however widespread, as herbaceous genera occur in tribes which are typically woody, such as *Bouvardia* in the tribe *Cinchoneae*. The climbing habit is represented by herbaceous or shrubby twiners, such as *Manettia*, or hook-climbers, as *Uncaria*, lianes climbing by means of branches which are converted into stiff woody hooks, or scramblers, such as our native Cleavers (*Galium Aparine*), which cling to surrounding vegetation by stiff recurving hairs. Epiphytes occur, as, for instance, in a group of Malayan genera (*Myrmecodia*, etc.), small shrubby plants with a tuber-like stem developed from the swollen hypocotyl which is permeated by irregular galleries in which ants make a home. The exact relation of ants to plant is not clear, as the characteristic hollowed tuber is developed when the plants are grown from seedlings to which ants have no access. Swollen hollow internodes inhabited by ants occur just beneath the inflorescence in species of *Nauclea*, *Duroia* and others; the position suggests that the ants protect the flowers from undesirable visitors. Species of *Duroia* and *Remijia* have flask-shaped swellings on the leaf-base which are regularly inhabited by ants.

The stipules stand either at the side of each leaf-base and thus lie between the leaves (interpetiolar), or in front of the leaf-base between stalk and axis (intrapetiolar). They are sometimes united with each other and the leaf-stalk forming a sheath. In some genera of the tribe *Gardenieae* they form a conical cap protecting the apex of the stem, and are thrown off with the growth of the bud. They vary much in form and are sometimes split into bristle-like structures (as in *Pentas*), each of which ends in a resin-secreting gland; the base of the stipules (*Coffea*, *Gardenieae*, etc.) may bear slender glands, the copious excretion from which often covers the younger parts of the shoot. Most noteworthy are the characteristic stipules of the *Galieae* which are leaf-like, the stem apparently bearing whorls of leaves at the nodes; the stipules are generally distinguished by the absence of buds from the axils, though "stipular shoots" have been demonstrated in *Galium* (an occurrence with which we may compare the development of spines in the axil of the small stipules in *Damnacanthus*); there may be one or several between the true leaves; this variation in number is explained by union or splitting in various degrees of the normal pair belonging to each leaf.

Solitary terminal flowers occur in genera which have large white or bright-coloured flowers, as in *Gardenia*, *Randia* and others; this is often associated with a pair of lateral flowers forming a dichasium. Generally the inflorescence may be referred to a dichasial type or consists of decussate panicles. Conspicuousness is attained in the tribe *Naucleae* by aggregation of the small-flowered dichasia into dense globose heads. In these heads the inferior ovaries of the flowers are often completely united, as in *Sarcocephalus*, and a similar aggregation occurs in other cases, as in *Morinda*, and in the union of two flowers of a pair which may extend so far that the two have a common calyx, as in the Australian *Pomax*.

A different method of increasing the attractiveness of the inflorescence is more or less characteristic of certain genera (*Mussaenda*, *Warscewiczia* and others)—in one or more flowers of the inflorescence one or rarely more of the calyx-segments become leaf-like and white or brightly coloured; the parts affected are the one or two anterior segments of the flowers on the first lateral branches and the phenomenon recalls the enlargement of the outer petals on the outer flowers of the umbel in certain Umbelliferae (e.g. *Heracleum*) and the characteristic development of ray-florets in the Compositae. In other cases enlargement of the sepals occurs only after fertilisation and serves as a means of distribution of the fruit, as in *Alberta*, *Nematostylis* and others.

The aestivation of the corolla supplies characters for distinction of tribes; most general are the valvate and contorted. Zygomorphy occurs in some genera, the flower tending to become bilabiate, as in *Henriquezia* (Amazon province). The stamens are generally inserted at or near the throat of the corolla, more rarely at its base when they are sometimes united below, or throughout their length; the anthers usually dehisce by longitudinal slits, though sometimes by apical

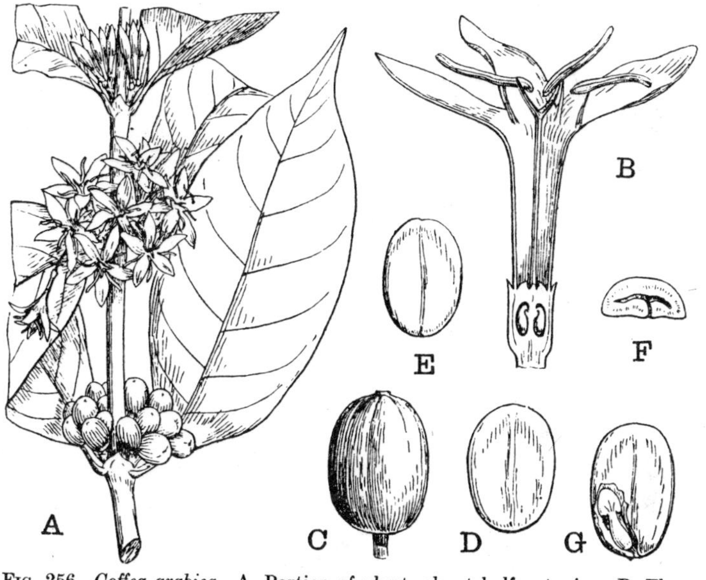

FIG. 256. *Coffea arabica*. A. Portion of plant about half nat. size. B. Flower in vertical section, × 2½. C. Fruit. D. Seed. E. Seed after removal of parchment-like testa. F. Horny endosperm cut across. G. Seed cut open to shew embryo lying towards the base of the endosperm; the radicle is inferior. C–G × 2. (A, B after Bentley and Trimen.)

pores, as, for instance, when the filaments are united. Hetero-styled flowers with stamens of different length occur (*Oldenlandia* and others). In zygomorphic corollas the stamens may be of unequal length and bent.

Exceptions to the completely inferior ovary occur in certain genera, as the Australian *Synaptantha*, where it is half-inferior, and in *Gaertnera* (tropical Africa and India) and *Pagamea* (Brazil, Guiana), where it is superior. In most cases the ovary is chambered, but *Gardenia* has a unilocular ovary with two to many parietal placentas. The development of the placenta is very varied. Most frequently it forms a thickening on the median septum or in the inner angle of the loculus (as in *Cinchona*, fig. 257, B); this may

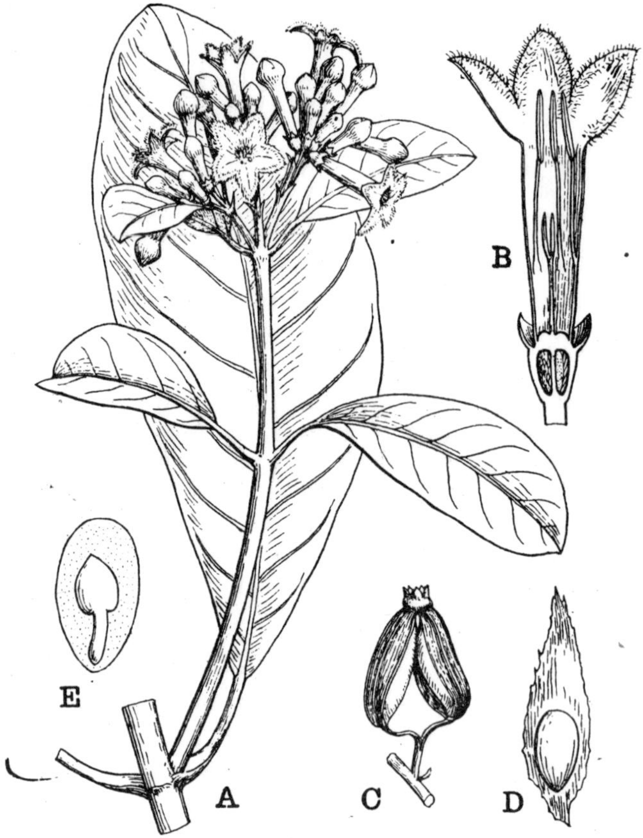

FIG. 257. *Cinchona calisaya.* A. Flowering shoot, slightly reduced. B. Flower in vertical section, × 2½. C. Fruit opening septicidally from below, × nearly 2. D. Winged seed, × 5. E. Seed (without wing) opened to shew embryo surrounded by endosperm, × 13.

broaden to become T-shaped in transverse section, and the arms may further become inrolled, bearing ovules on the outer or both faces; peltate or stalked spherical placentas also occur, as in the *Oldenlandieae.* The ovules are sometimes sunk in a fleshy development of the funicle, as in the one-ovuled chambers of *Pavetta*, etc.

Unisexual flowers (fig. 258) are obviously reduced from bisexual forms; generally rudiments of the non-functional members persist, the male flowers having a style but no ovary and the female having stamens the anthers of which contain no pollen-grains; occasionally in dioecious species, as in *Anthospermum*, no trace of the other sex remains. Often the male and female inflorescences are remarkably different, the male flowers being usually crowded together while the female are solitary and terminal.

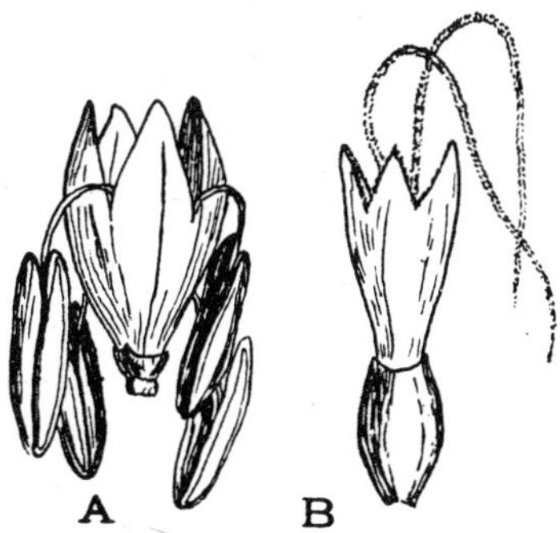

FIG. 258. Flowers of *Coprosma*. A genus with about 60 species, many in New Zealand, with small inconspicuous dioecious flowers. A male, B female. Enlarged. (After Le Maout and Decaisne.)

Pollination by insect-agency is indicated in some cases by the large long-tubed white or bright-coloured flowers, in others by the aggregation of smaller flowers into conspicuous inflorescences; also by the general development of the epigynous nectar-secreting disc in the form of a cushion, ring or cup, and the widespread occurrence of heterostyly. In our native *Galieae* the small flowers are more or less proterandrous; as the flower opens the stamens rise up, shed their pollen and bend outwards before the two stigmatic style-arms become separated (see fig. 259, A); the scanty nectar is sipped by small short-tongued insects which carry the pollen in passing from flower to flower. The longer-tubed flowers of *Asperula* contain more nectar and are visited by larger insects which in withdrawing the proboscis after probing the flower remove the pollen from the anthers converging round the mouth of the tube; in visiting a second flower some of the pollen is deposited on the stigmas which stand close together in the tube below the anthers and mature simultaneously with the stamens. In absence of insect-visitors self-pollination occurs by pollen falling on to the stigmas.

The strongly zygomorphic flowers of *Posoqueria* (tropical America) have an explosive mechanism comparable with that of some Papilionaceous plants; the pollen is dehisced into a chamber formed by the close approximation of the anthers which are in a state of tension when the flower opens; the touch of the insect's proboscis frees the tension and the pollen is scattered in a cloud over its body. The long tubular flowers of the American *Manettia* are visited by humming-birds.

The fruit shews important differences affording useful characters for the subdivision of the family. It is dry or fleshy, in the former case generally dehiscing either septicidally (fig. 257, C) or loculicidally, or being indehiscent,

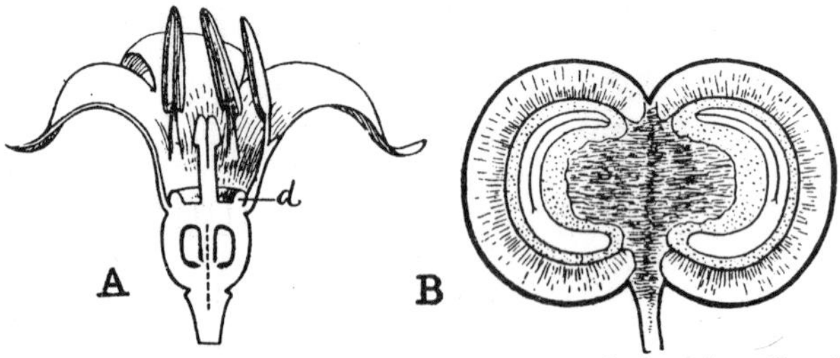

FIG. 259. *Rubia tinctorum*. A. Flower in median vertical section; the stigmas have not yet separated; *d*, nectar-secreting disc. B. Fruit in similar section, shewing the curved embryo, surrounded by endosperm, in the seed. Enlarged. (After Baillon.)

sometimes separating into one-seeded portions. Fleshy fruits are generally indehiscent. The calyx-limbs are often persistent and sometimes become enlarged and aid in distribution. Sticky fruits, due to the presence of small stiff recurving hairs, occur in the *Galieae* and some members of the *Anthospermeae*; the stem and leaves are often also provided with similar hairs so that portions of the plant bearing fruits are broken off and carried away. Fleshy fruits are distributed by aid of animals.

The seeds are generally small; in dry fruits they are sometimes more or less winged, especially in the tribe *Cinchoneae* (fig. 257, D). *Henriquezia* is exceptional in having a large two-valved woody capsule containing a few (four in each valve) large winged seeds recalling the fruit and seeds of Bignoniaceae, a resemblance which is further borne out by the open two-lipped flower. The straight, more rarely curved (as in *Rubia*, fig. 259) embryo lies in a more or less abundant fleshy, rarely cartilaginous (as in *Coffea*, fig. 256, F) endosperm, the flat cotyledons lying face to face. In some members of the tribe *Guettardeae* there is no endosperm.

The great majority of the species are tropical. The tribes *Galieae*, *Anthospermeae* and *Oldenlandieae*, which are remarkable for the prevalence of the herbaceous habit, pass into the temperate zone in which many of the genera are specially developed. Species of *Galium* reach the Arctic zone or high elevations on tropical mountain ranges. The genus *Nertera*, which consists of small low-growing herbs, spreads along the Andes from northern South America to the Strait of Magellan, and occurs also in Tristan d'Acunha, New Zealand, Australia, the Sandwich Islands and on the mountains of the Malay Archipelago. The allied genus *Coprosma* has also a similarly wide distribution in the south temperate and antarctic area. *Guettarda speciosa* is a widely-distributed strand-plant occurring round the Indian Ocean from Africa to Malacca, also in the Malay Archipelago, North Australia and the Polynesian Islands.

The family contains a large number of monotypic genera estimated by K. Schumann as nearly half the total number, but, on the other hand, there are several very large genera, such as *Galium* (300 species), *Oldenlandia* (200 species), *Psychotria* (nearly 1200 species) and others.

Two subfamilies are recognised, differing in the number of ovules in each carpel. They are subdivided by Schumann, in the *Pflanzenfamilien*, into twenty-one tribes as follows.

Subfamily I. CINCHONOIDEAE. Carpels with numerous (rarely few) ovules.

A. *Fruit dry.*

(*a*) Flowers solitary or in decussate panicles or cymose.

Condamineae. Woody plants with showy regular flowers, corolla-segments valvate in bud; ovules placed horizontally. Mainly tropical American.

Oldenlandieae. Mainly herbs with small regular flowers, valvate corolla-segments and vertically-placed ovules. Widely distributed, mainly tropical. *Oldenlandia* has more than 200 species.

Rondeletieae. Generally trees or shrubs with regular flowers; corolla-segments imbricate or contorted. Mainly tropical American, a few tropical Asiatic. *Rondeletia*, 60 species, tropical America.

Henriquezieae. Large trees with showy, generally zygomorphic flowers and large woody capsules with a few flat winged seeds. A few species in tropical South America.

Cinchoneae. Trees or shrubs with generally regular flowers. Ovules ascending; seeds numerous, winged. Tropics, mainly New World. *Cinchona*, about 40 species in the Andes of South America; the bark of several species, now widely cultivated in the tropics, is rich in the alkaloid quinine. *Bouvardia*, 30 species, Central America; several are grown as decorative plants.

(*b*) Flowers crowded in heads.

Naucleae. Trees or shrubs, sometimes twiners or hook-climbers (*Uncaria*) with regular flowers. Tropics, mainly of the Old World.

B. *Fruit fleshy.*

Mussaendeae. Trees or shrubs, more rarely herbs. One calyx-segment is sometimes enlarged and showy. Corolla-segments valvate in bud. Tropics, more numerous in the Old World. A few are epiphytic.

Gardenieae. Shrubs or trees. Corolla-segments contorted, rarely imbricate. Tropics. *Randia* (150 species). *Gardenia* (100 species, Old World tropics); *G. florida*, a native of China, is widely cultivated in hot-houses for its strongly scented flowers.

Subfamily II. COFFEOIDEAE. Carpels with a solitary ovule.

A. *Ovule pendulous with micropyle directed upwards; radicle superior.*

Vanguerieae. Vangueria, 50 species, mainly tropical Africa; *V. edulis* is cultivated in the tropics for its edible fruit.

Alberteae. Knoxieae. Guettardeae. Chiococceae.

B. *Ovule ascending, micropyle directed downwards; radicle inferior* (fig. 256, G).

(*a*) Corolla-segments contorted.

Ixoreae. Tropical trees and shrubs with regular flowers. *Coffea*, 30 species in the Old World, mainly Africa. *C. arabica* (fig. 256), the original coffee-plant, a shrub, native in Abyssinia and east tropical Africa, is widely cultivated in the tropics of both hemispheres. Liberian coffee is obtained from *C. liberica*, native in west tropical Africa; this species has shewn itself more resistant to the attacks of the disease caused by a parasitic fungus, *Hemileia vastatrix*. *Ixora* (200 species); *I. ferrea*, Iron-wood, yields a very hard wood.

(*b*) Corolla-segments valvate.

α. Ovules attached at the base of the ovary.

Psychotrieae. Generally shrubs or trees with bisexual flowers. *Psychotria*, 600 species in the tropics of both worlds, and other large tropical genera. *Uragoga ipecacuanha* is a herb, the branching closely knotted root of which yields the well-known drug. *Hydnophytum* and *Myrmecodia* (tropical Asia to Australia) are ant-inhabited epiphytic shrubs. *Paederiae*, mainly Asiatic.

Anthospermeae. Shrubs, undershrubs or herbs, with polygamous or dioecious flowers. Several genera are temperate, others reach the antarctic area, such as *Nertera* (*N. depressa* is a small low-growing herb cultivated for its red fruit), and *Coprosma* (fig. 258). *Coussareae*, mainly Brazilian.

β. Ovules attached to the septum.

Morindeae. Trees and shrubs. *Spermacoceae*. Herbs and small shrubs.

Galieae. Herbs with leafy stipules. Includes several north temperate genera—*Sherardia* (British); *Crucianella*; *Asperula* (80 species, a few in India and Australia); *A. odorata* (Woodruff) is British, rich in coumarin. *Galium* (British) has 300 species, and occurs in all parts of the world except Australia. *Rubia tinctorum* (Madder) (fig. 259) and other species contain alizarin and purpurin in the root and, before the introduction of aniline dyes, were widely cultivated.

Rubiaceae are very closely allied to Caprifoliaceae and the two families were combined by Baillon; the absence of stipules is not a universal distinction and it is impossible to cite any single distinctive character either floral or vegetative of universal application. In practice, however, there is no difficulty in distinguishing the two; thus *Sambucus*, in which stipules are present, is distinguished from Rubiaceae by its pinnately compound leaves. As indicated under the order, Caprifoliaceae may be regarded as a distinct group of genera in which the tendency to zygomorphy already foreshadowed in Rubiaceae becomes definitely worked out; in Caprifoliaceae also the pistil is generally tricarpellary, and there is an indication in some genera of the suppression of carpels and ovules which has become a constant character in the more advanced families of the order. The few genera of Rubiaceae in which valvate aestivation of the corolla is combined with a bicarpellary ovary with a single pendulous ovule in each chamber (e.g. *Knoxia, Pentanisia*) approach Cornaceae, which however are distinguished by polypetaly and absence of stipules. On the other hand, genera with many-ovuled cells suggest an affinity with

Loganiaceae which have opposite stipulate leaves and regular isostemonous sympetalous flowers, but is at once distinguished, apart from anatomical characters, by a superior ovary. This strong resemblance has led some botanists, for instance Warming, to regard the order as an epigynous continuation of the Contortae. A remarkable resemblance to the Bignoniaceae is shown in the small tribe *Henriquezieae*, with a more or less zygomorphic corolla, semi-inferior ovary and large woody two-valved capsules containing winged seeds.

Family II. CAPRIFOLIACEAE

Flowers bisexual, regular or zygomorphic, generally pentamerous with often reduction in the number of the carpels (2–5). Calyx and corolla present with parts united. Calyx generally small, and five-toothed or lobed. Stamens inserted on the corolla-tube, with generally introrse anthers. Ovary inferior, one- to five-chambered, with one to many pendulous ovules in the inner angle of each chamber. Styles free or united. Fruit a berry or a one- to several-stoned drupe; a capsule in *Diervilla*. Seeds with a fleshy endosperm and generally small straight embryo.

Generally woody plants with decussate usually exstipulate leaves and cymose often showy flowers.

Genera 11; species about 400; mainly north temperate.

The shrubby habit is the most common; examples of this are seen in *Sambucus*, as *S. nigra* (Elder), *Viburnum*, as *V. Lantana* (Wayfaring tree) and *V. Opulus* (Guelder-rose), *V. Tinus* (Laurustinus), *Symphoricarpus* (Snowberry), *Diervilla* (*Weigelia*); *Linnaea borealis* is a low-growing slender creeping evergreen shrub, native of the colder parts of the north temperate zone. Climbing shrubs characterise the section *Caprifolium* of *Lonicera* (Honeysuckles). Herbs are rare but are represented by *Sambucus Ebulus* (Danewort) and the small genus *Triosteum*. The leaves are generally entire, sometimes lobed, as in the Guelder-rose; *Sambucus* is exceptional in having pinnate leaves. The opposite leaves of a pair may unite, embracing the stem, as in species of *Lonicera*. Though the family is generally contrasted with Rubiaceae as being exstipulate, stipules not infrequently occur, as in species of *Sambucus*—where there are sometimes more than four side by side at the node—and in some cases—*Viburnum* and *Leycesteria*—they become nectar-secreting organs.

The difference between *Sambucus* and the rest of the family already noted in the leaves is emphasised in the anatomical structure. Generally the bulk of the wood consists of prosenchymatous tracheids with bordered pits, and the vessels are mainly of the scalariform type, while in *Sambucus* the wood-fibres bear simple pits and the scalariform type of vessel is rare. The origin of the primary periderm is variable even in the same genus.

The inflorescence shews varying degrees of development. A regular cyme is seen in *Leycesteria formosa*; in species of *Lonicera* the central flower is often suppressed (fig. 262, D) forming a two-flowered cyme in which case the ovaries of the pairs of flowers become more or less united (compare *Naucleae* in Rubiaceae); in *Symphoricarpus* and *Diervilla* the lateral flowers are suppressed and the inflorescence has the appearance of a spike or raceme. Large repeatedly branched inflorescences, ending in cymes, occur in *Sambucus* and *Viburnum*. The bracteoles are generally developed.

The flowers of *Sambucus* and *Viburnum* are regular (figs. 260, 261), with a generally rotate short-tubed corolla—except the zygomorphic sterile flowers on the outside of the inflorescence of species of *Viburnum*, which recall the similar phenomenon in some Umbelliferae. Those of *Lonicera* are markedly zygomorphic with generally a long slender tube and a two-lipped limb, four of the five segments uniting to form the upper lip (fig. 262). A tendency to zygomorphy is noticeable in the other genera, thus the bell-shaped corolla of *Symphoricarpus* and *Linnaea* (also tubular-funnel-shaped) is regular with a tendency to zygomorphy. The posterior stamen is reduced or suppressed in *Dipelta* (China) and *Linnaea*, and the stamens are didynamous. *Sambucus* differs from the rest of the genera in having extrorse dehiscence of the anthers (fig. 262, E).

FIG. 260. Floral diagram of *Sambucus Ebulus*. (After Eichler.)

The number and development of the carpels shew great variability even in one and the same genus. In *Leycesteria* and some species of *Sambucus* there are five carpels (rarely more in *Leycesteria*) which are equally developed; four in *Symphoricarpus*, three in most of the remaining genera and two in *Diervilla*. The number of ovules is also variable; *Sam-*

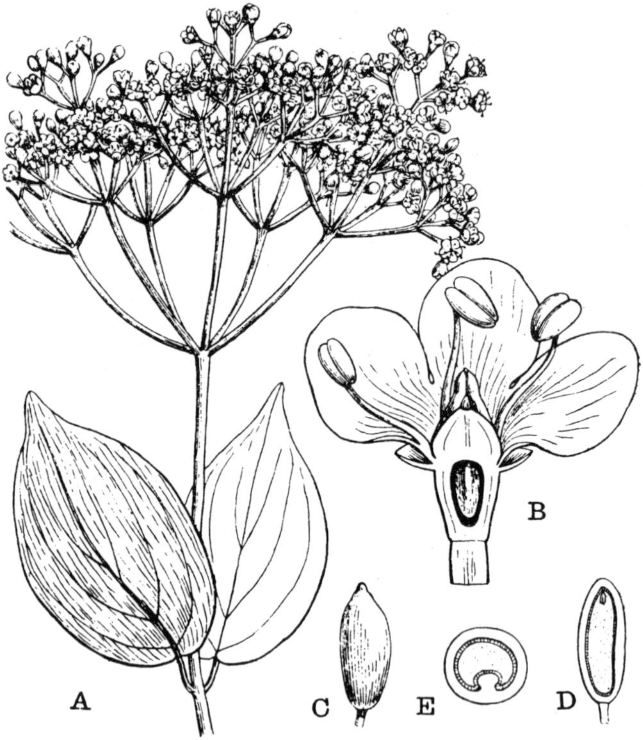

FIG. 261. *Viburnum villosum*. A. Flowering shoot, ¾ nat. size. B. Flower, in longitudinal section to shew the single functional chamber of the carpel with pendulous ovule, ×8. C. Drupe, ×1½. D, E. Drupe cut lengthwise and across, shewing the single seed. D×1½, E×2.

bucus, *Viburnum* and *Triosteum* have one in each chamber, while in *Leycesteria* they are numerous. The carpels are often unequally developed; thus in *Viburnum* two of the three original carpels become suppressed and the fruit is unilocular and one-seeded. In *Symphoricarpus* two chambers are one-ovuled and fertile, two contain several ovules and are sterile, and the fruit is two-seeded; similarly in *Linnaea* there are one fertile carpel containing one ovule, and two sterile carpels with several ovules.

The flowers are insect-pollinated. Adaptations to this end are the white or conspicuously coloured corollas, the aggregation of the smaller flowers, sometimes associated, as in *Viburnum*, with the specialisation of sterile specially attractive florets, the strong scent and the development of nectar. The most advanced type is represented by the *Caprifolium*

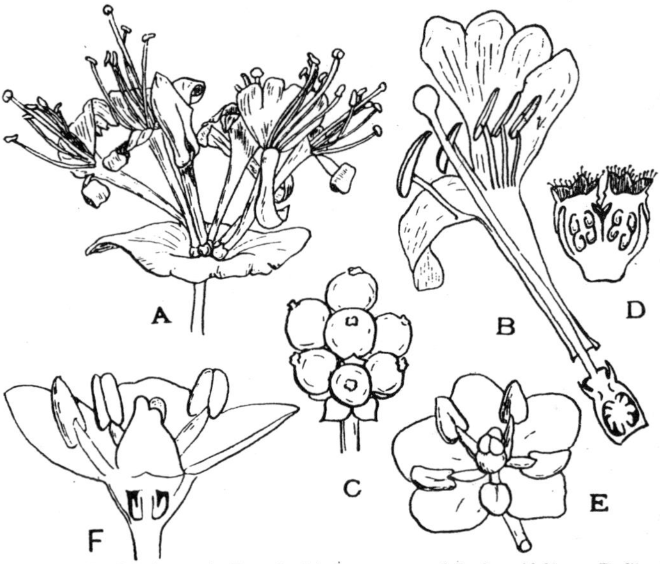

FIG. 262. A–D. *Lonicera*. A. Terminal inflorescence of *L. Caprifolium*. B. Single flower, enlarged; the corolla cut open, the ovary in vertical section. C. Cluster of fruit. D. Pair of flowers of *L. ligustrina* in vertical section, the corolla removed. E. Flower of *Sambucus nigra* enlarged. F. Same in vertical section. (A, F after Le Maout and Decaisne; D after Fritsch.)

section of *Lonicera*; with a bilabiate long-tubed corolla in the base of which nectar is secreted; the flowers are strongly scented at nightfall and attract night-flying insects, especially Sphingidae; the stigma is carried out beyond the stamens, thus preventing self-pollination.

The fleshy berry or drupaceous fruit is adapted for distribution by birds.

The species occur mainly in the temperate regions of the

northern hemisphere; in the tropics they are found usually only at higher elevations. The family is sparingly represented in the southern hemisphere by a few species of *Sambucus* (South America and Australia) and *Viburnum* (Andes), and the small genus *Alseuosmia*, confined to New Zealand, consisting of small evergreen shrubs with alternate leaves. Numerous fossil species of *Viburnum* have been described, especially in the Tertiary strata of North America.

Sambucus contains about 20 species and is very widely distributed; *Viburnum*, 120 species, mainly in temperate and subtropical regions of East Asia and North America, with three species in Europe; *Symphoricarpus*, 15 species in North America and Mexico, one in China; *Lonicera*, 180 species, widely distributed throughout the northern hemisphere, especially developed in Eastern Asia and the Himalayan region.

The close affinity of this family with Rubiaceae has already been indicated. Affinity with the polypetalous Cornaceae is indicated by *Viburnum* with simple opposite leaves, regular isostemonous flowers, one-ovuled ovary-chambers and drupaceous fruit.

Family III. ADOXACEAE

This family comprises only the single species *Adoxa Moschatellina* (Moschatel) (fig. 263), a small glabrous herb which is widely spread over the north temperate zone. It is a perennial with a creeping monopodial rootstock which bears a cluster of ternately divided radical leaves without stipules, and a simple two-leaved stem ending in a cymose cluster of five (or seven) small, green, sessile, regular, bisexual flowers. The terminal flower is tetramerous, the lateral are pentamerous and borne in decussating pairs. The simple floral envelope is deeply four- to five-lobed, and alternating with the lobes is a whorl of stamens, each of which is split to the base and bears a half-anther. The ovary is half-inferior and four- or five-chambered with a single ovule pendulous from the inner angle of each chamber; the short styles are equal in number to the ovary-chambers. Below the perianth is a small shallowly two- (terminal flower) or three- (lateral flowers) lobed involucre which has been regarded as a calyx, but which Eichler interpreted as representing the union of the pair of bracteoles or the bract and two bracteoles respectively; it persists in the fruit around the one- to five-seeded drupe. The compressed seeds contain a small embryo at the apex of a copious endosperm.

The flowers are pollinated by small insects which are attracted by the greenish-yellow colour, the musky smell, and the nectar secreted by a fleshy ring surrounding the bases of the stamens.

Insects crawling over the small inflorescence bring their feet and proboscis into contact at one time with the anthers, at another with the stigmas and thus effect cross-pollination (H. Müller).

The affinity of *Adoxa* is doubtful; widely different positions have been assigned to it—Saxifragaceae, suggested by the half-inferior ovary and perhaps by the superficial resemblance of the flower to such forms as *Chrysosplenium*, a comparison which does not seem warranted; and Araliaceae, which it recalls in its inconspicuous flowers, with a half-inferior isomerous ovary with solitary pendulous ovules. It is however probably best placed in its

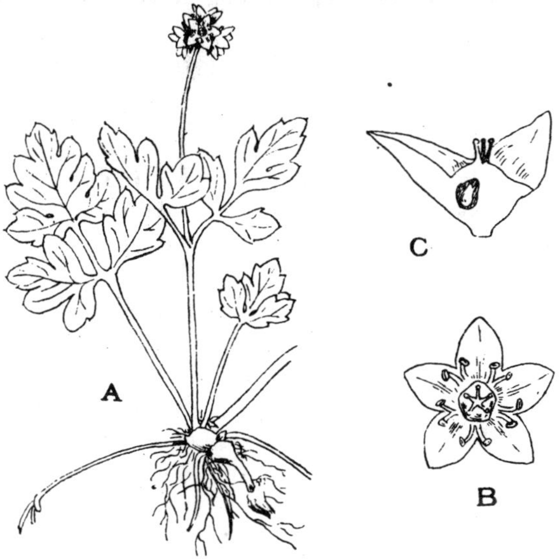

Fig. 263. *Adoxa Moschatellina.* A. Portion of plant in flower, slightly reduced. B. Flower. C. Same in vertical section. B and C enlarged. (After Fritsch.)

present position, and regarded as a derivative from the sympetalous tetracyclic type represented by the Rubiales. Eichler referred the floral structure to Caprifoliaceae, but the absence of a calyx, and the fission of the stamens associated with the general habit of the plant, seem sufficient to remove it from that family.

Family IV. VALERIANACEAE

Flowers bisexual or unisexual by abortion, irregular with no plane of symmetry. Calyx represented in the flower by an epigynous ring, becoming enlarged in the fruit. Corolla generally tubular and five-limbed, often saccate or spurred at the base. Stamens 1–4, attached to the corolla-tube and alternating with its segments; anthers dehiscing introrsely. Ovary of three united carpels typically three-chambered but one only fertile and containing a single pendulous anatropous ovule; style simple, ending in a simple or two- to three-partite stigma. Fruit one-seeded; embryo straight, completely filling the seed; radicle directed upwards.

Herbs, more rarely shrubs with opposite exstipulate leaves and numerous small flowers in cymose inflorescences.

Genera 10; species about 350; chiefly north temperate and Andine.

The plants are small annuals, as in *Valerianella*, e.g. *V. olitoria* (Lamb's Lettuce), a small glabrous cornfield plant with simple narrow leaves and minute pale lilac flowers. The majority are perennial, peristing by an underground rhizome, as in species of Valerian (*Valeriana*); some alpine species of this genus are acaulescent with leaves in a close rosette and a long thickened tap-root; in *V. tuberosa* (mountains of Mediterranean region) the plant persists by means of a tuberous adventitious root, borne each year on a short axillary shoot on the tuber of the preceding year, as in certain *Ophrydeae* (see vol. I). The aerial stem when branched is dichasial, or by suppression of the main axis dichotomous, and the same character is repeated in the inflorescence. The leaves are simple, as in *Valerianella*, *Centranthus* and others, or more or less pinnately divided, as in *Valeriana officinalis*; in others, as in *V. dioica*, the radical leaves are entire and the cauline pinnatifid.

The many-flowered inflorescences often pass over into monochasia; the flowers are sometimes closely crowded into heads, as in *Plectritis* (North America, Chile) and *Nardostachys* (Spikenard). The latter is exceptional in having an obvious five-limbed calyx, otherwise the calyx is late in development; its protective function, so far as concerns the flower, is usually transferred to the pair of bracteoles which stand close up to the flower, or mutual protection is effected by aggregation of the flowers in a head. The corolla crowns the top of the ovary, and is tubular or funnel-shaped; the tube is regular or produced at or near the base into a short sac or a spur of varying length in which nectar is secreted; the five (rarely

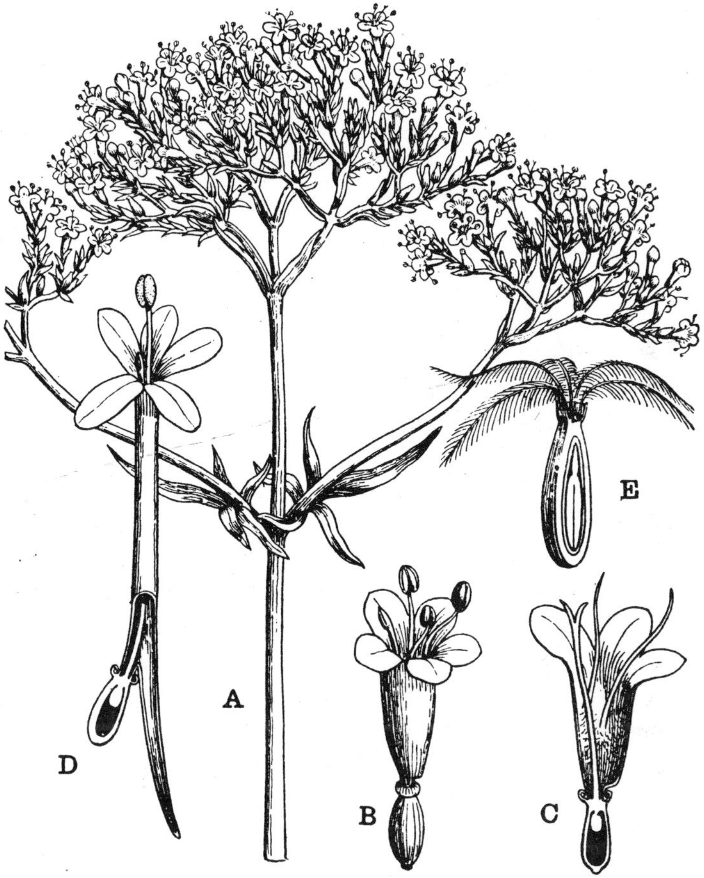

FIG. 264. *Valeriana officinalis.* A. Inflorescence, slightly reduced. B. Flower in male stage, the stigmas still unexpanded, ×4. C. Flower in vertical section in later stage with expanded stigmas; the anthers have fallen. D. Flower of *Centranthus ruber*, the lower portion in section; earlier stage, ×4. E. Fruit of *V. officinalis* in vertical section shewing seed with embryo, ×4. (B–E after Höck.)

three or four) limbs are equal, or bilabiate as in *Fedia* and *Centranthus*; they are imbricate in aestivation.

A progressive modification of the flower is traceable in the different genera. The least modified type is represented by the two small temperate Asiatic genera, *Patrinia* and *Nardostachys*, where the flower is almost regular. The calyx is five-toothed, or in *Nardostachys* obviously five-limbed, and does not become enlarged in the fruit; the corolla is generally regular (exceptionally spurred), the posterior stamen only is suppressed and the two sterile ovary-chambers are well-developed (fig. 265, D). In *Valerianella* (north temperate) (fig. 265, B) the outer (on the side towards the sterile bracteole) anterior stamen is also suppressed, the calyx is variously

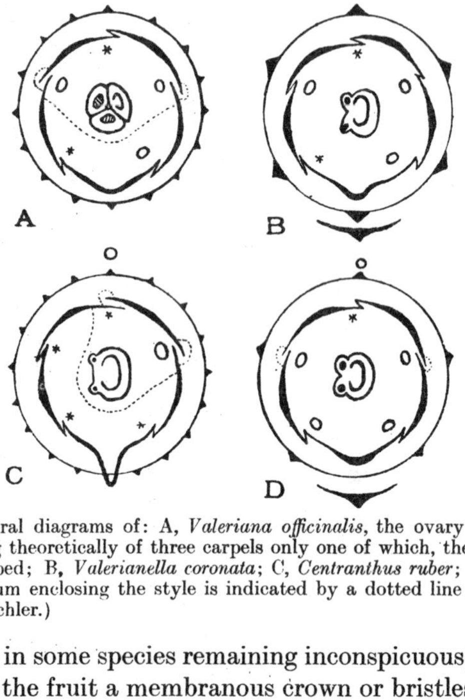

FIG. 265. Floral diagrams of: A, *Valeriana officinalis*, the ovary is shewn as consisting theoretically of three carpels only one of which, the right-hand, is developed; B, *Valerianella coronata*; C, *Centranthus ruber*; D, *Patrinia*. The septum enclosing the style is indicated by a dotted line in A and C. (After Eichler.)

developed, in some species remaining inconspicuous, in others forming in the fruit a membranous crown or bristles or hook-like structures; it never forms a pappus; the sterile ovary-chambers also shew various stages of development; the corolla is regular and funnel-shaped.

Fedia (a monotypic Mediterranean genus) has an inconspicuous two- to four-toothed calyx which does not enlarge in the fruit, a long tubular corolla with a short sac above the base, and a two-lipped limb; both anterior stamens are suppressed, but the two sterile ovary-chambers are still well developed. In the large genus *Valeriana* (figs. 264, 265, A) the calyx forms an epigynous ring in the flower which in the fruit unrolls to form a feathery pappus, the corolla-tube is regular or saccate, there are three stamens, as in *Valerianella*, and the ovary is one-chambered, the sterile chambers being reduced to small protuberances. *Centranthus* has a similar calyx, but the corolla is spurred and the reduction in the androecium reaches a maximum, one stamen only being present; the sterile ovary-chambers are reduced to nerves (figs. 264, D, 265, C). In the last two genera the corolla-tube is divided by a septum into two compartments, one of which encloses the style; the division is a low one in *Valeriana* but reaches to the throat of the tube in *Centranthus*; it is indicated by a dotted line in the floral diagrams.

The flowers represent an advanced stage of adaptation to insect-pollination, which may extend to dimorphism. H. Müller has indicated an interesting series in species of *Valeriana*. *V. officinalis* (Cat's Valerian, All-heal) is proterandrous (fig. 264, B, C); the small flowers are rendered conspicuous by aggregation, and the nectar in a small pouch near the base of the short tube is accessible to insects with moderately short probosces; the anthers, covered all round with pollen, in the first stage of the flower, or the three outspread stigmas, in the second stage, are touched by the feet and under surfaces of insects creeping over the inflorescence or by the heads of insects sucking nectar. In *V. dioica* cross-pollination is ensured by dioecism; there are four different forms of flowers borne on different individuals, (1) male flowers with no trace of a pistil and with large corollas, (2) male flowers with a rudimentary pistil and small corollas, (3) female flowers with evident traces of anthers and still smaller corollas and (4) female flowers from which anthers have almost entirely disappeared and with smallest corollas of all. Intermediate stages are found in *V. montana*, which has large- and small-flowered individuals, the former being bisexual, the latter functionally female, as the anthers though present are empty of pollen, and *V. tripteris*, which differs in that the large flowers are functionally male, the style bearing no stigmatic papillae.

The pappus or membranous or wing-like developments aid distribution of the fruit by air-movements; the bristly crown or hooks in species of *Valerianella* are effective by clinging to fur or feathers, and the pappus of *Valeriana* and *Centranthus* will also act in the same way.

More than 200 of the species are included in the genus *Valeriana*. In the Old World the chief development is in the Mediterranean region, and the family is scarcely represented south of the equator; in the New World there is a strong development of the genus *Valeriana* on the Pacific side, especially in the Chilian Andes and extending southwards.

The rhizome and roots of many of the perennial species contain an ethereal oil with a sharp bitter taste and penetrating smell, which from its strong action on the nervous system has a medicinal value.

Spikenard, one of the most prized unguents of the Romans and eastern nations, is procured from the young shoots of the Himalayan *Nardostachys jatamansi*; the generic name is derived from the spike-like appearance given to the shoot by the persistent stiff fibres of the older leaves. The young leaves of some annual species, as of *Valerianella*, e.g. *V. olitoria* (Lamb's Lettuce), are used as a salad.

Centranthus (Mediterranean region) contains the familiar garden species *C. ruber*, which has become naturalised on old walls and chalk banks in the south of England.

Family V. DIPSACACEAE

Flowers small, bisexual, generally medianly zygomorphic and aggregated into dense involucrate heads; each flower is enveloped by an epicalyx formed by the union of the pair of bracteoles. Calyx small, variously developed. Corolla tubular or tubular-funnel-shaped, with a five- or four-lobed limb, lobes imbricate in bud. Stamens four, sometimes fewer by abortion; anthers introrse. Carpels two, but the inferior ovary unilocular, with a single anatropous ovule pendulous from the apex; style filiform, stigma simple or bifid. Fruit dry, one-seeded, enclosed in the epicalyx and often crowned by the persistent calyx-limb. Embryo straight, surrounded by endosperm, radicle superior.

Annual or more often perennial herbs, rarely shrubs, with opposite exstipulate leaves.

Genera 9; species about 155; mainly Mediterranean, but spreading eastwards, southwards and northwards.

Though the plants are generally herbaceous, a lower prostrate persistent portion of the stem sometimes becomes woody, as in *Scabiosa lucida* and others. In some steppe-inhabiting species of *Scabiosa* the whole plant becomes

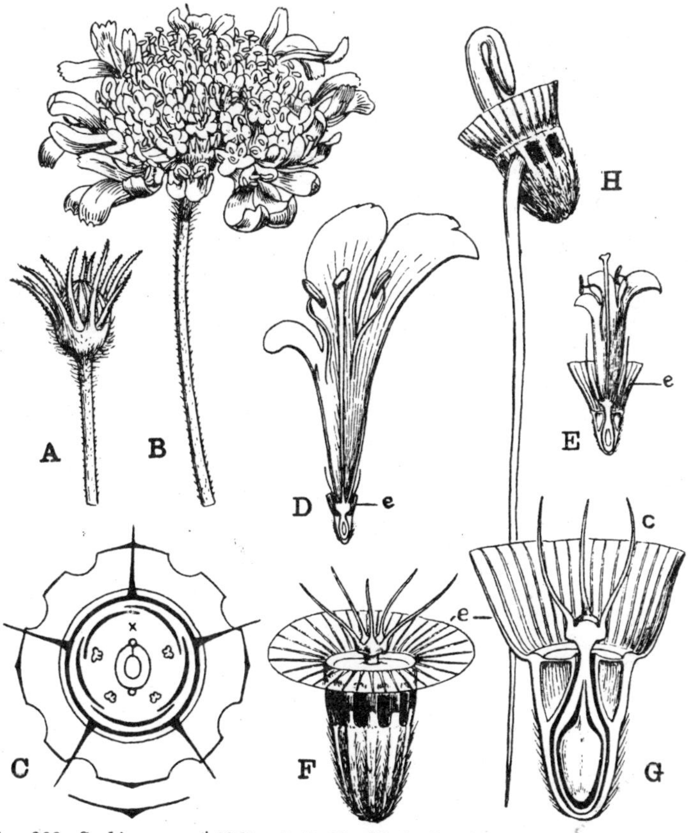

FIG. 266. *Scabiosa graminifolia.* A, B. Head in bud and flower, slightly reduced. C. Floral diagram. The two outer rings represent the epicalyx, indicated by *e* in the figures. D an outer, E an inner floret in vertical section, $\times 2\frac{1}{2}$. F. Fruit, $\times 5$. G. Fruit in longitudinal section, the epicalyx, *e*, is more ascending than in F; *c*, persistent calyx-teeth. H. Germination.

densely hairy; instances of an interesting adaptation to seasonal changes are seen in *S. pulsatilloides* (Granada) and to a less degree in other species, where in the first year the plant forms a dense leaf-rosette, which is protected from the dry hot summer by a grey felt-like hair-covering; this is absent from the leafy flower-bearing stem produced during the damp period of the following spring. The leaves are variable in form, being entire, cut, lobed or pinnatifid; they may differ on the same plant, as in *S. Columbaria*, where the narrow radical leaves are entire or divided and the cauline pinnatifid.

The erect biennial stem of the Teasels is angular, and hairy or prickly; the leaves are generally toothed or cut, and connate forming water-receptacles round the stem.

The epicalyx shews considerable variation in form in the different genera and species; it has generally an angled or ribbed or furrowed lower portion surrounding the ovary and the fruit when ripe; it may be truncate above, as in Teasel, but more often is provided with a membranous crown (fig. 266, F, G) sometimes bearing bristles or teeth and aiding the distribution of the enclosed achene by wind- or animal-agency. The numbers of angles, ribs, teeth, etc., are in relation to the number of limbs of the calyx or corolla (equal or double, etc.). The calyx and corolla may be regarded as typically pentamerous, though often actually tetramerous, the latter condition being, according to Eichler, due to union of two members (cf. *Veronica*). A fifth, posterior, stamen is always absent, and of the four stamens the two anterior are often smaller; sometimes, as in *Morina*, represented merely by barren rudiments. The calyx-limb varies in form, being sometimes cup-shaped or lobed, toothed or ciliate; it may fall after fertilisation, but generally persists in the fruit (fig. 266, F, G). The corolla-limb is often more or less two-lipped or the lobes are unequal. The irregularity is sometimes much enhanced in the outer flowers of the head (as in *Scabiosa*), which are also larger than the inner, indicating the differentiation into ray- and disc-florets which becomes specialised in the Compositae. According to Payer, the study of the development of the flower in *Scabiosa* indicates the presence of two carpels, but the ovary is always one-celled, with no trace of the sterile carpels which were characteristic of the Valerianaceae; the stigma, however, is often forked.

The aggregation of the flowers and the floral structure point to insect-pollination, details of which have been studied in several species. Nectar is secreted by the top of the ovary, and through the widening upward of the corolla-tube is often accessible to short-tongued insects. Self-pollination is avoided by the projection of the stigma beyond the stamens, as in *Morina*, or usually by marked proterandry, as in *Dipsacus*, *Knautia*, *Succisa*, *Scabiosa*; in *Knautia arvensis* this has led to dicliny, individuals occurring in which the flowers have all more or less aborted anthers which do not open.

The epicalyx becomes thickened and leathery, protecting the fruit, the pericarp of which is thin and easily ruptured at the apex; the seed has a thin membranous testa, and the embryo is surrounded by a thin or thick layer of fleshy endosperm. In germination the radicle protrudes at the top of the fruit and bends downwards, generally piercing the membranous wing of the epicalyx or penetrating one of its symmetrically arranged thin portions or furrows and pinning the whole to the earth by the thickened base of the hypocotyl (fig. 266, H). The oblong or ovate cotyledons are then drawn out of the fruit. In *Succisa australis* a small tooth-like projection at the base of the hypocotyl fixes the narrow rim of the involucel and holds it down while the cotyledons are drawn out.

The family is developed mainly in the Mediterranean region, especially the eastern portion, spreading through Western Asia to India and the Himalayas, and, with a few representatives, southwards into tropical Africa. It also spreads northwards into Central, and to a less extent Northern Europe. *Dipsacus sylvestris* is our wild Teasel; *D. fullonum* (Fuller's teasel), distinguished by its hooked bracts, is probably a cultivated form of *D. sylvestris*. Our native Scabious are included in three genera, formerly regarded as sections of *Scabiosa*, namely *Succisa* (Devil's-bit Scabious), *Knautia*, *Scabiosa* (*S. Columbaria*).

Order 2. *CAMPANULALES*

Flowers tetracyclic, bisexual, or unisexual by suppression, regular or zygomorphic; typically pentamerous, with isomerous stamens and generally fewer carpels. Anthers converging and often laterally united (syngenesious). Ovary inferior (rarely more or less superior), several-celled with $\infty - 1$ ovules in each cell or one-celled with a single ovule.

Generally herbs, more rarely woody plants, often containing latex-vessels or oil-passages. Flowers generally racemose with a tendency to aggregation in heads.

The distinguishing feature of the order is the simple but remarkably effective pollen-presentation mechanism. This consists in the lateral approximation or coherence of the anthers to form a tube into which the pollen is ejected, and the development of the style into a brush by which the pollen is swept out at the top of the tube. The subfamilies Campanuloideae, with free anthers, and Lobelioideae, with laterally united anthers, of the family Campanulaceae represent respectively simple and more elaborate types from this point of view; in the great family Compositae the syngenesious character is constant and the mechanism shews a wide variation in detail. The two small and geographically restricted families Goodeniaceae and Stylidiaceae have more elaborate highly specialised pollination-mechanisms. The family Campanulaceae shews considerable variety in floral structure; in the less advanced subfamily the flowers are regular with free anthers and an isomerous gynoecium or reduction to three or two carpels; in the more advanced the flower is strongly medianly zygomorphic with syngenesious anthers and a bicarpellary pistil; there is also a tendency to reduction in the number of ovules. The two small families have a zygomorphic bicarpellary pistil. The fruit in these and in Campanulaceae is, with few exceptions, a capsule, shewing considerable variety in mode of dehiscence, and containing numerous to few seeds; the epigynous character is not fixed, though inferior in the great majority it may be half-inferior to superior. Again, though the inflorescence is often lax there is a tendency to aggregation of the flowers, which in some genera form compact spikes or heads, sometimes surrounded by an involucre of bracts. In Compositae the general type of flower, so far as regards number and relative position, is remarkably constant; the always inferior ovary, though formed of two carpels, is unilocular and one-ovuled and the fruit one-seeded and indehiscent; the capitular inflorescence is also a constant character. Within the limits of the capitulum there is considerable variety in the form of the floret, which is regular or more or less zygomorphic, in the differentiation of the florets for reproductive or attractive purposes, and also in the specialisation of the calyx, the protective function of which has been transferred to the bracts of the involucre, for seed-dispersal.

We have noted under Rubiales biological tendencies in connection with inflorescence and flower similar to those which have become more or less characteristic of Campanulales. The aggregation of the flowers into dense heads, which reaches an extreme development in Dipsacaceae where the head is subtended by an involucre of bracts, is associated with loss of the protective function of the calyx and its development as an organ for seed-dispersal. But a marked distinction is the cymose character of the heads in Rubiales as contrasted with the racemose capitulum in Campanulales, and the resemblances between the two orders are to be regarded as indicating convergence of development rather than direct relationship. A clue to the origin of the Campanulales may be found in the conflicting views which have been put forward regarding the position of the family Cucurbitaceae (q.v.) and the order may represent a sympetalous development of a stock allied to Passiflorales in which a tendency to sympetaly and epigyny occurs.

Family I CAMPANULACEAE

Flowers generally bisexual, regular or medianly zygomorphic; usually pentamerous. Sepals generally free and open in aestivation; petals generally united below, aestivation typically valvate. Stamens free or united, sometimes partly united with the corolla-tube; anthers introrse. Carpels 2–5, united; ovary inferior or half-inferior, very rarely superior, generally 2–5- (rarely 6–10-) celled, rarely one-celled, with generally numerous anatropous axile, rarely parietal ovules; style simple, often with pollen-collecting hairs. Fruit a capsule; rarely berry-like. Seeds generally small with a fleshy endosperm and a straight embryo.

Annual or perennial herbs or undershrubs, rarely shrubs or trees, with usually alternate, exstipulate leaves, and often showy flowers, which are solitary or arranged in more or less complicated inflorescences. Latex is generally present in segmented sacs.

Genera 61; about 1500 species; mainly temperate and subtropical.

The plants are annual herbs, as in *Specularia*, a small genus, mainly north temperate, represented in England by *S. hybrida*—a small plant with an erect or recumbent stem occurring in cornfields and dry soils—some species of *Campanula* and others. Others are biennial, with generally a thick fleshy tap-root, as in species of *Campanula*, e.g. *C. Rapunculus* (Rampion), *Michauxia* (Asia Minor) and others. Generally, however, they are perennial, in which case the main axis bears a rosette of leaves and the flower-bearing shoots arise laterally, as in *C. rotundifolia* (Hare-bell), or the main axis bears flowers, as in *C. Trachelium*, *glomerata*, *persicifolia*, etc.; the lateral shoots bear leaves and an inflorescence, the lower portion persisting generally for one or two years as a rhizome; in *Canarina*, a native of the Canary Islands, one or more of the roots become thick and tuber-like. In other cases the main shoot perishes each year and the new lateral shoots become separate individuals, as in our British *Lobelia Dortmanna*, which roots in gravelly mountain lake-bottoms, the short main axis developing long slender stolons which form the new shoots. The aerial stems are generally erect, sometimes twining, as in species of *Codonopsis* (natives of the mountains of Central and Eastern Asia) and *Cyphia* (South Africa). Shrubby forms also occur, sometimes of remarkable form, as in *Campanula Vidalii*, a native of the Azores, and several small South African genera which have sometimes a heath-like habit (as *Roëlla*, *Rhigiophyllum*). A group of genera belonging to the subfamily Lobelioideae, native of the Sandwich Islands and other Pacific Islands, are shrubs or small trees, and the large tropical American genus *Centropogon*, of the same subfamily, consists of shrubs which are sometimes climbers. The large genus *Lobelia* comprises herbs, more rarely shrubs; a remarkable development is shewn by the Tree Lobelias, which are a feature of the high mountain flora of tropical Africa. These reach 15 ft. in height and have a stout, sometimes woody, cylindrical stem bearing a dense crown of long narrow leaves and ultimately forming a tall terminal spike of flowers. The lower part of the stem is marked with the scars of the fallen leaves.

The leaves are generally spirally arranged, but whorled or opposite in *Canarina*, alternate or opposite in *Wahlenbergia*, opposite in *Campanumaea* and others. The radical leaves often differ from the cauline, as in *Campanula rotundifolia*. The leaves are usually entire or toothed, rarely pinnately cut.

A general anatomical character is the presence of articulated laticiferous tubes which occur in almost all the organs. In the stem they are generally limited to the inner region of the phloem, more rarely they are found in the outer portion or in the cortical parenchyma; they are infrequent in the pith. Deviations from the typical stem-structure occur in many species of *Campanula*, and in species of allied genera. In the simpler cases internal phloem is present, in the more complicated there are one or more rings of distinct collateral or sometimes concentric bundles, which may even unite to form a closed inner ring. These internal bundles generally shew secondary growth in thickness.

The flowers are usually in racemes, spikes or heads, which in the subfamily Campanuloideae have a terminal flower, generally absent in the other subfamilies. In *Jasione* and *Hedraeanthus* the outer bracts form an involucre. Solitary terminal flowers are characteristic of some small genera (*Cyananthus*, *Roëlla*) and are also found in species of *Campanula*, and other genera; solitary axillary flowers also occur.

Lateral flowers generally have a pair of bracteoles, in the axils of which branching may take place developing dichasia which pass over into monochasia. The small, somewhat peculiar, genus *Pentaphragma* (India and Malaya), which has also very asymmetrical leaves, has a terminal dorsiventral cyme rolled up at the apex closely resembling an inflorescence of Boraginaceae or Hydrophyllaceae.

The family falls into three subfamilies distinguished especially by floral characters as follows.

Subfamily I. CAMPANULOIDEAE. Flowers regular; anthers generally free (figs. 267, 268).

The sepals, petals and stamens are usually five in number, but not infrequently more (6 to 10), as in *Michauxia* and *Canarina* (generally six), sometimes fewer, as in *Wahlenbergia* (often three or four). The sepals may be swollen at the base as in many species of *Campanula*, and may sometimes bear lateral appendages. The corolla is generally bell-shaped, sometimes very open, as in *Campanula persicifolia*, *C. carpatica* and others, more rarely funnel-shaped (*Merciera*) or rotate (*Specularia*), or the petals are almost or quite free, as in *Michauxia* and *Phyteuma*; the union of parts may be greater in the bud than in the open flower, and

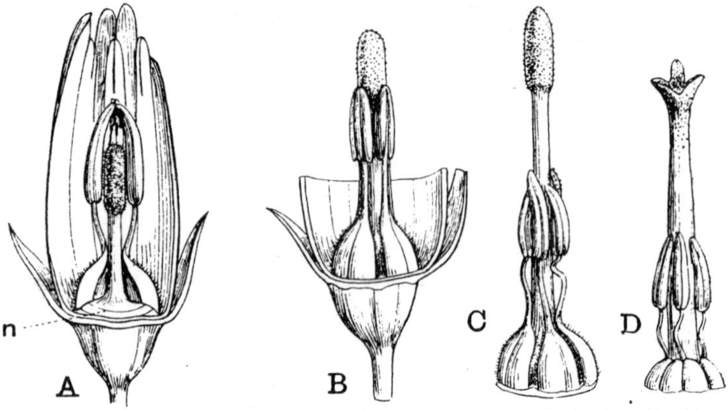

FIG. 267. Flower of *Campanula pusilla* to illustrate method of pollination. A. Young bud in longitudinal section, the unopened anthers surround the brush of the style; *n*, nectary, protected by the arching bases of the stamens. B. Stamens and pistil of a bud about to open; the style-brush is covered with pollen from the anthers. C. Stamens and style in first or male stage. D. Same in second or female stage. (After H. Müller.)

the method of separation is often very characteristic, as for instance in *Phyteuma*, where the petals may remain united at the apex, becoming free below. In the cleistogamic flowers of *Specularia* there are no petals.

The stamens are often broadened at the base, forming a hollow chamber above the nectar-secreting epigynous disc, and the anthers are free. The relation of the parts is admirably adapted to pollination by insect-agency. Thus in *Campanula* in the bud (fig. 267, A) and the early stage of the flower the appressed stigmas form a cylinder, the outer surface of which is covered with long hairs; in the bud the anthers surround this brush and shed their pollen upon its hairs (fig. 267, B); as the flower opens (fig. 267, C) the stamens wither and fall back into the flower, while the columnar style presents a brush thickly covered with pollen which is rubbed off by the bodies of humble-bees and other insects which probe between the broad bases of the stamens for nectar. In the next stage (fig. 267, D) the collecting hairs have withered and the three stigmas separate and expose their receptive upper faces. In absence of cross-pollination, self-pollination may be effected by the bending back of the stigmas to pick up grains of pollen still adhering to the style. In *Phyteuma* the long strap-shaped lobes of the corolla cohere for a time in the upper portion and form a narrow tube up which the pollen is pushed by the lengthening style and projects in a mass above the mouth; in the second, female stage, the style bearing the three now expanded

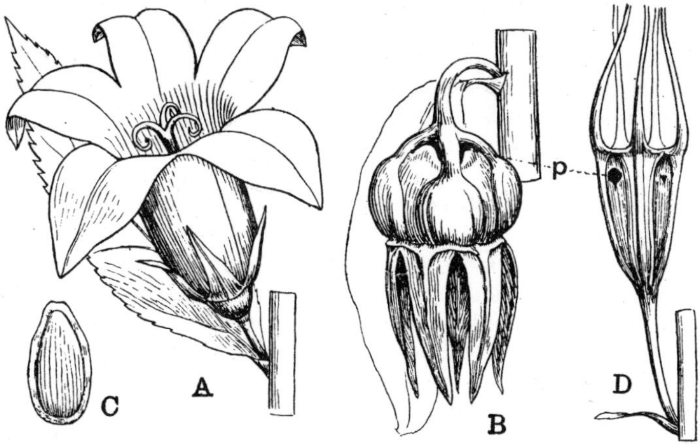

FIG. 268. A. Flower of *Campanula latifolia*, slightly reduced. B. Fruit of same, slightly enlarged. C. Seed of same, × 10. D. Fruit of *C. Rapunculus*, × 3. *p*, valve.

stigmas escapes from the corolla; the small flowers are aggregated into heads or dense spikes, their aggregation rendering them conspicuous; the method of pollination closely resembles that of a Composite.

The resemblance to the Compositae is still more marked in *Jasione* (Sheep's-bit), in which the small flowers are closely arranged in terminal heads surrounded by an involucre of bracts. The nectar is easily accessible even to short-tongued insects, as the corolla is cleft to the base into narrow linear lobes, and the thin staminal filaments are free. The stamens cohere at the base of the anthers to form a ring round the style, which in the first stage elongates, bearing the brush densely covered with pollen above the corolla-lobes, and afterwards, when pollen and hairs have disappeared, displays a two-lobed stigma. Larger insects will come into contact with many flowers at one visit, and the small size of the flowers is thus compensated for by the union of a large number, sometimes considerably over 100, in a single head.

The number of carpels varies. When two in number they are generally median, when three the odd one is generally posterior; when there are five they are alternate with the stamens (*Platycodon* and other genera) or opposite (species of *Campanula*, *Wahlenbergia* and others). The ovary is generally two- to five-chambered (rarely 6 to 10). The ovules are generally borne on thick, sometimes stalked, central placentas. In the small South African genus *Merciera* the one-celled or incompletely two-celled ovary contains four basal anatropous ovules and in another small South African genus, *Siphocodon*, the ovules are pendulous from the top of the three-celled ovary. In the subtribe *Wahlenberginae* the ovary is often elongated conically above the line of insertion of the calyx, being more or less half-superior, as in *Wahlenbergia*, or sometimes quite superior, as in *Cyananthus* (mountains of Central and East Asia).

Subfamily II. CYPHIOIDEAE. A small group of herbaceous plants with zygomorphic flowers and stamens with filaments sometimes united, but the anthers are free. Includes *Cyphia*, with about 20 species in Africa, mainly south (some have twining stems), and a few small West American genera.

Subfamily III. LOBELIOIDEAE. Flowers strikingly zygomorphic (fig. 269), rarely unisexual, pentamerous with reduction to two in the pistil. The odd sepal is anterior but by twisting of the flower-stalk through 180° (resupination) becomes posterior in the open flower. The petals are united into a tube, which is, however, often split more or less to the base, rarely are they quite free. The filaments of the stamens are more or less united and the anthers join laterally to form a tube (syngenesious), as in Compositae. The style bears a crown of hairs which in the first or male stage of the flower is at the bottom of the anther-tube; by lengthening of the style the pollen is pushed out of the tube and detained above it by the stiff hairs on the top of the anthers (fig. 269, C). After the removal of the pollen the two stigmas project from the tube and fold back to expose their receptive surfaces (female stage, fig. 269, D, E). The long-tubed flowers of some species of *Siphocampylus* (a large tropical American genus) are probably pollinated by humming-birds. In *Lobelia* (fig. 269) and other genera the ovary is often half-inferior or almost superior.

The fruit is generally a many-seeded capsule, the form and mode of dehiscence of which vary considerably in different genera. Thus in *Campanula* it opens by three to five small lateral valves below the calyx; when the fruit is erect, as in

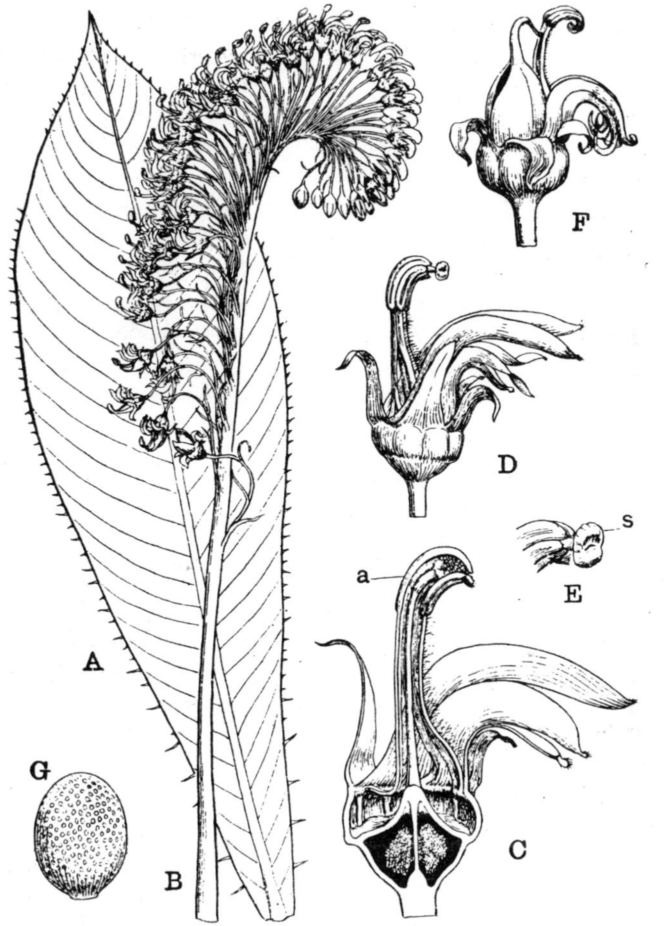

FIG. 269. *Lobelia assurgens.* A, B. Leaf and inflorescence, slightly reduced. C. Flower in vertical section, in male stage, × 2½; the pollen is being pushed out of the anther-tube, *a*, by the brush of hairs below the stigma. D. Flower in later, female stage, the stigma has opened, × 1½. E. Tip of anthers with spreading stigma, *s*; a tuft of pollen-detaining hairs is seen on the lowest anther; × 3. F. Fruit beginning to open (with remains of flower), × 1½. G. Seed, × 25.

C. Rapunculus (fig. 268, D), the valves are close under the calyx-lobes, when the fruit droops, as in *C. rotundifolia*, *C. latifolia* (fig. 268, B) and others, the valves are at the base of the fruit and therefore uppermost in space; thus in both cases some jerking of the fruit is necessary to allow the escape of the small seeds. In *Specularia* the long slender capsule opens by three slits in the upper part. On the other hand, in *Jasione* and *Wahlenbergia* it dehisces by valves, equal in number to the carpels, above the calyx-lobes. The capsule of *Lobelia* also opens loculicidally by two valves above the persistent calyx (fig. 269, F). Rarely, as in *Canarina* and a few other genera, the pericarp is fleshy and the fruit a berry. In *Sphenoclea*, a genus with only one species, a herbaceous marsh-plant widely distributed in the tropics, the two-celled capsule dehisces by means of a broad lid below the sepals.

The Campanuloideae are with few exceptions natives of the temperate zones reaching northwards into the frigid zone; many are alpine. The principal centre of development is the Mediterranean region. The largest genus, *Campanula*, contains about 230 species in the temperate regions of the northern hemisphere, especially the Mediterranean region, a few extending into the arctic area. It is represented in the southern hemisphere by *Wahlenbergia* (100 species), a very widely distributed genus which occurs also in north temperate regions; *W. hederacea*, a Western European species, occurs in the west and south of the British Isles.

Lobelioideae are mainly tropical and south temperate, but also occur in the north temperate zone. *Lobelia* has about 200 species, generally distributed in hot and temperate regions; two species occur in Europe, both of which are British—*L. Dortmanna*, which inhabits the margins of lakes in North Europe and North America, and *L. urens*, a West European (West France and Spain) species found rarely on heaths from Sussex to Cornwall.

The family supplies many well-known garden plants, especially of the genera *Campanula* (Canterbury Bell, etc.) and *Lobelia*; the very common dwarf blue species is *L. Erinus*, a native of the Cape. The fleshy roots of some species and the berries of *Canarina* and other genera are eaten; a few are used medicinally; the chief is *Lobelia inflata*, the "Indian tobacco" of North America, which contains a volatile alkaloid, lobeline.

Family II. GOODENIACEAE

A small family of 11 genera and about 300 species, mainly Australian. The plants are herbs or small shrubs without latex, with simple exstipulate leaves and flowers solitary and axillary or in somewhat loose cymose or racemose inflorescences. The flowers are bisexual, generally zygomorphic, and pentamerous with reduction to two in the pistil. The tube of the corolla is generally open on one side, and the petals have often broad membranous wings. The stamens are usually free from the corolla; the anthers are introrse and free or united laterally. The flowers are proterandrous; the pollen is passed into a collecting cup at the end of the style before the flower opens, the cup then almost closes, leaving a narrow opening covered by hairs. The style bends down and stands in the mouth of the flower, visitors to which become dusted with the powdery pollen. Pollen is continually forced through the narrow opening by growth of the stigma-lobes, which finally emerge to receive pollen in turn from visitors from another flower. The form of the stigma varies considerably in different genera.

The pistil consists of two median carpels; the ovary is one- to two-chambered, generally inferior or half-inferior, rarely quite superior. The ovules are generally anatropous and ascending. The fruit is usually a capsule dehiscing by valves, sometimes a drupe or nut. The straight embryo is generally surrounded by a fleshy endosperm.

The family is Australian, but species of one of the largest genera, *Scaevola*, have spread to the Pacific Islands and the tropical coasts of both hemispheres; the fruit of this genus is drupaceous. It is allied to the subfamily Lobelioideae of Campanulaceae, but is distinguished by absence of latex, the anatomical structure of the stem, which has frequently an extra-fascicular cambium, and by the further specialisation of the pollination-mechanism in the development of the characteristic pollen-collecting cup.

Family III. STYLIDIACEAE

A small Australian family of three genera, with about 120 species (of which more than 100 are included in *Stylidium*) of herbs or small shrubs, sometimes moss-like in habit. Latex is absent. The leaves are exstipulate and generally narrow and grass-like, often forming a basal rosette from which springs the elongated axis bearing a raceme of flowers. The flowers are bisexual or unisexual by abortion and have a pentamerous calyx and corolla, the sepals are generally free, the petals united below with five limbs, four of which are generally similar, while the anterior (labellum) is often different in form and bent downwards. There are two stamens which are united with the style to form the "column." The extrorse anthers dehisce transversely. The two carpels are median; the ovary is inferior and two-celled or more or less completely one-celled by disappearance of the septum, and the placentation ranges from axile to free-central. Two short stigmas alternate with the anthers. The flowers are proterandrous and evidently entomophilous. In many species of *Stylidium* the column is sensitive to touch; in its normal position it is bent forwards and springs backwards when touched, after which it gradually resumes the original position. The fruit is a capsule, generally opening septicidally. The usually numerous seeds have a delicate coat and contain a very small embryo in a fleshy endosperm.

The stem-structure is peculiar. The primary vascular bundles remain separated by parenchymatous tissue which later becomes sclerenchymatous. In species with an elongated stem where secondary growth in thickness takes place, new bundles are developed outside the primary ring, recalling the method of secondary growth in *Dracaena* and *Yucca*. An interesting anatomical peculiarity is the oblique position of the tall epidermal cells in many species of *Stylidium* giving the appearance in section of a several-layered epidermis; this is associated with the prolonged apical growth of the young leaves[1].

In some species of *Stylidium* carbohydrate is stored in the form of inulin.

The peculiar habit and anatomical structure, and above all the high degree of specialisation in the androecium and pistil, distinguish the family from other members of the order.

Family IV. COMPOSITAE

Flowers (florets) in heads (capitula) or short spikes, generally sessile and either subtended by scale-like bracts, or bracts are absent; the head surrounded by a few- or many-leaved involucre of bracts. Florets of a head bisexual or unisexual (monoecious or dioecious), or the outer (ray-florets) female or asexual; generally 5-merous and regular or medianly zygomorphic. Calyx absent or represented by the pappus in the form of an epigynous ring, or hairs, bristles or scales, which persist in the fruit. Corolla gamopetalous, regular, bilabiate or ligulate. Stamens isomerous, inserted on the corolla-tube; filaments generally free; anthers two-celled, and united laterally into a tube, dehiscing introrsely by longitudinal

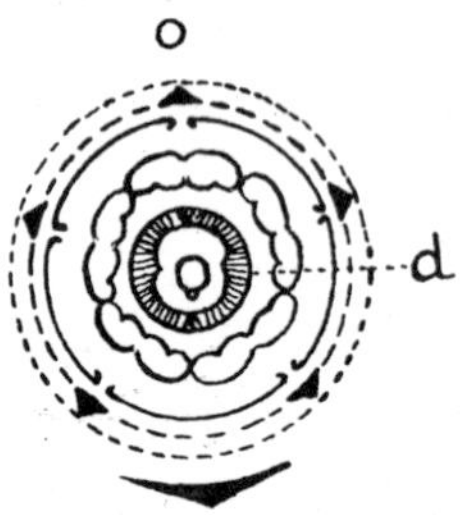

FIG. 270. Diagram of a bisexual tubular flower; *d*, disc. (After Eichler.)

[1] See Burns, G. P. "Beiträge z. Kenntniss d. Stylidiaceen." *Flora*, LXXXVII, 313 (1900).

slits; connective generally prolonged above the anthers; stamens usually absent from the female flowers. Ovary inferior, unilocular, formed from two median carpels, containing a single anatropous ascending ovule with one integument; style slender, in the fertile flowers divided at the top into two limbs which are stigmatic on the inner face and, at least in the bisexual flowers, bear hairs on the outer face or at the tip; in the male flowers generally undivided. Fruit a one-seeded inferior achene (cypsela), often crowned by the persistent pappus. Seed erect, filling the fruit, the coat uniting with the pericarp; endosperm absent; embryo straight with a short inferior radicle and flat or semicylindrical, sometimes involute cotyledons.

Herbs or shrubs, rarely arborescent, with generally alternate, more rarely opposite, exstipulate leaves.

Genera over 800; species about 14,000. A very large cosmopolitan family, containing about one-tenth of the total number of flowering plants.

The plants shew a remarkable diversity in habit. In the world-wide genus *Senecio*, the largest in the family, with about 2500 species, are included annual and perennial herbs, such as our British species *S. vulgaris* (Groundsel) and *S. Jacobaea* (Ragwort), climbers, as the Eastern Asiatic *S. scandens*, and various tropical and South African species, undershrubs or shrubs, as in species in the Mediterranean region, South Africa, Canary and Sandwich Islands, fleshy herbs or shrubs, as in the section *Kleinia* (mainly Cape) and trees with a simple or slightly branched stem bearing dense crowded heads of large leaves, as in the tree Senecios of the mountains of tropical Africa, Madagascar and St Helena. Other species on the Andes of South America have creeping tufted stems, sometimes woody at the base, and here also occur shrubby species of heath-like habit with closely crowded narrow leaves. Marsh-plants occasionally occur in the family but true water-plants are rare[1].

[1] Hutchinson, J., in *Gardeners' Chronicle*, LIX, 305 (1916), figures four aquatic herbs widely separated in systematic position and geographical distribution (species of *Bidens* from N. America, *Cotula*, S. Africa, and *Pectis* and *Erigeron*, Mexico). They are small herbs with slender stems, and having much cut submerged leaves.

A tap-root is frequent, as in Dandelion, Sow-thistle and others, and may bear adventitious buds by means of which rapid vegetative propagation is ensured. In some cases the roots are tuberously thickened or bear tubers, as in *Dahlia*. In others, as perennial species of *Helianthus*, the plant multiplies by stem-tubers produced on the rhizome; those of *H. tuberosus* (Jerusalem Artichoke) are edible. Effective vegetative propagation is ensured by short stolons, as in *Bellis perennis* (Daisy), or the long runners of *Achillea Millefolium* (Milfoil). The stem is most generally herbaceous, sometimes undeveloped, giving a scapigerous habit, the flowering stem rising from a radical rosette of leaves, as in Dandelion, and some Hawkweeds (*Hieracium*), or erect and more or less branched, as in *Cichorium* (Chicory) and *Lapsana* (Nipplewort); the herbaceous stem may become shrubby below, as in *Artemisia*, *Tanacetum* (Tansy) and others. The suffruticose or shrubby habit is also frequent and several widely separated genera are arborescent; the shrubby and arborescent habits are characteristic of island- and mountain-floras; examples are three small genera confined to St Helena, *Wilkesia* (Sandwich Islands) and *Astemma* (Andes of Quito); also the small subtribe *Tarchonanthinae* (Africa and Madagascar). The large genus *Mikania* (chiefly tropical America) consists mostly of twiners, herbaceous or shrubby, and the South American *Mutisia* includes shrubby climbers, the midrib of the leaf ending in a simple or branched tendril. A section of the large American genus *Baccharis* has a leaf-like winged stem, on which the leaves are often more or less suppressed; its species are natives of the Campos of Southern Brazil. In some xerophytic species the branches become converted into thorns, as in *Proustia* (Andes) and others.

The leaves are generally alternate, though not infrequently crowded in radical rosettes. Opposite leaves are characteristic of many species and genera, and are the rule in the large tribe *Heliantheae* (Sunflower, etc.); whorled leaves occur exceptionally. Although simple, the leaves shew great diversity in size, form and depth of incision. In the tree Senecios they are very large and undivided. In xerophytic species they may be small and scale-like, or, as in *Hoplophyllum*, a small genus of the South African Karroo, converted into thorns. An ericoid habit with needle-like rolled leaves is characteristic of many genera and even subtribes, such as *Relhaninae* and *Athriocinae* (mainly South African). Narrow parallel-veined leaves of monocotyledonous appearance are characteristic of a few genera, as *Corymbium* (South Africa) and *Schlectendalia* (Brazil), with the habit of an *Eryngium*. The venation is pinnate and the blade is entire or more or less deeply cut between the lateral nerves. Though typically exstipulate, auricles are sometimes present at the leaf-base, as in *Senecio* and other genera; and decurrent leaves are widely distributed through the family, sometimes rendering the branches winged, as in *Onopordon Acanthium* (Cotton Thistle).

There is also great variety in the indumentum, often associated with the nature of the habitat, from more or less complete absence of hairs, as in Dandelion, Daisy and others, to the thick felt characteristic of alpine or otherwise xerophytic plants, as *Gnaphalium*, *Filago* and others. Many species of *Senecio* have a loose cottony indumentum; stellate hairs are frequent in the Hawkweeds (*Hieracium*). In *Helianthus* and allied genera the hairs are surrounded at the base by silicified cells which after the fall of the hairs render the surface rough, as is characteristic of many Boraginaceae. Glandular hairs also occur, often rendering the plant sticky, as in *Senecio viscosus*, a local plant in Britain on waste dry ground. The tribe *Cynareae*, which includes the various genera of thistles and thistle-like plants, is characterised by development of bristles and spines on the edge and surface of the leaves and often also on the stem.

The tribe *Cichorieae* is characterised by a system of laticiferous vessels which accompany the phloem-tissue; they are formed by the disappearance of the transverse walls from longitudinal rows of cells and are also freely joined by cross-unions. In the other tribes oil-containing passages occur in the cortex outside the vascular bundles, running from the root through the stem and generally continued into the leaf.

The carbohydrate reserve-material inulin is dissolved in the cell-sap of the roots and tubers of many members of the family, as *Helianthus tuberosus* (Jerusalem Artichoke), *Inula Helenium* and others.

The capitulum is an indefinite inflorescence, the flowers opening in acropetal succession, the youngest being in the

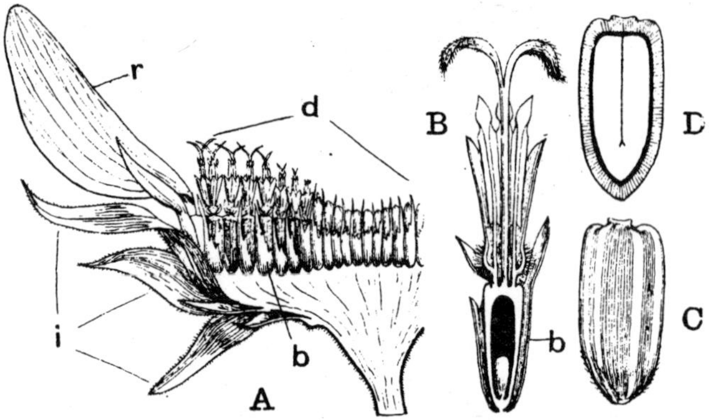

FIG. 271. *Helianthus annuus*. A. Portion of head in longitudinal section. B. Disc-floret in vertical section, × 3. C. Fruit, × 1½. D. Fruit in vertical section. *i*, bracts of involucre; *b*, bract subtending a single floret; *r*, ray-floret; *d*, disc-florets.

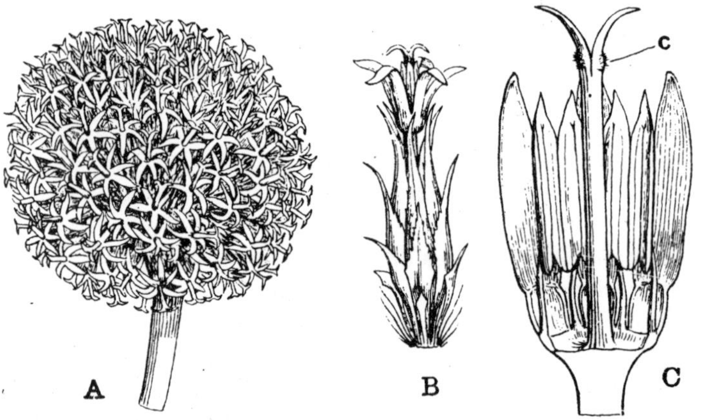

FIG. 272. *Echinops globifer*. A. Compound head, slightly reduced. B. A single head, × 1½. C. Floret, the corolla partly cut away, × 6; *c*, collecting hairs on style.

centre (fig. 271, A). It conveys the impression of, and is to some extent biologically analogous to, a single flower. The involucre of bracts performs the protective function of the calyx in the bud and also subsequently in cases where the heads close periodically, and ultimately in the fruiting stage. The ray-florets play the part of an attractive corolla and the disc-florets the parts of the stamens and pistil. The number of florets in a head varies enormously from several thousands, as in the huge heads of some Sunflowers, to a single flower, as in *Echinops* (fig. 272) and other genera, or in individual species of certain genera, as *Vernonia uniflora*. Single-flowered heads are however generally associated in secondary heads, as in *Echinops*, or in close corymbs. In few-flowered heads the number of florets is often fairly constant.

The shape of the floral or fruit-bearing axis also varies widely. It is usually flat or slightly convex, but may be elongated and spike- or cone-like, as in *Rudbeckia* (fig. 275) and *Speilanthes*, or, on the other hand, deeply concave, as in *Epaltes* (fig. 274). These two extremes are connected by intermediate forms. In *Xanthium*, where the heads are unisexual, the axis grows up round the pair of female flowers which become quite buried, only the styles emerging through an apical aperture (fig. 277, C). The form of the head is correspondingly varied. The axis is solid, or, as in *Matricaria*, *Tragopogon* and others, hollow; in the Sunflower it is filled with loose spongy tissue, in the true Artichoke (*Cynara Scolymus*) it is fleshy. In many of the tribes or subtribes each floret stands in the axil of a bract—the so-called scales or paleae—as *Heliantheae* (e.g. Sunflower, fig. 271), *Anthemidinae*, in others, as *Helenieae*, *Calenduleae* (e.g. Marigold) these bracts are absent. After removal of the flowers or fruits the receptacle is often smooth but in other cases variously pitted or tubercled at the places of insertion of the florets or fruits (fig. 279, E).

The involucre may consist of a very large indefinite number of spirally arranged bracts (200 or even more), or the number is comparatively small and sometimes very definite; thus in the climbing genus *Mikania* (fig. 273) there are four. In genera where the heads are few-flowered and united into secondary heads, the primary heads may have very few bracts while the secondary head has an independent involucre; thus in the subtribe *Angianthinae* of *Inuleae* (mainly Australian) the very small heads are one- to few-flowered, with involucres of two or few bracts; the secondary heads have a common involucre and take the form of a normal capitulum or are elongated into cylindrical spikes giving the plant a remarkable appearance. In the arborescent genus *Wilkesia* (Sandwich Islands) a true involucre is absent and its place

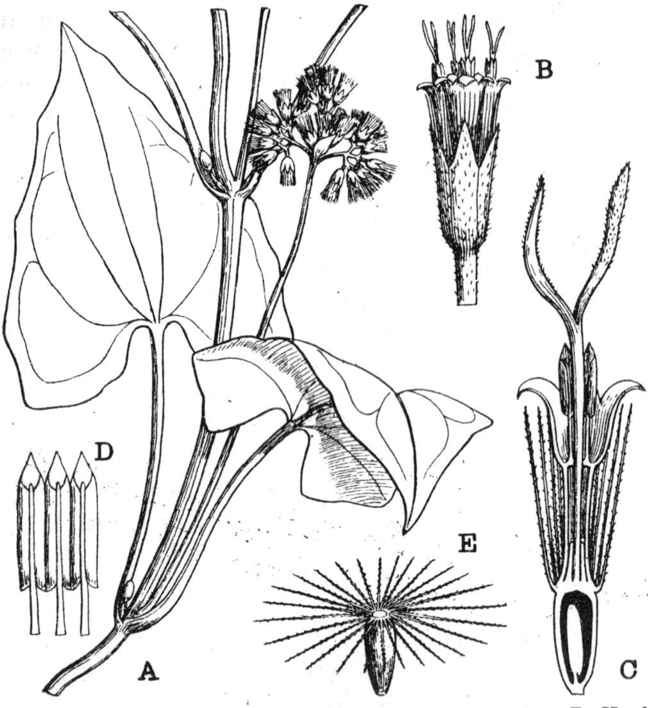

FIG. 273. *Mikania scandens*. A. Portion of plant, ⅔ nat. size. B. Head of flowers, × 4; note the involucre of four bracts. C. Disc-floret in vertical section, × 10. D. Three stamens, × 14. E. Fruit surmounted by pappus, × 6.

is taken by a whorl of true floral bracts, that is those which subtend the individual florets. This is assumed to be the case by a comparison with the allied genus *Argyroxiphium*, in which a true involucre is followed by a series of female flowers, inside which is an inner "involucre" formed from the true floral bracts and enclosing the bisexual flowers. In *Eriothrix*, a monotypic genus with the habit of *Lycopodium*, from Bourbon, there is no true involucre, but the small heads stand at the end of the branchlets, the small closely-crowded narrow leaves of which amply protect the few-flowered heads. Where the involucral bracts are arranged in several series, differences often occur between the members of each series, generally the outer are shorter than the inner; the outer may be so different as to suggest a distinct outer involucre, as in *Senecio*, where, in addition to the upper series or inner involucre, there is in some species an outer involucre of smaller bracts.

The consistence of the involucral bracts also shews great variety. They are leaf-like in Sunflower, tough and leathery in the true Artichoke (*Cynara*), or more or less membranous or scale-like. Frequently they become dry and scarious at the tip or margin or form large dry-membranous white or brightly coloured attractive outgrowths simulating ray-florets as in *Xeranthemum* (Mediterranean), and the Everlastings (*Helichrysum*). In the Carline thistle (*Carlina vulgaris*) the outer bracts are leafy and spreading with spiny-toothed margins, the inner are long, narrow, scarious, coloured and shining, forming against the green background the attractive element of the head. The stiff, bristle-like edges of the bracts of Knapweed (*Centaurea*) and the stiff barbed point in Burdock (*Arctium*) are aids to scattering the fruits. In the section *Calcitrapa* of *Centaurea* the bracts are prolonged into strong stiff somewhat branched spines.

The heads are sometimes solitary, either terminal or axillary. In some cases where a scape springs from a radical leaf-rosette the plant bears only one head, as in some Hawkweeds, e.g. *Hieracium Pilosella* (Mouse-ear Hawkweed). Generally, however, the heads are associated in inflorescences, frequently paniculate, but also racemose and cymose, and often corymbose in form. Reference has already been made to the secondary heads formed by association of a number of small heads and often possessing a common involucre.

The form of the flowers in the head and the distribution of the sexes are also very varied. In the simplest case all the heads are uniform and the individual florets of each head have the same form and the same function; they are then styled homogamous and probably represent the original form from which others have been derived. In the homogamous head the florets are regular as in species of *Senecio*, *Eupatorium* (Hemp-agrimony), or two-lipped (many members of the tribe *Mutisieae*), or ligulate, as in *Cichorium* (Chicory). An early stage in differentiation is represented by cases where, as often in Dandelion, the corollas of the outer florets shew a difference in colour, in this case red on the underside, or are, as in many *Mutisieae*, better developed or shew slight differences in form, thus indicating a transition to the development of distinct ray-florets. In the monotypic Australian genus *Schoenia* all the flowers are bisexual, but only the outer are fertile. In the heterogamous head the disc-florets are bisexual and either fertile or functionally male, while the marginal florets are female and more or less different in form from the disc-florets (rudiments of stamens may be present, as in *Tussilago*, Coltsfoot); in a further stage the ray-florets become merely attractive organs, a pistil is present but no fruit is formed, or the pistil is more or less completely suppressed and the flower becomes structurally and functionally asexual. In a limited number of genera the separation of the sexes is complete, and male and female flowers are borne on distinct heads, either on the same plant, as *Xanthium* (fig. 277) (where the male heads are found at the ends of the branchlets and the female lower down), *Ambrosia*, and other genera of the same tribe, or on distinct plants, as in the large American genus *Baccharis*. *Petasites* (Butter-bur) is subdioecious, the separation of the sexes being incomplete.

Its usual functions having been relegated to the involucre, the calyx is represented by a rudimentary structure or becomes fully developed only in the fruit, and functions as an organ for aiding dispersal. In a few cases there is no trace of a calyx, as in the subtribe *Ambrosiinae* (*Xanthium*, etc.); generally however the calyx develops as a small ring-like swelling round the base of the corolla, which most frequently appears after the corolla and stamens, the position of the sepals being indicated by five thickenings or swellings at points alternating with the petals and opposite the stamens. This type recalls the reduced calyx of Umbelliferae and many Rubiaceae. In most cases a further development takes place to form a pappus, which is only fully developed when the fruit is ripe. The whole edge develops to form a toothed, notched or divided crown, or the pappus takes the form of scales or bristles corresponding in number and position to five sepals (or fewer), as in many members of the *Heliantheae*, for instance *Bidens* (fig. 278, B), where there are two to four barbed bristles, or *Gaillardia*, where there are five (or more) dry

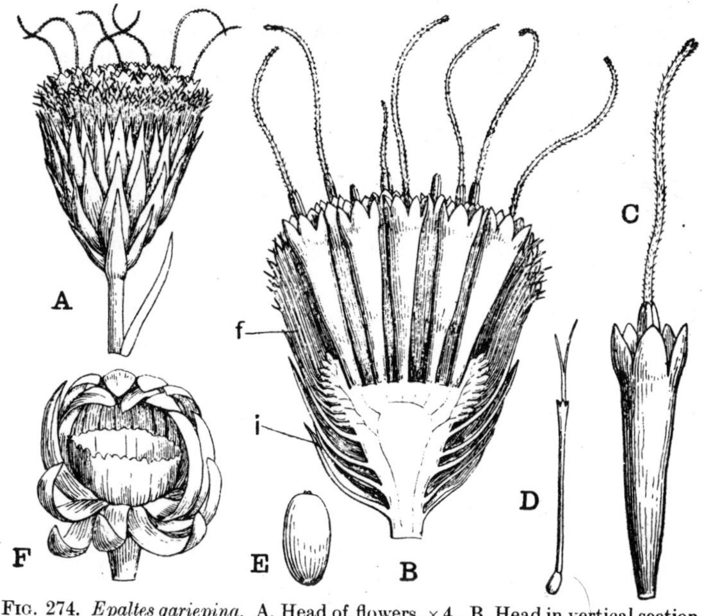

FIG. 274. *Epaltes gariepina*. A. Head of flowers, × 4. B. Head in vertical section, × 2½; *i*, involucre. Several series of female flowers, *f*, are arranged on the outside of the concave receptacle, the large disc-florets have both stamens and pistil but do not produce fruit; they occupy the floor of the receptacle. C. A disc-floret, × 12. D. A female (fertile) floret, × 12. E. Fruit, × 22. F. Receptacle surrounded by involucre of bracts after the fruits have fallen, × 2½.

membranous awned scales; or there are an indefinite number of bristles or hairs (fig. 273, E), sometimes forming several concentric series, often differing in nature, or crowded into a dense tuft. The outgrowths are purely epidermal developments or the underlying cell-layers also contribute to their formation.

It seems more in accordance with comparative morphology to regard the pappus as a development of the calyx, but it has also been regarded as representing merely trichomes or emergences. Much has been written as to the true homology of the pappus and the discussion has been summed up by Eichler (1875), who regards it as a modified calyx, and recently by Small (1919), who concludes from its general trichome-structure that it has no relation to a true calyx. Where, as in many *Vernonieae* and *Cynareae*, there are a large number of hairs or bristles arranged in several series of varying size, the less important series may represent merely epigynous outgrowths.

The regular corolla has a shorter or longer tube passing gradually or suddenly into a broader cylindrical funnel, or it is bell-shaped opening with five teeth. The two-lipped corolla has an upper lip representing two and a lower lip representing three segments; in rare cases the upper lip represents one and the lower lip four segments (as in the South American *Barnadesia*). The true ligulate corolla is deeply split on the posterior side and prolonged on the anterior into a strap-shaped limb made up of all the segments, as indicated by the five teeth at the apex. This may be compared with the arrangement of the segments in *Teucrium* (Labiatae), where the upper lip is deeply cleft and the two lobes are bent towards the lower lip which thus becomes five-lobed. In the so-called falsely ligulate corolla the strap is only formed of three petals, and represents a much-developed lower lip, the upper lip being slightly developed or suppressed, as in *Calendula* (Marigold) (fig. 279, C).

The nervation of the corolla-limb is peculiar. Generally the main nerves do not run, as might be expected from a comparison with the mid-vein of a leaf, to the apex of the segment but towards the base of the notch between two segments (commissural nerves), below which the nerve forks sending a branch right and left to unite below the apex of the segment with the corresponding branch from the nerve on either side (fig. 279, C); in some cases a midrib is also present, but is not connected with the commissural nerves. In Sunflower the broad petals have numerous parallel nerves which anastomose freely.

The stamens are attached to the corolla-tube and alternate

with its segments. The filaments are with few exceptions free; in *Silybum* and other members of the tribe *Cynareae* they are united to form a tube. In the anemophilous subtribe *Ambrosiinae* the anthers are quite free or only loosely united (*Xanthium*, fig. 277); but the cohesion in the family generally is not very strong as the anthers may separate when the flower withers. The base of each anther-cell is sometimes rounded (tribe *Eupatorieae*), but is often produced into a short or long tail, which may unite with the adjacent tail of the next anther, and serve to protect the nectary round the base

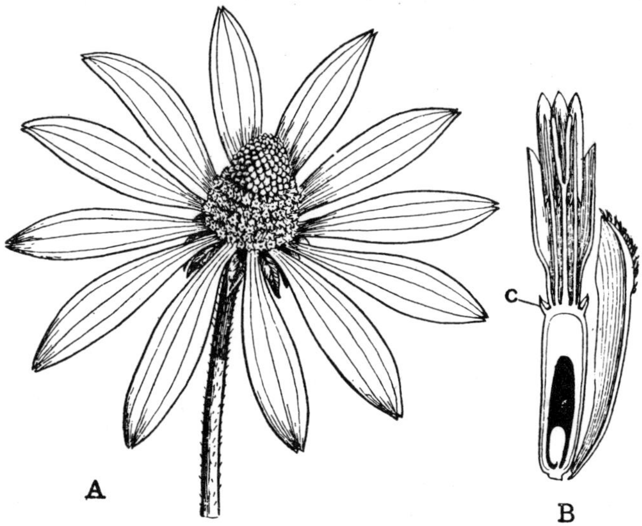

FIG. 275. *Rudbeckia pinnata*. A. Head of flowers, ⅔ nat. size; the disc-florets are bisexual and arranged on a convex receptacle. B. Disc-floret with subtending bract, in vertical section, ×4; *c*, scale-like pappus.

of the style from access of water, and also more effectively to close the end of the staminal tube and thus avoid waste of pollen (see Small, p. 37). The connective is with few exceptions prolonged above the anther into a broad blunt appendage. The filaments are sometimes sensitive, as in *Centaurea* and others, and shorten on being touched (fig. 276).

The ovary is formed from two median antero-posterior carpels which unite at the edges; the single ovule springs from the floor of the chamber, slightly below the centre, in front of the middle line of the anterior carpel (figs. 271, 275, B).

Except in asexual flowers a long cylindrical style is present, surrounded at the base by a ring-like or shortly tubular nectary—the nectary is absent in many male flowers. The style generally divides at the apex into a pair of longer or shorter style-arms on the inner face of which are borne the stigmatic papillae; on the outer side are the pollen-collecting hairs, at first appressed but afterwards spreading. The form of the style-arms and the distribution of the stigmatic papillae and the collecting hairs shew remarkable differences in detail which afford characters for distinguishing the various tribes and also genera. In species of *Arctotis* the style is sensitive to touch.

The remarkable success of the Compositae, as evidenced by its great numerical preponderance in genera and species over other families, the abundance of many of its species, and its world-wide distribution, is at any rate in part due to the admirable adaptation of its flowers for cross-pollination by a great variety of insects. H. Müller points out that Umbelliferae alone rank with Compositae in regard to variety of insect-visitors, but while in Umbelliferae the nectar lies exposed to the rain on the epigynous disc and is thus suitable only for short-tongued insects, in Compositae it is secreted at the base of a narrow tubular corolla, and, as it accumulates, rises into the wider part of the corolla, where it then becomes accessible to the most short-tongued insects, but is protected by the anthers from rain. It is thus accessible not only to short-tongued insects but may be sucked by bees and Lepidoptera, and observation shews that most Compositae are visited to a greater extent or even principally by the most specialised orders of insects[1].

In addition to the plentiful supply of easily accessible nectar and its protection from rain, other factors are the close association of the flowers and the possession of a pollen-mechanism which ensures cross-pollination in the event of insect-visits. Aggregation renders the flowers more conspicuous, and this may be increased by the florets being directed outwards, as in *Centaurea* (Cornflower, Knapweed), or by the ligulate character of the corolla which becomes exaggerated towards the margin of the head, or by the development of distinct ray-florets, or sometimes, as in *Carlina*, by an inner series of conspicuous involucral bracts.

The pollen-mechanism is similar to that already described for Lobelioideae, but shews greater elaboration and variety. The

[1] For an elaborate account of pollination-methods in this family, see H. Müller *Fertilisation of Flowers*.

coherent anthers form a hollow cylinder, and dehisce introrsely, filling the cylinder with pollen before the flower opens. The style-arms are at first closely applied face to face in the lower part of the pollen-cylinder, and, as the style grows, brush the pollen out of the tube by means of the hairs on their outer surface; ultimately they separate and expose their stigmatic surfaces. The hairs form a simple ring round the style at the base of the arms, as in *Centaurea* (fig. 276) or *Cirsium*, or are aggregated in a tuft at the ends of the arms, as in *Achillea* and *Chrysanthemum*; in

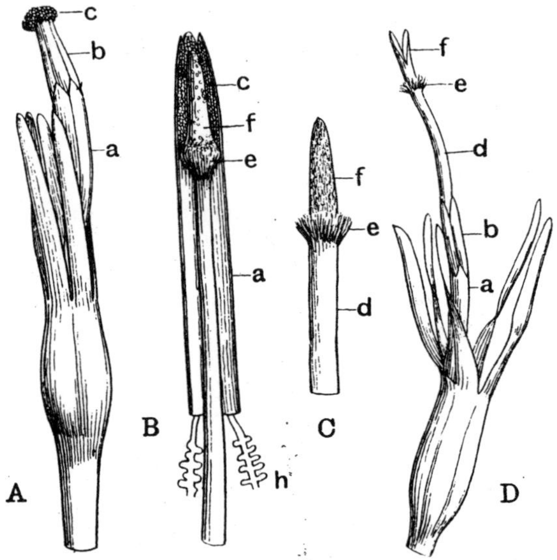

FIG. 276. *Centaurea Jacea.* A. Upper part of flower in first stage, ×6. B. Section of anther-cylinder before pollen has been shed, ×9. C. End of style removed from anther-cylinder, ×9. D. Upper part of flower in second stage after removal of the pollen; the style has elongated and carried up the now expanding stigmas (*f*). *a*, anther-cylinder; *b*, appendages of anthers which at first close the top of the anther-cylinder; *c*, pollen; *d*, style; *e*, ring of sweeping hairs; *f*, stigmas; *h*, teeth on filaments which guard the nectar. (After H. Müller.)

these cases they sweep the pollen before them as with a brush. In others, as Dandelion or *Leontodon*, the ultimately projecting surface of the style is covered with hairs, among which the pollen-grains become entangled and are thus carried out (see also fig. 271, B). In some cases, as in Dandelion, in absence of insect-visitors self-pollination is rendered possible in bisexual flowers by the style-arms ultimately curving backwards till their papillae come in contact with the pollen; but the general perfection of the pollen-mechanism is illustrated by the fact that the power of self-pollination has to a great extent been lost. On the other hand, in some species with inconspicuous heads, such as *Senecio vulgaris* (Groundsel), where the small heads have no conspicuous marginal florets and are very sparingly visited by insects, self-pollination takes place regularly; the pollen-grains swept out by the hairs at the tip of the style remain partly on the edge of the stigmas and partly fall upon their inner faces when they diverge.

Small has studied in detail the pollen-presentation mechanism and the correlation between the form of the anther-tube, as determined by the presence and degree of development of the prolongations of the connective above the anther and of the basal appendages, and the position and character of the hairs on the style-arms. He regards these as the expression of a tendency to economy of pollen, the appendages providing for a sufficient length of pollen-tube without encroaching on the polleniferous portion. An additional feature is the frequent irritability of the anther-tube in response to a touch. This had long been known in several genera, *Centaurea* (fig. 276) and others; when touched by an insect's proboscis the filaments contract and by drawing the anther-tube downwards squeeze some pollen out at the upper end at exactly the right moment (fig. 276, A). The list was considerably extended by Juel; and the phenomenon was shewn by Small to be of very general occurrence, some form of irritability being observed in 253 out of 360 species and varieties examined (i.e. 70 per cent.). The list of irritable species includes all the tribes and the majority of the subtribes of the family. In a few cases the pollen was merely presented at the mouth of the anther-tube on touching, but in the majority there was also a lateral movement of the anther-tube, generally towards the touch; in a few cases (*Gerbera* and *Perezia*) the pollen was extruded with almost explosive rapidity. Irritability of the style has been recorded only in three genera of the tribe *Arctotideae* (South Africa). The mechanism is described in detail by Small[1]. The closely appressed style-arms emerge in the first (male) stage of the floret covered with pollen and if touched move quickly in the direction of the touch. Recovery is rapid and irritability is regained in less than half-a-minute.

The marked proterandry of the florets, associated with the centripetal development in each head, favours crossing between separate inflorescences. In cases where the development of the bisexual florets proceeds slowly from the margin inwards, an insect alighting on the head at the margin will, in early stages of the head, visit pollen-bearing florets only, and in later stages of

[1] *New Phytologist*, XIV, 216 (1915).

the head will visit florets in the second or female stage before it reaches those in the male stage (as in *Bellis*, *Chrysanthemum*, etc.); in many *Cynareae* (as *Carduus*), however, the florets open in such quick succession that the head is for a time purely male and afterwards for some time purely female. In the comparatively few cases where the florets are all unisexual, the outer and first visited are female and the inner functionally male, as in *Calendula* (Marigold) (fig. 279), or the two kinds occur in separate heads, thus ensuring cross-pollination.

A further development of the condition referred to in the Groundsel, where the inconspicuous flowers depend almost entirely on self-pollination, is illustrated in *Artemisia* and allied genera, in

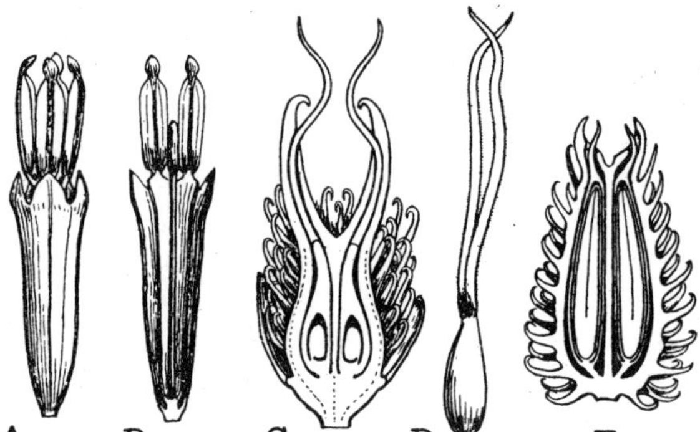

A B C D E

FIG. 277. *Xanthium orientale*. A. A male flower, × 12. B. The same in vertical section. C. A female head, in vertical section, the two flowers are completely enveloped in the receptacle, × 4. D. Female flower, × 4. E. Head in fruit in vertical section, slightly enlarged. The female receptacle is covered on the outside with stiff hooked spines. (C, D, E after Baillon.)

which many species have become adapted for wind-pollination. Delpino has traced the steps by which this may be assumed to have occurred, through suppression of the corolla of the female flower and elongation of the style, the pollen at the same time becoming dry and powdery. The small subtribe *Ambrosiinae* (*Xanthium*, fig. 277) is completely anemophilous, the flowers and frequently the capitula are unisexual; the corolla of the female flowers is more or less suppressed and the stigmas are much elongated; the functionally male heads are small and pendulous, the corolla of the male florets has a short tube, the anthers are slightly associated or quite free, the pollen is often quite smooth, and there is no nectary. Goebel finds evidence of an insect-pollinated ancestry in the frequent spinose thickening of the extine of the pollen-grain and in the hairiness of the rudimentary stigma of the male flower.

The fruit is rarely somewhat fleshy and varies in shape from cylindrical to inverted pyramidal or ovoid or compressed; the pericarp is traversed by two to five nerves, which may project as ribs or wings. There are numerous contrivances for ensuring distribution. During the ripening of the fruit the head is enclosed by the involucral bracts which when the fruit is ripe stand erect or bend right back; this movement may be checked by damp weather, which would be unfavourable for scattering the fruits. In *Sonchus oleraceus* (Sow-thistle) the fruits are pressed out by the bending in of the bracts, the lower parts of which become ultimately applied to the receptacle, and Hildebrand has described a similar method in the large globose heads of *Silybum Marianum* (Milk-thistle). The hollow broad club-shaped peduncle of *Zinnia*, *Arnoseris* and others favours movement of the ripe heads by the wind. The wind also scatters the fruits from the drooping heads of *Artemisia*, in which the fruits are detained in still air by the withered perianths which are pressed together between the involucral bracts. In *Gorteria*, a small South African genus of annual herbs, the few-flowered terminal heads are enveloped in the fruit by spine-pointed bracts, and are distributed without breaking up; the spines serve to anchor the head in the soil, and a single seed germinates, the root growing downwards through the hole formed by the detachment of the head from the peduncle, and the stem upwards between the bracts; as the fruit does not leave the head the pericarp remains thin.

Direct aids to distribution by the wind are found in the chaffy character of the small fruits, as in many *Anthemideae*, or a covering of loose woolly hairs on the fruit itself (*Tarchonanthus*) or on the bract (*Eriocephalus*); or more generally in the parachute development of the pappus, which also favours distribution by water, as the bubbles of air imprisoned between the hairs act as a float. The variety in the form of development of the pappus supplies characters for the distinction of genera or smaller groups of species; the hairs spring directly from the top of the fruit, as in the Thistles, or are carried up on a slender beak-like development of the fruit, as in Dandelion; or the individual hairs may be smooth, or rough or feathery, sometimes, as in *Tragopogon* (fig.

278, A), they are woven into a delicate web by transverse connections. In other cases, as *Gaillardia*, the pappus consists of a whorl of broad light scales. Wing-like developments sometimes occur on the sides of the fruits, as in *Verbesina* (two-winged) or *Tripteris* (three-winged) or the bract subtending the female flower forms an enveloping wing, or, as in the female flowers of *Zinnia*, the flat ligular corolla persists in the fruit as a broad wing. The fruit of *Diotis maritima* (Cotton-weed), which is widely distributed on the shores of the Mediterranean and of the Atlantic from Britain (rare) to the Canaries, floats by means of a spongy development of the persistent corolla-tube, which grows downwards and surrounds the greater part of the fruit.

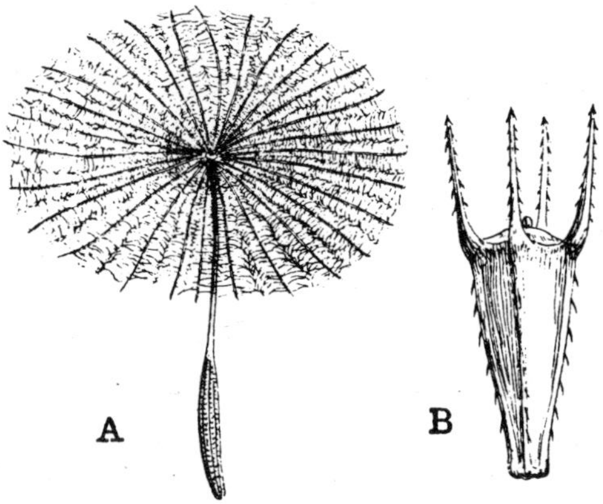

FIG. 278. Fruit of A, *Tragopogon major,* slightly reduced; B, *Bidens cernua,* × 5.

In other cases spine- or hook-developments ensure distribution by means of the fur of mammals or the feathers of birds. Such are the few stiff many-barbed pappus-bristles of *Bidens* (fig. 278) and allied genera, the hooked spines covering the outside of the involucral bracts of *Xanthium* (fig. 277, E) or the familiar stiff spreading hooked tips of the bracts of Burdock, which cause the whole head to be carried off. In *Tragoceros*, a small Mexican genus allied to *Zinnia*, the corolla of the female flower persists in the fruit and becomes bent back, forming a stiff simple or double hook. In *Siegesbeckia orientalis*, a species widely distributed in the warmer regions of the whole world, the involucral and smaller floral bracts bear sticky glandular hairs to which individual fruits adhere and are thus transported. A very exceptional case is afforded by the small tropical American genus *Milleria*, where a large bract encloses the small head, containing only one female flower, and becomes fleshy in the fruiting stage and is presumably attractive to birds.

In some genera the fruits are polymorphic on the same head, as is well seen in the common Marigold (*Calendula officinalis*) (fig. 279), where the outermost are much elongated, with small spines on the back, often beaked at the tip, and with a process

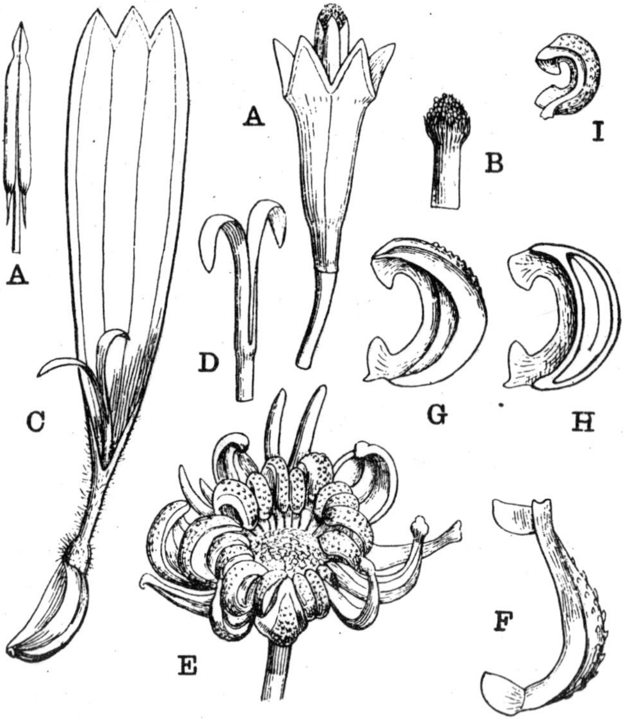

FIG. 279. *Calendula officinalis.* A. Disc-floret, pollen-bearing, × 8. A′. One anther from same. B. Infertile stigma of disc-floret, × 16. C. Ray-floret, pistillate, × 8. D. Top of style of same, × 16. E. Head of fruits, the sterile disc-florets have dropped, slightly enlarged. F. An outer fruit. G. A fruit of the second type. H. Same in longitudinal section. I. One of the innermost fruits. F–H nearly three times nat. size.

near the base; the next set have broad wings inrolled at the margin and are without a beak and the innermost are much incurved, often forming a ring, and have also narrow involute wings. Pethybridge has recently noted a difference in shape between a few of the outer fruits and the rest, in the head of *Picris echioides*.

Many members of the family are of economic or medicinal value from the presence of ethereal and fatty oils, resins and bitter principles. Among those of medicinal value are *Anthemis nobilis* (Chamomile), *Arnica montana*, species of *Artemisia* (Wormwood) and *Tussilago* (Coltsfoot). The seeds of some *Heliantheae* are rich in a fatty oil, especially Sunflower (*H. annuus*), which is cultivated for this in various parts of Europe, Egypt and India. Insect-powder is prepared from *Chrysanthemum roseum* and *carneum*; *Adenostemma tinctorium* and *Carthamus tinctorius* (Safflower) yield colouring matters. The young leaves of many species are eaten as salad; such are Lettuce (*Lactuca sativa*), Endive (*Cichorium Endivia*), Dandelion, etc. The young shoots of *Scolymus hispanicus* and *S. maculatus* are eaten in the south of Europe as asparagus, and the roots of species of *Scorzonera*, the tubers of *Helianthus tuberosus* (Jerusalem Artichoke), and the heads of *Cynara Scolymus* (Artichoke) as vegetables. The ground roasted root of *Cichorium Intybus* is Chicory, the common adulterant of coffee. Ornamental plants are numerous, *Dahlia*, *Aster* (including the Michaelmas Daisy), *Ageratum*, *Coreopsis*, *Zinnia*, *Helianthus*, *Calendula*, *Tagetes* (French Marigold), *Gaillardia*, *Helichrysum* (Everlastings or Immortelles) and many others.

The family is subdivided into 13 tribes included in two groups.

I. TUBULIFLORAE (including the Labiatiflorae). Laticiferous vessels absent, schizogenous oil-passages frequent. Corolla of the disc-florets not ligulate. Tribes 1–12.

II. LIGULIFLORAE. Anastomosing laticiferous vessels present, oil-containing passages very rare. Corolla of all the flowers ligulate. Tribe 13, *Cichorieae*.

Tribe 1. *Vernonieae*. Heads homogamous; corolla tubular and regular with five narrow lobes, never yellow; anthers at the base sagittate, blunt, or more rarely short-tailed, with filaments inserted high above the base; style generally deeply divided into two long semi-cylindrical pointed limbs; pappus usually setose and copious. Herbs, shrubs or small trees with generally alternate leaves. Mainly American, especially tropical; less numerous in Africa (many species of *Vernonia*) and Asia; a few in Australia; not represented in Europe. *Vernonia* (700 species); *V. cinerea* occurs throughout the tropics.

Tribe 2. *Eupatorieae*. Heads homogamous; corolla tubular and regular with generally five short teeth, never pure yellow; anthers blunt at the base, basifixed; style-arms long, semi-cylindrical, sometimes club-shaped or flattened at the tip, with collecting hairs on the outside, also on the inside of the limbs above the stigmatic surfaces, which form a pair of marginal lines; pappus often of 5 to ∞ bristles or of scales, sometimes absent. Herbs or shrubs. Leaves opposite or alternate, sometimes whorled. The large genus *Eupatorium* (over 800 species) has a few representatives in the Old World; *E. cannabinum* (Hemp-agrimony) is British. *Ageratum conyzoides* is a widely distributed weed throughout the tropics. *Mikania* (fig. 273), 200 species mostly Brazilian, some twining.

Tribe 3. *Astereae.* Heads heterogamous with ray-florets female or neuter and disc-florets bisexual or male, rarely homogamous; corolla of all the florets or of the disc-florets regular; anthers as in tribe (2); style-arms flat, stigmatic surfaces forming a pair of strongly marked marginal lines above which is a shorter or longer portion densely covered on the outside with collecting hairs; pappus various. Mostly herbaceous perennials, or sometimes annuals, but a few southern or insular genera are shrubby, as *Felicia* (South Africa), *Olearia* (Australasia), *Tetramolopium* (Sandwich Is.). Distribution world-wide but more in the temperate than in the tropical zones; about twice as many species in the New World as in the Old. *Solidago Virgaurea* (Golden-rod) occurs throughout the north temperate zone; many species are aggressive colonists, such as *Bellis perennis* (Daisy), a European species which has become naturalised in North America, and American species of *Aster*, *Erigeron* and *Solidago* similarly in Europe. *Erigeron canadensis*, a common North American weed, has become widely diffused throughout the world. British genera, ***Aster***, ***Erigeron***, ***Bellis***, ***Solidago***. ***Olearia***, **100 species, Australasian.** ***Conyza***, **60 species, tropical.** ***Baccharis***, **350** species, tropical American, has dioecious or polygamo-dioecious heads.

Tribe 4. *Inuleae.* Heads heterogamous or homogamous; corolla of all, or of the disc-florets only, regular, with four or five short teeth or lobes, usually yellow. Anthers tailed; style-arms various, without terminal appendages; pappus usually setose with simple or plumose bristles. Habit various; leaves generally alternate, often with woolly hairs, sometimes heath-like. Widely distributed. *Tarchonanthus*, a few south and tropical African species of small dioecious trees with leaves smelling of camphor. British genera, ***Inula*** (100 species, Europe, Asia, Africa), ***Pulicaria*** (30 species, mainly Mediterranean), ***Gnaphalium*** (150 species, temperate and subtropical), *Filago* (12 species, north temperate of Old World). *Leontopodium alpinum* is Edelweiss. *Raoulia*, mainly New Zealand, forms thick cushion-like growths on the mountains ("Vegetable sheep"). *Helichrysum* (Everlastings), 350 species, especially South African. *Odontospermum* *pygmaeum* (Rose of Jericho) is a small shrubby plant in which the involucral bracts close over the fruiting head in the dry and rapidly open in the damp, to allow the fruits to escape under circumstances favourable for germination. *Epaltes* (fig. 274), a small genus in the warmer parts of both hemispheres; several series of female flowers occur on the outside of the head; the disc contains bisexual but usually sterile florets.

Tribe 5. *Heliantheae.* Heads heterogamous, with female or neuter ray-florets and fertile or sterile bisexual disc-florets; more rarely homogamous. Involucral bracts often imbricated in several rows and more or less herbaceous without scarious edges; florets subtended by scale-bracts; corolla of disc-florets regular with five or rarely four short lobes, generally yellow, of female and neuter florets ligulate and trimerous, of the same colour as the disc-florets; anthers blunt or rounded at the base or acute, without tails, basifixed; a crown of collecting hairs is present on the style-arms, and is sometimes continued downwards, as in *Helianthus*; pappus various but not hairy, generally of two or three rigid awns or scales corresponding with the principal ribs of the fruit. Usually rather coarse herbs or shrubs, rarely trees, with usually opposite leaves. A large tribe, mainly American, though some of the American genera are represented in the Old World, such as *Bidens* (100 species), widely distributed in temperate and tropical countries (two British). Some American species have become widely naturalised, such as *Galinsoga parviflora*, a small annual herb with small yellow heads, native of Peru, which has spread widely in the warmer and temperate parts of the world and is found in cultivated fields and by road-sides in Surrey and Middlesex; and *Xanthium spinosum*, probably of South American origin, but now widely distributed. A small number of genera are restricted to limited areas in the Old World. There are 10 subtribes. The *Ambrosiinae*, as already noted, are anemophilous forms with small few-flowered heads and reduced flowers; they are mainly American but *Ambrosia* and *Xanthium* occur also in the Old World. *Zinnia*, 15 species, North America and Mexico; *Z. elegans* is a garden plant. *Helianthus*, 100 species, American; *H. annuus* (Sunflower) is a native of Mexico; *H. tuberosus* (Jerusalem Artichoke) of North America. *Coreopsis*, 100 species, mainly American. *Dahlia*, nine species, Mexico. *Rudbeckia* (fig. 275), 30 species in North America.

Tribe 6. *Helenieae.* Resembles the last tribe, but floral bracts are not present on the receptacle. Mainly Mexican and Pacific North American. *Helenium, Gaillardia, Tagetes* are met with in gardens.

Tribe 7. *Anthemideae.* Resembles *Heliantheae,* but the involucral bracts have a scarious margin. Pappus absent or reduced to a ring or cup. Generally herbs, sometimes shrubs, with generally alternate often pinnately divided leaves. Includes two large subtribes; *Anthemidinae,* with floral bracts on the receptacle, and *Chrysantheminae* without floral bracts. The former are mainly Old World: *Anthemis,* 130 species, Europe and Mediterranean, has three British, including *A. nobilis* (Chamomile); *Achillea,* 130 species, north temperate, has two British—*A. Millefolium* (Yarrow) is widely distributed in north temperate regions. To the second subtribe belong *Matricaria,* 50 species, mostly in Europe (two British), Africa and Western Asia; *M. suaveolens,* a North American species, has recently spread throughout Britain; *Chrysanthemum,* 200 species, mainly north temperate Old World, several widely naturalised; British are *C. segetum* (Corn Marigold), *C. Leucanthemum* (Ox-eye Daisy), *C. Parthenium* (Fever-few); *C. indicum* and *sinense,* natives of China and Japan, are the source of the cultivated Chrysanthemums; *Artemisia,* 200 species, north temperate zone, four British; *Tanacetum* (Tansy) (British).

Tribe 8. *Senecioneae.* Flowers homogamous or more frequently heterogamous. Involucre generally of a series of similar bracts, sometimes more or less united below into a cup, with frequently an additional outer series of shorter scales; receptacle generally without floral bracts; style and anthers as in *Heliantheae,* but pappus hairy. Herbs, shrubs or sometimes trees, with generally alternate leaves. Widely distributed. *Tussilago* (north temperate Old World, introduced in North America) (Coltsfoot). *Petasites* (14 species, north temperate); *P. hybridus* is British. *Arnica. Doronicum,* 25 species, north temperate Old World. *Cineraria,* 25 species, mainly South African. *Senecio,* about 2500 species, world-wide; *S. cruenta,* a native of the Canary Islands, is the source of the greenhouse Cinerarias.

Tribe 9. *Calenduleae.* Heads heterogamous, with female, generally ligulate, ray-florets and sterile regular yellow or orange disc-florets; floral bracts absent; anthers pointed at the base (fig. 279, A′); pappus absent. Herbs or shrubs with generally alternate leaves. A small tribe, mainly South African. *Calendula,* 20 species, Mediterranean; *C. officinalis* is the African Marigold of gardens.

Tribe 10. *Arctotideae.* Heads heterogamous, with ligulate, female or sterile ray-florets; anthers blunt-pointed at the base; pappus absent or if present not hairy. A small tribe, mainly South African, also mountains of tropical Africa. *Arctotis. Gazania.*

Tribe 11. *Cynareae.* Thistle group. Heads homogamous, or with neuter, rarely female, not ligulate ray-florets. Involucre of many series, increasing in length inwards, often thorny; receptacle covered with numerous bristles; anthers generally tailed; style thickened beneath the arms and bearing the collecting hairs there (fig. 272, C), or with a dense crown of collecting hairs beneath the arms; pappus generally hairy. Generally herbs with alternate prickly or spiny leaves. The chief centre is the Mediterranean region. *Echinops* (fig. 272) (70 species, Europe, Asia, north tropical Africa) has one-flowered heads grouped in secondary heads. *Carlina*, 20 species, Canary Islands and Europe to Central Asia; *C. vulgaris* (British, Carline-thistle). *Arctium*, six species (British), Europe, temperate Asia (Burdock). *Carduus*, 100 species, Europe (British), Asia, West Africa, the true Thistles. *Cirsium*, north temperate (British). *Onopordon*, Europe (British), N. Africa, West Asia, (Cotton-thistle). *Cynara*, 11 species, Mediterranean; *C. Scolymus* (Artichoke). *Silybum* (Milk-thistle), West Europe (British), Mediterranean. *Serratula tinctoria* (Saw-wort) (British) yields a yellow dye. *Centaurea*, 500 species, mostly north temperate, especially Mediterranean; five British; *C. nigra* (Knapweed), *C. Cyanus* (Blue-bottle or Cornflower), *C. Calcitrapa* (Star-thistle). *Carthamus tinctoria* (Safflower), Mediterranean, the flowers yield a red dye.

Tribe 12. *Mutisieae.* Heads homogamous or heterogamous. Ray-florets bilabiate or absent, rarely ligulate; disc-florets regular with a deeply divided limb, or bilabiate. Mainly developed in the South American Andine region. *Mutisia*, 60 species, South American, generally shrubby climbers with leaf-tendrils. *Gerbera*, 40 species, perennial herbs, warmer parts of Africa and Asia, one in Tasmania.

Tribe 13. *Cichorieae.* Liguliforal. Heads homogamous; corolla ligulate, truncate and five-toothed; anthers sagittate at the base; style-branches slender, papillose. Achenes usually narrow or flat, and sometimes produced into a slender beak bearing the pappus, which has usually one or more rows of simple or plumose bristles. Herbs, rarely shrubs; trees only in two allied genera, *Dendroseris* (Juan Fernandez) and *Fitchia* (South Sea Islands); with generally yellow flowers. Widely distributed, but especially Old World, and less numerous in the southern hemisphere. Many genera in the Mediterranean region; others are localised in Western America and Mexico. *Cichorium*, eight species, north temperate Old World, especially Mediterranean; *C. Intybus* (Chicory) is British, *C. Endivia* (Endive), Mediterranean. *Lapsana*, nine species, north temperate; *L. communis* (Nipplewort), British. *Crepis*, 200 species, mainly north temperate Old World (British). *Hieracium* (Hawkweeds), 400, mainly north temperate and andine, a remarkably polymorphic genus (British). *Hypochaeris* (Cat's-ear), 50 species, temperate (British). *Leontodon* (Hawk-bit), 50 species, Europe (British), Central Asia, Mediterranean. *Taraxacum* (Dandelion), 25 species, mainly temperate and cold zone of Europe (British) and Asia. *Lactuca*, 100 species, mainly Old World (British); *L. sativa* (Lettuce). *Sonchus* (Sowthistle), widely distributed in the Old World (British), one in New Zealand. *Tragopogon* (Goat's-beard), 40 species, Europe (British) to Central Asia, Mediterranean; *T. porrifolius* (Salsify) has edible roots. *Scorzonera*, 100 species, from Central Europe and the Mediterranean to Central Asia; the roots of *S. hispanica* are eaten like Salsify.

As regards the relative antiquity of the tribes of Compositae, Bentham considered that the *Heliantheae* contained the more primitive characters[1], such as the more leaf-like outer bracts of the involucre, the bracts subtending the flowers more normally developed and more firmly attached, the calyx-limb less transformed, consisting frequently of persistent teeth or awns directly continuous with the ribs of the ovary, and the anthers in some genera less firmly united. On the other hand, the *Cichorieae* suggest the most recent development in that the uniformity of structure of flower and fruit is greater than in any other tribe, the pappus is of a type shewing the least resemblance to a calyx-limb and the receptacular scales when present are among the least like ordinary bracts. In his classical essay on the Compositae Bentham suggests the course of development of the various tribes from a primitive form having regular gamopetalous flowers with an inferior ovary, isomerous and probably pentamerous calyx, corolla and stamens, a bicarpellary pistil but with the ovary reduced to a single cell with one erect ovule, the seed without endosperm and enclosed in an indehiscent pericarp. Such a flower-form resembles that of present-day Rubiaceae, but with a reduction in the gynoecium, and might well be an offshoot of the same group or plexus to which Rubiaceae and allied families owe their origin. By gradual consolidation the bracts have crowded round the condensed flowers and usurped the functions of the calyx-limb which has become reduced or transformed for the new duty of fruit-dispersal; the corollas have become contracted or the outer ones variously developed in form and colour to assist in the process of cross-pollination, and the anthers have become

[1] Hutchinson has recently expressed the same opinion.

united, as in the Lobelioideae section of Campanulaceae, and the styles modified to assist in the discharge of the pollen. The reduction of the bisexual to the unisexual flower may have preceded or followed some or all of these changes. To account for the present day distribution of the various tribes, Bentham suggests that several of these changes had taken place at a very early period, previously to the disruption or stoppage of communication between what are now the tropical regions of the globe, and that several important modifications existed, such as (1) the regular and uniform tubular development of the corolla, accompanied by more or less suppression of the inner bracts and of the normal calyx-limb and substitution of a pappus; (2) the reduction of the corolla-limb, attended frequently by a sexual dimorphism, and oblique development of the outer female corollas; and (3), perhaps at a later period, the uniform unilateral development of the whole of the corollas, accompanied usually by a suppression of the inner bracts and conversion of the calyx-limb into the pappus. From the first of these forms would have sprung the *Eupatorieae* in America, the *Vernonieae* (probably in the New World, but early spreading into the Old World), the *Cynareae* in the northern and the *Mutisieae* in the southern hemisphere; from the second form would have arisen, first, the *Heliantheae*, originating probably in the New World and spreading into the Old; secondly, the *Helenieae* in America, and the *Anthemideae* in the Old World; and thirdly the *Astereae*, the *Senecioneae* and the *Inuleae*. From the third form have developed the *Cichorieae*.

More recently Small has suggested that *Senecio* was the first genus of the Compositae to come into existence and that from it are derived directly or indirectly all the other genera of the family. He maintains that this theory is supported by the details of floral morphology, physiology and cytology, as well as by the geographical distribution of the genus. But the genus *Senecio* represents a stage of specialisation already well advanced and it seems probable that many less specialised forms existed from which the present day Compositae have arisen. Bentham's indication of the nature of these forms seems to offer a more philosophical explanation.

Remains of the Compositae, mainly fruits, are recorded from Tertiary deposits extending as far back as the lower Oligocene. They indicate the presence of Compositae resembling those of genera existing at the present day.

REFERENCES

BENTHAM, G. "Notes on the Classification, History and Geographical Distribution of Compositae." *Journ. Linn. Soc. (Bot.)*, XIII, 1873.

HOFFMANN, O. Engler and Prantl, *Die naturlichen Pflanzenfamilien*, Teil IV, Abt. 5, 1894.

LAVIALLE, P. "Recherches sur le développement de l'ovaire en fruit chez les Composées." *Ann. Sci. Nat. (Bot.)*, sér. 9, XV, 1912.

SMALL, J. "The Origin and Development of the Compositae." *New Phytologist*, reprint, 1919. Includes an extensive bibliography.

HUTCHINSON, J. *Gardeners' Chronicle*, LIX, 305 (1916).

GOEBEL, K. *Trans. and Proc. Bot. Soc. Edinburgh*, XXVI, 60 (1913).

PETHYBRIDGE, G. H. *Irish Naturalist*, XXVIII, 25 (1919).

REFERENCES TO GENERAL WORKS

In the Historical Introduction (vol. I) reference is made to some of the general works on the classification of flowering plants. The following may be added as bearing generally on the group or on large sections of it.

BESSEY, C. E. "The Phylogenetic Taxonomy of Flowering Plants." *Ann. Missouri Botanical Garden*, II, 109. (1915.)

EICHLER, A. W. *Blüthen-Diagramme.* 2 vols. (1875–1878.) The system developed by Engler is based on that of Eichler.

ENGLER, A. and PRANTL, K. *Die Pflanzenfamilien. Nachträge*, I–IV. (1897–1915.) Supplementary to the main work and supplying references to later literature. A new edition of the *Pflanzenfamilien* by Dr Engler, continued by Dr H. Harms, is in progress.

ENGLER, A. *Das Pflanzenreich.* A series of Monographs of the families of Flowering Plants, in course of publication. (1900– .)

ENGLER, A. *Syllabus der Pflanzenfamilien.* Ed. 11. Edited by Dr L. Diels. (1936.)

HALLIER, H. "Provisional Scheme of the Natural (Phylogenetic) System of Flowering Plants." *New Phytologist*, IV, 151. (1905.)

HUTCHINSON, J. *The Families of Flowering Plants.* 2 vols. (1926, 1934.)

LOTSY, J. P. *Vorträge über Botanische Stammesgeschichte.* Vol. III. (1911.)

PAYER, J. B. *Traité d'Organogénie de la Fleur.* (1857.) Text and Plates. A description of the structure and development of the flower, arranged systematically.

VAN TIEGHEM, PH. and COSTANTIN, J. *Éléments de Botanique.* Ed. 5. (1918.)

WARMING, E. *Frøplanterne (Spermatofyter).* (1915.) A revised account of the Flowering Plants as given in his *Handbook of Systematic Botany*; translated by M. C. Potter. (1895.)

WERNHAM, H. F. "Floral evolution with particular reference to the Sympetalous Dicotyledons." *New Phytologist*, XI. (1912.)

WETTSTEIN, R. R. v. *Handbuch der Systematischen Botanik.* 2 vols. Ed. 3. (1923, 4.)

WILLIS, J. C. *Flowering Plants and Ferns.* Ed. 4. (1919.) A handy descriptive dictionary of families and genera.

SOLEREDER, Hans. *Systematic Anatomy of the Dicotyledons.* Translated by L. A. Boodle and F. E. Fritsch. 2 vols. (1908.)

CHALK, L. "The Phylogenetic Value of certain Anatomical Features of Dicotyledonous Woods." *Annals of Botany.* N.S. I, 409. (1937.)

MUELLER, H. *The Fertilisation of Flowers.* English edition by D'Arcy Thompson. (1893.)

KNUTH, P. *Handbook of Flower Pollination.* English translation by J. R. Ainsworth Davis. 3 vols. (1906–1909.)

WODEHOUSE, R. P. *Pollen Grains: their Structure, Identification and Significance in Science and Medicine.* (1935.)

SERNANDER, R. *Entwurf einer Monographie der europaischen Myrmecochoren.* (1906.) Deals with the subject of dissemination by ants.

MORTON, F. *Die Bedeutung der Ameisen für die Verbreitung der Pflanzensamen.* (1912.) Reprinted from *Mitteilung. Naturwiss. Vereins Universität Wien,* 1912. A résumé of the subject with a list of papers which have appeared since Sernander's monograph.

Some references are given in the text. Numerous others will be found in the general works cited, especially the *Pflanzenfamilien* and *Pflanzenreich.* For further references the volumes of Just's *Botanischer Jahresbericht* may be consulted under the special heading dealing with the systematic study of flowering plants, and since 1918, the American periodical, *Botanical Abstracts.*

APPENDIX

p. 83. POLYGONACEAE. R. A. Laubengayer ("Studies in the anatomy and morphology of the polygonaceous flower," *American Journal of Botany*, XXIV, 329; 1937) reviews previous work and concludes from his studies of the anatomy of the flower that the trimerous whorled condition is fundamental and that the 5-merous or 4-merous condition as seen in *Polygonum* and other genera is not a spiral development but due to suppression in development or fusion of parts. Also that the flower is more or less perigynous, not hypogynous.

p. 94. In *Pflanzenfamilien*, Ed. 2 (Harms), XVI, C, 291 (1934), Pax and Hoffmann suggest a somewhat different arrangement, the main difference being that we must regard the Chenopodiaceae with the closely associated Amarantaceae, not the Phytolaccaceae, as representing the origin of the group. The following diagram expresses this view:

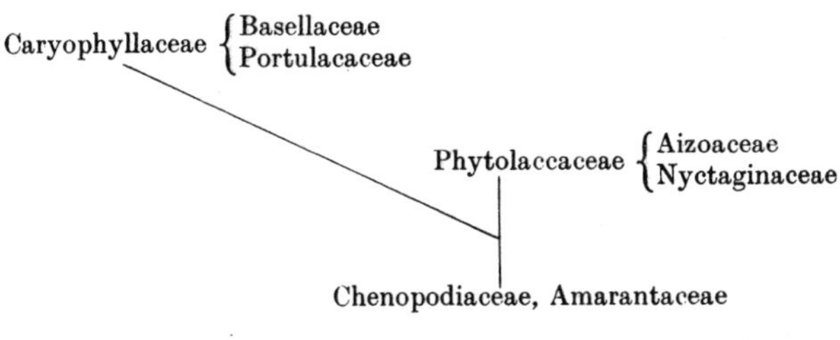

The student must not be bewildered by variation in views on phylogeny which, though of interest, are largely expressions of opinion.

p. 95. CHENOPODIACEAE. The numbers of genera and species given are from Ulbrich's monograph of the family in Edition 2 of the *Pflanzenfamilien*. *Chenopodium* has 250 species, *Atriplex* 120, and *Obione*, *Suaeda* and *Salsola* 100 each, *Kochia* 80 and *Corispermum* 50. The great majority of the genera contain a few species only; 42 are monotypic.

p. 108. NYCTAGINACEAE. Pollination. The attractive character of the flower, often due to a contrast between the white or coloured bracts and the perianth, the long tube, scent, and presence of nectar, suggest insect-pollination. When, as in *Mirabilis*, the flowers open in the evening night-flying insects are responsible. The flowers are protogynous but after anther-dehiscence by a rolling up of the style and stamens the stigma is brought in contact with the anthers rendering possible self-pollination.

Humming-birds may be agents of pollination, as in some species of *Bougainvillea*. Cleistogamous flowers occur in some species.

p. 114. AIZOACEAE. The arrangement of the genera followed in the text is that of Pax in the *Pflanzenfamilien*, Edition 1. In their revision of the family in Edition 2 Pax and Hoffmann recognise six tribes: (1) *Gisekieae*, including one genus *Gisekia* with an apocarpous pistil, formerly included in Phytolaccaceae. The remaining tribes have a syncarpous pistil. (2) *Orygieae*, formerly included in Molluginoideae. (3) *Limeae*, including a few genera removed from Phytolaccaceae. (4) Mesembryanthemeae, including the *Aizoeae*. (5) *Mollugineae*, with two subtribes *Mollugininеae* and *Sesuviinae*. (6) *Tetragonieae*, including *Tetragonia*. The subdivision is based on the apical or basal position of the ovule, presence or absence of an aril, and the length of the funicle.

The huge genus *Mesembryanthemum* has in recent years been intensively studied by several workers whose results are not altogether in agreement. They agree, however, in the division of the genus into a large number of smaller genera. For an account of their work see Pax and Hoffmann in the *Pflanzenfamilien*.

p. 116. Family BASELLACEAE. A small family, with 5 genera and about 20 species, mainly tropical American and West Indian, of glabrous perennial twining herbs with alternate simple often fleshy leaves. Flowers small, regular, bisexual, in many-flowered spikes or panicles, with a pair of bracts forming an involucre below the perianth. Perianth white or reddish of 5 members more or less united at the base. Stamens 5, opposite the perianth leaves and united to them at the base. Ovary free, unilocular, with one basal erect campylotropous ovule and a generally 3-armed style. Fruit fleshy, indehiscent, surrounded by the persistent perianth. Seed with an annular or spirally rolled embryo and generally sparse endosperm.

Basella alba, probably a native of tropical Asia, is widely cultivated in warm countries as a vegetable.

p. 155. MENISPERMACEAE. A. C. Joshi (in *Journ. of Bot.* LXXV, 96; 1937) finds in *Cocculus villosus* a fourth carpel frequently present, rarely fertile and generally a reduced structure and bearing no ovules. The carpels are borne spirally on the floral axis. In early stages the carpels are 2-ovuled, one arising at a slightly lower level than the other and proceeding as far as the development of the megaspore-mother-cell, but then becoming crushed by the growth of its companion ovule, and seen in the open flower merely as a scale beneath the functional ovule. Joshi cites this as evidence for reduction in the number of carpels and ovules in the Menispermaceae.

p. 319. Family CEPHALOTACEAE. A monotypic family comprising one species, *Cephalotus follicularis*, in South-west Australia, growing in damp humus-sand. It is a small herb perennial by a rhizome which bears a rosette of entire foliage-leaves beneath which is a rosette of modified pitcher-leaves adapted for the capture of insects. The stem ends in an inflorescence bearing short few-flowered crowded lateral cymes of small white flowers. The flowers are 6-merous; a simple persistent perianth of 6 sepals followed by two alternating perigynous whorls of 6 stamens each and a whorl of 6 free carpels, each containing one erect anatropous ovule with two integuments. When ripe the carpel separates with its seed which contains a straight embryo lying in a copious endosperm.

The genus is distinguished from the Crassulaceae by absence of petals and its pitcher-leaves.

INDEX

When the number is in black type a figure
will be found on the page indicated